Modern Actuarial Theory and Practice

Second Edition

D1481423

Modern Actuarial Theory and Practice

Second Edition

Philip Booth
Cass Business School, City University, London

Robert Chadburn
BPP Actuarial Education, Abingdon, Oxon

Steven Haberman
Cass Business School, City University, London

Dewi James
James Brennan & Associates, London

Zaki Khorasanee
Cass Business School, City University, London

Robert H. Plumb
Cass Business School, City University, London

Ben Rickayzen
Cass Business School, City University, London

CHAPMAN & HALL/CRC

A CRC Press Company
Boca Raton   London   New York   Washington, D.C.

Library of Congress Cataloging-in-Publication Data

Modern actuarial theory and practice / P. Booth ... [et al.].
   p. cm.
Includes bibliographical references and index.
ISBN 1-58488-368-5 (alk. paper)
1. Insurance–Mathematics. I. Booth, P. (Philip), 1964-

HG8781.M63 2004
368′.01–dc22
                                    2004056100

Visit the CRC Press Web site at www.crcpress.com

© 2005 by Chapman & Hall/CRC

No claim to original U.S. Government works
International Standard Book Number 1-58488-368-5
Library of Congress Card Number 2004056100
Printed in the United States of America 1 2 3 4 5 6 7 8 9 0
Printed on acid-free paper

Foreword

The publication of the first edition of *Modern Actuarial Theory and Practice* in 1998 marked an important step in the emergence of new actuarial textbooks in the English language. Although there had been many advances in the theory and practical application of actuarial science, and considerable development in thinking about the way in which the subject should be presented and taught, there had been relatively few new actuarial textbooks. A decade or so ago the U.K. actuarial profession (the Faculty of Actuaries and the Institute of Actuaries) stopped relying on textbooks for educating new generations of actuaries and began to develop its own core reading around the defined syllabus objectives, relying on commercial tuition providers and the universities to develop more tailored tuition materials to assist students in preparing for the examinations.

Nevertheless, there remained (and remains) a great need for textbooks, in particular to support the teaching of actuarial science in the universities, where there have been an increasing number of first-degree and post-graduate diploma programs in the fields of actuarial science and financial mathematics. It is also important for the actuarial profession to ensure that the fundamentals of actuarial science and its applications are accessible, both to current and future generations of actuaries, in permanent and codified form, through the availability of high-quality textbooks.

There is a growing need for textbooks to support the internationalization of actuarial education, in particular to cover the application of actuarial principles and techniques in the major practice areas of the profession. The International Actuarial Association (IAA) insists that all member associations should have in place by 2005 an actuarial education process that satisfies the requirements of the IAA Core Syllabus and Guidelines, which the IAA adopted in 1998. Two key elements of the Core Syllabus are the subjects *Investment and Asset Management* and *Principles of Actuarial Management*, which form the subject matter of *Modern Actuarial Theory and Practice*.

Mutual recognition of professional qualifications within the E.U. is already a fait accompli, following the passing of the Higher Education Diplomas Directive in 1988. The actuarial profession has embraced this concept actively, with mutual admission to professional bodies and a determined

effort to harmonize the qualifications of actuaries throughout the E.U.
Like the IAA, the European Actuarial Consultative Group (the Groupe
Consultatif) has approved a core syllabus for the education of a fully
professional actuary, which all member associations are expected to have
implemented in full by 2005, in cooperation with other relevant institutions,
including the universities, which in many countries play a dominant role
in the actuarial education process. The core syllabus is annotated with
references to suitable textbooks to cover the various topics. Prominent
among these references — for several of the subjects in the core syllabus — is
Modern Actuarial Theory and Practice.

Internationally, there are many good textbooks addressing the
fundamental mathematics of actuarial science and associated disciplines.
However, there is relatively little teaching material dealing with the practical and professional applications of actuarial science. *Modern Actuarial
Theory and Practice* has helped to fill that gap, providing an introduction
to the practical applications of modern developments in actuarial science
across a range of practice areas. The material is presented, to the greatest
extent possible, in generic terms, so that it will be accessible and useful to
an international audience.

In the years since the publication of the first edition there has been
considerable development of thinking in the actuarial profession to incorporate ideas and theories from the rapidly growing field of financial
economics. Market-consistent valuation methods have taken center stage
and the International Accounting Standards Board is working towards
the introduction of International Financial Reporting Standards for pension
funds, insurance companies, banks and other financial institutions that
will be based on fair value of assets and liabilities. The authors have
updated the first edition to reflect these developments and have also
introduced a new section of the book on actuarial applications in the fields
of health and long-term care. The pensions section is also now more
balanced between defined benefit (DB) and defined contribution (DC)
arrangements, reflecting the trend in the UK and in many other countries
away from DB and towards DC.

The new edition of *Modern Actuarial Theory and Practice* will offer valuable
additional study material for those preparing for the Core Applications
examinations of the U.K. actuarial profession, and similar-level studies in
other countries, as well as being a convenient general textbook for every
qualified actuary to have on their bookshelf.

I congratulate the authors on bringing the material together for this book,
and hope that it will be widely used by students and by qualified actuaries
around the world.

Chris Daykin CB, MA, FIA
Chairman, Education Committee of the Groupe Consultatif

Introduction

This book is specifically aimed at final-year undergraduates, M.Sc. students, research students preparing for an M.Phil. or Ph.D. degree, and student actuaries preparing for the actuarial examinations of various professional bodies. Practising actuaries may also find the book a useful guide to current thinking, methodologies, and models. Although all of the authors are U.K.-based, the book has been written with the intention of presenting the principles of actuarial science (rather than the details of practices specific to certain jurisdictions). Thus, various illustrative examples have been taken from the U.K. and the U.S.

The book describes the traditional areas of actuarial activity, but in a manner which highlights the fundamental principles of actuarial theory and practice, as well as their economic, financial, and statistical foundations. The commonalities and synergies between the different practice areas are emphasized. The last 25 years have seen a significant shift in thinking and in approach — from deterministic methods, with implicit or explicit margins to protect against variability, to stochastic methods and a full recognition of the nature of risk. This shift is reflected in the presentation of the material in this book.

The second edition updates the first edition with changes in actuarial thinking, technique, and application over the last 5 years. Specific changes relate to Part IV, on *Pensions*, which has been extended to provide a balanced treatment of defined-benefit and defined-contribution pension plans. Also, a new part has been added to reflect the growing significance of actuarial applications in the fields of disability, health, and long-term care insurance.

At the time of publication, changes in actuarial education are being implemented in several countries. As Chris Daykin mentions in his Foreword, both the International Actuarial Association and the European Actuarial Consultative Group have approved core syllabuses which member associations are expected to have fully implemented by 2005. It has been one of our objectives to ensure that the content and structure of the book match the philosophy of these developments on the international front, as well as the manner in which actuarial qualifications are changing and are likely to change in the future.

The detailed structure of the book is as follows.

Part I, *Investment*, comprises Chapter 1 to Chapter 5. Most actuaries are employed by intermediary companies working in the financial services sector. These companies have both financial assets and financial liabilities. A crucial part of the role of the actuary is the management of asset liability risk, which involves both the management of investments and the joint management of assets and liabilities. Although the book takes an "institutional" approach, Chapter 1 stands out in giving an overview of financial intermediaries and their economic functions and in looking at how they might evolve over time. In doing this, the chapter brings out the similarities and differences between the institutions discussed in detail in the later chapters and points out that the techniques and methodologies used by one type of institution could be exported to other types that perform the same kind of economic function.

Chapter 2 looks at the investments available to financial institutions. Each of the main types of investment is analyzed from economic and risk perspectives. Techniques of valuation are discussed. Those who are familiar with the principles of investment can omit this chapter without loss of continuity.

Chapter 3 examines the principles of asset liability management and the management of investment risk from a conceptual viewpoint. It then applies these ideas to different types of financial institution and compares and contrasts the investment policy of these different types of institution. International differences are also discussed and rationalized in terms of the principles presented earlier in the chapter.

Chapter 4 develops a number of different approaches to the measurement of investment risk, including utility theory, coherent risk measures, and the use of shortfall constraints. The discussion also considers the critical conceptual (and practical) trade-off between risk and return.

Chapter 5 builds on the discussion in Chapter 4. Actuarial science is often described as a mixture of judgement and application of scientific techniques. Chapter 5 discusses some of the scientific techniques that are used to determine investment policy in terms of portfolio selection and asset allocation. Investment problems can be so complex that simulation and numerical methods are more useful than analytical techniques. Chapter 5 goes on to discuss the subject of stochastic investment modeling and considers its application to asset allocation within a financial institution, in the presence of liabilities.

Following on from this general analysis of the management of assets, Part II, *Life Insurance*, considers actuarial management in respect of the major types of liability structure, describing the process of financial intermediation through life insurance and other long-term insurance-based contracts.

The basic operation and general nature of life insurance are described in Chapter 6. The aim is to give a brief description of the types of product available and how they are distributed, with particular examples from the U.S. and the U.K. The chapter also introduces the nature of the risks faced by life insurers, as a preface to the more detailed discussions in Chapter 10 and Chapter 11.

Chapter 7 and Chapter 8 are devoted to the operation of nonparticipating and participating long-term insurance contracts, again with reference to the U.S. and the U.K. Chapter 8 examines the different profit distribution methods that are in common use inernationally, i.e., the contribution method, the uniform reversionary bonus method, revalorization methods, and the accumulating with profits method. Simple numerical examples are given in order to illustrate the financial operation of these benefit structures. The two chapters include a full discussion of approaches to reserving and their financial effects. Chapter 9 describes the process and rationale for the solvency regulation of long-term insurance business, using the E.U. approach as an example. The significant financial effects of such regulation on the reserves of life insurers and, hence, upon profit emergence are discussed for the main types of contract. Again, simple numerical examples are given.

Having built up a sound understanding of the financial operation of a life insurance business, the reader confronts the financial risks carried by life insurers. Chapter 10 discusses the various methods and techniques that are available to control such risks, including underwriting and reinsurance.

Chapter 11 discusses the role of the actuary in the control of life insurance company risks, based on the concept of the "control cycle". Product pricing is also discussed. The use of appropriate modeling techniques is promoted as an aid to the actuary in preparing his or her advice. In particular, whole-office modeling techniques are thoroughly described. Both stochastic and deterministic approaches are considered, and the chapter concludes with warnings regarding the risks that may arise from modeling errors.

Part III, *General Insurance*, comprises Chapter 12 to Chapter 16. The aim is to demonstrate ways in which the profitability of general insurance companies and the security of the policyholders can be safeguarded by the application of scientific procedures for the classification and assessment of risks and the establishment of adequate premium bases and reserves. The focus is on the practical aspects of understanding the underlying processes. The range of perils to which general insurance contracts are exposed is much broader than a typical life insurance contract. The general insurance market aims to indemnify policyholders in the event of loss or injury due to a wide range of perils. The scope of general insurance policies and the variety of clients mean that, frequently, the data required to measure pricing and profitability are limited. This leads to significantly greater parameter estimation error in general insurance actuarial models than is the case with life insurance. Careful policy design and management can help to control some of the uncertainty, and retrospective premium rating (or profit sharing) provides a means of absorbing the problem of estimation error.

In most free markets the price achieved for a given risk is more a function of the supply and demand of insurance capacity than of a theoretically ideal level. The insurance market is no exception. Swings in market prices, arising inter alia from a mismatch between supply and demand, are a fundamental feature of general insurance.

Chapter 12 provides an introduction to general insurance markets and contracts. In order to gain an appreciation of the dynamics of a general insurance company's operations, readers will need an understanding of the underlying accounting framework, and this is introduced in Chapter 13.

Chapter 14 emphasizes the importance of understanding the "true" underlying risk premium. It then examines the reasons for differences between that and the premium charged in order to understand the contribution of each class of business to the overall profitability of the insurer.

For commercial insurers, the reinsurance market provides access to additional capital and assistance with the development of products for new and unusual risks. Chapter 15 discusses some of the ways in which this market operates. Reinsurance is a swiftly changing area of operation. Brokers and insurers develop products with characteristics similar to derivatives in order to compete with other financial institutions, such as banks, which are designing insurance-type products and offering hedging opportunities. The chapter closes with consideration of such alternative methods of risk transfer.

Chapter 16 sets out the processes that lead to a claim entry on an insurer's books. It describes some basic methods which may be applied in projecting claims to an ultimate level and, hence, in the estimation of claims reserves. The availability of increased computing power and developments in statistical modeling have led to a proliferation of regression techniques for modeling the behavior of the claims run-off triangle. Some of these are discussed briefly.

Chapter 17 to Chapter 20 constitute Part IV, *Pensions*, and they provide a general introduction to the subject of pension provision.

The extent and type of private-sector pension provision in developed countries appears to be driven by two main factors. First, those countries with state schemes providing a high level of replacement income, such as Germany, tend to have less-developed private provision; those with a low level of state benefit, such as the U.S., have well-developed sectors. Second, countries which give tax advantages to certain types of pension plan are more likely to have this type of provision.

Nonetheless, certain major characteristics are found in most systems, and these are highlighted in the coverage here. They include: aspects of benefit design, such as the provision of retirement income and dependants' benefits in defined-benefit pension plans; the types of financing in prefunded pension plans; the nature of actuarial modeling; and the general principles of investment to meet the future liabilities of pension plan members.

Chapter 17 provides an overview of pension plans, describing the main features of defined-benefit, defined-contribution, and hybrid plans. For prefunded defined-benefit plans, actuarial modeling plays a role of critical importance in the determination of levels of funding and in the assessment and management of financial risk. Chapter 18 provides a detailed and in-depth study of this subject, reviewing the nature of cash-flow projections, the role of the actuarial valuation, the range of funding (or cost) methods that have been proposed and implemented, and the different approaches to the valuation of plan assets. Chapter 18 also briefly discusses unfunded plans and how their costs may be managed. For defined-benefit pension plans, a subject of equal importance to funding is the choice of investment strategy, and this is discussed in Chapter 19. After a brief review of the principal characteristics of the asset classes available, we consider the investment objectives for different types of plan and the theoretical and practical aspects of stochastic asset liability modeling.

In many industrialized countries, there is a trend towards greater pension provision through individual accumulation policies or defined-contribution plans. In both cases, the individual member may be responsible for critical decisions affecting how much is contributed and how it is invested. Chapter 20 looks at these individual choices, presenting deterministic and stochastic approaches to benefit projections in the accumulation phase, preretirement. We also consider postretirement choices for the pensioner in the decumulation phase — for example, whether and when to buy an annuity and the opportunities presented by income drawdown.

Part V considers actuarial theory and practice for *Health Insurance*. It comprises an introductory chapter and four further chapters, each devoted to one of the principal types of insurance coverage. This approach has been adopted in preference to considering the underlying processes of underwriting, pricing, reserving, and claims management because these processes vary markedly in practice between products and between markets for a particular product line.

Chapter 21 provides an introduction to health insurance and considers the principal contract types, contrasting these with life insurance. It looks at some of the external factors that affect health insurance. It considers the key contract features and the risks facing the insurance provider.

Chapter 22 deals with income protection insurance, a class of product that provides income to the insured while they are unable to work because of illness or injury. Product design, underwriting, pricing, claim management, and reserving are discussed. The section on pricing methodology includes a brief review of some of the most widely used approaches, with a focus on the multiple-state modeling framework.

Chapter 23 follows the structure of Chapter 22 but deals with critical illness insurance, under which a benefit is payable upon the first occurrence or diagnosis of one of a number of specified medical conditions. It notes that in many countries (but excluding the U.S.) this has proved to be a very popular type of insurance product.

Chapter 24 considers long-term care insurance, again using the same principal headings as Chapter 22 but with a discussion also of alternatives to prefunded insurance products. The relationship between the insurance market and State provision is particularly important for this type of product, and the chapter includes a brief overview of different international examples.

The same template is used again in Chapter 25, which considers private medical insurance products. These contracts reimburse the insured, wholly or partially, for the cost of receiving medical treatment. The short-term nature of these products and the indemnity characteristics of the benefits provided mean that this type of insurance has similar features to the general insurance products described in Part III. They are unlike the other three types considered in Part V (i.e., income protection, critical illness, and long-term care insurances), which are more akin to life insurance because of their long-term nature.

The authors would like to acknowledge the positive and helpful comments received on the first edition from the many students whom they have taught in recent years and who have helped us to refine the material. We would also like to thank colleagues for their help and support and our partners for their forbearance. In particular, Rachel Rickayzen provided sterling editorial assistance with certain sections of the book.

The Authors

Professor Steven Haberman is Professor of Actuarial Science and the Deputy Dean of Cass Business School, City University. He graduated in mathematics from the University of Cambridge, qualifying as a Fellow of the Institute of Actuaries in 1975, and obtained his Ph.D. and D.Sc. in actuarial science from City University. He has worked at Prudential Assurance and for the Government Actuary's Department, and has been a member of the Council of the Institute of Actuaries. He has written over 125 papers on a wide range of topics, including mortality and morbidity models, annuities, insurance pricing and pension mathematics. His papers have won research prizes from the Institute of Actuaries and the Society of Actuaries (U.S.). He is co-author of four books, including most recently for CRC Press: the second edition of *Modern Actuarial Theory and Practice* and *Actuarial Models for Disability Insurance*. He is one of the Founding Editors of the *Journal of Pension, Economics and Finance*.

Professor Philip Booth is Professor of Insurance and Risk Management at Cass Business School and Editorial Director of the Institute of Economic Affairs. He has previously been Associate Dean of Cass Business School and also worked as a Special Advisor to the Bank of England on financial stability issues. He has published widely in the fields of social insurance, regulation, real estate finance, and investment. His papers in the real estate finance field have won prizes from the European Real Estate Society and form the *Journal of Real Estate Investment Mathematics*. He is Editor of *Economic Affairs* and Associate Editor of the *British Actuarial Journal*.

Zaki Khorasanee is a Senior Lecturer in Actuarial Science at the Cass Business School, London. He qualified as a Fellow of the Institute of Actuaries in 1991 after six years experience in pensions-related work for consulting firms. Since joining academia in 1992, his research papers on pension topics have been published in actuarial journals and he was awarded a PhD in actuarial science in 1999. He is currently responsible for teaching a postgraduate pension module corresponding to Subject ST4 of the actuarial syllabus.

Dewi James runs a legal and actuarial consultancy, James, Brennan & Associates. His principal business interests include the development of techniques for the efficient management of reinsurance assets (encompass-ing legal structures, actuarial techniques and web-based technologies for deploying efficient solutions), winding up defunct insurers and Dynamic Financial Analysis. Dewi has been chief actuary of London based reinsurers, a consultant with a major accounting firm and a civil servant responsible for insurance supervision in the UK.

Ben Rickayzen is a Senior Lecturer in Actuarial Science at Cass Business School, City University. He graduated with a first class honours degree in Mathematics from the University of Nottingham following which he joined the actuarial consultants, Bacon & Woodrow. He worked in their pensions consultancy and qualified as a Fellow of the Institute of Actuaries in 1990. He joined City University in 1994 and his main area of research has become health insurance. He has published papers on topics in both income protection and long term care insurance.

Robert H Plumb is a Lecturer in Actuarial Science at the Cass Business School, London. He qualified as a Fellow of the Institute of Actuaries in 1968. After 37 years experience in the life and health insurance business, he joined academia in 2000. He is currently responsible for teaching a postgraduate pension module corresponding to Subject STI of the actuarial syllabus.

Robert Chadburn is a tutor with BPP Actuarial Education, helping to run distance learning porgrammes for the UK actuarial profession in Statistics, Contingencies, Modelling, Life and Health Insurance.

Before 2000, Robert was a Senior Lecturer in the then Department of Actuarial Science and Statistics at The City University in London, teaching a borad range of actuarial subjects. He became a principal examiner for the UK actuarial profession during this time. His research activities particularly included mortality estimation and life insurance modelling.

Robert obtained his actuarial training in Life Insurance with the NFU Mutual in Stratford-upon-Avon, and has a PhD in Population Genetics from the University of Liverpool.

List of Chapters

Part I: Investment

Part II: Life Insurance

Part III: General Insurance

Part IV: Pensions

Part V: Health Insurance

Contents

Part II: Life Insurance

Part III: General Insurance

Part IV: Pensions

Part I

Investment

Chapter 1

The Widening Scope of Actuarial Theory and Practice

1.1 Introduction

Theory and practice often develop in a particular field as a result of *longitudinal* research. Research that throws more light on the validity of the efficient market hypothesis, and research into the relationship between risk factors and claims distributions under employee liability insurance contracts would be examples of empirical longitudinal research. The development of better claims reserving methods based on the chain ladder technique would be an example of applied theoretical longitudinal research. These three examples of longitudinal research all develop the discipline of actuarial science in a linear fashion; they help actuaries solve old problems more effectively. However, there are times when *latitudinal* developments are important, too. For various reasons, activities in both the commercial and academic world that are similar in principle often become separated in practice, and there is little traffic between the two areas of activity either in terms of the development of theory or the application of practical techniques. This has been the case, for example, in the finance world, where the professions involved in banks and in non-bank financial activities, such as insurance and pension fund management, have generally remained separate. There are various reasons for this, including regulatory reasons. There may also be practical reasons for the separation. For example, it may be the case that when computational facilities were not as freely available as they are today, the priorities of insurance and banking professionals were rather different despite the fact that the problems they were solving had many similar characteristics. The most obvious difference was the long-term nature of insurance contracts and the labour-intensive computational techniques that were required to value and price insurance contracts. In recent years, latitudinal developments in actuarial theory and in finance theory have begun to overcome the separation between the different disciplines.

In the last 30 years there has been increasing involvement by actuaries in general insurance, health insurance, and long-term care insurance. There has also been much more non-actuarial involvement in the management of life insurance and pension funds. Actuarial professional examinations have broadened to take account of new subject areas and also to take account of the contributions that non-actuaries have made to the development of actuarial theory. More recently, the relationships between banking practice and theory and actuarial practice and theory have become better understood. These relationships are summarised by Booth (1999) and are discussed in greater detail in this chapter. In this regard, the Institute of Actuaries Research Committee and the Royal Bank of Scotland financed research into the application of actuarial techniques in bank-loan pricing and this led to publications by Allan et al. (1998) and, Booth and Walsh (1998, 2001). The intrinsic similarities between bank-loan pricing and insurance pricing are drawn out in these publications and a bank-loan pricing model is developed. Since that time there have been further developments in both theory and practice that have brought actuarial methods and banking theory and practice closer together. For example, some of the ideas in value-at-risk modelling, developed in the banking sector, are now applied in insurance company capital management. In summary, it could be said that, from a theoretical perspective, although there are some issues in insurance that are different from those that need to be addressed in banking (for example, the equivalent of a "bank run" is rare in insurance — although not impossible) the similarities between the discipline of banking and finance and that of actuarial science and insurance are greater than the differences.

Given these observations, it is worth tracing the economic roots of the differences between insurance companies, banks and other forms of financial institution; this will help us better to distinguish the intrinsic differences and similarities between the sectors and disciplines and those differences and similarities that have arisen as a result of institutional pressures. In turn, as market structures evolve, it should help us to understand how the frontiers of actuarial science as a theoretical and practical discipline could evolve. In general, the authors of this text would expect that risk management techniques will become closer in the banking and non-banking sectors and that the various risk management professions and academic groups will cross the sectors to a greater extent than is hitherto apparent. This analysis helps to provide the context for the received ideas and techniques discussed in later chapters.

The remainder of this chapter is structured as follows. First, we discuss the process of financial intermediation, as it is an understanding of that process that gives rise to our understanding of the fundamental functions of different financial institutions. Then we consider the economic similarity between bank products and insurance products. Finally, we will consider some of the implications of the blurring of the boundary between banks and non-banks. It may be thought that, in this chapter, we do not give enough attention to the practical differences between insurance and other financial activities; that is quite deliberate. The remaining chapters of *Modern Actuarial Theory and Practice* consider the details of actuarial science as a theoretical and practical subject. This chapter looks to the future and tries to communicate the similarities between different parts of the financial sector to help provide an understanding of some of the new frontiers of actuarial science. In general, in this chapter we refer to the commercial and retail lending and deposit taking functions of a bank. Investment management and broking functions are not "banking" functions in the economic sense.

1.2 Financial Intermediaries: Their Role in Resolving the "Constitutional Weakness"

Households wish to have a secure method of saving. Those households and companies that wish to borrow or invest need a secure source of funds. In general, households wish to lend short so that their assets can be easily liquidated, although this may be true for only part of a household's portfolio. Firms need to finance their activities through longer-term borrowing and do so through bank loans, bonds, and equity finance. Hicks (1939) described that feature of an unintermediated financial system, where households wish to lend short and firms wish to borrow long, as a "constitutional weakness." Financial intermediaries resolve this weakness because they create a financial liability of the form a saver wishes to hold and then invest in assets which form the liabilities of firms (we will ignore borrowing by the government sector, although many of the same issues regarding the term of borrowing and lending still apply; see DMO (2000) for a discussion of the factors that the U.K. government takes into account when determining its borrowing needs). The financial liabilities created by firms (such as bonds, bank loans and shares) and which become the financial assets of financial intermediaries, enable firms to borrow long term and provide households with the required liquidity. Securities provide liquidity for the ultimate saver because they can be traded on secondary markets. Bank lending creates liquidity for the ultimate saver as a result of the intermediation function of banks in managing liquidity, based on the assumption that not all households will want to liquidate their savings at the same time—an application of the "law of large numbers" principle. Banks are able to lend to firms, and households make deposits with banks. The proper management of a banking business ensures that there are sufficient liquid assets within the bank to meet day-to-day claims on assets by households, but it is still possible for the majority of the bank's assets to be tied up in more illiquid forms of lending.

There is a practical difference between the way in which banks and non-banks facilitate the creation of liquidity in the financial system. There is a tendency for banks to create liquidity through the taking of deposits and lending activity, and there is a tendency for non-banks to be involved in the creation of liquidity through their use of securities markets. But this division is not clear cut. Banks securitise bank loans (see below) and non-banks invest in highly illiquid private equity and real estate ventures, sometimes by granting loans.

The creation of liquidity is just one of the functions of financial intermediaries. More generally, financial intermediation can be seen as the process through which the savings of households are transformed into physical capital. It can be understood as a chain. At one end of the chain we have households giving up consumption and saving. They then save these funds through financial institutions or intermediaries, such as banks, pension funds, and insurance companies. These institutions then either lend directly to corporations or purchase securities issued by corporations, thus buying assets that offer a financial return. Corporations then use the money raised from borrowing or from the issue of securities to purchase physical capital or provide financial capital. The returns from capital are then passed back down the chain, through paying returns to the holders of securities (or paying interest on bank loans), and the institutions that hold securities or that have made loans then pay returns to their savers on their savings products. There has been a tendency in most developed countries for banking sectors to shrink relative to non-bank sectors (see Davis and Steil (2000)). This is often described as a process of "disintermediation," which is a quite inappropriate phrase to use to describe this trend: there is simply a move away from one type of intermediation (using banks) to another type (using non-banks).

1.3 Functional Approach to the Analysis of Intermediaries

All financial institutions or intermediaries, such as banks, insurance companies and pension funds, hold financial assets or claims on firms (i.e., they lend to firms) to meet financial liabilities that are issued to households (for example, bank deposits or insurance policies). It is because of the fundamental nature of the balance sheet of financial institutions, with exposure to financial assets and financial liabilities, that they have a particular risk profile that is not shared by non-financial firms. It is the management of those kinds of risk that dominates much of the discussion in the remainder of this book. In the non-bank sector, actuaries have tended to be predominant in managing those risks.

In general, actuarial texts and actuarial examinations have taken an "institutional approach" to studying the financial risks, looking separately at non-life insurance, life insurance, and pension funds (and occasionally banks). It is also possible to study financial institutions from the perspective of the functions that they perform, a so-called "functional analysis." A functional analysis of financial intermediation can be found in Bain (1992), Blake (2000), Booth (1999), and Crane et al. (1995). Such an approach is sometimes helpful in understanding the economic nature of the risks that are underwritten by different institutions and is also helpful in understanding the developing links between different institutions and how pricing and risk management practices could be transferred between different institutions performing intrinsically similar functions. We follow that approach below.

Financial intermediaries must "add value," otherwise in a market economy they would not exist. We can go beyond the suggestion of Hicks (1939), that financial intermediaries resolve the constitutional weakness of an unintermediated financial system by allowing households to lend short and firms to borrow long, by describing the following functions that financial intermediaries or financial institutions perform:

Risk transformation. Financial intermediaries transform risk by risk spreading and risk pooling; lenders can spread risk across a range of institutions. Institutions can pool risk by investing in a range of firms or projects.

Risk screening. Financial intermediaries can screen risk efficiently (this helps deal efficiently with information asymmetries that are often said to exist in financial markets). It is more efficient for investment projects to be screened on behalf of individuals by institutions than for all individuals to screen the risk and return prospects of projects independently. If investment takes place through institutions, all the investor has to do is analyse the soundness of the institution and not of the underlying investments.

Risk monitoring. Financial intermediaries can also monitor risk on a continual basis. Banks can monitor companies that have borrowed from them when deciding whether to continue lending. Purchasers of securities (particularly of equities) can monitor by exercising voting rights (including selling shares on a takeover).

Liquidity transformation. Financial intermediaries ensure that assets that are ultimately invested in illiquid projects can be transferred to other savers in exchange for liquid assets. As has been noted above, this happens both in the securities markets and through the functions of banks.

Transaction cost reduction. Financial intermediaries provide convenient and safe places to store funds and create standardised and sometimes tax-efficient forms of securities. Furthermore, financial intermediaries facilitate efficient exchange. People who have surplus capital do not need to incur the search costs of finding individuals who are short of capital, as an intermediary forms a centralised market place through which exchange between such individuals can take place, albeit indirectly.

Money transmission. Banks facilitate the transmission of money assets between individuals and corporations for the purpose of the exchange of goods. The role of money is discussed by Meltzer (1998), and its role is crucial in explaining the special function of banks in the chain of intermediation. Banks could just involve themselves in money transmission (narrow banks) without being involved in the chain of intermediation. However, in most banking systems the roles of money transmission, bank lending and financial intermediation go hand-in-hand.

Asset transformation. Insurance companies and pension funds, and sometimes banks, are involved in "asset transformation," whereby the financial liability held by the institution is of a different financial form from the asset held (for example, insurance company liabilities are contingent due to the insurance services they provide, yet the assets are not subject to the same type of contingency). Actuaries tend to work in institutions that insure contingencies and provide asset transformation functions in this way. In fact, their skills are useful in any financial intermediary that is managing financial assets to meet financial liabilities. Although actuaries have tended just to work in institutions that sell contingent products, that is now changing.

The above are the basic functions of financial intermediaries. Fundamentally, the risks inherent in financial institutions arise from their functions. There are significant practical differences between the functions of banks and those of non-banks, which explains their separate development hitherto. It is helpful to understand more about their specific functions in order to help us to understand how the functions of financial intermediaries are becoming closer.

1.4 Intermediating Functions of Banks

The traditional commercial and retail functions of a bank tend to involve risk spreading, risk screening and monitoring, liquidity transformation and the provision of money transmission services. The business of a bank can be conducted in such a way that asset transformation is not undertaken. Therefore, it is not an intrinsic function of a bank. However, a bank might undertake floating-rate borrowing and fixed-rate lending or may offer mortgages with prepayment options. Both of these arrangements are forms of asset transformation, but both can be either avoided or hedged, so that the risk is taken on by another form of financial institution.

The risks that are underwritten involve credit risk — the risk that financial loss will arise from the failure of a counterparty.

1.5 Intermediating Functions of Insurers

The primary role of insurers is asset transformation. The assets held by insurers are fundamentally different from the liabilities created by the insurer: the insured's asset (and hence the insurer's liability) becomes activated on the occurrence of a certain contingency.

One of the roles of those who provide capital for the insurer is to bear the risk of financial loss from the asset transformation function. Insurers will generally also perform investment risk pooling, screening and limited monitoring functions. In the U.K., the total value of assets of life insurance companies is about £800 billion. Investment functions are a major aspect of an insurance company's business. One would not expect a life insurance company to be involved in money transmission functions. Also, whilst insurance companies tend to invest in liquid, securitised, assets, their liabilities do not form liquid assets of households as such. Therefore, they do not tend to provide liquidity transformation functions.

1.6 Intermediating Functions of Unit Trusts and Mutual Funds

Unit trusts and mutual funds[1] generally perform pure investment functions and hold securities (some mutual funds invest directly in real estate); defined-contribution (DC) pension funds have similar functions. As such, they perform risk screening, risk spreading, and also limited monitoring functions. In markets for non-securitised investments, mutual funds also provide liquidity. They allow large numbers of investors to pool funds that are invested in illiquid markets, such as direct real estate, and the intermediary can allow buyers and sellers to trade units without having to deal in the underlying investments. There are limits to the extent to which this function can be performed: if there are not equal number of buyers and sellers of the units, then the investments underlying the units have to be sold by the fund's manager: in some mutual funds there are redemption clauses which allow a moratorium before units can be redeemed.

Money-market mutual funds (see Brealey (1998) for a further discussion of their role) also provide money transmission services. They will normally invest in securitised loans that form a portfolio in which investors buy

---

[1]Whilst there are technical differences, both unit trusts and mutual funds are open-ended investment funds with the unit holder being exposed to all the investment risk.

units. In some cases, such funds are often simply used as savings vehicles (thus providing risk screening, spreading, and pooling functions), but such funds are also often used for money transmission, with unit holders making payments by transferring units to those receiving payment using a cheque book. Their development will be discussed further below.

In general, mutual funds pass investment risks back to unit holders, although there will be a reputational risk from not performing intermediating functions, such as risk screening, effectively. DC pension schemes do not provide asset transformation, unlike defined-benefit pension funds. The former are generally just savings vehicles providing no insurance function. However, they may sometimes provide insurance functions, e.g., guaranteeing investment returns, annuity rates or expenses; where such guarantees are given, they are normally underwritten by an insurance company.

1.7 Banks, Insurance Companies and Pension Funds: Fundamental Similarities and Differences

The money transmission function can be regarded as distinct from the other functions of financial intermediaries, as it does not involve the mobilisation of capital as a factor of production. In an economy with no savings or investment and no borrowing or lending, money transmission would still be required. It is perhaps this function that makes banks intrinsically different from other financial institutions, although, as has been discussed above, other intermediaries are also beginning to perform money transmission functions (note that in the case of money-market mutual funds, they combine money transmission with investment). The money transmission functions, as performed by banks, gives rise to a liquidity risk, or the risk of a "run," which does not arise to the same extent (or at least is easier to manage) in other financial institutions that do not perform this function.[2]

From the functional analysis of financial intermediaries we see that the "special" role of insurance companies and pension funds is that they face insurance risks due to the difficulty of estimating the amount and timing of liabilities and because of the contingent nature of the liabilities. This arises from the asset transformation function of non-banks. However, whilst this difference between banks and non-banks seems evident from the functional analysis of financial intermediaries, banks face many of the same kinds of risk from the credit-contingent nature of their assets as

---

[2]A run arises when depositors observe the behaviour of other depositors and try to liquidate their deposits more quickly because they fear that a solvent bank will run short of liquidity. Whilst this might be a distinct banking problem, it could be argued that actuarial skills could be used to model this behaviour statistically.

non-banks face in respect of their liabilities. Underwriting credit risk through granting bank loans can be seen as a form of asset transformation not dissimilar from the asset transformation in which non-banks partake. Furthermore, the other functions of the different financial institutions (risk spreading, risk monitoring, etc.) are broadly similar, at least conceptually.

In principle, there is a prima facie case for assuming that many of the same solvency and risk management techniques could be used for both banks and non-banks. If we were to classify the institutions by the intermediation risks that they undertake, rather than by whether they are banks, insurance companies, pension funds, etc., then we would see many similarities which may otherwise be obscured.

1.8 Bank Loans, Credit Risk and Insurance

The relationship between the functions and risks borne by banks and those borne by non-banks is further demonstrated when one considers the fundamental nature of a bank loan contract. Consider a bank that grants a loan to a risky client (e.g., a mortgage loan). The price, or interest rate, for that loan could be regarded as being made up of a risk-free rate of interest, an interest margin that covers the expected cost of default (sometimes known as the default premium), and an element that will provide the required return on capital for the bank, given the risk of the loan (sometimes known as the risk premium). The bank could turn this risky loan into a (more or less) risk-free loan in at least two ways. The first way would involve the bank obtaining credit insurance for that loan from an AAA-rated credit insurer. The second way would involve ensuring that the borrower himself takes out insurance. This insurance could be of two forms. The borrower could insure the loan itself (e.g., as with mortgage indemnity insurance, where an indemnity against loss is purchased) with an AAA-rated insurer. Alternatively, the individual could use an insurance policy that insured the kind of contingencies that may lead to default (disability, unemployment, etc.). The first method of insurance, whereby the bank purchases insurance, leads to a direct link between the banking and the insurance sector (direct risk transfer). The second method of insurance, where the borrower insures, leads to an indirect link. There are other ways by which the bank could pass the credit risk to an insurer. The loan could be part of a securitisation (see below). In this case, the bank would issue securities backed by this loan, and insurance companies could purchase the underlying securities. The bank could also purchase credit derivatives (with the risk being passed through to an insurer using a "transformer" vehicle). This array of potential connections between the banking and insurance sectors is, of itself, a rich topic for research.

There is a sense in which, where the loan is not insured in any way, the bank is taking on a risk-free loan and simultaneously "self-insuring" the loan. Thus, a risky loan can be regarded as a risk-free loan plus credit insurance, as far as the bank is concerned.[3] It is quite clear that the pricing, reserving, capital setting, and risk management techniques that would be used in an insurance company writing credit insurance for similar types of loan should be of direct relevance to the bank (and those used in the bank for setting the interest rate on the loan should be of relevance to the insurance company). Indeed, the risk factors that are of relevance to the bank are exactly the same as those that would be relevant to the insurance company. These forms of transaction have been conducted for decades (indeed, arguably, for centuries), yet it is only recently that techniques have begun to cross the divide between the insurance and banking sectors (see the references quoted in the Section 1.1).

The above example illustrates one situation where banking functions and insurance functions are intrinsically linked, in terms of the financial functions of the business. Both banks and insurance companies underwrite contingencies. In banks, these contingencies are directly linked to credit events. In insurance markets, a wider range of contingencies is under-written, but the range does include credit contingencies.

Another example of the similarities between the risks underwritten in the banking and insurance sectors arises from bank loan securitisations, a subject to which we have already alluded. Consider the situation where a bank has made mortgage loans to a group of individuals. We have already noted that the credit risk could find itself underwritten by the insurance or the banking sector and where it is underwritten by the insurance sector, this can be done in at least three contractual forms. However, now consider the situation where the risk is not insured but remains with the bank and where the bank then securitises the mortgages. There are at least two ways in which the risk attached to those mortgages can find itself being taken by the insurance sector. First, a life or non-life insurance company could buy the securities, with all the risk being passed through to the purchaser of the securities. Through the purchase of securities with credit risk attached, insurance companies have, for at least 150 years, taken on credit risk on the assets side of their balance sheet that has very similar characteristics to that taken on by banks.[4,5] Also, the securities could be marketed to investors as risk-free securities and the credit risk insured with a non-life insurance company or a

---

[3]Of course, this is not the legal position, merely a way of expressing the underlying financial relationships.

[4]Although it should be noted that there are some differences between the credit risk implied by a bank loan and the credit risk implied by a corporate bond, these differences are not so great when comparing bank loans with securitised bank loans.

[5]It should also be noted that, outside the U.K., it is common for insurance companies to invest directly in mortgages, and it was also common in the U.K. up to the 1960s.

credit enhancement provided by the bank. The idea of the bank providing a credit enhancement is analogous in theory (and to a large degree in practice) to the purchase of insurance by the bank in order to protect the purchaser against credit risk, except that, where an enhancement is provided, the bank rather than an insurer accepts the risk. Exactly the same risks should be considered and priced when providing such credit protection for the purchaser of the securities as a non-life insurance company would take into account when selling credit insurance for securitised loans.

1.9 The Evolving Relationship Banking and Insurance

There are at least four implications of the blurring of the boundaries between banks and non-banks that are worthy of further consideration. Three of these issues (the transfer of techniques between the sectors, corporate integration between banking and non-bank entities, and the development of regulation) will be considered briefly in this section. The fourth (the development of products that link the bank and non-bank sectors) is discussed separately in more detail in Section 1.10. References are given to further literature so that the interested reader can investigate these issues in greater detail.

There are some techniques used in the analysis of banking problems that have features similar those of the techniques used by actuaries in non-bank financial institutions described in the later chapters of *Modern Actuarial Theory and Practice*. For example, techniques for analysing expected credit risk losses by considering default probabilities and loss given default (see Altman (1993)) are similar to techniques used in Bowers et al. (1997) to analyse insurance claims by looking at claims frequency and claims size using compound probability distributions (see Part III). Techniques of stochastic investment modelling used in the life, non-life, and pensions industries (see Part I for a more detailed discussion of stochastic models, and see Part II to Part IV for the applications) have much in common with value-at-risk models used to ascertain the risk of investment portfolios in banks. As has been mentioned, the work by Allan et al. (1998) and Booth and Walsh (1998, 2001) used actuarial techniques, commonly applied in life insurance (e.g., see Part II) to develop pricing models for bank loans. In recent years a number of actuaries have crossed the boundary between banks and non-banks in the U.K. and a number of finance experts have crossed the boundary in the other direction.

The processes of risk management known as stochastic modelling, dynamic solvency testing, deterministic risk-based capital setting and deterministic solvency margin provisioning that are used in the non-bank (mainly insurance) sector (see Part II and Part III) all have analogues in the banking industry. Those analogues are value-at-risk modelling, stress testing, setting capital based on risk-weighted assets, and simple capital-to-asset (or to liability) ratio setting. Such techniques are described in their different contexts by Bessis (2002), Booth (1999), Harrington and Niehaus (1999), Joint Forum (2001), Kim et al. (1999), KPMG (2002), Mina and Xiao (2001), and Muir and Sarjant (1997). Kim et al. (1999) suggest how the value-at-risk method, which is commonly used to manage risk in the trading books of banks and applied in packages such as Riskmetrics, can be extended to deal with long-term asset modelling problems, such as those encountered by pension funds. It would be expected that common approaches (and, indeed, a common language) to describe such risk and capital management techniques might arise. Indeed, such a common language is already developing in official documents (e.g., FSA, 2002) and in commercial and quasi-academic documents (e.g., KPMG, 2002). However, there are practical differences between the bank and non-bank sectors that mean that the details of approaches to risk management, capital setting and so on may well remain different in their practical application and in terms of the detail of model structures that are used. With regard to this point, though, it should be noted that the practical differences between models used in pension fund and non-life insurance applications, which are necessitated by the practical differences between the risks in those two sectors, are probably as great as the differences between models used in the banking and non-life insurance sectors.

The blurring of boundaries between banks and insurance functions is partly a function of, and partly a cause of, corporate integration between the bank and the non-bank sectors. It is discussed further in Genetay and Molyneux (1998). This issue formed one of the main focuses of the 1999 Bowles Symposium, "Financial Services Integration: Fortune or Fiasco" (see *North American Actuarial Journal*, 2000, 4(3)).

The development of bancassurance groups is one aspect of corporate integration. Fabozzi et al. (1994) describe three forms of bancassurance. They are:

- The creation of insurance or banking subsidiaries. An example here is the creation of a banking subsidiary (Egg) by the Prudential Corporation.
- Interpenetration of markets arising as a result of a merger. An example in the U.K. is the creation of a banking/insurance group by the merger of the Halifax with Clerical Medical and General Life Assurance.
- The use of cooperative arrangements. These may take many forms, but could involve banks being "tied agents" for insurers or banks administering banking products provided by insurance companies. This latter form does not necessarily lead to the same organisation actually providing insurance and banking products, but a bank may market both banking and insurance products provided by two different companies.

These developments may have various economic results. For example, they may lead to economies of scale in marketing and distribution; there may be risk reduction through diversification of a business; there may also be a development of joint techniques of risk management across the group (in fact, this is another form of economy of scale). Genetay and Molyneux (1998) suggest that the main objective of bancassurers is success in selling insurance products to their bank customers, although they also identify the objective of diversifying earnings sources. There are other corporate developments, not strictly related to the relationship between banks and non-banks, that are worth noting in passing. In the U.K., supermarkets are increasingly offering banking services (frequently in partnership with established banks). Here, the supermarket makes use of its customer relationships in order to sell banking products. Also, non-financial companies often set up insurance "captives" that are, in effect, wholly owned subsidiaries of a company that carry insurance risk placed within the captive by the parent company.

Whilst the bank and non-bank sectors are becoming less distinct and might be underwriting many of the same risks, this has no necessary implications for the regulation of the two sectors. It is often suggested that, as the bank and non-bank sectors move closer together, they should be regulated according to the same principles. This view is misplaced. Any decisions regarding regulation should be based on sound economic principles. The roles that banks play in the payments system as a result of the particular way in which banks provide money transmission functions may lead to negative externalities (or systemic risk) from bank failure (see Jackson and Perraudin (2002) for a brief discussion, with the issues being covered further in the references contained within that paper). The systemic nature of the banking sector may justify a different degree of regulation or a different approach to regulation in the bank and non-bank sectors, even if they face the same risks.

Corporate integration leads to the development of complex groups, a trend that does lead to legitimate reasons for regulators to take an interest in the blurring of banking and insurance activities. If banks and non-banks become so inextricably linked that the failure of a non-bank could lead to the failure of a parent bank and then to the failure of the payments system, a more sophisticated approach to regulation may be needed that takes account of this problem. This issue is discussed in Joint Forum (2001) and also in Thom (2000), one of the papers in the Bowles Symposium referred to above.

Discussions of the different approaches to regulation and the reasons for regulation in the banking and insurance sectors appear in Booth (2003) and Wood (2003).

1.10 Some Examples of the Evolving Product Links between Banks and Non-banks

It is not surprising, given the fundamental similarities between bank and non-bank financial intermediaries, that links between the sectors are growing. Some aspects of those links are discussed in Section 1.9 and will not be discussed further. However, it is worth noting and discussing further how the development of products is further blurring the boundaries between and further eroding the distinction between the functions of banks and non-banks. These developments are likely to continue, so that relationships between the bank and non-bank sectors will continue to evolve. In particular, we will consider further the development of money-market mutual funds and the development of the credit derivatives and credit insurance markets.

Hirshleifer (2001) discusses the blurring of boundaries between banks, other financial institutions, and the financial markets in terms of the development of financing through capital markets (i.e., financing corporate activity through the issue of securities rather than through bank loans) and the securitisation of bank loans. The securities so created can then be purchased by non-banks which then carry the credit risk. This development is an important one and, to a large degree, is reflected in the decrease in relative importance of the banking relative to the non-banking sector (the process often referred to, inappropriately, as disintermediation — see above). However, the development of money-market mutual funds that also perform money transmission functions takes the process of the blurring of boundaries between the bank and the non-bank sectors a step further.

The development of money-market mutual funds relates most closely to the distinct intermediation functions of banks because money-market mutual funds perform money transmission functions. Individuals can transfer units to another individual using a check book. In the U.S. from 1994 to 1998 there was a 121% growth in money-market mutual fund holdings. The value of U.S. retail money-market mutual funds was $834 billion in September 2004 (source: Investment Company Institute, http://www.ici.org). Money-market mutual funds are unitised mutual funds that invest in a portfolio of secure, short-term, securitised, liquid assets (held by an independent custodian if they take the form of U.K. unit trusts). This development challenges the unique role of banks in money transmission. This is not to say that banks will not continue to provide money transmission functions — perhaps even through marketing money-market mutual funds. But the use by bank customers of money-market mutual funds could lead them to do so in a way that is not intrinsically different from the functions provided by other financial intermediaries. Money-market mutual funds are not intrinsically different from other forms of mutual funds, in the way that bank deposits are "different" from securitised investments. Therefore, it is of interest to ask questions such as who would bear the risk of a "run" (i.e., investors wanting to liquidate money-market assets more quickly than the underlying assets can be called in — often defined as "liquidity risk") or the risk of securities losing their value due to default (due to inadequate "risk screening," or simple bad luck).

With traditional bank deposits, banks keep capital to guard against the second eventuality and ensure that their capital and their assets are sufficiently liquid to guard against the first. In the case of money-market mutual funds, both risks would be borne by the unit holder, as the value of the units, determined by the market value of the securities in the second-hand market, could fall below the par value of the securities.

If the securities had to be liquidated quickly by mutual funds, in order to redeem units, then unit values could fall, even if the investments were still regarded as totally secure; in the same way that credit risk manifests itself in a fall in the value of securities when lending is securitised, liquidity risk would also manifest itself by a fall in the value of securities. However, it would be the unit holder who would bear these risks. Limits may be put on liquidations in order to prevent these problems; if this happened, it could impair the money-market mutual fund's function as a money transmission vehicle. But the unit holder will have to balance the lower interest spreads that can be obtained from money-market mutual funds (because of lower costs, the elimination of capital requirements, and the absence of deposit insurance) against the higher risk that results from liquidity risk and credit risk being passed back to the unit holder.

The securitisation of bank loan books to provide securities purchased by money-market mutual funds (and other non-bank financial institutions) leads to a change in the intermediating functions of banks. The securitisation of banks' loan books can be regarded as a method of providing the benefits of raising money through the securities markets to companies (or individuals) that are not sufficiently large borrowers for the direct issue of securities to be cost effective. This could be an important development, because it could lead to the separation of the risk screening and of the risk monitoring and pooling functions of financial intermediaries. These processes have already developed to a significant extent in the U.S., with most mortgages originated in the U.S. being securitised. Banks remain a major provider of services, including screening borrowers for creditworthiness and administering loans; they can then securitise a book of loans or issue floating rate notes against a book of loans, so that savers can invest in such loans through money-market mutual funds or through non-bank institutions rather than through bank deposits. Money-market mutual funds and non-bank institutions can then purchase floating rate notes or near-cash instruments and provide the risk pooling services. The optimal sharing of risk between the originating bank and the purchasers of the securities has to be determined. For example, if risk monitoring is more efficiently carried out by a bank, then banks can provide credit enhancements to the securities, carry the first loss on default, etc.; this will ensure that the bank has an incentive to monitor loans and take appropriate action on default.

The possibility that the risk of the securities backed by the loans can be borne by a number of different parties shows how similar banking and insurance functions are when it comes to underwriting credit risk. The holder of the securities (e.g., the purchaser of money-market mutual fund units) may not bear the ultimate credit risk, but the risk could be insured with a credit insurer by the bank that originated the loans. Alternatively, the bank could offer guarantees or credit enhancements providing exactly the same cover itself. It is clear, when seen in this way, that the whole process of granting risky bank loans is no different in principle from that of underwriting credit risk insurance. The risk can be, and frequently is, packaged and repackaged to be divided in different ways between the mutual fund customer, insurance companies, the originating bank and other banks.

Thus, the development of money-market mutual funds leads to further blurring of the boundaries between banks and non-banks. It has also led to transactions developing between the bank and non-bank sectors in respect of the guaranteeing and insuring of credit risk. The development of the credit insurance and credit derivatives market over recent years has had the same effect. In fact, insurance companies have underwritten credit insurance through indemnity products, personal lines, and trade guarantees for many decades. However, recent developments in the credit derivatives and credit insurance markets have created more complex links between banks and non-banks and have also facilitated rapid, large-scale risk transfer. The different credit transfer product lines that are offered by insurance companies and banks can hardly be distinguished in terms of their functions and basic characteristics. FSA (2002) and Rule (2001a,b) provide a more in-depth discussion of these markets. Credit derivative contracts can appear very much like credit insurance contracts. Indeed, the writer of a credit derivative may reinsure certain aspects and, in markets that do not require the regulatory separation of banks and insurers, credit derivative and credit insurance contracts can be virtually identical.

The simplest form of credit derivative, the credit protection swap would involve the payment of a fee by the entity requiring protection, in return for a contingent payment by the counterparty, on the occurrence of a specified credit event (e.g., the failure of a loan). The events that will cause a payout on a credit swap contract may relate to a group of loans that the bank has made, or to a group of similar loans. The credit default swap is generally tradable and may relate to the value of a group of traded loans so that it can be marked to market. There is a close analogy with credit insurance. If the same bank purchased insurance it would pay an up-front premium in return for compensation on the occurrence of the insurable event. Both credit derivatives and credit insurance would leave the seller of the investment with a potential credit risk liability. However, with credit insurance, it will always be the case that the bank insures the loans that it has actually made, rather than obtaining protection in respect of a group of similar loans. Thus, as with securitisation and the development of money-market mutual funds, these products bring banks and insurance companies together.

A further erosion of the special functions of banks and non-banks has arisen with the development of Defined Contribution (DC) pension schemes (see Chapter 3 and Part IV). In such schemes, there is virtually no asset transformation or insurance risk, although independent insurance provision might be made. However, in principle, the savings element of a DC pension scheme simply involves risk screening and pooling, and these functions could be performed by any financial intermediary, including a bank. Indeed, such products are provided by banks, but they tend to be managed by a separate investment management arm of a bank, rather than by the arm that deals with commercial and retail lending and savings. It should be noted that, in most countries, tax rules normally prevent such savings being cashed before retirement, and they are, therefore, of longer term than is normal for banking products. However, this is not an intrinsic feature of such a product, merely a constraint on its design imposed by governments. In fact, in many countries, DC pension products are provided directly by banks and are linked to straightforward bank savings accounts.

1.11   Conclusion

On examining many of the intrinsic functions of banks and non-banks, we find notable similarities. In particular, credit risk underwritten by a bank has very much the same characteristics as insurance risk. On looking at the fundamental economic functions of financial intermediaries, we do see differences between banks and non-banks, but these differences are perhaps not as "clear cut" as has generally been assumed. Even if certain functions are different, there are certainly significant similarities between different types of financial institution.

Financial innovation is leading to the development of products that further blur the distinction between banks and non-banks. Products offered by different institutions can provide similar benefits to customers and generate similar risks that have to be managed within the institutions. Historically, the division between bank and non-bank financial institutions was drawn much more sharply than it can be today. These developments in financial intermediation should lead to greater competition and further innovation in the financial sector as product providers and purchasers concentrate on the economic functions provided by products and not on the particular institution that happens to be the originator of the products. This form of financial integration may lead to risk management techniques changing and/or becoming more alike in banks and non-banks.

The evolving nature of the practical ways in which financial institutions are performing their intermediation functions and the increasing risk transfer that is taking place between different forms of financial institution allows us to see that the intrinsic differences (as opposed to the institutional differences) between bank and non-bank financial institutions are not as great as they once were (or were perceived to be). The divide between bank and non-bank institutions in professional and commercial life might have arisen because of the important nature of some particular differences between their functions. It might also have arisen because of market frictions arising from regulation, tax policy or the roles of professions. The important role that the money transmission function of banks played in the payments systems has led regulators to treat banks as "special," and this may continue. But it is clear that the evolution of the roles of banks and non-banks has implications for actuaries at the professional and academic level, who will increasingly benefit from looking latitudinally at financial problems involving risk. Such an approach should not reduce the academic rigour of actuaries' work but will allow them to solve a greater range of practical, theoretical and academic problems.

The remaining chapters of this book concentrate on received actuarial theory and the main areas of current practice. The book takes an "institutional" approach to analysis, looking at life and non-life insurance and pensions. The purposes of this introductory chapter have been to communicate the intrinsic economic functions of financial intermediaries (including insurance entities) and to illustrate how these functions may evolve over time. We conclude this chapter by highlighting the point that the techniques used by particular institutions discussed in the later chapters could be used in other institutions that perform the same kinds of economic function.

References

Allan, J. N., Booth, P. M., Verrall, R. J. and Walsh, D. E. P. (1998). The management of risks in banking. *British Actuarial Journal* 4(part IV), 707–802.

Altman, E. (1993). Valuation loss reserves and pricing commercial loans. *Journal of Commercial Lending* (August), 8–25.

Bain, A. D. (1992). *The Economics of the Financial System*. Blackwell, U.K.

Bessis, J. (2002). *Risk Management in Banking*. Wiley.

Blake, D. (2000). *Financial Market Analysis*. Wiley, U.K.

Booth, P. M. (1999). An analysis of the functions of financial intermediaries. Paper for Central Bank seminar on disintermediation, Bank of England.

Booth, P. M. (2003). Who should regulate financial institutions? *Economic Affairs* 23(3), 28–34.

Booth, P. M. and Walsh, D. E. P. (1998). Actuarial techniques in risk pricing and cash flow analysis for U.K. bank loans. *Journal of Actuarial Practice* 6, 63–111.

Booth, P. M. and Walsh, D. E. P. (2001). Cash flow models for pricing mortgages. *IMA Journal of Management Mathematics* 12, 157–172.

Bowers, N. L., Garber, H. U., Hideman, J. C., Jones, D. A. and Nesbitt, C. J. (1997). *Actuarial Mathematics* 2nd ed., Society of Actuaries, Illinois, U.S.

Brealey, R. A. (1998). The future of capital markets. In: *VII Annual Meeting of the Council of Securities Regulators of the Americas (CONASEV)*, Lima, Peru.

Crane, D. B., Froot, K. A., Mason, S. P., Perold, A. F., Merton, R. C., Bodie, Z., Sirri, E. R., and Tufano, P. (1995). *The Global Financial System: a Functional Perspective*. Harvard Business School.

Davis, E. P. and Steil, B. (2001). *Institutional Investors*. MIT Press, U.S.

DMO (2000). Debt management report 2000/2001. Her Majesty's Treasury, London, U.K., www.dmo.gov.uk.

Fabozzi, F. J., Modigliani, F., and Ferri, M. G. (1994). *Foundations of Financial Markets and Institutions*. Prentice Hall International, U.S.

FSA (2002). Cross sector risk transfers. Financial Services Authority discussion paper, May 2002, www.fsa.gov.uk.

Genetay, N. and Molyneux, P. (1998). *Bancassuranc*. Macmillan, U.K.

Harrington, S. and Niehaus, G. (1999). *Risk Management and Insurance*. McGraw-Hill, U.S.

Hicks, J. (1939). *Value and Capital*. Oxford University Press, U.K.

Hirshleifer, D. (2001). The blurring of boundaries between financial institutions and markets. *Journal of Financial Intermediation* 10, 272–275.

Jackson, P. and Perraudin, W. R. M. (2002). Introduction: banks and systemic risk. *Journal of Banking and Finance* 26, 819–823.

Joint Forum. (2001). Risk management practices and regulatory capital: cross-sectoral comparison. Bank for International Settlements, 13–27, www.bis.org.

Kim, J., Malz, A. M. and Mina, J. (1999). LongRun — technical document. Riskmetrics Group, New York, U.S.

KPMG (2002). Study into the methodologies to assess the overall financial position of an insurance undertaking from the perspective of prudential supervision. European Commission, Brussels, Belgium.

Melzer, A. A. (1998). What is Money? In: Wood, G. E. (Ed.). *Money, Prices and the Real Economy*. Edward Elgar, UK.

Mina, J. and Xiao J. Y. (2001). Return to Riskmetrics — the evolution of a standard. Riskmetrics Group, New York, US.

Muir, M. and Sarjant, S. (1997). Dynamic solvency testing. Paper presented to the Staple Inn Actuarial Society, London.

Rule, D. (2001a). The credit derivatives market: its development and possible implications for financial stability. *Financial Stability Review*, No. 10, Bank of England, London, U.K., 117–140, www.bankofengland.co.uk

Rule, D. (2001b). Risk transfer between banks, insurance companies and capital markets: an overview, *Financial Stability Review*, No. 11, Bank of England, London, U.K., 137–159, www.bankofengland.co.uk

Thom, M. (2000). The prudential supervision of financial conglomerates in the European Union. *North American Actuarial Journal* 4(3), 121–138.

Wood, G. E. (2003). Too much regulation? *Economic Affairs* 23(3), 21–27.

Chapter 2

Investments and Valuation

2.1 Introduction

This chapter provides an introduction to investors and investments, and to the valuation and risk analysis of individual investments. We discuss the general characteristics of different types of investment; methods of valuation are introduced; the economic factors affecting investments are discussed, and their risk characteristics are analysed. Where appropriate, the analysis of different investment categories is developed in the context of actuarial techniques that are commonly applied to the analysis of both assets and liabilities of financial institutions. Because this chapter is an introductory chapter, discussing issues briefly that can be covered in much more detail, a considerable number of references is given to standard investment texts. These references include Web sources. Most of the Web resources are U.K. based but have links to international sites covering similar areas of investment. The investment texts are a mixture of international and U.K. texts.

2.2 Cash Instruments

As far as long-term institutional investors such as life insurance companies and pension funds are concerned, cash and cash instruments will form a relatively small part of their investment portfolios. Non-life insurers, with their shorter term liabilities, will tend to hold a higher proportion of assets in cash. Whilst it is necessary to understand the nature of the various cash instruments, we will not spend a great deal of time on their analysis.

2.2.1 General Characteristics

Cash deposits have no risk of capital fluctuation and the interest earned depends on the level of short-term interest rates in the economy. Thus, whilst providing capital protection, cash instruments do not give any guarantee that a particular rate of return will be earned over any given substantial period. Cash instruments may take the form of deposits (similar to bank deposits often used by personal investors) or of tradable cash instruments. The life of the tradable securities is often short (less than a year), but the important characteristic is the frequency with which interest rates change to take account of changes in short-term interest rates. The rate of interest on money-market cash instruments will not be fixed for more than a year and will normally be adjusted more frequently, to reflect changes in short-term interest rates. There are cash instruments available that have a long maturity time but on which interest rates are frequently adjusted to reflect short-term interest rates.

2.2.2 Specific Cash Instruments and Valuation Issues

Institutions, just like personal investors, can put money on deposit with banks. The money may be placed for a fixed term, at a fixed rate of interest, for periods such as 1 week, 1 month, 3 months or "on call" (immediate access). The 3-month interbank interest rate provides a good proxy for the general level of short-term interest rates. It smoothes out abrupt changes that may arise in much shorter term rates due to day-to-day shortages and excesses of liquidity.

Treasury bills are not widely used investment instruments by U.K. non-bank institutions, but they are used in other countries. Such bills are normally issued by the central bank on behalf of the government. Being short term, they normally have high levels of security relative to other investment instruments, even in countries with a high national debt. Typically, the quoted yields from Treasury bills are simple rates of discount per annum. Thus, for a given yield d, the price P of a 3-month Treasury bill (maturing in 91 days) is

$$P = 100(1 - d \times 91/365)$$

The effective rate of interest per annum from the transaction would be i, such that:

$$P(1 + i)^{91/365} = 100$$

This could then be compared with the effective rate of interest from other transactions of a similar length.

Floating-rate notes have many of the characteristics of short-term bonds, except that interest is normally paid quarterly and set at a margin over a given short-term interest rate. Floating-rate notes are issued by corporations and governments and have various terms to redemption. The notes are tradable and, despite their coupon being related to short-term interest rates, they need not necessarily be priced at par (the redemption level). There are a number of reasons for this, including the fact that the interest rate paid on floating rate notes lags short-term rates slightly, the existence of a credit spread reflecting risk and dealing costs, and illiquidity (making the notes slightly less attractive than shorter term instruments).

Cash instruments are available with a range of different types of credit characteristics. Floating-rate notes and other cash instruments can be secured or unsecured, they can be guaranteed by third parties, or they can be backed by "receivables" arising from a process of securitisation. For example, a bank may make mortgage loans and then issue securities, the cash flows on which are financed by the cash flows from the mortgage portfolios.

2.2.3   Risk Characteristics

Cash instruments offer complete capital security, subject to credit risk. They do not offer income security — their return varies according to the level of short-term interest rates. In the long run, higher inflation will tend to lead to higher short-term interest rates and, in this sense, cash instruments may provide some inflation protection. Over short to medium time periods, however, cash rates can lead or lag inflation significantly. This is partly as a result of the role of cash rates in the management of monetary policy. In the U.K., during the 1970s, real cash rates were highly negative (-15% after tax at their lowest). This was a result partly of the cash market structure at the time and partly of the way in which monetary policy was being managed. In the 1980s, swings in real cash returns were substantial; but, since the beginning of the 1980s to the present time, medium-term real returns from cash have been more in line with very long-term historical norms. For further analysis, see Table 3.4.

2.3   General Characteristics of Conventional Bonds

The terms "fixed interest bond" and "fixed interest security" can effectively be used interchangeably. A security is a contract that gives the lender of money particular rights in relation to the borrower. In particular, a *fixed interest security* normally provides the lender with a series of fixed coupon (or interest) payments and a capital repayment at a specified time or times.

The fixed interest security market is very dynamic, and a very wide range of securities with different characteristics are available; for a discussion of these the reader should refer to a standard investment textbook, such as that by Bodie et al. (2002). We shall consider, here, some of the basic types of fixed interest security. Most government bonds provide a series of fixed coupon payments at regular intervals (half yearly in the U.K. and U.S.A.) and a single maturity payment at a specific time. Some fixed-interest securities can be redeemed between two dates at the option of the borrower. A convertible loan stock allows the lender to choose whether to take redemption at a specified time or convert the stock, at that time, into another stock or into equity (the terms of conversion being set at the time of issue of the original stock). Sometimes, the capital of a loan stock will be repaid over a number of years (the total coupon payment necessarily reducing, as capital is repaid).

A pure discount or zero coupon bond provides no coupon payments and, hence, is simply bought for the capital gain. In a number of fixed interest markets, "strips" are available that enable investors to buy securities which have the characteristics of zero coupon bonds. Strippable bonds allow the coupon and capital payments from a bond to be separated from each other and traded separately so that investors can purchase the right to receive any individual payment from the bond.

The various terms and conditions of a bond will be stated at issue. The issue price will, ultimately, be determined by supply and demand in the market and the rate of return that investors require; the factors that determine the rate of return will be discussed in Section 2.4.4. In the following subsections, we consider particular aspects of bond investment and analysis.

2.3.1 Government Bonds

In a given country, the most secure form of fixed-interest security is likely to be issued by the government. Governments have the power to tax, borrow, and print money to service debt, a power not possessed by the private sector. In most developed countries, inflation is the most serious problem that holders of government debt are likely to face, although this problem itself is not insignificant. Most government fixed-interest securities are interest-bearing stocks, paying regular coupons denominated and fixed in a given currency. The securities are generally listed and can be traded on the stock exchange.

There is normally a wide variety bonds with different maturity terms issued by governments. Government issues tend to be large and liquid. For example, in the U.K. there are currently 15 government bonds in issue with an outstanding issue size greater than £6 billion, with maturity times ranging from 1 year to 29 years. Up-to-date details of U.K. government bonds in issue can be found at www.dmo.gov.uk; regular annual and quarterly reports of the U.K. gilt market are issued by the Debt Management Office and published on this Website. Further variety in the U.K. market is provided by "bond stripping." Individual coupon and capital payments can be traded. This means, for example, that an investor can purchase just the capital payment from 4.25% Treasury 2032, effectively holding a 27-year zero coupon bond. The market for stripped bond payments is less liquid (only about 1.5% of that gilt is held in stripped form).

2.3.2   Corporate Bonds

Corporate bonds provide an alternative investment medium to government bonds. For reasons described below, corporate bonds tend to offer higher rates of return than government bonds.

Different types of traditional corporate bond (mortgage debentures, debentures, guaranteed loan stocks and unsecured loan stocks) have varying degrees of security. Their level of security does, of course, affect their yield. The above types of corporate bond, as well as government bonds, would normally be classed as domestic bonds. That is, they are bonds issued by a domestic borrower and generally traded on a domestic stock exchange and denominated in a domestic currency.

Most large corporate bond issues are issued in the form of eurobonds. The eurobond market is an international capital market, where large amounts of capital can be raised, free from the regulation of a particular sovereign government. The term "eurobond" is perhaps a misnomer — such bonds could perhaps be better termed "international bonds." Eurobonds are issued by an international syndicate of banks, on behalf of a borrower in a particular country (although the borrower may be a multinational corporation) to investors throughout the world. The bond may be in any currency. Thus, a German-based multinational corporation may issue a eurodollar bond, of which a U.K. investor could purchase part. Whilst domestic bonds are normally of the traditional form, offering fixed coupon and capital payments, the eurobond market is particularly innovative. A number of different types of bond exist to suit the varying needs of investors.

As with cash instruments, corporate bonds may be issued with a range of third-party guarantees and/or be backed by receivables in the form of a "securitisation." Corporate bonds may also have complex option characteristics. For example, issuers may have the ability to "prepay" bonds (make capital payments early) or investors may be able to demand repayment on the occurrence of particular events such as a credit downgrade.

2.3.3   Bond Valuation

Except where options exist, to either the investor or issuer, the valuation of bonds is a relatively simple matter. All payments are known and the present value of the known payments can be discounted at the investor's required rate of return. The rate of return from a bond for a given price can be found by solving the equation of value for the yield. Conventions as to how yields are quoted vary across bond markets, depending mainly on the frequency of coupon payments.

If an investor purchases a bond, redeemed for £100 in one installment, paying an annual coupon at rate D and there are n years to redemption (where n is an integer), the equation of value linking the purchase price to the coupon and redemption payments is:

$$P = Da_{\overline{n}|} + 100v^n \tag{2.1}$$

where $a_{\overline{n}|}$ is the present value of a series of annual payments of one in arrears for n years.

The equation can be solved to calculate a price for a given yield to maturity or yield for a given price. The yield will be an effective yield per annum in this case. Variations of this formula, where there is a non-integer term to redemption, are discussed in Adams et al. (2003a). Where coupon payments are paid annually (e.g., in continental Europe or on eurobonds) annual effective yields are quoted. Where coupons are paid half yearly, yields per annum convertible half yearly tend to be quoted (i.e., twice the yield per half year). If i is the effective yield and $i^{(2)}/2$ the yield per half year, then $(1 + i^{(2)}/2)^2 - 1 = i$. It is important that a given investor compares yields quoted on the same basis.

Two important concepts in bond analysis are volatility and duration. Volatility measures the sensitivity of bond values to interest rate changes. If yields rise (fall) then bond prices will fall (rise). Volatility can be defined in the following way:

$$\text{Volatility} = -\frac{1}{P}\frac{dP}{di}$$

where P is price or present value and i is the yield to maturity. (This is often assumed to be an annual effective yield, although it does not have to be.)

Duration or discounted mean term is the weighted average of the times to payment, where the weights are the present values of the payment. It can therefore be expressed as follows:

$$\text{Duration} = \frac{\sum_{r=1}^{r=n} t_r C_{t_r} v^{t_r}}{\sum_{r=1}^{r=n} C_{t_r} v^{t_r}}$$

where $t_1, t_2, \ldots t_n$ are the times of the various payments from the investment.

It can be shown that

$$\text{Volatility}(1 + i) = \text{Duration}$$

where *i* is the effective yield per annum to maturity if volatility is calculated using the effective yield per annum. Thus, volatility and duration are proportional to each other and their ratio is approximately equal to unity. In general, volatility and duration increase with the term to redemption of a bond (although there are exceptions to this general rule); they increase as the coupon rate reduces and increase as the yield reduces. We will use the concepts of volatility and duration in Chapter 4. The concepts can be applied more widely to any set of cash flows, relating either to assets or liabilities.

2.3.4   Economic Analysis

We will deal, first, with factors that affect prospective rates of return from a bond, i.e., which affect the general level of gross redemption yields available in the bond market. The economic factor that most critically affects long-term bond yields is the investor's expectations of future inflation. The current level of inflation and also the monetary policy being followed are important data that help determine long-term expectations of inflation. However, the main factor that investors would take into account would be the "credibility of the monetary policy regime," i.e., the extent to which institutions have been developed that are likely to give rise to low and stable levels of inflation. In general, such institutions would be expected to be independent of the political process and have a clear target for inflation (or an intermediate variable that directly affects inflation) and an effective process for delivering that objective. Credibility is difficult to assess by anything other than subjective analysis. However, for an analysis of these issues see Goodhart (2003). Issing (2004) reviews the implementation of the euro. As he makes clear, the monetary regime enjoys credibility but does not have a specific inflation target.

Table 2.1 shows the level of long-term bond yields at various times, when different levels of current inflation pertained. Investors require compensation for inflation and, in most circumstances, a positive expected return after inflation; thus, any increase in expected inflation will tend to put upward pressure on bond yields. In addition, if the outlook for inflation becomes uncertain, as is often the case at high rates of inflation, investors may require an inflation risk premium (i.e., a higher return to compensate for the increased uncertainty of the real value of payments). High current inflation does not necessarily imply high inflation expectations.

Interest rates on short-term bonds will be particularly affected by government monetary policy. Monetary policy is normally adjusted through changes in short-term interest rates. This has an inevitable effect on short-term bond yields. Long-term bond yields need not necessarily move in the same direction as short-term interest rates because a tightening

Modern Actuarial Theory and Practice

Table 2.1 U.K. government bond yields and current inflation for selected years

Year	Retail price index inflation (%)	Long-dated government bond yields (December, %)
1976	15.1	15.0
1979	17.2	14.7
1983	5.3	10.5
1987	3.7	9.5
1990	9.3	10.6
1994	2.9	8.6
1997	3.6	6.3
2001	0.7	5.0

Source: Data from Barclays Capital (2003) and Financial Statistics published by National Statistics

(loosening) of monetary policy may lead to improved (worsened) prospects for inflation, which may lead to lower (higher) long-term bond yields despite higher (lower) short-term interest rates.

As well as being affected by domestic monetary policy, short-term interest rates are also affected by international economic news. Governments may, for example, raise interest rates in order to protect the value of the national currency, particularly if they are pursuing a formal or informal exchange rate target. If short-term interest rates in other countries change, this may affect domestic interest rates; this effect is likely to be most apparent when exchange rates are more closely integrated, so that domestic and foreign currency bonds form more perfect substitutes. Long-term bond yields will also be affected by long-term yields in other world markets. Again, the relationships will be closer the more closely integrated are the exchange rates between the currencies of issue. For example, one sees a much closer relationship between yields from German and French bonds (where both are denominated in euros) than between yields from U.K. and Japanese bonds. For a regular update of developments in international markets, readers could consult the *Bank of England Quarterly Bulletin* at www.bankofengland.co.uk.

On the supply side, any significant increase in government bond issues, due to an increase in the public sector borrowing requirement, will put upward pressure on bond yields. New funds will have to be attracted into the bond market to accommodate the increased supply; this can only happen if prices fall and yields rise. If the total outstanding government debt becomes too high, investors may be concerned that the government will be unable to service the debt. This may lead to increased yields, because of a perceived default risk. Alternatively, investors may believe that the debt will be devalued because of the effects of inflation. An increase in corporate issues may also put upward pressure on bond yields, as more funds have to be attracted into the market. The supply of funds to investment markets will also be important in determining the general level of interest rates. Thus, countries with higher savings ratios are likely to be able to sustain lower long-term interest rates, although, again, this will depend on the degree of openness of the economy.

Analysts will also consider a number of other factors when considering the rates of return which should be available from bonds. Growth in national income, producer prices, unemployment and the balance of payments position are examples of these other factors. However, these economic indicators are most valuable for the information they provide on the likely future progress of the more important economic variables discussed above.

To compensate for their greater risk of default on capital or interest payments, lack of marketability, and higher dealing and research costs, corporate bonds will normally offer higher yields than government bonds of the same currency. This yield differential will vary according to the risk characteristics of the particular issues being considered and the general strength of the corporate sector. Opinions on the credit risk of a bond can be found from rating agencies. For example, Standard & Poors (www. standardandpoors.com) rates bonds from AAA (extremely strong potential to meet obligations) to D (payment in default). The credit rating will be based, amongst other things, on the company's financial position, its management quality, the risk of the business, and its operating performance. The credit spread (the difference between the yield on a bond with credit risk and that on a risk-free bond) will depend to a large degree on the credit rating. However, average credit spreads will also change with the general economic and business outlook. It should not be thought that all government bonds are risk free. For example, at the time of writing, the Standard & Poors rating for Italy (AA) is less than that for some private companies issuing bonds denominated in euros. The ratings criteria are different for governments than for companies, for example concentrating on debt : GDP ratios, tax levels relative to taxable capacity, and so on.

The analysis of corporate bonds could involve similar principles to the analysis of non-life insurance liabilities (see Part III). In theory, a probability can be attached to each payment. In the event of default, one can think in terms of a probability distribution of losses given default. Some of the basics of analysing losses in such a framework are presented by Bessis (2002). A more detailed technical discussion of techniques of corporate bond analysis is presented by Kao (2000) and Jackson and co-authors (Jackson et al., 1999; Jackson and Perraudin, 1999). The latter two articles consider the idea of a "transition matrix" between different credit ratings, applying ideas that are familiar to actuaries in the form of "exposed to risk" and "multiple state modeling." Bonds with different credit ratings will have different credit spreads and different values for a given payment stream. Transition matrices are a way of modelling the probability that a bond will migrate from one credit rating to another. A review of credit conditions in sovereign and corporate bond markets appears in the *Bank of England Financial Stability Review*. This is published twice per year and can be found at www.bankofengland.co.uk. The International Monetary Fund (IMF) produces a global financial stability report that includes an analysis of credit conditions (www.imf.org).

Where Can I Learn More about Monetary Policy and Government Bonds?

In most countries, the government is the single biggest borrower. In the U.K., outstanding government debt is about 35% of GDP. Analysis of the government bond market is an important aspect of institutional investment and analysis of monetary policy is important for understanding the government bond market. Some sources of information on monetary policy are given below. They are U.K. orientated, but all countries will have similar sources of information (central bank Websites, including the ECB and so on).

A clear discussion of theory and practice of monetary policy can be found in Miles and Scott (2002). Friedman et al. (2003) contains four papers on the theory and practice of the relationship between monetary policy and inflation, inflation and the real economy, and the operational framework within which a central bank can deliver a stable price level.

The minutes of Bank of England Monetary Policy Committee (MPC) meetings, which sets short-term interest rates in the U.K., and the subsequently published *Inflation Reports* and *Quarterly Bulletins* can be found at www.bankofengland.co.uk. The quarterly bulletins often include the text of speeches given by members of the committee or by the Governor or Deputy Governors of the Bank of England. The MPC is "shadowed" by the Institute of Economic Affairs Shadow MPC, which monitors decisions taken by the MPC and explains its own thinking on monetary policy. The minutes of those quarterly meetings can be found at www.iea.org.uk. More recently, *The Times* newspaper has introduced a Shadow MPC. Its meetings are reported in *The Times* on a monthly basis. *The Economist*, the *Wall Street Journal* (including its regional variations), the *Daily Telegraph* and the *Financial Times* have commentaries on bond markets and financial market developments more generally.

The annual and quarterly reviews of the U.K. government bond markets can be found at www.dmo.gov.uk, together with a wealth of other information on government bond markets (particularly index-linked markets). There are Web links from here to gilt-edged market makers, as well as to international government bond Websites. These links, combined with www.imf.org and www.oecd.org, would be a good starting point for the analysis of other government bond markets.

2.3.5 Risk Characteristics

Three major risk characteristics of government bonds will be considered: inflation risk, reinvestment risk, and the risk of a portfolio dropping in value, relative to liabilities, if an investor is "mismatched." With corporate bonds, as with some government bonds, there is also a default risk. A conventional bond provides an investor with a guaranteed rate of return to maturity. That rate of return from conventional bonds is fixed and defined in nominal terms. Conventional bonds do not provide investors with protection from the erosion of the purchasing power of money by inflation. One example of this problem was in the U.K. in the 1970s; when inflation became much higher than investors had anticipated, bond investment actually provided investors with negative real rates of return. For example, in the period from 1972 to 1979, long-dated U.K. government bonds gave an average real return of −6.8% (calculated from Phillips and Drew (1985, 1995)). It would generally be felt that the risks of such a negative real return are much less now that inflation is low and stable in most OECD countries. However, for long-term investors, such as pension funds, the risk of a long-term change in the inflation environment should not be discounted.

If an investor matches assets to liabilities, by purchasing a bond portfolio that is very similar in duration to that of the liabilities, there will be little investment risk. However, it will often be the case that the liabilities will have too high a duration to be matched with appropriate assets, given the range of maturities available and the pattern of payments from fixed-interest bonds. If interest rates at which coupons can be reinvested fall below expectations, the investor may not achieve the required yield over the period of the investment. A potentially more serious situation would arise if the maturity dates of the bonds in the portfolio were shorter than the length of the liabilities. In this case, both coupon and maturity payments would have to be reinvested at future, unknown, rates of interest. A rigorous treatment of this problem appears in Chapter 5.

If an investor has a portfolio of long-term bonds but liabilities that are short term, then there is a risk that an increase in market interest rates will reduce the value of the bond portfolio relative to the liabilities. Volatility can be used to measure the likely fall in value of a bond portfolio as a result of a rise in the general level of yields. This problem of mismatching can also occur in the case of a personal investor. For example, a policyholder with a unit-linked personal pension fund may be close to retirement. The fund may be invested in a unit-linked bond fund, which in turn may be invested in longer term bonds. A rise in the general level of yields will cause the unit fund price to fall; the likely extent of this fall can be found by measuring the volatility of the investments in the fund. It should be mentioned that any rise in the level of bond yields would be likely to be offset somewhat by an improvement in annuity rates for the policyholder,[1] so that the resultant effect on the post-retirement income to the policyholder would be somewhat mitigated (see Booth and Yakoubov (2000)).

As well as the above risks, corporate bonds also carry the risk of default. Such default risk can be managed by portfolio diversification and an analysis of the companies issuing the bonds that make up the portfolio; rating agencies will often be used. An undiversified portfolio could lose a significant part of its value if a company defaults.

2.4   General Characteristics of Index-Linked Bonds

Index-linked bonds were issued in the U.K. largely as a response to the problem of sustained high inflation in the 1970s. The first issue was in 1981. Issues of index-linked bonds have also been made in Australia and, more recently, in the U.S. and in a number of other countries. The major characteristics of index-linked U.K. government bonds are as follows. Both coupon and capital payments are guaranteed by the U.K. government and linked to the retail price index. The payments are linked to the index with an 8-month time lag and, therefore, index-linked U.K. government bonds do not give complete inflation protection. Most non-U.K. issues have only a 3-month time lag. Because of the inflation protection they offer, index-linked U.K. government bonds should not require an inflation risk premium in times of high or variable inflation. Thus, index-linked U.K. government bonds should enable the government to fund part of its debt more cheaply, with greater certainty about the real-terms cost of funding debt, while providing investors with a risk-free investment which gives a (more or less) guaranteed real yield to maturity. In general, the other important characteristics of index-linked U.K. government bonds are those that have already been described for conventional U.K. government bonds. However, it is worth noting that, because of the high prospective nominal capital gain, index-linked U.K. government bonds tend to have a low initial yield. The characteristics of index-linked U.K. government bonds make them attractive to institutional investors with real liabilities.

There are relatively few private-sector issues of index-linked bonds, and those that have been made have a small issue size and are relatively illiquid. Many of the private sector index-linked bonds that have been issued have been issued by organisations such as housing associations or privatised utilities, where regulation or other circumstances are likely to lead to the

---

[1] As rates of return that institutions could earn increase, there will tend to be downward pressure on annuity prices because of an increase in the rates of return an insurer would use to price annuities.

organisation receiving income related to increases in the retail price index. The Debt Management Office at www.dmo.gov.uk has detailed information about private and international issues with links to appropriate Websites. International issues do not all have the same coupon payment frequency, and a few do not index the coupon to the retail price index. The valuation of different types of index-linked bond is covered by Deacon and Derry (2004) and to a more limited extent in Adams et al. (2003a). We will just cover valuation in the U.K. context here.

2.4.1 Index-Linked Bond Valuation

Because it is the real, rather than nominal, rate of return that is of interest to the investor, we value the real payments from the stock at a real rate of return. This rate of return is guaranteed if the bond is held to maturity, notwithstanding the inflation lag. We will make a number of assumptions. The nominal coupon rate is D per annum, paid half yearly, linked to the increase in the retail price index with an 8-month lag. The real rate of return is j per annum convertible half yearly. The rate of inflation is constant at rate f per annum. There are exactly n years to run to maturity and the maturity payment is 100, linked to the increase in the retail price index with an 8-month lag. The valuation formula ignoring tax is then

$$P = \frac{I_{t_1}/I_{t_0}}{(1+f)^{1/2}} \left(\frac{D}{2} a_{\overline{2n}|} + 100v^{2n} \right)$$

calculated at rate of interest $j/2$, where I_{t_1} and I_{t_0} are the values of the retail price index 8 months before the payment of the first coupon and 8 months before issue respectively. A full derivation of this formula appears in Adams et al. (2003a), where more complex valuation cases are also discussed. It is more important, in this context, to understand the elements in the formula. If the index-linked bond were indexed without a time lag, then the $(I_{t_1}/I_{t_0})/(1+f)^{1/2}$ term would not appear in the formula. The formula would, therefore, be like that for the valuation of a conventional bond, with the nominal payments (before indexing) being discounted at a real rate of return. This rate of return would then be achieved, regardless of the rate of inflation, throughout the life of the stock. The existence of the indexation lag means that the price for a given real yield, and hence the real yield for a given price, will vary somewhat with the rate of inflation f. This variation in the real yield, as future inflation varies, is greater for shorter term than for longer term stocks (see Table 2.3).

2.4.2 Economic Analysis

In theory, because index-linked U.K. government bonds provide almost complete protection from inflation, their real yields should not be affected significantly by changes in inflation. In reality, however, a change in the outlook for inflation may change investors' perception of the risk characteristics of index-linked U.K. government bonds. Thus, investors may be more attracted to index-linked bonds when inflation is high and the outlook for inflation is uncertain. Given their limited history in most countries, which has coincided with relatively stable economic circumstances, it is difficult to tell whether index-linked bonds have provided this "safe haven" in practice.

Real yields from index-linked U.K. government bonds will be affected by the level of government borrowing and by the balance between savings and investment in an economy. Other things being equal, a high level of government borrowing will lead to a general increase in real yields from all investments. Conversely, a high level of savings in an economy will put downward pressure on all real investment yields. In general, index-linked U.K. government bonds can be regarded as the most secure real investment and, therefore, we would expect their returns to be somewhat less than the expected real returns from most other real investments. Index-linked U.K. government bonds can, therefore, be used as a "benchmark" with which other real investments can be compared. The level of real yields from index-linked U.K. government bonds remained in the band 3.25% to 4.25% for the first 15 years after they were issued, but real yields declined sharply in the late 1990s and have remained at around 2% since.

The levels of index-linked gilt yields from a representative stock and of average long-dated conventional gilt yields in the years 1993, 1996, 1999 and 2003 are shown in Table 2.2. It should, of course, be noted that the index-linked gilt yield is a real yield, over and above inflation. The difference between the nominal yield from conventional U.K. government bonds and the real yield from index-linked U.K. government bonds could be regarded as a first-order approximation to the market's expectation of future inflation (see Section 2.4.4).

Table 2.2 Index-linked gilt yields and long-dated conventional gilt yields for selected times

Year	Long-dated U.K. government index-linked bond yields (%)	Long-dated U.K. government conventional bond yields (%)
1993	3.0	6.5
1996	3.6	7.7
1999	2.0	5.4
2003	2.1	4.4

2.4.3 Risk Characteristics

Index-linked U.K. government bonds bear some inflation risk because of the indexation lag. If the rate of inflation increases in the 8 months before a payment is received, compared with the rate of inflation that was assumed in the valuation, then the real value of that payment is affected; see Figure 2.1.

It is the level of inflation between points A and C that determines the monetary amount of a payment, but it is the level of inflation between points B and D that is used to calculate its real present value. An increase in inflation can, therefore, decrease the real return from an index-linked stock. This effect can be quite significant for a short-term index-linked stock. Indeed, an index-linked stock with only 8 months to redemption does not provide any inflation protection at all; all the payments are known and fixed. The average real yields from index-linked bonds of different terms, at two different rates of assumed future inflation on 3 March 2003, are shown in Table 2.3.

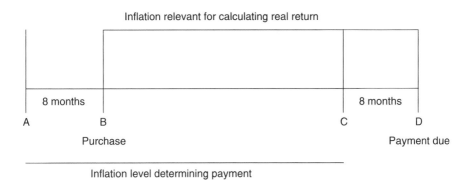

FIGURE 2.1

Table 2.3 Comparison of real yields from representative index-linked bonds at different assumed rates of future inflation, 4 October 2004

Index-linked bond term	Real yield to maturity (%)	
	Assuming 0% future inflation	Assuming 5% future inflation
≤5 years	2.20	1.39
5–15 years	2.10	1.79
>15 years	1.89	1.73

Source: *Financial Times*, 5 October 2004

For someone who invests in long-term index-linked bonds, the indexation lag is not a serious problem, unless inflation rises significantly. There is, of course, a risk that the expected nominal return will not be achieved if inflation falls. For this reason, index-linked bonds should not generally be used by investors seeking to achieve a particular nominal rate of return, unless the risks are understood.

Ignoring the problem of the 8-month lag, the real yield from an index-linked bond is guaranteed if it is held to maturity. However, if the level of real yields increases and an index-linked bond is not held to maturity, then the real holding-period return will be below the real yield at purchase. Indeed, it is possible for the real holding-period return from a portfolio of index-linked U.K. government bonds to be negative. Thus, a short-term investor should not hold a portfolio of long-term index-linked U.K. government bonds, unless the risks are understood and accepted. An investor with longer time horizons, who holds a portfolio of short-term index-linked bonds, will have to reinvest the proceeds at unknown future real rates of interest. Such an investor will, therefore, be exposed to a possible reduction in the level of real yields available. It should be said that, because real investment yields are likely to be less volatile than nominal investment yields, such reinvestment risks may be less serious with index-linked bonds than for conventional bonds. However, the sharp decline in real investment yields in the late 1990s (see Table 2.2) should be noted.

2.4.4   Estimating Market Expectations of Inflation Using Market Information

The existence of index-linked bonds allows us to develop methods to estimate the market's expectation of future inflation by comparing real yields to maturity from index-linked bonds with nominal yields to maturity from conventional bonds. Deacon and Derry (1994) defined the methods of doing so, and these methods are also presented more formally than here in Adams et al. (2003a). Scholtes (2002) also has an interesting discussion of the derivation and application of market-derived measures of inflation expectations.

Three methods of estimating inflation expectations can be used. The first of these is the so-called "simple approach." Here, it is just assumed that the difference between the nominal yield from conventional government bonds and the real yield from index-linked government bonds is an estimate of the market's expectations of future inflation. The underlying assumption of the calculation is that the expected real return from a conventional bond should be equal to the real return from an index-linked bond of the same term to maturity. Considering the data in Table 2.2, for example, the estimate of inflation expectations would be 2.3% for 2003. This is a useful, first-order approximation to inflation expectations. However, there are three reasons why it should not be relied upon where an accurate assessment is required.

It is possible that there should be a risk premium from conventional bonds because they carry an inflation risk, so that the expected real yield to maturity from conventional bonds and the expected real yield to maturity from index-linked bonds are not equal.

Secondly, the real yield to maturity from index-linked bonds is not independent of the expected rate of inflation because of indexation lags; this was defined by Deacon and Derry (1994) as the "problem of internal consistency."

Thirdly, the conventional bond and index-linked bond used (or the indices used) to estimate inflation expectations will have different durations. Even if the bonds used have the same term to redemption, the different pattern of payments from the conventional and indexed instruments means that they represent yields over different time periods. If yields curves are steeply sloping, then this problem invalidates the whole analysis because the difference between conventional and index-linked bond yields would include a yield curve as well as an inflation effect.

To deal with the first problem, it is necessary to obtain an estimate of the inflation risk premium which investors will require from conventional bonds. It may also be necessary to take account of other reasons why expected real yields may differ between the index-linked and conventional markets (e.g., there may be an illiquidity premium in the index-linked bond market, or the two markets may be "segmented" from each other with little trade between them, preventing expected yields from equalising). It is extremely difficult to address this problem, but it may be safe to assume that fluctuations in any risk premium will be second order in magnitude compared with fluctuations in inflation expectations.

It is not difficult to deal with the second problem: that of "internal consistency." Formal equations of value can be set up for the present value of the index-linked and conventional bonds and solved for the real yield from index-linked bonds and the nominal yield from conventional bonds. An appropriate relationship between those yields and expected inflation can then be solved. This involves simple financial mathematics techniques.

The third problem can only be solved by estimating full yield curves of both forward and spot interest rates from both index-linked and conventional bonds. The relationship between those two yield curves will then provide estimates of "forward" and "spot" inflation expectations. The n-year spot inflation expectation estimate would be the estimate of the market's expectation of geometric average inflation over the next n years. The n-year forward inflation expectation estimate would be the estimate of the market's expectation of the rate of inflation in the year n years from now. Such estimates are complex to derive from first principles, but could, in any case, be obtained from standard investment data sources. Spot and forward inflation expectations, for example, can be found at www.bankofengland.co.uk.

Information about market expectations of inflation could have a number of uses for actuarial practitioners. For example, where liabilities of a financial institution depend on inflation (either directly of indirectly), estimates of market expectations of inflation provide an independent estimate of inflation that can be used for projecting liabilities. In U.K. pension funds, pension increases are linked to inflation in each year with a "cap" or "ceiling." Market-derived forward expectations of inflation can be used to determine by how much pensions might be expected to increase in each future year. Deferred pensions are linked to inflation. Again, increases are capped, but the cap is activated if compound average inflation over the period of deferment is greater than 5% per year. Thus, spot market expectations of inflation could be used to determine the expected value of future pension increases. Spot inflation expectations, derived from market data, are shown in Table 2.4.

Published data on inflation expectations derived from market data are not available before 1985. The limitations of index-linked government bond data preclude the collection of reliable data before that time. It is notable that inflation expectations declined significantly over the period. From 1979, there was a strong downward trend in actual inflation, with each peak being lower than the preceding peak and each trough lower than the preceding trough. The institutional mechanisms were also strengthened, leading to greater "credibility" in the monetary policy regime. This started with the development and implementation of the "Medium Term Financial Strategy" in the early 1980s and continued with the inflation targeting regime and the granting of some independence in the timing of changes in interest rates to the Bank of England. Credibility was further enhanced when the Bank of England was granted full operational independence. Whilst inflation expectations did not fall to the inflation target of 2.5% immediately, they had fallen to very close to that level by June 2001. The Bank of England was given full operational independence and responsibility for meeting the inflation target in May 1997. Between the end of April 1997 and mid June 1997, 10-year inflation expectations fell by 0.7%.

Table 2.4 Ten-year spot inflation expectations

Date	Ten-year spot inflation expectation (%)
30 June 1985	6.9
30 June 1991	6.0
30 June 1995	4.7
30 June 1997	3.4
30 June 1999	3.2
30 June 2001	2.6
30 June 2003	2.5

Source: Bank of England

2.5 General Characteristics of Foreign Currency Bonds

The particular characteristics of a foreign currency bond issue will depend on the issuer and the type of bond (domestic bond, eurobond or foreign bond). In general terms, an investor purchasing a foreign currency issue will receive a series of payments which are fixed in terms of the foreign currency. These could be held to meet foreign currency liabilities, in which case the investment will have similar risk characteristics to those of a sterling bond investment held to meet sterling liabilities. Alternatively, foreign currency bond investment may be carried out by funds with predominantly domestic currency liabilities. In this case, exchange rates and exchange rate risk will have to be considered and analysed (see below).

2.5.1 Valuation

For an institution with foreign currency liabilities, which invests in foreign currency bonds for matching reasons, the valuation issues are no different from those described above for domestic currency bonds. An investor with domestic currency or real liabilities, investing in foreign currency bonds, has other considerations in mind — in particular, exchange rate movements.

High long-term nominal bond returns may exist in markets where investors expect a long-term depreciation of the currency. In the valuation, investors must allow for any expected depreciation or appreciation of the currency. The mathematics of allowing for expected currency movements in the valuation of foreign currency bonds can be found in Adams et al. (2003a). Generally, if a foreign currency is expected to appreciate by $x\%$ per annum against the domestic currency and the foreign currency return from a foreign currency bond is $i\%$, then the expected return in the domestic currency is approximately $(x+i)\%$.

2.5.2 Economic Analysis

The only additional consideration for the economic analysis of foreign currency bonds is that of exchange rate fluctuations. In other respects, the economic factors affecting yields from foreign currency bonds are the same as those affecting yields from domestic currency bonds, although some foreign government bonds may be regarded by investors as not free of the risk of default (see Section 2.3.4). Exchange rate movements can be a very important influence on the domestic currency return from a foreign currency bond. The fundamental factors that affect exchange rates will be discussed in Section 2.8 on international equity investment.

Table 2.5 Ten-year government bond yields (4 October 2004) in selected countries

Country	Ten-year government bond yield (%)
Germany	4.03
Italy	4.20
Japan	1.54
Great Britain	4.89
U.S.	4.18
Brazil	7.69[a]

[a]Redeemable 2010.
Source: *Financial Times*, 5 October 2004

It is of interest to look at bond yields in different markets and relate differences between them to the economic issues affecting bond yields discussed in Section 2.3.4: see Table 2.5.

Both Italian and German government bonds are issued in the same currency. There is a yield spread between them and this will reflect the extra credit risk attached to the Italian government (or, possibly, a risk of EMU break up). As a result of its high stock of outstanding debt, Italian government bonds are only AA rated. Japanese government bonds have particularly low yields, relative to those of the other countries: most analysts expect persistent deflation in that country; there is also a very high level of saving. U.K. government bond yields are slightly higher than those in the Eurozone. However, a full decomposition of the yield curve does not find a consistent pattern of lower yields in the Eurozone. U.S. government bond yields are lower than those in the U.K. This may reflect low inflation expectations and the relatively high liquidity of the U.S. bond market, but also possibly reflects low current short-term interest rates (which will always have some impact on longer term yields, even if they are not expected to persist for more than 2 or 3 years). The Brazilian bond is denominated in euros. It should be noted that this has a maturity time of 6 years not 10 years. Because the currency of issue is euros, euro rather than Brazilian domestic currency inflation is irrelevant when valuing the bond. The reason for the yield spread is the credit risk attached to the Brazilian government bond, it is rated B+ by Standard & Poors.

2.5.3   Risk Characteristics

We can distinguish between three types of investor: those with foreign currency liabilities (such as non-life insurers conducting business abroad); those with real liabilities (such as pension funds); and those investors with domestic currency liabilities or who are investing in foreign currency bonds in order to maximise the domestic currency return on an investment fund.

For an investor with foreign currency liabilities, the risk issues are the same as for a domestic investor investing in domestic currency bonds (see Section 2.3.5). For an investor with real liabilities, the greatest risk arises from inflation in the foreign currency. Unanticipated inflation in the foreign currency is likely to lead to a depreciation of the foreign exchange rate and a reduction in the real return. The investor may be protected from inflation in the domestic currency, as this may be compensated by an appreciation of the foreign currency. An investor concerned primarily with domestic currency returns has a straightforward exchange rate risk.

This risk could be managed by the use of forward exchange markets. In either case, there is also the risk of a real exchange rate movement, i.e., a movement in exchange rates not related to relative inflation rates. Such exchange rate movements can be particularly significant in the short term, but they can also be lasting. There is an analysis of the evidence for different explanations of exchange rate movements in Miles and Scott (2002).

2.6 General Characteristics of Equity Investment

In the U.K. and U.S.A., the issue of equity capital is the most important method by which companies raise long-term capital to finance their operations. In the U.K., a particularly large proportion of institutional investment takes place in equities (see Chapter 3). The holders of the equity capital of a company own the company. After all other commitments have been met, equity holders are entitled to a share in the residual net profits of a company. Some of the residual net profits will be retained for further investment and the remainder paid to equity capital holders by way of a dividend. In the event of a winding up because of insolvency, equity holders have no rights to repayment of capital until all other claims on the company have been met. Therefore, it is possible for equity capital holders to lose all of their investment (i.e., obtain a minimum return of −100%). It is very unusual for owners of corporate bonds to lose all of their capital (i.e., if the company defaults and gives rise to a loss, given default of 100% on its debt capital). Equities have a so-called "put option value": given the feature of limited liability, it is not possible for an equity investor to obtain a return of less than −100%, even if the company has a negative net value. The put option value grows as the company becomes more highly geared (has a higher level of debt finance), because it is more likely that the company will suffer a "shock" that will lead it to have a negative net present value if it has a higher level of debt.

It is generally expected that the dividends declared on equity investments will tend to increase; in the long term, they will tend to increase at about the same rate as the company's profits increase. It is not unreasonable, in the long-term, to expect company profits to be unaffected in real terms by inflation. Thus, profits and dividends should grow to compensate the investor for inflation in the long term.

Equities are traded on the stock exchange, and their capital value, to a large extent, will reflect investors' expectations of future profits and dividends. Capital values can be volatile and equities are, therefore, not suitable as a short-term investment for most institutional investors. Because, in the long term, capital values should also rise to compensate the investor for inflation, one of the most important characteristics of equities is that they can provide the investor with protection from long-term inflation. The value of an equity investment, calculated at a particular rate of return, is a subjective matter because an investor does not receive a fixed series of payments.

2.6.1 Equity Valuation

We will begin by discussing a fundamental approach to valuation. An equity provides the investor with a dividend stream. A starting point in equity valuation could, therefore, be the valuation of the expected dividend stream: the so-called "discounted dividend model." If the investor requires a rate of return of r per annum, and an equity is expected to pay a dividend annually of d_1 at the end of the first year, d_2 at the end of the second year, and so on (in practice, dividends would be paid half yearly), then the present value of the dividends, V_0, would be

$$V_0 = \sum_{j=1}^{\infty} \frac{d_j}{(1+r)^j}$$

If the dividends are expected to grow at a constant compound rate g per annum from an initial level of d_1, then:

$$V_0 = d_1 \sum_{j=1}^{\infty} \frac{(1+g)^{j-1}}{(1+r)^j}$$

Therefore:

$$V_0 = \frac{d_1}{r-g} \quad \text{if } g < r \tag{2.2}$$

This formula is known as the "Gordon growth model" formula. It can be seen that it provides a straightforward way of valuing an equity investment.

A two- or three-stage discounted dividend model could be used, whereby one rate of dividend growth could also be used in the near future, followed by a second, medium-term rate, followed by a long-term growth rate, assumed to continue into perpetuity. These models are easy to apply from a computational point of view using basic actuarial mathematics techniques. However, the difficulty with equity valuation is the subjective nature of the process of estimating future dividends and dividend growth.

We can compare the subjective valuation of an equity with the market price when taking a decision to buy or sell. If the market price of an equity is P_0, then, adapting Equation 2.2, we can say

$$P_0 = \frac{d_1}{r - g} \quad \text{if } g < r \tag{2.3}$$

where P_0 is the market price of the equity. In this equation, d_1, r, and g could reasonably be said to represent estimates of the next dividend, future dividend growth and the required return by marginal investors in the market. In general, the required return r could be decomposed into the risk-free rate of return r_f and the equity risk premium r_p.

There is legitimate dispute about the usefulness of the discounted dividend model in investment analysis. In general, fundamental analysts are more likely to use an earnings approach to valuation. Taking Equation 2.3 and dividing both sides by the earnings per share e_1, we obtain

$$\frac{P_0}{e_1} = \frac{d_1/e_1}{r - g}$$

The left-hand side of this equation is the price/earnings ratio and the right-hand side could be regarded as the variables that should determine the price/earnings ratio. The analyst can compare the actual price/earnings ratio with the price/earnings ratio believed to be reasonable, based on the analyst's own estimates of expected dividends and earnings, future dividend growth, and required rates of return.

Which method of valuation is more appropriate? In theory, discounted dividend models use all the information necessary to ascertain the value of an equity. However, in practice, the variables that are used in the discounted dividend approach are difficult to estimate. Methods of valuation that concentrate directly on earnings and the immediate prospects of the company are often preferred.

A chain of causality can be traced from the key variables which analysts study to future earnings and dividend growth. For example, costs and prices determine profit margins; profit margins and sales determine return on capital; return on capital, tax, and capital structure determine net return on equity; net return on equity and profit retentions determine future profits growth; and future profits growth and dividends determine value. Thus, in theory, it should be possible to translate the key variables that analysts observe into the components of the discounted dividend formula. However, if this is difficult to do in practice, it might limit the practical applicability of the discounted dividend approach. Fundamentally, the same variables are required for the price/earnings valuation approach. However, it is possible to rearrange this formula so that the focus is on return on equity rather than on future dividend growth. Some would argue that the discounted dividend model did not predict either the "dotcom" boom or the fact that equity prices rose in the U.K. after the government increased taxes on equity investment, to remove 20% of the cash flow for pension fund investors.[2] Supporters of the discounted dividend model would argue that this is simply an issue of speed of response to external shocks and that their approach has been more than vindicated by the collapse of "dotcom" companies and the subsequent decline in the equity market more generally.

This debate cannot be resolved in this brief introduction to investment. However, we will note that, fundamentally, an equity is valuable in the long term for its cash flows alone. Methods of valuation focusing on cash flows should at least be used as a check on other methods of valuation and analysis. Other techniques of valuation, such as "economic value added," do not value cash flows directly, but instead fundamentally dissect a company for its economic potential in a way that can be related to either the discounted-dividend or earnings-based approaches to valuation. More detailed discussion and empirical evidence on the different methods of valuation can be found in Lofthouse (2001) and Bodie et al. (2002).

As has been noted, a great deal of information is required for equity analysis, as subjective estimates of the key variables are required. Much of the information for equity valuation will be found from company reports and accounts. Methods of analysing this information are discussed by Bodie et al. (2002). Accounting and investment information about companies themselves can be found from proprietary sources, such as Hemscott at www.hemscott.net. Research on the issue of analysts' forecasts and the determination of equity valuations is discussed by Panigirtzoglou and Scammell (2002).

So far we have described *fundamental analysis*, where analysts will look at the financial prospects for the company to try to predict future profits and dividends, evaluate the efficiency of the company, and assess the company's ability to continue as a going concern. Analysts may also use *technical analysis*. Technical analysts believe that a share price may depart from its fundamental value for significant periods. They further believe that the sign and magnitude of this departure can be inferred by studying charts of past

---

[2]In the July 1997 Budget, the U.K. government removed tax credits on dividends paid to pension funds, thus reducing, at a stroke, expected future cash flows from dividends.

share price movements. Such techniques are described in greater detail by Bodie et al. (2002) and Lofthouse (2001). Analysts may well use a combination of fundamental and technical analysis. Technical analysts are unlikely to use chart patterns alone, but are also likely to consider whether there are fundamental reasons for the development of the chart patterns. Fundamental analysts may believe that a particular share is over- or under-priced according to fundamental factors, but may determine the timing of sell and buy decisions after considering price patterns described in charts.

2.6.2 Economic Analysis

In the discussion here, we will concentrate on the fundamental factors that affect equity values. We would expect that the value of an equity would be most critically affected by expected dividend growth and interest rates. In general, a reduction (increase) in interest rates will lead to an increase (reduction) in the present value of the income stream and to a general reduction in equity values. There are, therefore, interconnections between equity markets and bond markets. However, equity values should be unaffected by the outlook for inflation, as an increase in inflation should, in the long term, lead to an increase in expected dividend growth, as well as an increase in investors' required returns. Therefore, we might expect equities to maintain their real value in times of inflation. Thus, if the general level of interest rates in bond markets rises due to a rise in the market's anticipated rate of inflation, this may not affect equity values in the same way as it would affect bond values. Changes in long-term real rates of interest are much more important for equity markets. For this reason, a valuation model with variables expressed in real terms may be most appropriate: see Adams et al. (2003a).

There is evidence that increases in liquidity in the financial system, perhaps preceded by or caused by a relatively lax monetary policy, may impact on asset prices before impacting on inflation through the goods market. Thus, low short-term real interest rates and a loose monetary policy, all other things being equal, may lead to increases in asset prices not justified by fundamentals. Pepper (1994) provides a very detailed analysis of the effect of monetary policy on financial asset values that is still relevant today. There is much debate about the relationship between asset values and the monetary policy, which is summarised in Friedman et al. (2003).

A number of economic factors, other than inflation, need to be studied when considering the prospects for future dividend and capital growth. The general level of economic growth, increases in productivity, and changes in input prices are all important determinants of the key variable of return on equity. The equity market will often respond to a change in short-term interest rates for reasons other than those cited above. An increase in short-term interest rates may lead to lower expectations of future economic growth in the medium term, causing companies cash flow problems, possibly endangering their solvency and making investors more pessimistic about future levels of real interest rates. These factors may cause equity values to fall in their own right and may also make equities unattractive relative to bonds.

2.6.3   Risk Analysis

The major risk factors involved in equity analysis are evident from the discussion of economic issues above. Equities can provide the investor with long-term protection against unanticipated inflation, but large short-term fluctuations in capital values are evident. Because the income stream from an equity is uncertain, there are two major risks from equity investment. First, as has been mentioned, capital values can be quite volatile in the short term; second, the long-term income stream may prove to be lower than expectations (this will, in turn, probably lead to long-term capital value underperformance). These risk characteristics are critical in determining the role of equities in institutional portfolios (Chapter 3). It is worth noting that the idea of interest-rate sensitivity of bond investments, indicated by duration, can be applied to equity investments too. If there is a change in investors' required rates of return, or in expected dividend growth, this will affect the value of an equity investment. The longer that the cash flows from an equity are expected to be deferred (as a result of the company retaining profits for reinvestment and hoping to achieve long-term future dividend growth), the more sensitive to interest rate changes and expected dividend growth will be the value of the equity. Thus, low-growth, high-yield shares are sometimes said to be "defensive" and high-growth, low-yield shares are sometimes said to be "aggressive." Similarly, as has been noted, highly geared companies are likely to be more "risky" than less highly geared companies. A high level of debt on a company balance sheet both increases the risk of bankruptcy and increases the volatility of shareholders' funds.

If we accept that equities do provide investors with long-term protection against inflation, then their nominal return is likely to be more unpredictable than their real return. In a period of inflation which is lower than anticipated, the nominal return from equities may be lower than anticipated. Thus, holding a portfolio of equities to match long-term fixed liabilities would be risky. There is, first, the risk of disappointing real dividend and real capital growth. Second, there is also the risk of lower than expected inflation, leading to lower than expected nominal dividend and capital growth.

As with all investments, equity investments are at risk from tax changes affecting their value. However, there is probably a greater risk with equity investments, as their tax treatment depends on the corporate regime, the personal regime, and the institutional investor tax regime. The tax position of equity investments is often far from transparent, making it more prone to manipulation and instability. Of course, tax changes can be beneficial as well as detrimental to equity values.

It is sometimes helpful to divide risks into diversifiable and non-diversifiable risks. Diversifiable risks can be reduced by holding a diversified portfolio of equities. Such risks would include certain company-specific risks, such as the risks of mismanagement. Non-diversifiable risks cannot be reduced by holding a diversified portfolio of equities. These will include risks that affect the equity market as whole (e.g., the risk of tax changes). Techniques for exploiting the power of diversification are discussed in Chapter 4.

2.7   Real-Estate Investment

Investment in direct real estate involves the purchase of interests in land and buildings. Those interests may be complex. In the simplest case, the purchase of a freehold property, the investor owns the property and will lease it in return for a rental income. The rent will be reviewed at regular intervals to reflect market rental levels. An example of how interests can be split would arise when the freeholder assigns the lease in return for a consideration, the institution to which the lease had been assigned paying a fixed rental to the freeholder and itself receiving the right to receive the rent from the tenant.

There are a number of characteristics of direct real-estate investment that derive from the inherent physical nature of this asset type. Each property is unique and is not normally split into tradable units that could be regarded as analogous to shares. Therefore, direct investment in real estate is difficult for small investors. Because real estate is indivisible and because of the legal complexities and other costs surrounding the transfer of ownership, real estate is traded infrequently and is, therefore, regarded as rather illiquid. The absence of frequent deals in individual properties, together with the unique nature of each property, means that, at any particular time, it is difficult to ascertain a value. Valuation is a matter of subjective judgement. Real-estate indices are, therefore, difficult to construct, as regular objective valuations on individual properties are not available.

Unlike many other investment types, with real estate the physical asset has to be "managed" by the investor. The problems of maintenance, repairs, and insurance have to be dealt with. Rents have to be collected and negotiated. The investor has to allow for maintenance costs, management costs, refurbishment, redevelopment, and obsolescence. The complexities of real-estate law also increase both the costs of management and the transaction costs of investing in real estate. These various management costs are borne directly by the investor. The returns from real-estate investment come from the rental stream and any capital growth. In the U.K., rents are normally paid quarterly in advance, before deduction of any taxes. The terms and conditions of a lease are a matter for negotiation and, thus, vary from property to property. However, it is normal for rents to be reviewed regularly (normally, every 3 to 5 years), when they are either raised to the current level of market rents or maintained at the previous level, whichever is higher. This is the so-called "upward-only rent review" clause and is one example of many "embedded options" within real-estate investments (see Section 2.7.1). Rents related to business turnover, and breaks from rents in the early stages of leases, are additional terms that can be quite common. At the expiry of the lease, terms and conditions have to be renegotiated.

Two points should be noted from the above. First, capital values will depend on the level of rents and expected rental growth. Second, one would expect rents, and hence capital values, to be correlated with inflation, although imperfectly so. Real estate can be said to be a real investment that provides the investor with a long-term hedge against inflation. However, the value of real estate as an inflation hedge is something that is disputed. The evidence is reviewed by various authors, discussed in Brown and Matysiak (2000).

Investors can also obtain exposure to the real-estate market using real-estate "vehicles." These include equity vehicles, quoted on the stock market, such as real-estate investment companies (in the U.K.) and real-estate investment trusts (in the U.S.). Small investors who cannot afford the management costs of real-estate investment can obtain exposure though unit trusts, limited partnerships and managed funds. These vehicles are not considered further here. However, information on indirect methods of exposure to the real-estate market can be found from Chan et al. (2003), Hudson-Wilson (2000) and from papers posted on the Investment Property Forum Website at www.ipf.org.uk.

2.7.1 Valuation

To an actuary, the most obvious way to value real estate would be to value the expected rental stream using a discounted cash flow approach. In practice, however, traditional valuation methods use comparisons of initial yields with benchmarks for different types of real estate. Such methods may be effective in stable market conditions. The traditional methods of valuation and the comparisons with discounted cash flow methods are discussed in detail by Baum and Crosby (1995) and, from a more actuarial perspective, in Adams et al. (1999). The discounted cash flow method, described here, is presented in Adams et al. (1999).

Consider an investor who purchases a freehold property on the basis of the following assumptions:

V_0 is the value of the property
R_1 is the initial level of rental income per annum, paid in p instalments (net of the annual management costs described above)
g is the level of growth of market rents per annum (assumed constant)
n is the rent review period
r is the investor's required return

We will further assume that: the lease is a new lease or that a rent review has just been carried out; there are no tenant defaults, and that leases are renewed immediately upon expiry, on similar terms; refurbishment and redevelopment costs are ignored. It is not difficult to adjust the valuation formula, if the above assumptions are relaxed. The present value of the freehold property is the present value of the expected rents, at the investor's required return. If the lease is renewed on similar terms at the end of its life then the rental stream would continue in perpetuity. Therefore:

$$V_0 = R_1 \ddot{a}_{\overline{n}|}^{(p)} + \frac{R_1(1+g)^n}{(1+r)^n} \ddot{a}_{\overline{n}|}^{(p)} + \frac{R_1(1+g)^{2n}}{(1+r)^{2n}} \ddot{a}_{\overline{n}|}^{(p)} + \cdots$$

so that

$$V_0 = \frac{R_1 \ddot{a}_{\overline{n}|}^{(p)}}{1 - \left(\dfrac{1+g}{1+r}\right)^n}$$

where $\ddot{a}_{\overline{n}|}^{(p)}$ is calculated at rate of interest r.

In the U.K., where rents are payable quarterly, $p=4$. The investor's required return r would depend on the return available from other investments and the perceived risk of a particular property investment.

If real estate is considered to be a real investment, then it might be sensible to perform the discounted cash flow analysis in real terms, i.e., discount real rents (after adjusting for inflation) at a real rate of interest.

It is important when valuing real estate to take account of the particular characteristics of an investment. These characteristics will vary across different countries, as well as across different properties within a country. The characteristics of lease clauses available in particular markets may well change over time, including in response to legislation. In the U.K., there has been a tendency for lease terms to shorten in recent years. Most lease clauses have the "upward-only rent review" characteristics described above. This characteristic means that the formula presented above can be improved upon. The above formula values the cash flows from the investment calculated using the expected growth rate of rents. However, if a property is let with upward-only rent reviews, rents can increase if market rents increase, but they cannot decrease. This situation is as if the freeholder had a right to receive rents equal to the level of market rents but had a put option that allowed the investor to "sell" market rents for the current rent receivable from the property if market rents fall below that level. When this happens, the property is "over rented." A number of other particular lease clauses can also exist that can be important for valuation. For example, "break clauses" may allow a tenant to leave a property and "break" the lease if rents from a property are above the general level of market rents (or for any other business reason). Valuation of properties with option clauses is difficult. Despite the fact that options are the norm, they are very rarely valued explicitly. There is a developing literature on the valuation of options embedded in real-estate contracts. This is summarised in Adams et al. (2003b) and particular techniques are presented in Booth and Walsh (2001a,b).

Research papers and practical papers on various aspects of valuation and other aspects of real-estate investment can be found from the Investment Property Forum (www.ipf.org.uk), the Real Estate Research Institute (www.reri.org), and the Royal Institution of Chartered Surveyors (www.rics.org.uk). Data on current lease clauses and real-estate performance can be found from the Investment Property Databank (www.ipd.org.uk).

2.7.2 Economic Analysis

As with equity analysis, it seems reasonable to begin by considering the variables in the fundamental valuation formula. The theoretical relationship between real-estate values and different financial variables is discussed in detail in Adams et al. (1999). In theory, real-estate values should be most affected by changes in anticipated rental growth rates g and long-term interest rates that will affect an investor's required return r. The effect of inflation should be to increase both expected nominal rental growth rates and investor's required returns. The effect on a property value may, therefore, be broadly neutral. However, as has been mentioned, the effect of inflation on real-estate values is disputed. Economic factors which investors believe are likely to lead to higher real rental growth are also likely to lead to higher capital values. Thus, better expectations for economic growth are likely to benefit the real-estate market. Similarly, a shortage of supply of real estate, in a particular category, is likely to put upward pressure on rents, and hence capital values, in that sector.

If real interest rates fall in other investment markets, in theory at least, the required return of investors in the real-estate market should fall, so that capital values increase. The real-estate market can also be affected by the conduct of monetary policy, partly because development is often financed through the banking system. For example, the market is likely to be affected by a change in short-term interest rates. An increase in short-term interest rates is likely to induce a slowdown in economic growth. This will reduce the short-term demand for real estate and, therefore, reduce current levels of rents. At the same time, it will probably reduce expectations of future rental growth. These problems will often be compounded by two other factors. First, much real-estate development is financed by development companies borrowing at floating rates of interest; development companies may have difficulty servicing their debt if short-term interest rates rise (particularly if the slowdown in growth causes rented property to become vacant). Second, if the high interest rate policy follows a period when short-term interest rates were low, it is likely that the market will have become oversupplied with property at that time. Thus, there will be an "overhang" for the market to absorb, before the demand for new property increases.

2.7.3 Risk Analysis

The difficulty of ascertaining reliable real-estate valuations makes any statistical analysis of risk difficult. Real estate would appear to provide a good hedge against inflation on theoretical grounds, but the empirical evidence, as has been mentioned, is far from clear. Real estate is a long-term investment, which can be volatile in the short to medium term. The risks of using real estate as a short-term investment are exacerbated by its illiquidity, lack of marketability, and transactions costs. Real estate appears less volatile than equities in the short to medium term. However, this may simply be because real-estate performance indices use valuation-based data, rather than transactions prices, and valuations are affected by "valuation smoothing."[3]

As far as the longer term is concerned, real-estate investment is often regarded as being more secure than equity investment. The main reason for this greater security is that the rental stream is guaranteed by contract. Rents must be paid to the owners of the property before any dividends are paid to the shareholders. Also, the upward-only rent review clause prevents rents from falling during a lease term, at least in nominal terms. Furthermore, if a tenant becomes bankrupt, the owners of a property retain the physical property that can be re-let.

---

[3]This problem, whereby surveyors base estimated values on "historical comparables," combined with the general lack of timely data, is a serious impediment to serious analysis of property data. There is a discussion of this problem in Brown and Matysiak (2000).

2.8 International Equity Investment

The general characteristics of international equity investment are not inherently different from those of domestic equity investment. Therefore, only the additional considerations, necessary when investing in international equities, will be discussed here. An investor may be attracted to international equity investment in order to obtain higher returns than those available in the domestic market. The investor should analyse international markets and currencies, in order to try to identify those that will outperform the domestic market. In addition, the investor may try to reduce the risk of equity investment by diversifying the portfolio internationally. This should reduce the risk of both short-term volatility and long-term underperformance of an equity portfolio.

Against this, there are additional costs and risks of international equity investment. If direct portfolio investment is undertaken, then there will be difficulties in dealing with an information supply that may be inadequate and information presented in a different format than that used in domestic equity markets. Many international equity markets are small, and they tend to be quite volatile. In addition, there is the risk of adverse exchange rate movements (discussed below) and the possibility of additional costs in terms of taxation.

2.8.1 International Equity Valuation

There are few additional points to add to Section 2.6.1, other than to deal with the problem of currency fluctuation. If a discounted dividend approach were to be used, then the investment analyst would need to allow for expected dividend growth and expected currency movements. Given the "real" nature of international equity investment, other approaches may also be sensible — for example, considering expected real dividend growth and expected real movements in exchange rates.

When international equities are being analysed, the income stream may be less important than estimating future changes in capital values. Most foreign companies tend to retain a higher proportion of earnings than do U.K. companies, for reinvestment within the company. The dividend income can, therefore, be relatively low, but this should be compensated by higher capital growth in the long term.

2.8.2 Economic Analysis

We have already covered the factors that affect domestic equity returns (Section 2.6). Here, we need merely add some comment on exchange rate analysis (see also Section 2.5) on foreign currency bonds. In the long term, we would probably expect exchange rates to be affected mainly by relative changes in price levels in different countries. Thus, we would expect the exchange rate of a high-inflation economy to depreciate gradually. Other things being equal, we may also expect the exchange rate of an economy with strong productivity growth to appreciate over time.

In the short term, other factors must also be considered. Short-term real interest rates, the level of government borrowing and the balance of payments are all important factors determining exchange rates. In addition, exchange rates may be affected by speculation and appear to diverge from fundamental value for quite long periods of time. Analysts would need to consider the extent of this overvaluation or undervaluation when estimating future returns from international equity investments.

2.8.3 Risk Analysis

Again, most aspects of risk analysis have been covered in Section 2.5 and Section 2.6. We need only consider here some further difficulties with international equity investment, including foreign exchange risk. The fact that an equity is denominated in a foreign currency does not change the fundamental nature of the equity investment: it is still a real investment. However, there will be foreign exchange-rate risk factors. In the long term, an increase in the general level of prices in a particular country should lead to an increase in dividend levels and in the capital values of equities denominated in that currency. This is likely to be offset by a depreciation of the foreign exchange rate. If there is domestic inflation, the domestic exchange rate should depreciate, increasing the domestic currency value of the foreign currency investment, thus offsetting the fall in the purchasing power of money. Of course, insofar as long-term exchange rate movements do not reflect relative inflation levels, there is an additional exchange rate risk. In the short term, in particular, exchange rates can be very volatile and may not necessarily reflect obvious fundamental economic variables. There can, therefore, be considerable short-term foreign exchange risk from international equity investment.

Investors can hedge against short-term exchange rate movements using forward and futures markets, although this brings its own costs and risks. Most world equity markets (other than those in the U.K., the U.S.A., Japan and possibly Germany) are relatively small and tend to be volatile. This volatility in smaller, less-well researched markets can be used to the advantage of investors, in that markets may, arguably, become inefficiently priced or "good value." However, shares bought at a high level may well fall in value or shares may remain undervalued compared with the value indicated by fundamental economic and financial factors for some time.

If an institutional investor decides to buy international equities because it believes they are "underpriced", then they may have to be prepared to hold those equities for a considerable period before their price returns to reflect their fundamental value.

2.9   Derivatives

In this section we discuss some of the ways in which derivatives can be used to gain exposure to investment markets. Derivatives are investment instruments whose value is "derived" from the value of an underlying investment, or combination of investments, with similar characteristics. The most important derivatives, as far as institutional investors are concerned, are futures, options, swaps and currency forwards. It is only possible to deal with the issues very briefly here — readers who would like to pursue the matter in greater detail should refer to a text such as Hull (2002) and the references contained therein.

2.9.1   General Characteristics

A *forward* contract is a legally binding contract to deliver or take delivery of a given quantity of an underlying asset at a fixed price on a specified future date. A dollar currency forward, for example, would involve a contract for one party to deliver a specified quantity of dollars at a particular rate of exchange at a particular time to another party. Thus a currency forward contract involves a commitment to buy or sell a given quantity of foreign exchange at a pre-arranged price. Currency forwards are used to hedge foreign exchange risks. A currency forward contract can be used to protect an investor in foreign currency assets from an unanticipated fall in the value of the foreign currency.

A *future* is a derivative instrument that can be used to increase or decrease exposure to a particular investment market. It is a legally binding arrangement between two parties, whereby one side is obliged to buy and the other obliged to sell a particular amount of a particular asset, on a specified future date, at a specified price. The asset may be a specific asset, such as a certain number of shares of a particular company. Alternatively, the asset may be the value of a stock market index. Generally, the purchaser and seller of a future pay a returnable deposit, called a margin; this initial deposit is small relative to the value of the asset to which the contract relates. Any movement in the value of the asset in which the future is traded is reflected by a change in the *margin account*. Thus, the initial deposit is only a fraction of the value of the assets to be delivered at settlement, yet the investor is exposed to movements in the whole of the value of the deliverable assets.

A purchaser of a future might be trying to increase exposure to a particular market; however, unless the investor holds cash representing the value of the assets to be delivered, then the exposure to market movements from a small initial cash investment can be considerable. Futures and forward contracts are intrinsically similar. However, futures contracts can be traded, often on an exchange, whereas forward contracts are often held until the time of delivery. Settlement takes place for forward contracts at delivery, whereas for futures contracts it takes place on a daily basis, based on movements in the market price of the underlying asset — which is almost always tradable.

An option gives the purchaser the right to buy (a *call option*) or sell (a *put option*) a given amount of a particular asset at some future point at a given price. A *European option* gives the right to a purchase or a sale on a particular date and an *American option* on or before a particular date. Investors may buy or sell such options. An example of a call option would be the purchase of the right to buy 10,000 of a particular share, at a price of 150p, in 180 days. If the share price were currently 152p, then the option would have a relatively small value (the share can be bought in the market for 152p and, therefore, the right to buy it at 150p would not be particularly valuable). If the share price were currently 200p, however, then the option would have significantly greater value.

However, even a right to buy a share at the current market price will have some value; the option is a right, not an obligation. Therefore, when the option is exercised, the holder of the option will have the choice of exercising the option or buying in the market, whichever is cheaper. Again, for relatively small cash outlays, the holder of an option may be exposed to quite sharp market movements. A call option could become valueless at exercise if the value of the underlying assets falls below the exercise price; a put option could become valueless at exercise if the value of the underlying assets rises above the exercise price.

A swap involves two parties exchanging a set of cash flows on an agreed basis. For example, in an interest rate swap, one party may agree to pay the other party a series of fixed interest payments in return for a series of floating interest rate payments, based on short-term, market interest rates. Such an arrangement can be used to change the interest rate exposure (or effective duration) of a bond portfolio. It can also be used to allow companies to borrow in relatively cheap markets and then create swaps to produce the maturity structure of borrowing that is required. Such a swap effectively involves one party exchanging the cash flows from a cash instrument (see Section 2.2) for the cash flows from a fixed-interest bond (see Section 2.3). More generally, swaps can involve the exchange of any two sets of cash flows, such as payments in one currency for payments in another (e.g., U.S. dollars for Canadian dollars) or rents from a portfolio of retail properties for rents from a portfolio of industrial properties.

Derivatives are often available as standardised contracts that can be traded through exchanges. However, for long-term investing institutions, the generally short-term derivative contracts that are available through exchanges are not necessarily appropriate for risk management purposes. Tailor-made contracts can be traded "over the counter" frequently through investment banks. An example of a tailor-made contract useful for an insurance company would be a "swaption." A swaption is an option to buy a swap. These can be useful to help match guaranteed annuity options (see also Chapter 3 and Part II). The policyholders of an insurance company that has sold guaranteed annuities have an option to purchase an annuity at a fixed price or take a cash sum and purchase an annuity at open-market prices. If interest rates have fallen and annuity prices have risen, then it may be in the policyholders' interests to do the former. If the insurance company has invested on the expectation of interest rates not falling sufficiently to make it worth the while of policyholders taking the annuity at a fixed price, then the insurance company will make a loss, as the investment fund will be worth less than the (increased) value of the annuity. The insurance company can partially match those liabilities by purchasing a "swaption" that allows it to swap, at given time, a series of floating interest rate payments for a series of fixed interest rate payments. Thus, it can receive, at a fixed price, the fixed interest payments it needs to meet its liabilities. The insurance company can only partially hedge risks using swaptions because there are a number of aspects of the risk of guaranteed annuity contracts that cannot be hedged this way. For example, guaranteed annuity contracts contain an element of mortality risk and also equity price risk. If equity prices increase, then the value of the fund, with which an annuity can be bought on guaranteed terms, increases, thus increasing the cost of the guaranteed annuity option.

2.9.2   Valuation

The pricing of futures, options and swaps is a complex issue that it is not possible to cover in detail here. It is worth noting some general points. In general, we do not value derivatives as free-standing entities. We value them by relating their value to the price of the asset from which their value derives. For example, a future that enables an investor to purchase shares in a given company would be valued relative to the quoted market price of the shares in that company, taking account of any financial differences between exposure to the underlying asset itself and exposure to the derivative. Also, we generally use an "arbitrage-free" assumption. Normally, this involves valuing a combination of the asset itself (the price of which we know) and a risk-free asset (the value of which is easy to compute) which, when combined, has exactly the same financial characteristics as the derivative. In order to apply the arbitrage-free assumption, we need to be able to trade the derivative and the underlying asset in liquid markets, otherwise their prices may diverge. This arbitrage-free assumption allows us to use only risk-free interest rates when valuing derivatives. Sometimes, such as when valuing a forward, these assumptions are reasonable and give good results. On other occasions, such as when valuing options, the assumptions may hold reasonably well and give acceptable results in practice. However, there will be complex options and swaps (e.g., the long-term swaptions referred to above) for which the assumptions may not be reasonable. Either subjective elements will then come into the valuation method or more sophisticated methods of valuation need to be used or both.

We will give one example of a derivative pricing formula. This formula is only intended to be approximate but illustrates the principles enunciated above. In the U.K., the London International Financial Futures Exchange (LIFFE) recommends the following formula for calculating the theoretical price at time t for the FTSE 100 stock index future.

$$F_t = I_t\left(1 + \frac{i_\tau}{365}\right) - I_t y \frac{d^*}{d}$$

F_t = theoretical future prices at time t
I_t = index level at time t
i = risk-free rate of interest
τ = number of days until settlement
y = annual dividend yield on FTSE 100 index
d^* = dividend payments expected between time t and the settlement date
d = total dividend payments expected over the year ending on the settlement date

It is important to realise that only a proportion of the futures price is invested at the outset, representing the margin. F_t is the price to purchase the index at time $t + \tau$. The futures price depends most critically on the current level of the stock market index (as this is the asset to be delivered) and the level of dividends (which the investor will not receive, if the future is bought rather than the underlying asset). Because the contract is not settled until time $t + \tau$, the purchaser of the future has the advantage of being able to invest the money to be paid for that time period, even though the holder of the future has effective exposure to the underlying asset. Therefore the value is increased to allow for the interest that could be earned on the money invested by the purchaser of the future. In this formula, simple interest relationships are used, as are other approximations.

Futures and forward contracts have many characteristics in common and similar formulae are used in pricing.

The price of an option, which is the consideration paid for the right to buy or the right to sell an investment at a given future time, depends on the current stock price, the level of risk free interest rates, the exercise price (the price at which the stock can be bought or sold at the exercise date), and the volatility of the stock price. This latter term is important, because the option is a right to buy or sell — it is not an obligation. If the price of the underlying investment is volatile, then it is more likely that the investor will obtain an advantage by exercising the option. Further information on option pricing and the standard formulae for valuing options can be found in Rubenstein and Cox (1985) and in the paper by Black and Scholes (1973) or in current textbooks such as Hull (2002), Kolb (2002) and Adams et al. (2003a). American options are more difficult to price than European options and these texts cover the valuation of such options.

2.9.3   Risk Characteristics

The investment risks from derivatives arise from not using these investments for legitimate purposes. With regard to futures contracts, as has been explained, a small investment can provide considerable exposure to market movements. There is, therefore, the prospect of "gearing-up" an investment portfolio considerably. This should be avoided by monitoring the total market exposure of the investment portfolio and, where necessary, holding cash, or appropriately dated bonds, in order to deliver futures contracts.

A similar problem exists with a portfolio that includes options. A sharp downwards (upwards) market movement can render valueless a portfolio of call (put) options. Again, the effective investment exposure should be considered. If call options are held, the institution should consider holding the cash necessary to exercise the option. If put options are held, the institution should consider holding the stock to be sold if the option is exercised: if this is not done, a sharp market increase could wipe out the value of the option, without any benefit from an increase in the value of the investments.

Writing or selling options also carries risks. A company selling a call option may have to deliver stock to the holder of the option at a price considerably below the market price, if investment values have risen. This risk could be reduced by holding the stock. If put options are written and investment markets fall, then the writer of the option may have to buy stock at a price above the market price. It may, therefore, be sensible to reduce the holdings of that stock and hold sufficient cash to buy the stock in the event that the options are exercised.

When considering the risk of any derivatives position, it is important to consider the risk in the context of the overall portfolio. Risky positions in derivatives might be offset by equal and opposite risks in underlying assets — particularly if the derivatives have been bought for risk management purposes. On the other hand, risks might reinforce each other. A portfolio that is overweight in U.S. equities, holds U.S. equity futures, and has sold U.S. equity put options, for example, would lose three times over from a fall in the value of U.S. equities. On the other hand, if an institution were underweight in U.S. equities and had bought U.S. equity futures, the risk exposure of the portfolio might be exactly as was required by the institution; the institution may use the futures market to obtain such exposure to avoid the greater transactions costs of buying in the direct market. Similarly, an institution that has entered into a derivatives position as part of its asset portfolio might have an equal and opposite exposure in its liability portfolio. For example, a forward to buy Canadian dollars at a fixed price might be risky for a U.S. life insurance company with U.S. dollar liabilities, but if it has liabilities to pay pensions in Canadian dollars this might not be risky.

In any derivative contract, there always exists "counterparty risk." Counterparty risk is similar in principle to default risk on a corporate bond contract: it is the risk that the institutions with whom the derivative contract has been entered into defaults and, as a result, a loss is made. A number of institutional arrangements have been developed to reduce counterparty risk. These include the use of intermediaries to facilitate derivatives transactions, who will generally have a good credit rating and the use of "margin accounts" to prevent exposures from building up.

Further discussion of the uses and risks of derivatives in an actuarial context can be found in Kemp (1997) and The Faculty and Institute of Actuaries Manual of Actuarial Practice, GN 25, to be found at www.actuaries.org.uk. The latter document also includes professional guidance on the use of derivatives.

References

Adams, A. T., Booth, P. M., and MacGregor, B. D. (1999). Property investment appraisal. *British Actuarial Journal* 5(Part V), 955–982.

Adams, A. T., Booth, P. M., Bowie, D., and Freeth, D. S. (2003a). *Investment Mathematics*. Wiley, U.K.

Adams, A. T., Booth, P. M., and MacGregor, B. D. (2003b). Lease terms, option pricing and the financial characteristics of property. *British Actuarial Journal* 9(Part III), 619–636.

Barclays Capital. (2003). *Equity Gilt Study*. Barclays, London, U.K.

Baum, A. and Crosby, N. (1995). *Property Investment Appraisal*, second edition. Routledge, London, U.K.

Bessis, J. (2002). *Risk Management in Banking*, second edition. Wiley, U.K.

Black, F. and Scholes, M. (1973). The pricing of options and corporate liabilities. *Journal of Political Economy* 81, 637–659.

Bodie Z., Kane, A., and Marcus, A. J. (2002). *Investments*. McGraw Hill, U.S.

Booth, P. M. and Walsh, D. E. P. (2001a). The application of financial theory to the pricing of upward only rent reviews. *Journal of Property Research* 18(3), 69–83.

Booth, P. M. and Walsh, D. E. P. (2001b). An option pricing approach to valuing upward only rent review properties with multiple reviews, *Insurance: Mathematics & Economics* 28, 151–171.

Booth, P. M. and Yakoubov, Y. (2000). Investment policy for defined contribution pension schemes close to retirement: an analysis of the "lifestyle" concept. *North American Actuarial Journal* 4(2), 1–19.

Brown, G. R. and Matysiak, G. A. (2000). *Real Estate Investment a Capital Market Approach*. FT Prentice Hall, U.K.

Chan, S. H., Erickson, J. and Wang, K. (2003). *Real Estate Investment Trusts: Structure, Performance and Investment Opportunities*. Oxford University Press, Oxford, U.K.

Deacon, M. and Derry, A. (1994). *Deriving Estimates of Inflation Expectations from the Prices of UK Government Bonds*. Bank of England, Working Paper Series, No. 23. Bank of England, London, U.K.

Deacon, M., Derry, A. and Mirfendereski, D. (2004). *Inflation-Indexed Securities: Bonds, Swaps and other Derivatives* (2nd ed.), Wiley, U.K.

Friedman, M., Goodhart, C. A. E. and Wood, G. E. (2003), *Money, Inflation and the Constitutional Position of the Central Bank*. Institute of Economic Affairs, Readings, No. 57, Institute of Economic Affairs, London, U.K.

Goodhart, C. A. E. (2003). The Constitutional Position of the Central Bank. In: Friedman, M., Goodhart, C. A. E. and Wood, G. E. (Eds). *Money, Inflation and the Constitutional Position of the Central Bank*. Institute of Economic Affairs, Readings No. 57, Institute of Economic Affairs, London, U.K.

Hudson-Wilson, S. (Ed.). (2000). *Modern Real Estate Portfolio Management*. Frank Fabozzi Associates, Pennsylvania.

Hull, J. (2002). *Options, Features and Other Derivative Securities*. Prentice-Hall, U.S.

Issing, O. (2005), The ECB and the Euro — the first five years, *Occasional Paper No. 134*, Institute of Economic Affairs, Readings No. 134, Institute of Economic Affairs, London, U.K.

Jackson, P. and Perraudin, W. (1999). The nature of credit risk: the effect of maturity, nature of obligor and country of domicile. *Bank of England Financial Stability Review* (7), 128–140.

Jackson, P., Nickell, P. and Perraudin, W. (1999). Credit risk modelling. *Bank of England Financial Stability Review* (6), 94–121.

Kao, D. L. (2000). Estimating and pricing credit risk: an overview. *Financial Analysts Journal* (July/August), 50–66.

Kemp, M. H. D. (1997). Actuaries and derivatives. *British Actuarial Journal* 3(Part 1), 51–162.

Kolb, R. W. (2002). *Futures, Options and Swaps*. Blackwell, Oxford, U.K.

Lofthouse, S. (2001). *Investment Management*. Wiley, U.K.

Miles, D. and Scott, A. (2002). *Macroeconomics: Understanding the Wealth of Nations*. Wiley, U.K.

Panigirtzoglou, N. and Scammell, R. (2002). Analyst's earnings forecasts and equity valuations. *Bank of England Quarterly Bulletin* 42(1), 59–66.

Pepper, G. T. (1994). *Money Credit and Asset Prices*. The Macmillan Press.

Phillips and Drew. (1985). *World Capital Markets: A Review of the Last Decade*. Phillips and Drew.

Phillips and Drew. (1995). *Pension Fund Indicators: A Long-Term Perspective on Pension Fund Investment*. Phillips and Drew.

Rubenstein, M. and Cox, J. (1985). *Options Markets*. Prentice-Hall, U.S.

Scholtes, C. (2002). On market-based measures of inflation expectations. *Bank of England Quarterly Bulletin* 42(1), 67–77.

Chapter 3

General Principles of Asset Allocation

3.1 Introduction

In this chapter we introduce the general principles of asset allocation in respect of an institutional investment portfolio. We then apply those ideas to particular types of institutional investor. This chapter is both descriptive and analytical, and concentrates on the important concepts and principles that are used in the asset allocation process. In Chapter 4 and Chapter 5 we will look at mathematical techniques that can be used in the investment analysis and asset allocation process. It should be possible to rationalise the conclusions of the later chapters in terms of the discussion in this chapter.

Actuaries, whose role it is to manage risk in financial institutions, are often closely involved with asset allocation decisions. Investment policy can have considerable implications for the solvency and financial success of an institution. Thus, as well as considering actuarial theory and practice in relation to the liabilities of institutions, it is also important for actuaries to understand the theory and practice of asset allocation. Indeed, it is the interaction of assets and liabilities which is so important in determining investment risk, and it is on this aspect that we will concentrate in much of this chapter. We will begin by looking at the most important considerations in the asset allocation process: risk, return and institutional and governmental influences and constraints.

In Section 3.5 to Section 3.7 we apply those general principles to the specific examples of a pension fund, life fund, and non-life fund in the U.K. The most important guiding principles are drawn out and the points are illustrated by considering the actual asset distributions of U.K. pension funds and insurers. Actuarial guidance to the U.K. profession is also highlighted, as this provides a justification for some of the principles discussed. In Section 3.8 we discuss the considerations a personal investor might take into account when taking an investment strategy decision.

3.2   Investment Risk

The most important consideration for an institution is the degree of investment risk relating to an asset or asset class. For most institutional investors, investment risk is determined by the risk that the liabilities cannot be met. Therefore, it is important that investments are appropriate given the nature of the liabilities. Thus, investments should be determined so that the financial factors which would tend to cause the value of the assets to rise or fall also cause the value of the liabilities to react in a similar way. There are a number of aspects to the process of minimising investment risk by investing in assets that are appropriate for a given set of liabilities. The process can be summarised by the phrase, "matching assets with liabilities." As well as ensuring that assets are appropriate, given the liabilities, it is also desirable to reduce risk by diversification. This will help to reduce the risk of wholesale default and reduce asset value volatility.

If assets and liabilities are not properly matched, then risk will be taken on by those who ultimately bear the financial cost of investment failure. This will be with-profit policyholders in a mutual life company, shareholders and possibly with-profit policyholders in a proprietary insurance company, and a mixture of the shareholders of the sponsoring company and the members of a defined-benefit pension scheme. In the case of the pension scheme, it is not always clear who bears the investment risk. In a defined-contribution pension scheme, the member would always bear the risk. When insurance companies fail, there may be policyholder protection schemes, and so on, that bear some of the cost. Companies that remain solvent often fund such schemes by means of a levy.

3.2.1   Matching Asset and Liability Characteristics

Because an actuary would generally regard investment risk as the risk of not meeting liabilities, the nature of the liabilities is perhaps the most important consideration when taking asset allocation decisions. Where liabilities are defined contractually, such as in the case of the non-profit portion of a life assurance fund, it may be possible to develop clear guidelines when developing an investment policy designed to reduce risk. Liabilities that are fixed in money terms would be most appropriately matched by fixed money assets (such as bonds). Price index-linked liabilities would be better matched by investment in real asset classes. In countries where they are available, the most precise match would be index-linked bonds.

For many institutions, particularly pension funds and non-life (general) insurance companies, liabilities may not be specifically defined in money or index-linked terms. However, the nature of the liabilities may be such that changes in the general level of prices will tend to have a significant effect on the final money amount of the liabilities. Real asset classes would also be appropriate investments in respect of such liabilities, although other considerations are paramount for non-life insurance companies (see Section 3.7). In the case of some institutions, it is not always easy to define liabilities precisely. For example, the actuary responsible for a with-profit life fund may be charged with the duty of "ensuring that policyholders' reasonable expectations are met." In this case, there may be no written contract guaranteeing that benefits will be paid beyond a certain minimum level, but past experience leads policyholders to expect a certain stability of benefit payments, with bonuses following particular patterns in different stock market conditions. Investment policy must take this into account. Where the liabilities of the fund are not precisely defined, it can be said that choosing appropriate assets to match liabilities is not an exact science. The actuary must take into account the presumed preferences of policyholders, as well as any guaranteed benefits that have been promised. If options or guaranteed surrender values are provided as part of the policy terms, it is necessary to ensure that the investments are such that funds are available to meet the options if they are exercised by the policyholders. This will impose further restrictions on investment policy, and sometimes it will be difficult to match precisely options that have been written.

An example both of the problems of matching complex liabilities and interpreting the precise liability commitment arising from with-profit policyholders is given by the case of the Equitable Life Assurance Society. The society was closed to new business following a House of Lords judgement in 1999. Losses had been incurred as a result of selling with-profit policies that gave the policyholder the option of taking either the cash value of the policy or an annuity priced on guaranteed terms. As long-term interest rates decreased during the 1990s (see Table 3.3), so that market annuity prices increased, the guaranteed annuity became better value to the policyholder and had a greater cost to the society than the cash value. However, the society had purchased assets on the assumption that members would take the cash value of the policy at maturity (see Ballotta and Haberman (2003) for further discussion). Had the society purchased assets to match the annuity, and interest rates had then risen so that the cash value of the policy was better value, the society would also have made losses. In effect, the society had sold call options on fixed-interest securities to its members. Such options, written to mature over such a long time period, are difficult to match. They could only have been matched by purchasing long-term call options or, alternatively, "swaptions," which are relatively unmarketable investments and, at the time the policies were sold, were not available in the market at all. The solvency problems of the Equitable only became critical after a House of Lords judgement that defined "reasonable expectations" of the particular group of policyholders that had guaranteed annuity policies in a way that had not been anticipated by the company. The concept of reasonable expectations of policyholders that is so critical in determining the liability structure and, therefore, the investment policy in life insurance companies is discussed in detail in Shelley et al. (2002) and the references contained therein.

The Equitable example highlights the need to ensure that assets and liabilities are matched. However, it also highlights the need for the marketing and pricing side of the business to be "joined up" to the investment side.

3.2.2 Matching Asset and Liability Terms

Assuming that the correct asset classes have been selected, bearing in mind the nature of the liabilities, it is also important to ensure that the term of the assets is appropriate, given the term of the liabilities. If the term of the assets is too short, relative to that of the liabilities, the maturity proceeds from the assets will have to be reinvested at unknown future rates of interest. If interest rates have fallen below those assumed when premiums were calculated, then this may have implications for the solvency of an institution or the returns to shareholders and with-profit policyholders of an insurance company (or for the contribution rate, for a company that is financing a defined-benefit pension scheme). If the term of the assets is too long, then the assets may have to be sold, when the liabilities become due, at prices determined by a higher yield basis than that on which they were bought. Once again, in this situation, the asset proceeds may be insufficient to meet the liabilities. A more theoretical treatment of this issue appears in Chapter 5.

The length of the liabilities will also influence the asset allocation decision in other, less direct, ways. The short-term returns from some types of investment, such as real estate and equity, can be very uncertain, due to the inherent short-term volatility of such markets. If liabilities are short-term, therefore, such investments may be inappropriate. If liabilities are of a more long-term nature (such as most pension fund liabilities), unpredictability of short-term returns may be less important. It should be mentioned that there is considerable academic and practitioner debate about this issue (see also Section 3.2.5 to Section 3.2.7).

3.2.3 Matching by Currency

Liabilities may be denominated or fixed in different currencies. This should also be taken into account when determining investment policy. If, for example, life assurance policies have been written, with the benefits fixed and payable in U.S. dollars, dollar-denominated bonds may be regarded as appropriate investments a priori. Non-life companies tend to be more internationally diversified than life companies. Foreign currency deposits or foreign currency short-term bonds should generally be used to match the foreign currency liabilities of a non-life company.

Where liabilities are denominated in a particular currency but are not fixed in that currency, because they are inherently real liabilities, the issues are slightly more complex. In theory, a real asset does not lose its real characteristics simply because the payments are made in another currency. Similarly, in theory, the particular currency in which an inherently real liability is going to be paid is irrelevant. In both cases, if movements in relative currency values are determined by relative inflation rates, then the real value of the asset or liability calculated in any currency will be unaffected by relative currency movements. Thus, an institution with real liabilities (such as a pension fund) might be expected to be indifferent between real assets with returns paid in different currencies (for example, between U.K. real estate, U.S. real estate, etc.). The reality, however, is quite different. Whilst investors with real liabilities will tend to invest in overseas equities more readily than investors with fixed or short-term domestic currency liabilities, they will not normally fully diversify their investment portfolios internationally; this is often known as "home bias." There are a number of reasons for home bias. First, currency movements do not fit the purchasing power parity model, whereby currency values only change in response to changes in relative changes in prices in different currencies; thus, there can be significant real changes in exchange rates, particularly over short time periods. Second, there will often be institutional costs of diversifying into overseas investments; this is particularly the case for real-estate investment, where knowledge of foreign real-estate law and the use of international real-estate professionals would be required to diversify internationally. Overall, institutions with real liabilities will normally diversify, to some extent, into overseas equities and, to a more limited extent, into overseas real estate. The issue of home bias is discussed more fully by Davis and Steil (2001).

3.2.4 Diversification

Having chosen assets that are appropriate, given the nature of the liabilities, it may be possible to reduce further the risk of not meeting liabilities by diversifying between different types of inherently similar investment. If it has been determined that assets should be held predominantly in a particular group of asset classes (for example, in real asset classes), investment policy should be pursued in such a way that the pension fund or insurance company is not put at risk by the underperformance of one asset type or one investment within an asset category.

Therefore, it is important to ensure that the portfolio of assets held to meet a particular group of liabilities is diversified between different asset categories of a similar nature. The assets within any one category should, in turn, be diversified by sector where appropriate; in addition, the assets invested in any sector category should be further diversified. Thus, for example, if the nature of the liabilities were such that it had been determined that assets should be invested predominantly in real investments, then the diversification could take place as follows. At the highest level, the assets could be invested in index-linked bonds, real estate, international equities, U.K. equities, and possibly a proportion in cash. The real-estate investments, if the fund were large enough to justify investment in real estate, could be diversified between offices, shops, industrials, and warehouses, and each of those categories could be split between investment in central London, the Southeast, the Midlands, the north of England, Scotland, and so on. The international equity investments could be diversified geographically (by region and then by country within each region) and further diversified by industry sector within each country. U.K. equity investments could be diversified by industrial sector (for example, construction companies, retailers, banks, etc.). Where an investment holding in a particular asset category is not large enough to allow sufficient diversification, which may often be the case as far as real estate is concerned, collective investment vehicles can be used to provide diversification. For example, investing in investment trust shares is a method of diversifying a small portfolio of equities; this principle can also be extended to overseas investment. Investing in real-estate shares or through real-estate unit trusts (real-estate investment trusts in the U.S.) is one method of ensuring diversification whilst investing relatively small funds in the real-estate sector. The basic benefit of diversification is that it reduces the variability of investment returns because the factors that cause investment returns to vary in different markets will be less than perfectly correlated across all markets. This is analysed more formally in Chapter 5.

3.2.5 Short-Term Volatility Risk

A short-term investor will tend to consider risk in terms of the variability of short-term returns. There is considerable variability of short-term returns from real estate and equity investments. Whether short-term variability translates into long-term variability depends, to an extent, on whether the efficient markets hypothesis is valid. Short-term variability is mainly due to short-term market fluctuations, which may well rectify themselves in the medium to long term. Foreign equity markets, particularly smaller foreign equity markets, can be very volatile. This is illustrated in Table 3.1 and Table 3.2, which show the rates of return, in their

Table 3.1 Annual total return from the Hong Kong equity market

Year	Hong Kong equity market return (%)
1983	23.6
1984	51.9
1985	44.1
1986	45.7
1987	−7.2
1988	19.9
1989	5.9
1990	10.5
1991	52.1
1992	33.1
1993	130.4
1994	−30.3
1995	24.6
1996	41.8
1997	−19.1
1998	−8.4
1999	72.8
2000	−13.3
2001	−19.6
2002	−13.9

Table 3.2 Total annual return from U.K. equity market

Year	U.K. equity market total return (%)
1983	27.8
1984	35.8
1985	19.6
1986	27.7
1987	6.9
1988	10.4
1989	38.5
1990	−7.6
1991	22.4
1992	22.6
1993	30.2
1994	−6.3
1995	22.8
1996	17.7
1997	26.5
1998	15.7
1999	24.1
2000	−5.6
2001	−12.7
2002	−22.3

own domestic currency, from the Hong Kong and U.K. equity markets respectively. Both markets show considerable variability of returns, especially the Hong Kong market. The standard deviation of annual returns is considerably higher in the Hong Kong market (37.8%) than in the U.K. market (16.7%) over the period.

The extent to which short-term volatility of an equity portfolio can be reduced through diversification overseas has been thoroughly investigated in the literature. Solnik (1999) has done extensive work to find correlation coefficients between monthly returns from a number of world equity markets. The correlation coefficient between the U.S. and U.K. markets (1971 to 1998) was 0.52; that between the U.K. and Japanese markets was 0.35; and that between the U.K. and German markets was 0.44. All these correlation coefficients are significantly less than one; therefore, they appear to suggest that greater stability would be added to a portfolio by adding investments from overseas equity markets. However, before any firm statement can be made to this effect, one would have to consider the relative variance of returns in the different markets. Also, as financial markets become more integrated, one might expect the correlations between equity market movements to increase. However, Solnik finds no firm evidence of any established trend in this direction. The correlation coefficient between U.K. and Hong Kong equity returns, in the 20 years to 2002, shown in Table 3.1 and Table 3.2, is 0.59; it should be noted that this is annual data and based on domestic currency returns.

3.2.6   Risk of Long-Term Underperformance

In the longer term, as has been mentioned, institutional investors will be most concerned about their ability to meet liabilities. Having chosen investments that are appropriate, given the nature of the liabilities, the investor may increase the probability of achieving the long-term target return by diversifying between asset classes. There will be factors that will cause the long-term return from an asset class to deviate from that which was anticipated. These factors will often tend to be economic or political: some equity markets will perform better than others because of the economic or political climate prevailing in the country through any time period. By diversifying between asset classes and between different investment markets within an asset class, an investor can ensure that the whole portfolio is not affected by adverse long-term investment conditions pertaining to any one market.

It should be said, however, that long-term investment returns in different markets will tend to be positively correlated to some extent. This is because some of the economic factors that affect bond yields and the long-term returns from real investments will often prevail in different markets at the same time, often for a long period. These factors would often affect real-estate markets, equity markets, and bond markets at the same time, and also affect markets in different parts of the world.

3.2.7 Equity Volatility in the Short Term and the Long Term

In Section 3.2.5 and Section 3.2.6 we have distinguished between short- and long-term underperformance risk. However, we need to be careful before taking this distinction too far. Long-term conventional bonds are clearly an appropriate match for long-term liabilities. According to the discounted dividend model of equity valuation, the value of equities can also be regarded as the discounted value of their future income stream. Insofar as equity values fluctuate as a result of changes in risk-free interest rates (or even the risk premium in some cases) underlying their valuation, changes in equity values may not be detrimental to a long-term investor. In a valuation of assets and liabilities, if the interest rate sensitivity of the liabilities is approximately the same as the interest rate sensitivity of the equity investments in a discounted dividend framework (see Adams and Booth (1995) for more discussion of this), then the overall balance sheet will be unaffected by a change in equity valuations caused by a change in required rates of return.

Equities do not fluctuate in value purely because of changes in interest rates. In the discounted dividend model, they can also change in value because of changes in expected dividend growth. Such changes in value can be considerable. For example, a fall of 1% in the expected return on equity capital within a company could lead to a change of 20 to 30% in fundamental value.

In general, those who believe in efficient markets would not consider the long-term risk of equities to be less important than their short-term risk (except for the caveat about the impact of long-term interest-rate changes mentioned above). Those who believe that equity markets have a tendency to revert to the mean would argue that there is an element of "time diversification" to equity investment risk. Long-term risk, it is argued, is less than short-term risk, because equity values may move away from fundamental valuations in the short term but will return to them in the long term. There are further, more sophisticated arguments on both sides of this debate. However, as far as this chapter is concerned, it is sufficient to point out that it should not be thought that short-term equity risk can be ignored by long-term investors. Some actuarial investment models, such as the Wilkie model, described in Chapter 5, do assume that equity markets "revert to mean dividend yields" so that equity risk can be diversified over time. The work by Dimson et al. (2002) referred to in Section 3.3.1 is also a useful contribution to the debate on the relative importance of short-term and long-term equity risk.

3.2.8   Default Risk

The final aspect of risk that we will consider is the risk that the anticipated payments from an investment will not be received because of
default by the issuer. Normally, government bonds are regarded as being
risk free, as far as default risk is concerned, because a government
has the power to extract more taxes or print money in order to meet its
obligations. However, it should be noted that inflation and devaluation are
seen as very damaging options for governments that have overborrowed,
and default risk is certainly a consideration: given the increasing reluctance
of governments to devalue their debt by inflation. See also Chapter 1 and
publications such as the Bank of England's Financial Stability Review
(www.bankofengland.co.uk) and the IMF's Global Financial Stability
Report and other IMF publications (www.imf.org). It should also be
noted that inflation risk is always present with conventional government
bonds that.

Corporate bonds will carry a default risk because the issuer has no powers
of confiscation or powers to devalue the purchasing power of money
enjoyed by sovereign borrowers. When analysing corporate bonds, the
investor should consider the risk of default or use information provided
by companies, such as credit rating agencies: see Chapter 2. A higher yield
to maturity will be demanded the greater is the risk of default. The principle
of diversification to reduce risk can also be applied to investment in
corporate bonds. Although the factors that will cause different companies
to default will often be linked, particularly when economic conditions
are unfavourable, diversifying a company bond portfolio will reduce the
risk that the whole portfolio will be adversely affected by default.

3.3   Prospective Returns

For a given level of risk of not meeting liabilities, an institution will attempt
to invest in order to achieve the highest expected return over the period of
investment. As will be clear from the previous chapter, the returns from
many asset categories will have to be estimated by considering the expected
stream of payments generated by the investment. Both the valuation
techniques studied in the previous chapter and a historical analysis of the
returns generated by different investments are useful when estimating
prospective returns from investments.

3.3.1   Relationship between Risk and Return

The statement that an institution will attempt to obtain the highest return
for a given level of risk is not necessarily a particularly helpful one. It is

generally assumed to be the case that risk and return are related. If the risk from an investment is higher, then it is likely that the return will be higher also. Investors may be willing to trade a higher level of risk for a higher level of return. The preferences of different investors, with regard to the particular level of risk they are willing to accept for a given level of return, will be different, however, depending on the degree to which investors are risk-averse. The issue of trading risk and return is discussed in Chapter 4 and Chapter 5.

Investors will also have different perceptions of what is meant by risk. A short-term investor will not view risk in the same way as a long-term investor. A short-term investor may regard floating-rate notes as "risk free." However, they will be far from risk free for an investor who has to guarantee a given return over a long time period. Despite their capital value volatility, investments giving rise to a long-term fixed income stream may well be less risky for an investor with long-term liabilities.

A further complication is that we should not view the risk of a particular investment simply in terms of the risk of the investment not providing the required return, whether defined in real, nominal, or salary-related terms, over the desired time horizon. We should view the risk of a particular investment in terms of the marginal contribution it makes to the risk of the portfolio (see also Chapter 5). In the context of an institutional investor, that should be the marginal contribution of the investment to the risk of not meeting liabilities. Equilibrium models in capital market theory (e.g., see Adams et al. (2003)) are helpful in developing relationships between risk (defined in terms of the marginal contribution to the risk of a portfolio) and return, but these relationships rely on rather restrictive assumptions. In most practical situations, portfolio theory and stochastic investment modelling are necessary to ascertain the marginal contribution of a particular asset class to the risk of the portfolio.

Thus, the relationship between the return of an individual asset class and its risk is not necessarily a straightforward one. Nevertheless, in general, it can be said that if an asset class is regarded as having greater risk attached to it, then a higher return must be expected. Dimson et al. (2002) studied the risk premium available from equity relative to bond investments over the last century. Simply deriving the risk premium from historical data will not necessarily lead to accurate estimates either of the prospective risk premium at the current time or, indeed, the prospective risk premium at the beginning of that historical period. Equity investments may, in fact, have underperformed or outperformed expectations, as may bond investments if their real returns are considered. Indeed, a fall in the prospective risk premium will lead to a rise in the ex-post-recorded risk premium because it would cause equities to move onto a lower yield basis. Adjusting for these factors, Dimson et al. suggest that the range 2.5 to 4.0% would be a reasonable forward-looking arithmetic mean annual risk premium from equity relative to bond investments.

Dimson et al. also look at the risk of equity market underperformance over different time periods. They find a considerable risk of underperformance over 20-year periods and conclude that "short-term" underperformance of equities can be severe and last for decades.

The factors that affect the prospective returns from different investment categories were discussed in Chapter 2. In this chapter we again discuss the analysis of prospective returns, but concentrate on the factors that enable the investor to determine whether investments are "good value," given their risk to the investor. It is worth defining more precisely here what is meant by "good value." The efficient market hypothesis suggests that, at any time, investment prices should reflect all available information.[1] If markets follow the efficient markets hypothesis, a particular investment would not be "good value" at any time in the sense of risk-adjusted outperformance being likely. However, without entering the debate about the empirical validity of market efficiency, it can be said that, from the point of view of the individual investor, market efficiency is not a good starting point for the analysis of investments. Investment markets are composed of large numbers of individual and institutional investors, all of whom have different subjective preferences. At any time, a particular investor may regard a specific investment or class of investments to be "good value" given the needs of that investor. To illustrate by analogy, different supermarket own-brand baked beans may have the same price when the market for baked beans is in equilibrium, but a particular customer may have a preference for the beans of one supermarket rather than those of another. In the same way, equity and bond markets may be in equilibrium, but a particular investor may have a preference for bonds rather than equities, after an assessment of their relative value to that investor. It might also be added that the whole concept of market efficiency in equilibrium may not be helpful in describing markets in which new information is continually being discovered. Thus, regardless of one's view on the efficient market hypothesis, discussion of the assessment of the relative value of investments is important.

3.3.2 Returns from Bonds and Cash Instruments

If the payments from a domestic, government-guaranteed, bond can be regarded as being totally secure, and if the bond is to be held to maturity, then the prospective return can be calculated easily. A company fixed-interest security will be more risky than a government security and, therefore, will have a higher yield to maturity: the extent of that higher return will depend mainly on the risk and marketability of the particular

---

[1]The particular set of information that is assumed to be reflected would depend on the level of efficiency.

bond issue. In general, we can decompose the yield on government fixed-interest securities i into two components: the expected real rate of return j and the expected rate of inflation r. These two components are roughly additive, so that we can say

$$i = j + r$$

For most purposes, this relationship would be sufficiently precise, although the precise relationship, if returns are effective annual returns, is $i = j(1 + r) + r$.

An increase in either expected inflation or required real rates of return will lead to an increase in conventional government bond yields. At moderate and stable rates of inflation, investors may tend to demand a real rate of return of 2.5 to 3.5% from long-term conventional government bonds. Thus, when determining the adequacy of conventional government bond yields, this benchmark should be borne in mind. However, if the level of future inflation is perceived as unpredictable, then investors may demand a higher real rate of return, to compensate them for the extra inflation risk. Other factors, such as the level of government borrowing, will also affect the real rate of return that investors will require from government bonds.

At the short end of the conventional government bond market the real rates of return that investors demand will be influenced by the monetary policy being followed by the government. Monetary policy tends to be managed through short-term interest (cash) rates, which have an inevitable effect on short-term bond yields.

The effect of expected inflation on bond yields can be seen by looking at historical yields on fixed-interest bonds over time. In the U.K., the general trend of government bond yields in the postwar period was upwards until the early 1980s. At the beginning of the 1950s, the yield on long-term government bonds was about 3.5%; the level of yields gradually increased, as inflationary expectations increased, reaching around 17% in 1974. Since 1981, government bond yields have been on a declining trend, as inflationary expectations have fallen due to the successful pursuit of a monetary policy that did not accommodate inflation. In 1997, the U.K. followed a number of countries by establishing an independent central bank. This reinforced the credibility of the monetary policy regime and led to a further fall in inflationary expectations towards the Bank of England's target rate of 2.5%. As can be seen from Table 3.3, long-dated government bond yields fell from nearly 16% in 1981 to around 7.5% in 1996 and then to 4.7% in 2001, a level around which they have stayed since.

If a representative portfolio of government bonds were held, then a change in the yield basis could lead to negative returns being achieved in any particular year. In the postwar period, the real return from a representative portfolio of government bonds would have been generally

Table 3.3 U.K. long-dated government gilt yields, 1975–2002 end year

Year	Yield (%)
1975	14.8
1976	15.0
1977	10.9
1978	13.2
1979	14.7
1980	13.9
1981	15.8
1982	11.1
1983	10.5
1984	10.6
1985	10.5
1986	10.5
1987	9.5
1988	9.3
1989	10.0
1990	10.6
1991	9.8
1992	8.7
1993	6.4
1994	8.6
1995	7.6
1996	7.6
1997	6.3
1998	4.4
1999	5.3
2000	4.7
2001	5.0
2002	4.4

Source: data from Barclays Capital (2003)

positive in most years up until the late 1960s. For a number of years in the 1970s there were negative real returns, due to the yield basis rising in response to increasing inflation and high government borrowing (e.g., see data in Barclays Capital (2003)).

In general, one would expect government-guaranteed bonds issued by OECD countries to offer a lower return than fixed-interest securities guaranteed by a private company (see Section 2.3). The additional yield from a company bond would be determined by the credit risk attached to the company, the class of the company bond, and the marketability of the issue. The rate of return from a company bond can be decomposed into the risk-free rate of return offered by a conventional government bond of similar term and coupon and the credit spread that investors require from the particular company bond issue. The credit spread would depend on the likelihood that investors would receive the promised yield to maturity from the bond, the risk premium that investors required to compensate them for uncertainty, and a margin to compensate them for the lack of liquidity and marketability of corporate bonds. A benchmark level for the credit spread for a good-quality company bond would be around 0.5 to 1.5%.

In the U.K., index-linked government bonds offer investors a guaranteed real rate of return to maturity (ignoring complications such as the lag between indexing and the payment of coupon and capital payments). One would therefore expect that their prospective expected yield would be less than that from other real investments. The real yield on index-linked government bonds, as measured by the FT Actuaries All Stocks and Over 5 year Index-Linked Government bond indices, gradually rose from a level a little below 3%, when they were originally issued in 1981, to above 4.5% in September 1992. They then returned to a level of around 3% by 1997 and then reduced to just under 2%, a level at which they remained at the time of writing. In historical terms and in international terms, the level of 2% is very low for guaranteed real returns over a long time period. It is possible that institutional factors in the U.K. depress the level of real yields. A long-term benchmark real return from index-linked government bonds of around 3% might, therefore, seem reasonable. The particular levels of real yields at any one time will depend on the level of real yields available from competing investments, both in the domestic country and internationally, as well as the supply of index-linked government bonds and the levels of saving and borrowing in a country. The perceived relative risk of index-linked bonds and other real investment media, such as equities, will also be important.

Corporate index-linked bonds are available. The margins between real rates of return from corporate index-linked bonds and those from index-linked government bonds are roughly equal to the credit spread from conventional bonds of the same credit standing. However, in all countries, including the U.K., the corporate index-linked market is small in size and relatively illiquid. Further details of U.K. corporate issues can be found at www.dmo.gov.uk.

The return on short-term cash deposits will be heavily influenced by monetary policy: a lax monetary policy will generally be accompanied by low short-term real interest rates. In general, however, over a longer period, one would expect the level of short-term interest rates to be determined mainly by the level of inflation. It is not generally possible to keep interest rates artificially low, through the operation of a lax monetary policy, over a long period of time. Using Treasury bill data (e.g., in Barclays Capital (2003)), it can be seen that, in the postwar period, cash has generally yielded a small positive real rate of return of about 2 to 3% on average, except during the 1970s and in the immediate postwar period, when cash real returns were significantly negative. The 1970s was also a period when there were significant restrictive practices in the short-term deposit markets, reducing returns to investors. Except for the 1970s, the real

Table 3.4 Treasury bill returns and inflation 1975–2002

Year	U.K. Treasury bill returns (%)	U.K. retail price inflation (%)
1975	10.75	24.9
1976	11.34	15.1
1977	9.44	12.1
1978	8.06	8.4
1979	13.45	17.2
1980	17.17	15.1
1981	13.76	12.0
1982	12.38	5.4
1983	10.14	5.3
1984	9.55	4.6
1985	11.87	5.7
1986	10.95	3.7
1987	9.58	3.7
1988	11.01	6.8
1989	14.55	7.7
1990	15.86	9.3
1991	11.59	4.5
1992	9.47	2.6
1993	5.86	1.9
1994	5.40	2.9
1995	6.74	3.2
1996	6.16	2.5
1997	6.88	3.6
1998	7.92	2.8
1999	5.51	1.8
2000	6.22	2.9
2001	5.50	0.7
2002	4.12	2.9

Source: Barclays Capital (2003)

rate of return from cash deposits during the postwar period was around 2.5%, on average. Table 3.4 shows average Treasury bill discount rates and inflation rates for 1975 to 2002. The extent to which real short-term rates of interest were somewhat higher in the 1980s and early 1990s than in the 1970s can be seen from this table. The general reduction of nominal Treasury bill yields as inflation fell can also be seen from the table. Cash returns tend to be lower than the returns from other investments, except during periods when a very tight monetary policy is operating. For this reason, investors will not tend to invest in cash unless their liabilities dictate that they should or if they regard the outlook in long-term investment markets as very uncertain.

The rates of return from a foreign currency bond will depend on the rate of return expressed in the foreign currency and any exchange rate movement. Where there is freedom of movement of capital between bond markets, it could be expected that expected real rates of return from all bond markets would be approximately equal. A formal analysis of this proposition can be found in Solnik (1999). Thus, a benchmark real rate of return from foreign currency bonds of around 2.5 to 3.5% would be reasonable. However, a higher rate of return would be expected from bond markets in countries with high inflation, where there was concern about possible default or where the currency was perceived to be weak. The level of nominal bond yields in a particular currency will depend, to a large extent, on the level of inflationary expectations in that currency.

3.3.3 Returns from Equity and Real-Estate Investments

There is a significant amount of historical data on the past performance of U.K. equities. However, the observed long-term real rate of return from an equity index will depend critically on the starting and ending points of the data series. Thus, in common with all areas of investment analysis, the data require a degree of careful interpretation. Arguably, after the 1960s, equities were "re-rated" as their value as an inflation hedge became more widely appreciated by institutional investors. Also, short-term fluctuations in equity markets can be considerable, for the reasons discussed above: for example, there is a 3.5% difference between the average annual return from U.K. equities over 20 years, depending on whether the endpoint of the calculation is 1999 or 2002.

One would expect the real rate of return from equities, over the long term, to be greater than that from index-linked stocks, cash and real estate. This is partly because there is likely to be greater short-term capital value volatility of equity investments and also because there is less certainty about the future income stream. Historical evidence in the U.K. and a priori analysis are consistent with a long-term real return of between 5 and 6% from equity investments (such returns can be derived from the data in Barclays Capital (2003) over a long data period), although this is a highly subjective matter. However, this level of return provides a reasonable benchmark that can be used to value an equity dividend stream or that can be used in comparison with expected prospective rates of return when investment decisions are being taken.

As was discussed in Chapter 2, foreign equities are inherently real investments. Historically, the return from foreign equities has varied between investment markets. In general, it would be reasonable to expect that some foreign equity investment opportunities would give rise to greater returns than domestic equity investment opportunities, as the possibility of obtaining higher returns will probably be one of the motivations for foreign investment. However, investors will also invest overseas in order to obtain the benefits of diversification, and may be willing to do so even if the returns from investing overseas were no greater than those from investing domestically. Thus, in the long run, one would perhaps expect the returns from foreign equities to be about the same as those from domestic equities, or possibly greater in a well-managed portfolio that took advantage of undervalued investment markets. Table 3.1 illustrated the high level of short-term returns that are sometimes available from foreign equity markets — particularly from smaller markets.

The return from real-estate investment arises from the flow of rents from property and any increase in the value of property over the purchase price when it is sold. Against this income flow and capital appreciation have to be set the management costs and any costs of refurbishment. Real estate has reasonable security of income flow. The contractual obligations of rents are paid before any dividend payments can be made to the holders of a company's equities. Furthermore, in the U.K., lease contracts are often designed with an "upward-only rent review clause" so that, at each periodic review of rents (typically such a review takes place every 5 years), rents either continue at the level before the review or increase to the level of market rents (whichever is higher): see Chapter 2. Thus, property is both a real investment (because we would expect market rents to act as a hedge against unanticipated inflation) and provides an income stream with a fixed nominal floor, at least for the term of the lease.

Whilst real estate does provide a fixed nominal floor, we would expect rents to drift upwards over time. Therefore, it is probably closer to a real investment. Thus, it is useful to focus on the real return that might be expected from real-estate investments. In general, although real estate is regarded as less risky than equity investment, there are difficulties caused by transaction times and illiquidity. However, one would expect the lower risk to be combined with a lower real return. It is difficult to obtain a consistent long-term series of real-estate performance figures with which to assess a benchmark rate of return with any degree of confidence. However, on the basis of evidence that is available in the U.K., a long-term real return of around 4.5 to 5.5% from real estate may be regarded as a reasonable benchmark that could be used to assess the value of this asset class. The differences between the risks of different properties will lead investors' required rates of return to vary considerably between individual properties. Data for real-estate returns and inflation are shown in Table 3.5. The mean return from real estate over the period is 7.9% and the mean inflation rate is 3.2%.

3.3.4 Prospective Rates of Return and Investment Decision-Making

If, having taken all factors into consideration, an investment category appears to be (subjectively) good value, in that the rate of return it appears

Table 3.5 U.K. real-estate returns and inflation

Year	U.K. commercial real estate total returns (%)	U.K. retail price inflation (%)
1990	−8.5	9.3
1991	−3.1	4.5
1992	−1.7	2.6
1993	20.3	1.9
1994	11.9	2.9
1995	3.6	3.2
1996	10.1	2.5
1997	16.9	3.6
1998	11.7	2.8
1999	14.7	1.8
2000	10.5	2.9
2001	6.7	0.7
2002	9.7	2.9

Source: IPD U.K. Annual Index and Barclays Capital (2003)

to offer is high relative to historical norms and compensates for the risk to the investor, exposure to that category may be increased. However, timing will also be an aspect of the decision process. Any possible capital gain from a change in yield basis must be taken into account when deciding whether current market levels justify a change in investment strategy. A pension fund manager might not regard a bond yield that is 1% above the level that appears to reflect fundamentals as justifying a long-term switch into bonds. However, a short-term switch may be made if it was felt that a reduction in bond yields was likely to occur in the near future that would lead to the fund achieving a short-term capital gain. Conversely, an investment manager may regard a particular investment category as "good value" but not switch into that category if the investment manager believes that such value will not be reflected in market prices for some time to come. These issues of market timing are also tied up with whether or not the manager is actively managing the fund and whether it is believed that markets are efficient in the sense of reflecting all information in investment prices.

This leads on to a discussion of how to assess the relative value of different investments. We will consider the relative evaluation of conventional government bonds, index-linked government bonds and U.K. equities and consider some examples based on U.K. market data available at the time of writing.

3.3.5 The Real Yield Gap

Traditionally, measures such as the *yield gap* and *yield ratio* have been used as an indicator of the relative investment values of government bonds and equities. These measures may be crude but, nevertheless, they are helpful in indicating whether a particular investment category is good value relative to another. In this section we will look at the *real yield gap*, which can be used to compare the relative value of index-linked government bonds and equities.

The expected real rate of return from an equity investment can be approximated by

$$g_r + d_0/P \tag{3.1}$$

where g_r is the expected long-term real rate of growth of dividends and d_0/P is the historical dividend yield based on the last declared dividend d_0 and the current price P. This intuitive result can be found from the Gordon growth model of equity valuation, expressed in real terms, and is derived in Adams et al. (2003); in theory, the prospective, rather than the historical, real dividend yield should be used to give a more precise relationship. The real rate of return from an index-linked government bond is known with certainty and will be defined as j_{il}. Investors will require a risk premium for investing in equities, as index-linked government bonds offer a guaranteed real rate of return. Therefore, if an investor were to be indifferent between investment in index-linked government bonds and investment in equities, we would expect

$$j_{il} + r_p = g_r + d_0/P \tag{3.2}$$

where r_p is the risk premium required by the investor for investing in equities. As an indication of relative value, the investor can look at the real yield gap, defined by the historical dividend yield on equities minus the real yield from index-linked government bonds. Rearranging Equation 3.2:

$$d_0/P - j_{il} = r_p - g_r \tag{3.3}$$

Thus, the real yield gap should be equal to the risk premium minus the expected long-term real dividend growth rate from equities. We can consider the real yield gap as an indicator of relative value. Thus, it can be monitored over time; then, if it moves outside normal historical limits, it would suggest that equities were cheap or dear relative to index-linked government bonds. Alternatively, we can simply think of the real yield gap not in terms of whether or not a market correction might be expected, but as an indicator of the subjective value of equities relative to index-linked gilts for a particular investor. We can rearrange Equation 3.3 for the equity risk premium implied by the market pricing of equities and index-linked government bonds. If the equity risk premium is insufficient for a given investor, this suggests that equities are too dear for that particular investor.

Example 3.1

The yield on the FT Actuaries Index-Linked Government Bonds Over 5 Years Index is 2.3% (assuming 5% future inflation); the dividend yield on the FT Actuaries All-Share index is 3.9%.

1. Calculate the real yield gap.
2. Calculate the risk premium from equity investments if expected future real dividend growth is 1%.

Answer

1. The real yield gap is the equity dividend yield minus the real yield from index-linked government bonds, which is 3.9% − 2.3%, or 1.6%.
2. From Equation 3.3, the real yield gap should be equal to the risk premium minus the expected real growth rate of dividends; therefore: $1.6\% = r_p - 1\%$ and $r_p = 2.6\%$.

It is worth noting that this is at the low end of the range suggested by Dimson et al. (2002); see Section 3.3.1.

3.3.6 Relative Value of Index-Linked and Conventional Bonds

Subject to the complication of the inflation lag, the real return from index-linked government bonds is known. The expected real return from conventional government bonds is approximately

$$\frac{i - r}{i + r} = j \tag{3.4}$$

where r is the expected future rate of inflation (assumed to be constant) and i is the nominal rate of return from conventional government bonds. We will ignore complications caused by the quoting of U.K. government bond yields as nominal yields per annum convertible half yearly; quoted yields can easily be converted to yields convertible annually. At most values of r it is sufficient to approximate the expected real rate of return from conventional government bonds by $i - r$.

If $j = j_{il}$, then the expected real rate of return from conventional government bonds would be equal to the real rate of return from index-linked government bonds. We might expect a risk premium from conventional government bonds because of the inflation risk. There may be circumstances where such a risk premium would not prevail. The market as a whole will be made up of a range of investors with different liabilities, some of whom will regard conventional government bonds as more risky and some of whom will regard index-linked government bonds as more risky. Also, the larger supply of conventional government bonds may make conventional government bonds more liquid. In practice, the risk premium will vary with investors' perceptions regarding the stability of inflation and the credibility of the framework within which monetary policy is determined; there is further discussion of these issues in Scholtes (2002). Scholtes discusses the various different economic events that might have caused the risk premium to change over the last ten years or so.

We will proceed on the basis of there being no risk premium. In this case, index-linked and conventional government bonds will be fairly priced if

$$i - r = j_{il} \tag{3.5}$$

Example 3.2

The yield on the FT Actuaries Index-Linked Government Bonds Over 5 Years Index is 2.3% (assuming 5% future inflation); the gross yield on the FT Actuaries Fixed Interest long bond index is 5.1%. Comment on the relative value of index-linked government bonds and conventional government bonds to an investor, who requires no risk premium from either type of investment, if expected future inflation is 2.5% per annum.

Answer

From Equation 3.5, the expected real yield from conventional government bonds is 2.6% per annum (to a gross investor such as a pension fund). This is above the 2.3% real yield from index-linked government bonds. There could be a number of reasons for this:

1. Market expectations of future inflation may be higher than the government's long-term target of 2.5%.
2. Tax considerations would reduce the net real yield from conventional government bonds relative to that from index-linked government bonds for most investors.
3. Most investors may, in fact, demand an inflation risk premium from conventional government bonds.

There has been some discussion recently about the level of index-linked bond yields in the U.K., particularly in relation to the levels that have pertained in international markets (e.g., see recent Debt Management Office Annual Reviews at www.dmo.gov.uk and Scholtes (2002)). The level of yields will also depend on the issuing policy of the government-bond issuing authorities. It might be asked whether issuing authorities should issue conventional and index-linked bonds in such propor-tions that the real cost of borrowing is equalised using both types of bond. However, the issuing authorities may be constrained by institu-tional factors (such as the amount of government debt being issued) from issuing government debt in such a way that the expected cost of funding through real and nominal instruments is the same. They may also be constrained because the issuing policy has broader aims than simply minimising the cost of funding debt—such as diversifying the debt portfolio appropriately.

3.3.7 The Yield Gap

The yield gap is a traditional measure of investment value, when com-paring equities and conventional government bonds. However, study of the yield gap carries the difficulty that we are comparing two fundamen-tally different types of investment. The size of the yield gap will depend on the expected rate of inflation, which we would not expect to remain stable over time.

The expected nominal return from an equity investment, using the Gordon growth model, is approximately: $g + d_0/P$ where g is the expected long-term nominal growth rate of equity dividends (see the qualifica-tions mentioned in Section 3.3.5 and Adams et al. (2003)). If the expected long-term returns from equities and conventional government bonds were equal and investors required a risk premium r_{ep} from equities over conventional government bonds, then we would expect:

$$d_0/P + g = i + r_{ep} \qquad (3.6)$$

The yield gap, defined as the rate of return from conventional government bonds less the historical dividend yield from equities, would be

$$i - d_0/P = g - r_{ep} \qquad (3.7)$$

It should be noted that the expected growth rate of dividends g can be divided into a real and an inflation component. Thus, the yield gap can be seen to depend on expected real dividend growth rates, expected future inflation and the risk premium. If the yield gap is bigger (smaller) than the expected long-term growth rate of equity dividends less the risk premium for a particular investor, then it suggests that conventional government bonds are cheap (dear) relative to equities for that investor. If no risk premium is required for conventional government bonds over index-linked government bonds, then r_{ep} in Equation 3.7 should be equal to r_p in Equation 3.3 in equilibrium.

Example 3.3

The gross yield on the FT Actuaries Fixed Interest long bond index is 5.1%; the dividend yield on the FT Actuaries All Share index is 3.9%.

1. Calculate the yield gap.
2. Calculate the risk premium from equity investments if expected future real dividend growth is 1% and expected future inflation is 2.5%.
3. Comment on the relative valuation of conventional government bonds and equities.

Answer

1. The yield gap is the yield from conventional government bonds minus the equity dividend yield, which is $5.1\% - 3.9\% = 1.2\%$.
2. From Equation 3.7, the yield gap should be equal to the expected nominal growth rate of dividends minus the risk premium; therefore: $1.2\% = 3.5\% - r_{ep}$. The future nominal expected dividend growth rate is 3.5%, made up of the 2.5% future inflation assumption and 1% real dividend growth. Thus: $r_{ep} = 2.3\%$.
3. The risk premium from equities over conventional gilts is slightly less than that from equities over index-linked government bonds. This is consistent with there being a small risk premium from conventional over index-linked government bonds.

Investors may use yield gaps to compare investment categories and predict future market movements. Such yield gaps are a rather crude way of comparing value, and other techniques are discussed in Lofthouse (2001). Nevertheless, yield gaps are a starting point. More generally, yield gaps can be used by all investors to provide an indication of whether the risk premium in a particular market is high enough to compensate for the risk of that market given that investor's subjective assessment of the risk of that investment category. Thus, perhaps the most important application of "gap analysis" is not for predicting market movements but for ascertaining which investments provide the best subjective value for particular investors, given current yield levels.

3.3.8 Comparing Other Investment Categories

Similar techniques can also be used in the comparison of index-linked government bonds, conventional government bonds and equities with real estate. In the case of a real-estate investment, the real return can be approximated by the initial level of the rental yield plus expected real rental growth, less expected depreciation. The expected real return from real estate can then be compared with the expected real return from other investments. However, it is not necessarily appropriate to analyse properties with unusual lease terms or very long rent review periods in such a simplistic way.

Rules of thumb can also be developed for comparing foreign currency investments with U.K. investments. An attempt can be made to estimate the prospective real return from foreign equity investments by looking at the dividend yield, expected real dividend growth and expected real currency movements. The mathematical and economic bases for such rules of thumb are discussed in Adams et al. (2003) and in Solnik (1999).

Having discussed risk and expected return in some detail, we now move on to look at other factors investors should consider when determining the proportion of assets to devote to different categories of investment.

3.4 Institutional and Governmental Influences and Constraints

The other factors affecting asset allocation that will be discussed are the constraints imposed by marketing difficulties, the constraints imposed by law and taxation, and the need of institutional investors to fulfill the preferences of policyholders and the wishes of trustees. Most of these factors are specific to different groups of investors.

3.4.1 Marketability, Liquidity, and Transaction Costs

In general, an investment can be said to be marketable if there is a reasonable trading volume so that an investor can buy or sell a considerable quantity of the investment without affecting the market price significantly. The investment can be regarded as liquid if the transaction can be carried out and settled easily and quickly. However, definitions of liquidity vary: whilst an investment could not be said to be liquid unless it was realisable for cash quickly at the market value, many investors, particularly those in the bond markets, would also suggest that an investment is more liquid the closer it is to maturity.

Other than cash, government bonds are probably the most marketable and liquid investments of those we have discussed. The marketability and liquidity of corporate bonds and equities will depend, to a large extent, on the quality of the company, with market capitalisation of the issue being an important factor in determining marketability. Marketability of overseas equities will depend on the sophistication of the particular equity market in question, as well as on the size of the companies in which investments are being traded. Real estate is generally regarded as being unmarketable as a result of its highly specific nature. Real estate is also highly illiquid. It has a very long transaction time compared with most of the other major classes because of the necessity to transfer legal title in the physical asset. However, it is worth noting that private equity and venture capital vehicles can also be both unmarketable and have long transaction times associated with them.

Transaction costs tend to increase as liquidity reduces. The transaction costs for buying or selling government bonds are relatively low. In the case of corporate bonds and equities, transaction costs tend to be higher and, in addition, investors will have to consider the costs of research, which will form part of the costs of managing a portfolio of company securities. In general, the less sophisticated the market, the greater the dealing and research costs. Foreign equities will tend to have higher transaction costs, partly because investors from overseas incur additional costs that are not incurred in domestic markets. Real estate tends to bear very high transaction costs. For example, in the U.K., these transaction costs include 4% stamp duty on most commercial real-estate transactions.

Transaction costs are also important when considering a change in investment strategy. Transaction costs could prohibit, for example, an investor switching significant funds between asset classes because it was believed that there was a temporary mispricing of investments, particularly if one of these asset classes were real estate. Derivatives can sometimes be used to change exposure to asset classes quickly and with lower transaction costs. Also in the real-estate market, indirect vehicles can be used, such as real-estate investment trusts in the U.S. and property investment companies or managed funds in the U.K.

3.4.2 Legal Constraints

Legal constraints on investment policy may take many forms. In some countries investment is positively directed, or limitations are placed upon the amount of investment in any particular investment category. Such positive direction has often been the norm in E.U. countries, other than the U.K., Ireland and the Netherlands. Davis (2002) describes such controls as they have existed in other E.U. countries. Examples include insurance companies in Germany being limited to 30% of their investments in quoted shares and insurance companies in Italy being restricted to 20% in quoted shares. Currency-matching rules often limit the amount of overseas investment. In the U.K., regulations have tended to be relatively liberal, with restrictions generally forming part of a legislative framework intended to ensure that institutional investors are prudent, rather than enforcing specific action in particular circumstances. In the U.K. there are regulations requiring insurers to diversify their investments appropriately, particularly where free reserves are limited; insurers are also required to demonstrate that their solvency position is resilient to changes in financial conditions. Pension fund trustees are required to act in the best interest of members—this is normally interpreted as meaning their best financial interests. Specific regulation and its impact on investment policy will be discussed later in this chapter.

3.4.3 Taxation

Investors should always take the tax position of their fund into account when determining investment policy. Taxation may make certain investments advantageous or disadvantageous to particular investors. Institutions should, therefore, consider the net return from investments when taking investment policy decisions. The net return should be calculated at the institution's marginal rate of tax (for investors that do not pay tax this would, of course, be zero).

The particular types of differential tax treatment that can influence investment policy can be divided into three main groups:

1. The differential treatment of capital gains and income that can lead investors to seek investments that do not give rise to income.
2. The differential treatment of equities and bonds. This normally arises from the fact that returns to equity (profits after interest) are taxed in the corporation tax system but interest on bonds is taxed in the income tax system. In many jurisdictions, profits are taxed at a higher rate than interest on bonds and dividends paid on equities can often be "double taxed," in that tax can be levied on dividends even if the profits from which dividends are paid have previously been taxed. Until 1997, the U.K. tax system dealt with this problem through the granting of "tax credits" that non-taxpaying investors could reclaim in respect of dividends paid on equities; this system has now been abolished.
3. The differential tax position of domestic and overseas equities. This might arise because, where the corporation tax rate is higher overseas than in the domestic country, it is generally not possible for investors from a lower tax regime to reclaim the extra tax paid at source on overseas investments. This is a greater problem on equity-type investments than in the bond markets, where interest will often be paid gross and tax accounted for in the investor's home country.

Distortions in the tax system, such as those described here, have a primary impact, encouraging particular investors to invest in the instruments that are least highly taxed, given their own tax position. They can also have a secondary impact. For example, if institutions pay higher tax on income than on capital gains, then this raises the price (and hence lowers the gross yield) of investments that offer capital gains. Gross investors will then be attracted to those investments that are relatively less attractive to the taxpaying investor (i.e., those investments that pay a higher income). In this way, investment markets can become "segmented," with investors choosing between investments with similar risks simply because of their tax position.

3.4.4 Policyholder Preferences

Ultimately, institutions have to take into account the preferences and interests of personal investors who benefit from the provision of insurance and pension services by the institutions. The interests and preferences of personal investors may be reflected in a number of ways. Where contracts promise fixed benefits that are guaranteed, the design of contracts will, in the first instance, have to reflect the preferences of potential policyholders. Where returns are not guaranteed, so that there is a large degree of investment freedom for the institution, investments should be managed to ensure that the institution reflects the policyholder's relative preferences for stability, higher returns, inflation protection, etc.

3.4.5 The Use of Derivative Instruments

Derivative instruments can be used to overcome the institutional constraints and other difficulties discussed in this section. In particular, futures can be used effectively to implement an investment decision without recourse to dealing immediately in large quantities of the underlying asset; this may allow an institution to overcome some of the marketability and liquidity problems that have been discussed. It is worth noting, however, that there is not a well developed market in real-estate derivatives; thus, the difficulties of real-estate investment, mentioned above, cannot be overcome in this way. The duration and credit risk of a bond portfolio can also be altered using swap instruments, without dealing in the underlying assets.

Options could be used if it is believed that a particular event may adversely affect the value of certain shares in an equity portfolio. Options could also be used more generally to insure an investment portfolio (although such insurance carries costs) or to assist an insurance company in matching policy options and other complex policy clauses. For example, a number of insurance policies and investment contracts have been written in the U.K. that have guaranteed rates at which policyholders can purchase annuities when the policies mature (see also Section 3.2.1 and Section 3.6.2).

Some insurers have covered such options by purchasing "swaptions" in the "over-the-counter" derivative market. These swaptions allow the purchaser to exchange a floating interest income stream for a fixed-interest income stream should interest rates fall sufficiently to make the fixed stream better value. The swaption contract would, therefore, have a value to the investing institution in the same circumstances that the guaranteed annuity rate had a value to the policyholder. Options can also be used to hedge guaranteed surrender values and guaranteed equity products.

3.5 Asset Allocation Principles Applied to Pension Funds

In general, a fund will wish to maximise return for a given level of risk, where risk is defined in terms of variability of the funding rate or solvency level. The proportions invested in different asset classes that are regarded as a priori suitable will vary according to the trustees' and fund managers' expectations of future investment returns. In the U.K., the legislation that exists does not explicitly stop pension funds from investing in those asset categories that would, in any case, be deemed most suitable, although valuation standards do affect asset allocation.

3.5.1 Mechanism for Taking Investment Decisions

The precise mechanism for taking investment decisions will depend on the size and nature of the pension scheme. In the U.K., pension funds will have trustees and an actuary, and will normally have separate, clearly identifiable investment managers. The investment managers may be an "in-house" team, or the fund may be managed by the investment managers of a fund management group. If the fund is managed externally, then it may be a segregated fund managed independently of other pension funds, or it may be a pooled fund divided into units. The trustees are responsible for the investment policy of the pension fund and will lay down guidelines for the investment managers. These guidelines should be constructed after consultation with the investment managers and the actuaries. The investment managers should work within these guidelines. Trustees would not, in general, interfere with the day-to-day decisions of the investment managers.

3.5.2 Liability and Other Financial Considerations

In recent years there has been considerable debate within the actuarial profession about the role that financial economics should play in actuarial analysis. Some would argue that actuaries should use techniques derived from the academic discipline of financial economics to a greater extent. Others disagree, regarding those ideas as too abstract. There is a range of considerations that should determine the investment policy of a pension fund, and these considerations have to take into account the principles underlying corporate finance, financial economics and actuarial science. Whilst the principles that are important can be laid out clearly, there is room for legitimate debate about the validity of different theories and the assumptions that underlie those theories. In this section we will take account of relevant ideas from both financial economics and traditional actuarial science. We will also discuss some ideas from the field of corporate finance that should influence actuarial practice.

It has been suggested that, regardless of the liabilities of a pension fund, the asset allocation policy is irrelevant, as the pension fund is only one part of the balance sheet of a company. This is an extension of the proposition of Modigliani and Miller (1958), that balance sheet structure is irrelevant in determining the value of a firm. The proposition has been developed and qualified for pension funds by Black (1980).

The reasoning of those who take this position is shown by the following situation. If a sponsoring company holds equities in its pension fund, then the company (and ultimately its shareholders) must take the risk that the equities underperform within the fund but they would also get the benefit of any excess return. The company could then reduce the holding of equities in the pension fund and increase holdings of bonds. This would change the risk profile of the pension fund. However, because the shareholders of the company bear the ultimate risk of the pension fund, it is argued that the company could restructure its balance sheet, buying back its own equity and issuing bonds, thus increasing the gearing of the balance sheet exactly offsetting the changes in investment strategy in the pension fund. Furthermore, even if the company does not change its balance sheet in this way, the shareholders who own the shares of the company can rebalance their own portfolios to hold more bonds and fewer equities, if the companies in which they hold shares are "over geared." If a company reduces equity investment in its pension fund and buys back equity on its own balance sheet (financed by issuing bonds), then the company has fewer equity assets in the pension fund but less equity on its balance sheet. Its pension fund is (arguably) less risky, but its balance sheet is more highly geared. Thus, it is suggested that the liabilities of the pension fund, as such, are irrelevant in determining investment policy for the pension fund. The company should look at the capital structure of the firm as a whole, and the shareholders in companies with pension funds can always adjust their own portfolios in order to suit their risk profile if they feel that the risk taken on by the pension funds of companies in which they have invested is not optimal. We shall see below that this result is not incontrovertible. Nevertheless, it is an important point and it is worth illustrating with a numerical example.

The accounts of two simple companies after consolidating pension fund assets and liabilities into their main accounts are shown below. In the first case (ABC PLC) the pension fund is only invested in equities. For the second company (XYZ PLC) the pension fund is only invested in bonds but the company balance sheet is more highly geared. We assume that the value of the assets is represented by the present value of the cash flows from the fixed assets and that the market value of the equities represents the present value of the cash flows from the fixed assets less the value of the debt of the company for each company. Thus, the accounts represent the economic value of the companies rather than historic cost figures. We consider the situation where there is a fall in the value of cash flows from investment projects of 50% and a fall in the value of the equities held in a pension fund by the same proportion. In effect, this assumes that the equities held in a pension fund are not geared, but are exposed to exactly the same risks as those to which the company is exposed. This assumption is not material. The new balance sheet values, after the fall in the value of investment projects, are shown in brackets in each case.

ABC PLC

Assets

Present value of cash flows from fixed assets	100	(50)
Pension fund equities	20	(10)
Pension fund bonds	0	(0)
	120	(60)

Liabilities

Corporate debt	0	(0)
Corporate equity	100	(40)
Future pension liabilities	20	(20)
	120	(60)

XYZ PLC

Assets

Present value of cash flows from fixed assets	100	(50)
Pension fund equities	0	(0)
Pension fund bonds	20	(20)
	120	(70)

Liabilities

Corporate debt	20	(20)
Corporate equity	80	(30)
Future pension liabilities	20	(20)
	120	(70)

In both cases the value of the equity, or the shareholders' interest, has fallen by approximately 60% (from 100 to 40 in the case of ABC PLC and 80 to 30 in the case of XYZ PLC). It is implicit that the deficit in the pension fund caused by the fall in the value of the equities held by ABC, is made up by the shareholders in ABC PLC, as the value of the equity in ABC PLC has been written down to reflect the pension fund deficit. In XYZ PLC the pension fund is no longer exposed to equity risk, but the shareholders suffer the same fall in the value of their shares as a result of the fall in the value of cash flows from investment projects. This arises because XYZ PLC is more highly geared.

Thus, the ultimate bearers of risk can control equity risk at three levels, and if shareholders wish to access the risk premium from equities, then they can achieve this at the same three levels. The equity exposure of the company pension fund can be adjusted; the balance sheet gearing of the company can be adjusted; and the personal portfolios of the end investor can be adjusted (to change the proportion of equities and bonds held). This result suggests that the pension fund need not be too concerned about the liabilities of the pension fund, as such, and directors should consider the financial structure of the company as a whole. At this level, the theory of corporate finance suggests indifference between different investment policies for pension funds assuming that counter-balancing changes to the company balance sheet structure can be made. However, there are a number of other considerations that are, in fact, important:

- If the logic of the above argument is accepted (notwithstanding the points made below), tax considerations suggest that pension funds should, in fact, invest only in bonds. Equities have a relatively harsh tax treatment in a pension fund, in most jurisdictions. For example, in the U.K., since July 1997, equity returns have effectively been taxed in pension funds (through the unrelieved taxation of corporate profits), whereas bonds, cash, and real estate are not taxed. It can be argued that it is optimal to eliminate pension fund investment in equities and make appropriate capital structure changes to the company as a whole, so that the risk to the end shareholder is unchanged but the pension fund minimises its effective rate of tax on its investment returns.
- Pension funds are not a transparent aspect of company finance. It could be argued that shareholder value is maximised if pension fund liabilities are perfectly matched so that risks in the non-transparent aspects of the balance sheet are minimised as shareholders do not like opaque risks. Exley (2001) argues that if this argument is accepted, then bonds are the best match for pension fund liabilities. However, as we shall see below and also in Chapter 5, this is not necessarily the case.
- Pension funds cannot be regarded simply as an extension of the company balance sheet. They are designed to be independent funds managed by independent trustees in the best interests of their members and, as such, form part of the remuneration package of employees. Again, this suggests that, in many circumstances, the best approach for trustees to take is to match liabilities, as far as possible, to minimise risk to members. For

further discussion of risk measurement, see the methods introduced by Haberman et al. (2003a,b).

- There might be an element of "co-insurance," by which members of the fund bear some of the risk of underperformance. In these circumstances, the pension fund cannot simply be regarded as an extension of the corporation's balance sheet: the risk is shared between the company's shareholders and the fund members. This argument probably has less force today as a result of legislation and scheme design minimising the degree to which members' benefits, particularly early leavers' benefits, can be reduced. Also, in the past, in the U.K., increases in pensions to match inflation were discretionary (and might not be granted if the fund's assets underperformed), whereas the 1995 Pensions Act made such increases compulsory for future pension accrual.
- In many schemes, surpluses will be used to improve member's benefits, but the fund sponsor will make good any deficits. Thus, the fund sponsor has an incentive to invest in a minimum risk fund; to do otherwise would lead the company sponsor to have to bear the cost of underperformance without benefiting from outperformance.
- Failure to allow for bankruptcy risk and the costs of bankruptcy can invalidate the Modigliani and Miller theorem, and hence any conclusions for pension fund asset allocation that are based on that theory.

Overall, we would argue that there is a strong case for matching assets and liabilities in a pension fund, whilst bearing in mind the underlying corporate finance issues, in particular insofar as a discriminatory tax treatment applies. If we accept this view, then investment risk to a pension fund could be regarded as the risk of the funding rate varying, due to investment conditions rendering the existing funding rate insufficient to meet the liabilities. Equivalently, it could be viewed in terms of the risk of the fund having a deficiency with a given funding rate. In either case, it is matching the characteristics of assets and liabilities that is most important for minimising the investment risk of a pension fund.

The liabilities of a continuing final-salary pension scheme can be divided into three main types:

- Liabilities in relation to active members depend on unknown future wage levels and are very long term.
- Deferred pension liabilities, for members who have left the scheme, are generally linked to increases in retail prices, up to a maximum of 5% per annum, over the period of deferral (this is known as "limited price indexation" or LPI). Such liabilities can be regarded as price index-linked with a financial option to reduce the real payout if inflation is more than 5% per annum.

- Pensions in payment will often be fixed in monetary terms. However, they could be index linked as a result of the scheme rules, the discretionary action of the trustees, or as a result of the limited price indexation required under the 1995 Pensions Act.

As far as the active liabilities are concerned, the aim of the pension fund will be to protect itself from U.K. wage inflation. This would suggest investing in "real" asset classes that provide a long-term hedge against unanticipated price inflation, which is likely to be an important component of unanticipated wage inflation. No asset can be relied upon to be a hedge against real wage inflation. It could be argued that assets such as domestic equities, real estate, foreign equities, and index-linked bonds provide a long-term hedge against the unanticipated price inflation part of wage inflation. Exley (2001) argues that index-linked bonds are the best match for active member liabilities. However, portfolio theory suggests that, when there is no perfect match for a particular liability, a diversified portfolio of imperfectly matched assets might provide the lowest risk asset portfolio. We look at some empirical modeling in Chapter 5.

There is no conventional investment that exactly matches the option payout structure of deferred pension liabilities. However, investing in index-linked bonds could be regarded as a low-risk strategy. In the event of inflation being higher than 5% per annum compound, over the term of the liabilities a portfolio of index-linked bonds would generate a surplus.

As a fund matures, it will have a greater proportion of fixed liabilities made up of pensions in payment; the liabilities will also be shorter in term (although still long term, relative to the duration of most bond portfolios). The fund can reflect this in its investment policy, by investing a greater proportion of its assets in conventional bonds. If the trustees have an implicit or explicit commitment to grant discretionary increases to match U.K. price inflation, index-linked rather than conventional bonds may be regarded as a more appropriate match. A portfolio of index-linked bonds can be designed to be a very good match for pensions in payment liabilities that are index linked.

In general, the long-term nature of pension liabilities also means that a pension fund may be willing to sacrifice liquidity, in order to obtain higher returns. Properties with good tenant covenants, providing a long-term guarantee of a fixed rent level, with periodic reviews to market levels (see Chapter 2) may, therefore, make appropriate investments for pension funds. Liquidity may be important though, at least for part of the investments, in the event of unexpected calls on the scheme's assets (such as transfers). In addition, as a pension scheme matures, both the liquidity of investments and the need for immediate income will become more important considerations.

3.5.3 Legislation and Taxation Considerations

U.K. pension fund legislation is not particularly restrictive in terms of quantitative restrictions on asset holdings. Assets are held in trust and the investment powers of the trustees are listed in the trust deed; normally, the trust deed confers very wide powers. Trust law and the common law require trustees to act in the "best interests" of the fund members. (In practice, this has been determined to mean best financial interests of the members.) Trustees are expected to invest in the way that a "prudent person" would do.

Pension funds are limited as to the amount of "self-investment" they can undertake. They cannot invest in the capital of the employer (including loans, equity, and ownership of real estate) to an extent greater than 5% of the scheme's assets. In addition, if investments were over concentrated, the trustees could be regarded as being in breach of their trust deed. As far as contracted-out schemes are concerned, the concentration of investment rules limit the amount of assets the actuary can take into account, when signing the scheme's solvency certificate, if investments are over concentrated.

Pension law and regulation is a rapidly changing field, and it is worth discussing here one change in regulation that may affect the investment policy of pension funds. Pension funds will have to value their assets for the purposes of their company sponsor's accounts using fair value accounting concepts enshrined in FRS 17. This is based on international accounting standards. This accounting standard involves valuing pension fund liabilities at the rate of return underlying good-quality corporate bonds. Any deficit or surplus so revealed will be reflected within the company sponsor's balance sheet. This regulation may affect pension fund investment policy. Because schemes will be less able to change valuation bases to reflect changes in investment market values, pension funds may become more concerned with achieving lower short-term volatility of asset portfolios. They may, therefore, invest in assets that match the liability valuation benchmark (bonds with a good credit rating). We do not discuss here the issue of whether FRS 17 is an appropriate accounting standard (although it could be argued that it does not fully take into account all the financial aspects of a pension fund discussed in Section 3.5.2). However, if FRS 17 is regarded as implementing as standard practice an approach that is already good practice, then, whilst it may lead to changes in pension fund asset allocation, such a change may well have taken place in the long term in any case. It could also be argued that if the approach underlying FRS 17 is not good practice, then investment analysts will ignore FRS 17 pension fund deficits when valuing companies. The long-term impact of FRS 17 may, therefore, be limited. There are other aspects of regulation and market practice of which pension investment practitioners need to be aware (for example, those flowing from discussion that took place following the Myners Report (Myners, 2001)). However, we do not discuss those further here, as they do not affect asset allocation independently of the considerations discussed above.

As far as tax considerations are concerned, U.K. pension funds are exempt from U.K. income and capital gains tax. However, returns on equities held are effectively taxed at the corporation tax rate, with no relief now allowed for tax paid by the company on distributed profits. Pension funds in other countries generally receive similar treatment. The main effect of this is that it allows pension funds to purchase investments that taxed funds would regard as tax inefficient, such as investments that yield a high income. However, equities could be regarded as inefficient from the U.K. tax perspective. The change in the tax regime in 1997 (see Section 3.4.3) may have a long-term impact on pension fund investments, particularly when taken in conjunction with the increased understanding of the corporate finance issues discussed in Section 3.5.2. The tax discrimination against equities in a pension fund may also impact on how particular assets are held. For example, when real estate is held directly, returns are not taxed (although there is 4% stamp duty payable on any transactions). On the other hand, if real estate is held through quoted property investment companies, then the profits of the company will be taxed at the corporation tax rate.

3.5.4 Expected Investment Returns

Having taken into account the appropriate risk factors, the pension fund should invest to maximise expected returns, for a given level of risk. Different investors, though, do not view risk in the same way. Thus, the subjective risk-adjusted return for particular asset classes may be different for different investors, depending on their liability profile, liquidity restrictions, and so on. It could be argued that investors with long-term time horizons and liabilities that are not always well defined in legal terms will put a higher subjective value on equities than short-term investors. Some actuaries have argued that the long-term, downside risk of equities is limited and that the excess return from equities more than compensates for any small risk that might exist.[2] Others have argued that markets are inefficient (see Lofthouse (2001) for a discussion of the concept of market efficiency and an analysis of the

---

[2]This was the view taken by actuaries Robert Clarkson and Solomon Green at the *Faculty and Institute of Actuaries Finance and Investment Conference*, 2002. The proceedings were not published.

evidence), so that good analysis and timing can lead to long-term outperformance of a pension fund asset portfolio. A company may be willing to accept some long-term risk of underperformance of equity markets. In these circumstances, the expected long-term contribution rate will be reduced by investing in assets, such as equities, with a high expected return.

Changes in weightings between investment categories may be achieved gradually by changes in the way in which new money flowing into the scheme is invested, or by disinvestment from one investment category in order to increase investment in another category.

3.5.5 Size of Fund

A smaller fund may have additional restrictions on its investment policy imposed by practical considerations. For instance, a small fund is unlikely to be able to invest directly in real estate, although it may get exposure to the real-estate market indirectly through investment in real-estate companies, unit trusts, mutual funds or limited partnerships. In addition, the practical difficulties and the costs associated with investment in all but the largest foreign investment markets may well preclude direct exposure to them. Again, there are indirect methods of obtaining exposure to foreign investment environments, such as investment in multinational companies and investment in closed-end funds or investment trusts. These practical considerations will affect the investment policy of small and medium, independently managed or segregated pension funds. Many smaller pension funds may choose to invest in pooled investment funds. The pooled fund will then often be able to invest directly in real estate or foreign investment markets. However, the use of pooled funds does limit the discretion of the fund in terms of the particular investments chosen.

3.5.6 Marketability and Liquidity

Marketability and liquidity are not necessarily of prime importance for a pension fund, although their significance should not be dismissed. The liabilities of most pension funds are sufficiently long term so that, for the majority of assets, marketability and liquidity are not particularly important considerations. However, a mature scheme may require the ability to liquidate assets to pay pensions. Also, the possibility of transfers and other payments being made from a fund requires a degree of liquidity.

3.5.7 Summary of Considerations

The considerations we have discussed above generally lead U.K. pension funds to invest in real investments such as equities, real estate and index-linked government bonds. The anticipation of higher returns from equities is the factor that leads to the greater proportion of the fund being invested in this category. Conventional fixed-interest bonds may be held to match pensions in payment, where these are not index linked. There has been a reduction in the average level of U.K. equity investment by pension funds over the last 9 years, from 56.2% to 40.9%. This can be explained by the changed tax position of equities, the maturing of funds, the increased understanding of the risk of equity investments, greater understanding of the corporate finance issues discussed in Section 3.5.2, and the implementation of FRS 17 (and possibly of the Minimum Funding Requirement of the 1995 Pensions Act, which is now to be phased out). Nevertheless, U.K. pension funds remain relatively equity based, in terms of their investment policy.

These considerations are reflected in Table 3.6, which illustrates the average proportion of U.K. pension funds' assets in various categories at the end of 2003.

3.5.8 International Differences

For a number of reasons, the asset mixes of overseas pension funds differ from those of pension funds in the U.K. In this section, we discuss some of these reasons.

First, the importance of funded pension provision varies from country to country. In the U.S., U.K. and Canada, total pension fund assets as a proportion of personal-sector saving are much higher than in Japan, Germany and most of the rest of continental Europe (see Davis and

Table 3.6 Asset allocation of U.K. pension funds, 31 December 2003

Asset	Allocation (%)
U.K. equities	39
U.K. fixed-interest securities	12
U.K. index-linked bonds	9
Overseas equities	28
Overseas bonds	3
Real estate	6
Cash	3

Source: UBS Pension Fund Indicators, 2004.

Steil (2001) for a detailed discussion of international institutional assets and liabilities). There are a number of reasons for these differences in private-sector pension provision, including taxation incentives, the nature of benefits offered by schemes, legislative restrictions on schemes, and the extent of alternative provision (such as non-funded state or company arrangements). Bonds tend to constitute a larger proportion of pension fund assets in Canada, the U.S. and Japan than in the U.K. It could be argued that an important reason for this is the U.K.'s history of inflation. The proportion of assets held in the form of bonds decreased from over 40% in 1966 to 15% in 1988 (BEQB, 1991). The monetary framework in the U.K. is probably now as credible as, or more credible than, that in other OECD countries, and this is reflected in a reduced long-term inflation risk in the U.K. compared with the period from 1960 to 1993. As has been mentioned, in Section 3.5.7, there has been some reversal in the trend towards greater equity investment in the U.K. In the U.S., nearly 40% of assets are still held in cash or bonds; this reflects, in part, the supply of bonds and inflation history, but also legislation including a benefit insurance system (via the Pension Benefit Guarantee Corporation) with higher premiums for underfunded schemes, and asymmetric surplus distribution rules.

The proportion of assets held in equities is higher in the U.K. (67%) than in the U.S. (62%) or Switzerland (25%). Again, this will reflect the different inflation history in these countries but will also reflect regulation. The Netherlands is the only continental E.U. country that has funded pension provision comparable with that in the U.K. Bezooyen and Mehta (1997) point out that only 31% of Dutch funds were invested in equities (less than half the proportion in the U.K. at that time although the proportion in the Netherlands has now risen to 3%. Bezooyen and Mehta put this difference down to higher and more volatile inflation in the U.K., as well as to the ability at that time of U.K. funds to report using actuarially assessed (and therefore implicitly smoothed) asset values. Bezooyen and Mehta also suggest that peer pressure, related to performance measurement, drives asset allocation in the U.K. Historically, the tax system, too, favoured bond investment, rather than equity investment, to a greater extent in The Netherlands than in the U.K., although, as noted above, this changed in 1997.

3.6 Life Insurance Companies

The investment policy of a life insurance company will be determined to a large degree by its mix of business. The different types of life office business determine the nature of an office's liabilities. In general, a life office's liabilities will be made up of future non-profit liabilities, future with-profit liabilities (which, in turn, will have guaranteed and non-guaranteed portions), investment (or unit) linked liabilities, future expenses that are fixed in absolute terms, and future expenses that are expected to rise as the general level of prices in the economy increases. In addition, there will be shareholders' and policyholders' funds, the extent of which may have a significant effect on investment policy. More detail about the form of these liability types is given in Chapter 6.

Bearing in mind this wide variety of liability types, whilst it may be possible to determine an average distribution of assets for life insurance companies, it would be dangerous to put forward a typical distribution. We will now move on to discuss the appropriate distribution of assets for the various different liability types, as well as the other considerations a life company should take into account when determining investment policy.

3.6.1 Liability Considerations

Most well-established life offices will have a large portfolio of traditional non-profit and with-profit business. The non-profit business will be made up of contractual benefits that are normally fixed in money terms (although some contracts may have benefits fixed in retail price index terms) and long term. Therefore, it would be appropriate to match them with a portfolio of long-term, sterling, fixed-interest bonds — probably a mixture of government bonds and good quality corporate bonds. Any index-linked benefits (such as index-linked annuities) can be matched with index-linked bonds. It is unlikely that the office could exactly match the terms of the assets and liabilities, particularly if there had been a rapid expansion of new business. Whilst some non-profit business may be of a shorter term (for example, single-premium immediate annuity business), most non-profit liabilities will be very long term in nature (particularly annual premium business). Long-term fixed interest securities will, therefore, normally be used to minimise investment risk, although even the longest fixed interest bonds may have a duration less than that of the office's longer duration liabilities.

As far as the with-profit business is concerned, the liabilities can be split into two parts. The guaranteed benefits plus reversionary bonuses accrued to date can be treated in the same way as the non-profit liabilities. There will then be non-contractual liabilities, made up of expected future annual and terminal bonuses (see Chapter 8). The life office has much more discretion regarding investment policy in respect of such liabilities; nevertheless, certain rules of thumb should be applied. First, policyholders' reasonable expectations must be met, although "reasonable expectations" has never been defined with any precision in U.K. law (see also Section 3.2.1). However, the office will certainly be expected, in general, to provide a long-term return for the policyholder, on with-profit business, greater than that provided by non-profit business. Some protection from inflation will also be required. Second, policyholders will require greater predictability of returns from with-profit business than would be provided by, say, direct investment in the stock market; this will partly be provided through the management of the valuation of the life fund and the so-called "smoothing" of bonuses, but will also be managed through the investment policy. Third, the office will need to be able to compete with the bonus rates provided by other offices. These factors point to investment in a diversified portfolio of real assets for the non-guaranteed part of with-profit liabilities, including domestic equities, real estate, and some foreign equities.

When determining both the nature and term of the assets that should be held to meet various liabilities, the office should always bear in mind future premiums it expects to receive. Premiums, after deducting fixed expenses such as commission, can be deducted from future fixed liabilities. The investment policy appropriate for the remaining net liability should then be determined.

Insofar as unit-linked liabilities are concerned, a life insurance company will generally carry no investment risk. The liability of the life office is the value of the units in which policyholders have invested; this will be determined by the value of the investment funds, as long as the office has invested sufficiently in the fund to match the unit-linked liabilities. The split of assets between investment types will be determined by the unit funds in which policyholders have invested. In this respect, therefore, the investment policy of a life company will reflect the preferences of policyholders. However, very often a unitised fund will be described as, or will have the characteristics of, a "mixed fund." That is, the unit price would be determined by the value of a mixed portfolio of investments, determined by the investment managers. The investment managers should determine the mix of the fund after considering the risk/return preferences of the policyholders (insofar as this is possible) and by considering prospects in the various investment markets. The performance of unitised funds will generally be monitored closely by performance measurement agencies. There is, therefore, intense competitive pressure on fund managers to perform well over both the short and long term. This may induce pressure to structure funds close to the market or index average structure.

Unitised with-profit business (see Chapter 8) is less transparent than unit-linked business, and the investment considerations are closer to those surrounding with-profit business.

With regard to expenses, fixed money expenses, such as commission, can be regarded as a deduction from premiums. Increases in other expenses of managing the business are likely to be related to increases in the general level of prices. Such liabilities can only be estimated; nevertheless, they are liabilities against which reserves need to be held.

It could be regarded as appropriate to match such liabilities by investment in index-linked stocks.

3.6.2 Options and Guarantees

The existence of options and guarantees on policies makes the determination of the investment policy more difficult. In the U.K., surrender values are generally determined with reference to the current market value of assets backing the group of policies (often known as the "asset share" of a policy): options and guarantees on surrender are not normally given and the policyholder will bear the risk of movements in investment markets. However, a policy option that provides a policyholder with guaranteed payouts at various different stages of the policy (for example, in the form of guaranteed surrender values) cannot be matched using standard investments. If the insurance company invests in order to match the ultimate liability and interest rates rise, then more policyholders will surrender and invest their surrender proceeds at higher rates of interest. The assets that the insurance company holds will have fallen in value and may not match the surrender value. The first priority of the insurance company will be to follow a relatively conservative investment policy. The potential cost of the policy option can also be estimated using option pricing methods (see Beenstock and Brasse (1986)) and by looking at the historical behaviour of interest rates and of the reaction of policyholders in response to changes in interest rates; extra reserves can then be held against the potential liability. The insurance company may also consider investment in put options, which may give it protection against rising interest rates. Such options would generally have to be specially designed "over-the-counter" options because of the long-term and complex nature of the contracts.

Investment options and guarantees of this form are not the norm on U.K. insurance policies, although many unit-linked policies are offered that have investment guarantees (see Dodhia and Sheldon (1994) for a review of these; the range of policies has not changed greatly since the publication of that paper). In many continental European markets and in the U.S., guaranteed surrender values have been more common, although the Third Life Directive in the E.U. imposes a strong valuation basis on some of these options and this has reduced their prevalence. Also, in the U.K., companies have sold "guaranteed annuity options" (see Section 3.2.1). In general, these options can only be matched by "swaption" products (again, see Section 3.2.1) which are bought over the counter. Some companies, such as the Equitable Life in the U.K., have not hedged the risks of the guaranteed annuities. The losses subsequently caused by the guaranteed annuity option increasing in value as interest rates have fallen and longevity improved have caused financial difficulties for these companies; in the case of the Equitable Life, these financial difficulties were severe.

3.6.3 Free Reserves

We have determined that, a priori, the assets covering the non-profit and the guaranteed part of the with-profit liabilities should be invested in fixed-interest assets of the appropriate currency and term. The remainder of the with-profit assets would probably be invested in a diverse portfolio of real assets. We can now discuss the role of the free reserves, the presence of which may lead an insurance company to deviate from this norm. The free reserves should, of course, be invested in such a way that the statutory minimum solvency margin (see Part II for a discussion of statutory solvency regulations) is unlikely to be breached, if there are changes in financial conditions. An insurance company that was close to the solvency limit would, therefore, wish to invest cautiously.

An insurer that has a high level of solvency would generally wish to invest free reserves in real assets, in order to protect the real value of the shareholders' and policyholders' funds. These funds are also the working capital of the business, however, and they may well be used to support the expansion of new business. This will include setting up reserves for new business, meeting acquisition costs, and meeting other costs of policy development (including capital costs). The profitability of this use of funds would be estimated using the profit-testing techniques to be described in Chapter 6 to Chapter 8. The existence of significant free reserves also gives the insurance company greater investment freedom in respect of its other assets. It may be possible to invest a greater proportion of assets backing with-profit policies in investments from which a higher prospective return is expected (for example, equities). A decision could even be taken partially to mismatch the fixed liabilities. However, the solvency of the business and the risk that share-holders and policyholders wish to take should always be a paramount concern. It should always be borne in mind that shareholders and with-profit policyholders will bear the costs of any adverse fluctuation in investment values if the liabilities are not matched by term, type and currency.

3.6.4 Expected Returns

Whilst obtaining the maximum returns possible must always be subservient to matching the liabilities, the level of expected investment returns is an important issue for a life insurance company. Pegler (1948) suggested that the aim of the life office should be to invest its funds to earn the maximum yield. This was qualified by an explanation that "yield" meant expected yield, and that this consideration should be subject to one of security. The expected return on investment funds is of particular importance for three groups: shareholders, with-profit policyholders and unit-linked policyholders (to be dealt with in Section 3.8).

With-profit policyholders will generally require an expected return greater than that offered to non-profit policyholders. However, the guarantees and smoothing inherent in with-profit policies will probably mean that their expected return is somewhat lower than that which could be achieved, say, from investment in a unit-linked equity fund. The extra return, compared with that offered to non-profit policyholders, will arise from investment in equities and real estate.

3.6.5 Legislation

The investment policy of a life insurance company is also constrained (although to a relatively limited extent in the U.K.) by legislation. Further constraints are imposed by the professional duties of the actuary, which will be discussed in the next section. In the E.U., the main way in which the regulation of investment policy takes place, albeit indirectly, is through the valuation regulations. Only admissible assets can be taken into account in the valuation; non-admissible assets can be used for investment but are not to be taken into account when determining whether the insurer has an adequate degree of solvency.

All "approved securities" are admissible up to any limit. This includes all securities issued by the government or local authorities. Other assets can only be taken into account up to a specified percentage of the total liabilities. The percentages are not intended to be a significant constraint on investment policy, but exist to prevent over-concentration in any particular investment. Examples of the relevant percentages are:

5% in any single property
5% in listed shares and debentures in any one company (although not
    more than 2.5% in shares is admissible)
1% in unlisted shares in any one company
0.1% in options in any one company

There are no direct constraints on the proportion that can be invested in any particular investment category (for example, equities). Further constraints may be imposed implicitly through the valuation basis. Insurance companies have to apply particular valuation principles and also stress tests to test the resilience of their balance sheet to investment market movements. The stress tests encourage companies to match their fixed assets and liabilities by term and discourage them from investing in equities in respect of their fixed liabilities. If they do so, life insurance companies will have to hold additional capital. In particular, if a life company is close to its statutory minimum solvency margin, then it might reduce its investment in equities and increase the duration of its fixed-interest investments in order to reduce the extra reserves it needs to hold to meet the stress tests.

The valuation of liability regulations may also affect an insurance company's investment policy if the level of free assets is small. The rate of interest that can be used to value the liabilities is restricted to 97.5% of the reliable yield on assets. This is based on the running yield on equities[3] and real estate, which would normally be less than the yield from fixed-interest securities. It would be possible for an insurance company to increase the yield it could use when valuing the liabilities by switching assets out of equities and into fixed-interest securities. In many circumstances this may well be the prudent thing to do, although one should bear in mind that, depending on the nature of the liabilities, fixed-interest securities bear greater inflation risk. There is a very clear discussion of the recently introduced regulations in the UK in Muir and Waller (2003). However, the regulations are changing rapidly. Updates to the regulations can be found at www.fsa.gov.uk.

3.6.6 Professional Guidance to the Actuary

As well as legislation affecting investment policy, actuaries have a general professional duty to manage assets and liabilities prudently. This professional duty is undertaken within the framework of specific professional guidance in the U.K. Guidance Notes are issued to actuaries, some of which relate, in part, to the actuary's input into the asset allocation decision-making process. The Guidance Notes that relate to this aspect of life insurance business are as follows:

GN1 — Actuaries and Long-Term Insurance Business
GN5 — Actuaries Advising Long-Term Insurers in Countries Overseas
GN8 — Additional Guidance for Appointed Actuaries

Other Guidance Notes relating to life business do not have implications for investment policy. GN1 and GN8 are practice standard and GN5 is regarded as recommended practice.

---

[3]Although an adjusted earnings yield can be used.

GN1 (paragraph 6.5) makes clear that the investment policy is the responsibility of the directors. However, the actuary must pay due regard to the term of the assets and liabilities. The dangers of insolvency are increased where the company is transacting a large volume of non-profit business — particularly if there are guarantees. If the investment policy is inappropriate, bearing in mind the nature of the assets and liabilities, the Actuary should advise the company of the constraints necessary to protect the position of policyholders.

GN5 suggests that actuaries should apply the principles of GN1 when advising overseas companies. However, it is recognised that this would not always be appropriate. It is also recognised that it may be difficult to ascertain the value of the assets and, where this is the case, this should be mentioned in any report. Any U.K. actuary advising an overseas insurer should, of course, bear in mind local legislation.

As far as investment is concerned, GN8 draws the attention of the Actuary to the Prudential Sourcebook for insurers, published by the FSA. The company must pursue an investment policy or hold margins sufficient to ensure that the fund is protected from changes in financial conditions. The resilience test deals with this to some extent. However, the Actuary must also be aware of changes in currency values and of guarantees and options that are not specifically addressed by the resilience test. GN8 also draws the attention of the Actuary to the need to make provision for possible future increases in expenses; investment in index-linked stocks to the appropriate degree could assist in this regard.

3.6.7 Taxation

The taxation position of life offices is complex. In the U.K., a life office is charged tax on the higher of profits or investment income less expenses. In practice, this means that tax is charged on investment income, at a rate close to the national basic income tax rate. In addition, in the U.K., chargeable capital gains are also taxed. The taxation of life funds cannot be said to affect unduly investment policy to a great degree.

It should be noted that the pension fund part of the life office business is taxed on the basis described in Section 3.5.3. The investment decisions in respect of this segment of the business should reflect the taxation considerations mentioned in that section.

3.6.8 Marketability and Liquidity

The long-term nature of life insurance business means that marketability and liquidity considerations are not as critical as, for example, with non-life insurance business. If the liabilities allow, life offices can invest considerable assets in real estate. High marketability and liquidity can never be a disadvantage, however. It would be inappropriate for too large a proportion of a fund's assets to be invested in unmarketable and illiquid assets. Also dangerous would be too great a concentration in unquoted assets with values that are difficult to measure. Such investments would also often be unmarketable.

3.6.9 Summary of Considerations

The main considerations for the life office are the nature, term and currency of the liabilities. Thus, to a large degree, the investment strategy will depend upon the split between with-profit and non-profit liabilities. Where the office has investment freedom because of the nature of the liabilities, it will invest to maximise return, whilst bearing in mind the policyholders' risk preferences.

Regulation in the U.K. is not prescriptive, but the regulations concerning valuation of liabilities and asset admissibility do impose constraints on the life office. Actuaries, as well as others involved in the investment decision-making process, also have a professional responsibility to ensure that the interests of policyholders are met. There will be more discussion of unit-linked business in Section 3.9, with particular reference to personal pension provision.

Different life insurance companies will have investment distributions that vary widely, as a result of having different liabilities, different solvency margins, etc. The distribution shown in Table 3.7, therefore, cannot be regarded as typical for any particular type of life company, but represents the average for the industry in the U.K. It is noteworthy that over 80% of

Table 3.7 Asset allocation of U.K. long-term insurance companies, 31 December 2002

Asset	Allocation (%)
U.K. ordinary shares	31.4
U.K. fixed-interest securities	34.7
U.K. index-linked government bonds	3.0
Overseas equities	10.9
Overseas fixed-interest securities	6.8
Real estate	7.1
Cash	6.1

Note: Some figures are approximations as categories have been combined.
Source: National Statistics.

all the U.K. gilts held by long-term insurers have a maturity date greater than 5 years reflecting the long term of the liabilities.

3.6.10 International Differences

Elsewhere in the E.U., investment policy is regulated to a much greater extent than is the case in the U.K. In North America, regulation may be less important than in the E.U., but policy terms and conditions often play an important part in ensuring a greater concentration of investments in fixed interest securities.

The influence of regulation in the E.U. in both the pensions and insurance field is discussed in greater detail in OECD (2000). It should be borne in mind that there is an ongoing review of solvency regulation and that there may be further liberalisation and harmonisation of regulation within and between E.U. countries. There remain a number of restrictions on portfolio investment, such as maximum limits on equity investment and minimum proportions to be invested in bonds.

Davis and Steil (2001) analyse the breakdown of life insurance companies' portfolios on an international basis. At the time of the analysis, the average proportion invested in loans or domestic bonds in the countries studied was 61%; the average proportion invested in domestic equities was 23%. The proportion invested in loans or domestic bonds in the U.K. was 26% and the proportion invested in domestic equities 48%. This difference in investment policy was related to whether countries followed a "prudent person" or "quantitative restrictions" form of regulation of investment policy. It is also highly likely that the investment policy is also dependent upon the nature of the liabilities of the life offices. In the U.K., life offices have been used for equity-linked savings, through unit-linked and with-profit policies. This has not been the case in many other countries, where policies are more "protection orientated." In the U.K. market, more with-profit and unit-linked liabilities, backed to a greater extent by real assets, may also have been written to satisfy the demand of consumers wanting protection from inflation from the 1970s to the 1990s.

3.7 Non-life Insurance Companies

The main characteristic of non-life or general insurance liabilities is that they are much shorter and less predictable in size than life insurance and pension fund liabilities. International diversification of non-life business is also much more pronounced than that of life insurance and pensions business. Liabilities can, therefore, be in a number of different currencies. For many lines of business, in most countries non-life policies are generally short term and reserves are, therefore, smaller relative to premiums, and the interest earned on reserves may be less important than for life business. Traditionally, therefore, there has been less emphasis on invest-ment management in non-life companies. However, returns earned on assets can be a significant part of a company's profit, particularly on lines of business where profit margins are narrow. Some business lines also have long "run-off" periods (see Part III) and, in some countries, short-term inflation is high and volatile. In these circumstances, investment policy can also be very important. In addition, non-life companies often have large levels of free reserves that have to be invested.

There will be differences between the lines of business written in different countries, and also differences in the extent to which short-term inflation is a risk. However, there are no significant international differences between the principles followed by non-life insurers. The details of regulation and taxation (see Section 3.7.4 and Section 3.7.5) are, of course, different in different countries, although there are many similarities within the E.U.

3.7.1   Liability Considerations

The premiums received by a non-life company are held to meet future claims during the risk period. In general, most of the claims on most lines of business will be notified within the time of the contract (usually a year), and settled within that time or not long afterwards. The liabilities are, therefore, generally short term. The policy terms normally define a loss not in terms of a fixed money amount, but in terms of a replacement cost or some form of damages in reparation for a tangible or intangible loss based on the indemnity principle (see Part III). We would expect these amounts to increase as the general level of prices increases, although not necessarily at the same rate (and, depending on factors such as court settlements, claim amounts can increase considerably faster than general inflation). From year to year, when contracts are renewed, premiums will increase to reflect expected increases in money amounts of claims. However, in the invest-ment of the assets, the non-life insurance company should consider possible inflation risk over the period between the payment of premiums and the payment of claims.

Notwithstanding the above, it can be said that, in most countries, the general inflation risk for short-term non-life liabilities is not considerable. In the U.K. economy, over the course of a year or so, inflation is likely to be relatively predictable. However, in economies where inflation is higher or more volatile, the need for protection from unanticipated inflation is an important factor.

Thus, in the U.K., most non-life liabilities can be treated as if they were short term and nominal in character. However, a significant and possibly growing amount of liabilities could be much longer term. Even for certain common types of insurance policy, such as motor insurance, claims may well not be settled until some time after the end of the policy year; inevitably, these claims are the larger claims. For some types of insurance, claims may not even be made until several years after the end of a particular policy year. This is particularly notable in the case of industrial liability insurance, where industrial diseases will often not be identified and traced to the period of insurance until many years after the original onset of the disease. Thus, insurance companies have to hold a significant proportion of their reserves against outstanding claims and incurred but not reported claims. Uncertainties surround the timing and size of such outstanding claims which cannot be addressed through prudent investment policy. It is difficult to ascertain the expected number and size, the time of payment, and the inflation risk relating to these incurred but not reported claims However, we do know that, in general, the size of such claims is likely to increase where there is general inflation in the economy; in this sense, the longer term liabilities could be regarded as "real" liabilities.

This mixture of short-term liabilities and longer term liabilities of uncertain amount and duration makes it quite difficult to determine an appropriate investment policy. Cash and short-dated fixed-interest securities are certainly appropriate in respect of the majority of liabilities. The investment returns will be known over the term for which the liabilities are outstanding with a high degree of certainty. If desired, the expected investment returns can be allowed for in the premium calculations. With regard to the longer term liabilities, an inflation hedge would be provided by equities, real estate, and index-linked bonds.

3.7.2   The Place of Equities, Real Estate and Index-Linked Government Bonds in a Non-life Portfolio: The Role of the Solvency Margin

Equities have the benefit of providing an inflation hedge and, over the long term, they would be expected to provide a higher real return than fixed-interest investments. However, for a non-life company, they have the disadvantage of an uncertain real return in the long term and considerable short-term volatility. For these reasons, non-life companies have tended to invest relatively low proportions of their assets in equities, except in particular circumstances, which will be discussed below. Other than in these circumstances, a non-life company investing a high proportion of assets in equities would have a clear risk of insolvency if assets values were to fall.

Non-life companies with significant solvency margins may invest a considerable proportion in equities. If the size of the solvency margin is such that a reasonable adverse market movement is unlikely to take the solvency margin of the company below the statutory minimum solvency margin, then a proportion of equity investment can be justified. The risk of such equity investment is not reduced if the solvency margin is large — the shareholder bears the risk. However, if there is a large solvency margin, then the risk of bankruptcy, with the attendant costs and loss of franchise, is less. It may be thought that real-estate investment could be justified by the same criteria by which equity investment is often justified. However, it is normally felt that the poor marketability, poor liquidity and high transactions costs of real estate, combined with its low historical return compared with equity investment, makes it unattractive. The real-estate holdings of non-life companies consist mainly of their own offices.

Index-linked bonds provide a useful additional asset class for non-life insurers. However, they do not make up a large proportion of non-life insurance company assets. In the U.K., as has been mentioned, inflation is relatively predictable over the term of non-life insurance liabilities. In addition, it should be remembered that, due to the 8-month time lag in the indexation of payments from an index-linked government bond in the U.K., they would not provide significantly more inflation protection, over the term of the liabilities, than conventional government bonds or cash. Index-linked bonds in other investment markets tend to have shorter indexation lags.

As far as liabilities with a longer run-off period are concerned, index-linked government bonds may well provide an appropriate match. However, the nature of such liabilities is such that their amount and timing are very unpredictable, so that a non-life company may regard short-term investments as more appropriate.

The actual solvency margin of a non-life insurer is likely to be considerably above the level of the statutory minimum solvency margin. Free reserves effectively form shareholders' funds, and they need to be invested bearing in mind both this and the fact that the insurer's solvency margin must not fall below the statutory minimum. An appropriate mixture of equity and fixed-interest investment will normally be selected, the particular mix being determined mainly by the size of the actual solvency margin over the statutory minimum solvency margin and the policy of the owners of the company as interpreted by the directors.

It should be mentioned that the statutory minimum solvency margin would not be the only consideration related to the solvency position of the insurance company when setting investment policy. In fact, most companies hold capital of many times the statutory minimum margin. This is necessary in order to obtain a credit rating that is sufficiently high to allow the company to write good quality business. The lower that the actual solvency margin is, the more conservative the investment policy will have to be to obtain a given target credit rating.[4] Thus, the three

---

[4]The FSA is introducing risk based solvency standards for non-life insurance companies so that the required level of solvency will depend on the level of investment risk. It is not clear that this will influence a company's behaviour significantly given that the desire to obtain a particular credit rating is already an important constraint.

Table 3.8 Asset allocation of U.K. insurance companies (excluding long-term business), 31 December 2002

Asset	Allocation (%)
U.K. ordinary shares	12.3
U.K. fixed-interest securities	44.3
U.K. index-linked government bonds	1.3
Overseas equities	3.5
Overseas fixed-interest securities	22.8
Real estate	1.5
Cash and short-term assets	14.3

Note: Some figures are approximate because similar categories have been combined and some undefined categories
have been eliminated.
Note: Over 60% of U.K. government bond holdings are
short-term government bonds.
Source: National Statistics.

issues of investment policy, desired credit rating and solvency margin
should be considered strategically, in conjunction with each other, at
board level.

3.7.3 Foreign Currency Investment

On average, non-life insurers in the U.K. invest about 25% of their assets
in foreign currency investments (see Table 3.8). Non-life insurance
companies often operate on an international scale and underwrite risks
denominated in a number of different currencies. Therefore, it would be
appropriate for the non-life company to invest in foreign currencies to an
extent determined by the proportion of foreign currency liabilities. The type
of assets purchased would be much the same as the types of asset used
to match domestic currency liabilities (cash, short-term bonds, etc.).

3.7.4 Legislation

In the U.K., there is little explicit legislation directing the investment
policy of non-life insurers. Admissibility regulations are the same as for life
insurers (see Section 3.6): they simply enforce the degree of diversification
that would be followed, in most cases, by a prudent insurer. The statutory
minimum solvency margin requirement and the new risk-based capital
requirements might lead to lower investment in assets that tend to suffer
from short-term volatility (although the effect is unlikely to be significant, as
a prudent insurer would probably avoid high exposure to such investments
in any case).

3.7.5 Taxation Considerations

Non-life funds in the U.K. pay tax as traders in investments at the full rate of corporation tax. A non-life company would, therefore, pay tax on coupons from fixed-interest bonds and on capital gains from all investments including bonds. This leads non-life companies to invest in high-coupon bonds, which tend to have higher net yields on this basis. High-coupon bonds also have the advantage of providing a greater immediate income. All other things being equal, a non-life fund would prefer franked income (on which tax has already been paid and on which no further tax is paid by the recipient), given the high marginal rate of tax paid on investment income. However, the investment categories that would provide an investor with franked income (such as preference shares and equities) would tend to be judged by other considerations: in the case of preference shares, by resulting mismatch in duration; in the case of equities, by the risk attached to their returns.

3.7.6 Marketability and Liquidity

Marketability and liquidity have already been mentioned when considering the role of real estate and equities in a non-life company investment portfolio. Clearly, the liabilities of a non-life company are too short to consider non-marketable or illiquid assets. Thus, where a non-life company invests in equities, they are likely to be of the more marketable type. As has been mentioned, the only significant real-estate investment undertaken will normally be in relation to provision of office accommodation for the non-life company's own business use.

3.7.7 Expected Investment Returns

This consideration has been left until last, because investment is not the raison d'être of non-life insurance companies. Many of the related issues have already been covered. However, in a competitive non-life market, profits from investment income may be as important to non-life insurers as underwriting profits.

If the solvency margin of a non-life insurer were such that, even given a significant fall in equity values, the margin would not fall below the statutory minimum margin or the credit rating would not be reduced below its target, then the insurer might invest a larger proportion of assets in equities. Real estate would generally be ruled out on liquidity grounds. By investing a larger proportion in equities, the insurer may be able to obtain a higher expected return on assets; in addition, in competitive market conditions, the expectation of greater profits from investment may make it possible to keep premiums lower than those of competitors. Two further factors should be considered, however, before proceeding with such a strategy. First, as has been mentioned, the risk of a fall in equity values is passed to shareholders. Second, carrying a high solvency margin, merely to facilitate equity investment in a non-life fund, has a capital cost. Shareholders may reasonably ask why part of the excess solvency margin is not distributed and an investment policy more appropriate to the liabilities is not followed.

3.7.8 Summary of Considerations

Overall, the short-term nature of some parts of a non-life insurer's liabilities will be paramount in determining its investment policy. The existence of liabilities with long run-off periods may lead to a requirement for some longer term protection from inflation. Liquidity and marketability are, in general, paramount. A high level of free reserves may allow a non-life insurer to invest a greater proportion of assets in equities. However, whilst obtaining a higher investment return may be important, it should be subservient to the investment risk considerations. Many non-life insurers have significant foreign currency liabilities and, hence, may invest a corresponding proportion of assets in foreign currency investments. Overall, these considerations will lead a non-life insurer to invest a greater proportion of assets in short-dated domestic currency bonds, cash and foreign currency bonds than pension funds or life funds would. An average distribution of assets between categories for non-life insurers as at 31 December 2002 is shown in Table 3.8. It is noteworthy that 85% of assets held by non-life insurance companies consist of cash or bonds with a term to maturity of less than 5 years.

3.8 Personal Investments

In this section we consider some of the factors an individual, or a fund manager attempting to fulfill the preferences of an individual in structuring a fund, would take into account when determining an appropriate investment strategy for a unitised investment fund. The investment strategy may be at the discretion of the investment manager or of the unit holder; whoever takes the investment decision, an important feature is that the effects of the investment policy are transparent in any unitised fund vehicle.

3.8.1 Unit Trusts and Investment Funds

The main considerations for the investor are risk and return. A possible additional consideration, particularly for unit-linked life policies, may be the income yield. Generally, personal investors will use unit trusts (similar to mutual funds in the U.S.) as a savings or investment vehicle. Although they should generally be used as a medium- or long-term vehicle, there is often a concentration on the short- to medium-term performance of unitised funds. The variables that can be controlled, to give the desired investment profile are the investment mix within the fund, the fund manager (different managers may well have different investment styles), and the combination of funds with different investment objectives in which a personal investor invests.

When analysing the expected return from different funds and combinations of investment within a fund, the main considerations are the historical performance record and the likely future prospects for individual markets. This information will be used in order to form the investor's expectations of future returns. As far as risk is concerned, a number of factors may be considered. The individual, choosing between investment categories, may be most concerned about underperforming a particular rate of real return. Thus, the investor may implicitly adopt a decision criterion such as rejecting the investment if it fails to fulfill:

$$\Pr[R \geq r^*] \geq q$$

where R is the real return, r^* is some benchmark, and q is a predefined probability (say 0.95). Alternatively, the investor may regard short-term volatility of real returns from the fund as being the appropriate measure of risk (few funds would have sufficient history to produce a meaningful measure of volatility of long-term returns; and because of changes in style and managers, long-term risk measures are of doubtful value). Risk measures that can be used in a number of contexts, including in the context of personal investment, will be described in Chapter 4.

Fund managers may also determine, or be required to determine, investment policy to target risk-adjusted return measures. One such risk-adjusted return measure, known as the Sharpe index, uses such an approach. The Sharpe index is a summary measure that indicates the historical excess return per unit risk:

$$\text{Excess return per unit volatility} = \frac{r_p - r_f}{\sigma_p}$$

where r_p is the average return of the portfolio, r_f is the average risk-free return over the same period, and σ_p is the standard deviation of the portfolio return. Other risk-adjusted measures of performance are discussed by Adams et al. (2003).

In many circumstances, fund managers will have both risk and performance measured against a "fund-specific benchmark." Such a benchmark might be a relevant stock market index (for example, the FTSE 100 index if the fund is a U.K. equity fund, or the S&P 500 if it is a U.S. equity fund). Such an approach may be taken because the fund is an "indexed" fund that is specifically designed to match the performance of the stock market index (and risk is therefore defined in relation to the performance of the index). More generally, the approach may be taken because, when measuring both performance and risk, it is important to take into account the investment context. It would be unreasonable to expect a U.K. equity-based investment fund to outperform the risk-free rate of return if a representative index of the U.K. equity market had fallen by 20% over the period. In this context, measures such as tracking error will be used. These measure the average deviation (or the square root of the average of the squared deviations) of the fund performance from the index performance over a given period; see also Chapter 4.

So far, we have discussed how investment funds might consider their performance, relative to risk, in an ex post sense. Clearly there are lessons from ex-post analysis that will help fund managers determine asset allocation for investment funds and that will help personal investors choose unitised funds. If a fund has a volatile performance history, this is likely to be an indication of a particular management style. In addition, other lessons can be drawn from a forward looking analysis. For example, in a mixed fund of different investment categories, a higher weighting in bonds or cash is likely to produce lower long-term returns but less volatile performance. A higher weighting in equities will tend to produce the reverse. Diversifying the fund should produce higher long-term average returns for a given degree of risk (and hence a higher Sharpe index). There are other approaches to determining the ex-ante risk of a unitised fund. One forward-looking method of determining risk for a fund that is measured against a benchmark such as an index is the use of the "active money" concept. This measures how far a fund's investment policy is from the benchmark with which the fund is compared: the further an equity fund is, in terms of its asset allocation, from the index against which it is measured, the higher the tracking error is likely to be. Thus, active money, often described as the "index bets" taken by a fund manager is an ex-ante measure of risk. Formally, it can be described as the sum of the differences between the proportions of each investment that are held in the index and the proportion held by the fund.

A problem with the active money measure is that it does not take account of the risk of the index bets in terms of the volatility of the investments and the relationships between the returns from different assets. This is best illustrated by example. Take a situation where the benchmark for a European equity fund is the index weighting of the 40 biggest European stocks. The manager omits one stock completely but has a double weighting in another stock which has the same index weighting. If the omitted stock has a correlation coefficient of 0.9 with the stock in which there is a double weighting, then the risk taken would be negligible (for example, they may both be oil companies). On the other hand if the stock in which there is a zero weighting has a correlation coefficient of 0.2 with the stock in which there is a double weighting (for example, if one stock is an oil company and the other a retailer), then the fund manager would be taking a much more significant risk. Often, fund managers will use proprietary packages, with complex multi-factor models of equity markets and individual stocks, as part of the asset allocation and risk-monitoring processes for investment funds. More generally, there is a need to take into account a number of ex-ante and ex-post measures of risk and methods of risk management when determining asset allocation in an investment fund. Many of these issues are discussed by Kemp et al. (2000).

3.8.2 Personal Pensions

The considerations used for determining an appropriate investment strategy for a personal pension are similar to those described above for investment funds more generally, although it should be borne in mind that investment is over a longer time horizon. The investment decision may be taken by the policyholder, who will normally be allowed to choose the asset categories in which the fund should be invested. If the policyholder requests that the pension fund is invested in a mixed or balanced fund, discretion is effectively passed to the investment managers. In both cases, the considerations are broadly the same.

In theory, the most significant risk element for the personal pension policyholder should be the risk that the fund will be insufficient to purchase a pension that will maintain the pensioner's pre-retirement living standard. Therefore, it is the long-term inflation risk that is most important.

It is unlikely that the policyholder will analyse risk in a very sophisticated manner; however, a number of risk criteria could be used by the policyholder or fund manager, either explicitly or implicitly (see Booth (1995)). First, the pension-plan holder could consider

$$\Pr(A \leq a^*)$$

where A is the accumulation of the fund at retirement and a^* is the amount of fund necessary to purchase a pension of $x\%$ of salary at retirement (where a pension of below $x\%$ of final salary would reduce the investor's living standard significantly).

Other measures of risk that concentrate on downside risk could include the semi-variance of the ultimate accumulation, or the semi-variance of short-term returns (see Chapter 4). Alternatively, the variance of the ultimate pension (or of the cash sum necessary to buy the pension) could be considered. It is likely that a personal pension-plan holder would also be interested in the short-term volatility of investments: partly because it is a relatively easy quantity to measure and partly because investment categories that have high short-term volatility may well have a high degree of uncertainty attached to their long-term values. These considerations of risk are likely to encourage the personal pension-plan holder to invest in a diversified portfolio of real investments. The desire to increase expected return, to reduce the expected cost of providing a given pension, will tend to encourage the investor to diversify mainly between domestic equities, foreign equities and real estate. Index-linked government bonds may also play an important part in a portfolio.

As the investor moves closer to retirement, a lower short-term volatility of investments may be required. In theory, at least, this should lead to increased investment in index-linked and conventional government bonds, and possibly cash. From an asset/liability matching perspective, the most important criterion is to hold investments that will be well correlated with annuity purchase prices (possibly index-linked annuity prices). As long-term real interest rates fall, index-linked annuity prices will tend to rise (assuming constant mortality and expenses). If the investor holds a portfolio of index-linked bonds, then the value of those investments should rise to protect the investor from the rise in annuity prices. It is often suggested that investment policy in personal pension funds should be "lifestyled" so that investors move automatically towards a portfolio of index-linked stocks as they move towards retirement, thus matching the investment characteristics of the annuity that needs to be purchased at retirement. Office of Fair Trading (1997) was one such proponent. There is a critique of this approach and an empirical investigation of its success in Booth and Yakoubov (2000). There is other literature that supports lifestyling, such as Vigna and Haberman (2001); Haberman and Vigna, (2002). Personal-pensions products offered by insurance companies frequently do include lifestyling options.

Personal pension funds should also be considered in the context of wider asset portfolios held by individuals. Personal pension funds will tend to comprise just one of many sources of wealth that can be used to fund income in retirement. Other sources may include general saving, other investment funds and property (including the individual's own residence). If this is recognised, then asset allocation decisions within the personal pension fund should take place only after considering the wider investment and wealth portfolio of the individual. Tax considerations may well be important. For example, as a result of the fact that corporation tax cannot be reclaimed on equity investments in otherwise tax-free funds[5] (such

---

[5]The U.K. avoided such tax discrimination against equity investments until 1997, when the tax system was changed; see Booth and Cooper (2002) for an analysis of the results of this change, in the context of defined contribution pension funds.

as pension funds), it would be tax efficient to hold any equity investments an individual chooses to hold, as part of their wider portfolio, outside the pension fund and cash and bond investments within the pension fund. This may be sub-optimal on other criteria, as this would lead to the least liquid and most volatile investments being held in short-term investment vehicles, rather than in pension funds.

3.9   Conclusion

In this chapter, we have considered the most important factors affecting the investment policy of different types of institutional investor, with a view to explaining the very different investment policies followed by pension funds, life funds and non-life funds. We have also considered the factors affecting the investment policy of personal investors. The main factor affecting investment policy is the liabilities of the investor. Having been satisfied that liabilities are appropriately matched, the investor will take into account expected returns, liquidity and marketability, taxation and regulatory requirements.

References

Adams, A. T. and Booth, P. M. (1995). Sensitivity measures for equity investments. *The Journal of the Institute of Mathematics and its Applications to Business and Industry* 6(4), 365–374.

Adams, A. T., Booth, P. M., Freeth, D. S. and Bowie, D. (2003). *Investment Mathematics*. Wiley, U.K.

Ballotta, L. and Haberman, S. (2003). Valuation of guaranteed annuity conversion options. *Insurance, Mathematics and Economics* 33, 87–108.

Barclays Capital. (2003). *Equity Gilt Study*. Barclays, London, U.K.

Beenstock, M. and Brasse, V. (1986). Using options to price maturity guarantees. *Journal of the Institute of Actuaries* 113(Pt. I), 151–166.

BEQB (1991). The development of pension funds—an international comparison. *Bank of England Quarterly Bulletin* 31(3), 380–390.

Bezooyen, J. V. and Mehta, S. (1997). A comparative study of investment strategies for Dutch and UK defined benefit pension schemes. In: *Proceedings of the 1997 Institute and Faculty of Actuaries Investment Conference*.

Black, F. (1980). The tax consequences of long run pensions policy. *Financial Analysts Journal* 36, 21–28.

Booth, P. M. (1995). The management of investment risk for defined contribution pension schemes. In: *Transactions of XXV International Congress of Actuaries*, vol. 3.

Booth, P. M. and Cooper, D. R. (2002). The tax treatment of UK defined contribution pension funds. *Fiscal Studies* 23(1), 77–104.

Booth P. M. and Yakoubov, Y. (2000). Investment policy for defined contribution pension schemes close to retirement: an analysis of the "lifestyle" concept. *North American Actuarial Journal* 4(2), 1–19.

Davis, E. P. (2002). Ageing and financial stability. In: Aurebach, A. J. and Herrman, H. (Eds), *Ageing, Financial Markets and Monetary Policy.* Springer, Heidelberg.

Davis, E. P. and Steil, B. (2001). *Institutional Investors.* MIT Press, London, U.K.

Dimson, E., Marsh, P. and Staunton, M. (2002). *Triumph of the Optimists: 101 Years of Global Investment Returns.* Princeton University Press, U.S.

Dodhia, M. and Sheldon, T. (1994). Guaranteed equity products. Paper presented to the Staple Inn Actuarial Society, Staple Inn Actuarial Society, London.

Exley, J. (2001). Pension funds and the UK economy. Paper presented to the *Institute and Faculty of Actuaries Finance and Investment Conference*, Institute and Faculty of Actuaries, London, U.K.

GN1, Actuaries and Long-Term Insurance Business. *Guidance Note One of the Institute of Actuaries* (as revised December 2001).

GN5, Actuaries Advising Long-Term Insurers in Countries Overseas. *Guidance Note Five of the Institute of Actuaries* (as revised October 1998).

GN8, Additional Guidance for Appointed Actuaries and Appropriate Actuaries. *Guidance Note Eight of the Institute of Actuaries* (as revised December 2001).

Haberman, S. and Vigna, E. (2002). Optimal investment strategies and risk measures in defined contribution pension schemes. *Insurance Mathematics and Economics* 31, 35–69.

Haberman, S., Day, C., Fogarty, D., Khorasanee, M. Z., Nash, N., Ngwira, B., Wright, I. D., and Yakoubov, Y. (2003a). A stochastic approach to risk management and decision making in defined benefit pension schemes. *British Actuarial Journal* in press.

Haberman, S., Khorasanee, M. Z., Ngwira, B., and Wright, I. D. (2003b). Risk measurement and management of defined benefit pension schemes. *IMA Journal of Management Mathematics* in press.

Kemp, M., Cumberworth, M., Gardner D., Griffiths, J., and Sandford, C. (2000). *Portfolio risk measurement and reporting: an overview for pension funds.* Report by the Portfolio Risk Measurement Working Party of the Faculty and Institute of Actuaries, Faculty and Institute of Actuaries, London, U.K.

Lofthouse, S. (2001). *Investment Management.* Wiley, U.K.

Modigliani, F. and Miller, M. H. (1958). The cost of capital, corporation finance and the theory of the firm. *American Economic Review* 48, 261–297.

Myners, P. (2001). *Institutional Investment in the United Kingdom: A Review.* H.M. Treasury, London, U.K.

OECD (2000). *Private Pensions Systems and Policy Issues.* OECD, Paris, France.

Office of Fair Trading (1997). *Report of the Director General's, Inquiry into Pensions,* Office of Fair Trading, London, U.K.

Pegler, J. B. H. (1948). The actuarial principles of investment. *Journal of the Institute of Actuaries* 74, 179–211.

Scholtes, C. (2002). On market-based measures of inflation expectations. *Bank of England Quarterly Bulletin* 42(1), 67–77.

Shelley, M., Arnold, M., and Needleman, P. D. (2002). A review of policyholders' reasonable expectations. *British Actuarial Journal* 8(Part IV), 705–756.

Solnik, B. H. (1999). *International Investments*, 4th edition, Addison-Wesley, New York.

Vigna, E. and Haberman, S. (2001). Optimal investment strategy for defined contribution pension schemes. *Insurance, Mathematics and Economics*, 28, 233–262.

Chapter 4

Investment Risk

4.1 Introduction

This chapter makes the transition from the conceptual to the analytical and mathematical, as far as the subject of investment is concerned. Actuarial science involves the use of judgement in its applications. That is just as true in the area of investment as in any other area. Knowledge of the underlying conceptual issues is, therefore, no less important than knowledge of the mathematical techniques that are used in problem solving. However, there are analytical and mathematical techniques that can aid our understanding and aid the practical management of investment policy. In this chapter, we begin with a study of investment risk. This is followed in Chapter 5 by the introduction of various techniques that can be used in the asset allocation process.

Techniques employed when making asset allocation decisions, to be reviewed in Chapter 5, rely on the use of decision criteria. For example, implicitly or explicitly, the institution may attempt to maximise a function of expected returns and a risk measure. The decision criteria need to be related to relevant objective variables, such as the ultimate surplus of a pension fund or the returns on an asset portfolio. Asset allocation models will not always produce the same results if different risk measures are used. It is important, therefore, to have an understanding of the risk measures that are appropriate in theory and in practice. In this chapter, we discuss various different approaches to measuring investment risk and the appropriateness of different approaches in different contexts. We start by looking at the simple situation where an investor is interested in the terminal wealth arising from the investment decision.

4.2 Utility Theory and Risk Measures

Utility theory can be used as the basis for analysing investment risk. Daniel Bernoulli first used utility theory in the 18th century to explain the

"St Petersburg paradox." It appeared, from empirical observation, that gamblers were unwilling to stake money on a "fair game" which had a finite probability of a large loss. Bernoulli, in offering an explanation for this paradox, suggested that the determination of the value of wealth is not based on the amount of wealth, but on the utility that the amount of wealth provides. An investment corollary of this observation is that the value of an investment is not based on its expected payoff, but on the expected utility which may be yielded as a result of the various payoffs. From this, it is possible to develop reasonable measures of investment risk.

Two characteristics of utility functions that are normally regarded as necessary are

$$U'(X) \geq 0 \tag{4.1}$$

and

$$U''(X) \leq 0 \tag{4.2}$$

where X is the amount of wealth, $U(X)$ is the utility of wealth X, and $U'(X)$ and $U''(X)$ represent the first and second differentials of the utility function. The inequality in Equation 4.1 requires the investor not to prefer less wealth to more; the inequality in Equation 4.2 requires that the value we put on a given increment in wealth does not increase as the level of wealth increases.

Consider the utility function

$$U(X) = a + bX + cX^2 \qquad b < 0, \quad c > 0$$

In this case, the utility of wealth is a quadratic function of wealth. The restrictions on b and c enable the conditions in Equation 4.1 and Equation 4.2 to be fulfilled over relevant ranges of X. If we let the starting value of wealth be W and consider a random addition to wealth of A (arising from the accumulation of an investment), then the investment portfolio which maximises utility is that which

$$\text{Max } E[a + b(W + A) + c(W + A)^2]$$

i.e.:

$$\text{Max}[a + bW + bE(A) + cW^2 + 2cWE(A) + cE(A^2)]$$

With the quadratic utility function, the maximisation of utility depends only on the first and second moments of the random accumulation and, therefore, on the mean and standard deviation of the random accumulation or of investment returns.

Utility theory can, therefore, be used as a justification for the use of standard deviation of investment returns as a measure of risk.

Bernoulli suggested the use of a logarithmic utility function, so that

$$U(X) = \ln(X)$$

The log utility function has the attraction that, at all levels of wealth, a given proportionate increase in wealth is equally valuable. This might be regarded as a feature that reflects the preferences of most individuals.

Again, letting the starting value of wealth be W and the random accumulation of an investment portfolio be A, we are required to choose a portfolio which

$$\text{Max } E[\ln(W + A)]$$

Adapting the derivation by Markowitz (1991), it can be shown that this is approximately equivalent to maximising

$$\ln[W + E(A)] - \frac{\text{Var}(W + A)}{2[W + E(A)]} + \cdots$$

The higher order terms would include higher moments of the random accumulation A. This utility function would also lead to the investor using standard deviation as a measure of investment risk, at least over a significant range of investment returns (see Markowitz (1991)).

Thus, the two types of utility function described provide a case for the use of the standard deviation of investment returns as a measure of investment risk. This case is strengthened when it is appreciated that, if all portfolios have normally distributed investment returns, standard deviation can be the only relevant measure of investment risk which is of any significance because all higher moments of the normal distribution depend on the mean and standard deviation. Further arguments for the use of the standard deviation of investment returns can be proposed in terms of its practicality. Standard deviations have useful mathematical properties. If we can estimate the standard deviation of investment returns from all possible assets and the covariances of returns between the assets, then we can find the standard deviation of returns from any portfolio (see Chapter 5). Developing mathematical portfolio selection models can be very difficult if such models rely on moments of distributions of investment returns other than the first and second moments. Therefore, the use of standard deviation of returns as a measure of investment risk has a number of practical advantages and can be a justifiable approach if certain assumptions are made about the shape of investors' utility functions or the shape of the distribution of investment returns.

However, in many situations, such assumptions may be unrealistic. Investors may well not have smooth utility functions; they may be particularly averse to certain "downside" outcomes; they may have two or more distinct, joined segments of a utility function: one reflecting utility of wealth above a benchmark and one reflecting utility of wealth below that benchmark. Furthermore, returns from investments are unlikely to be normally distributed and could well be skewed. Difficulties such as this could be resolved by using utility theory, from first principles, in models of portfolio selection. This is often far from easy, because of a number of practical considerations. Therefore, it is, often useful to have summary measures of investment risk other than the standard deviation of investment returns. Some of these will be discussed in Section 4.3.

4.2.1   Relating Utility Functions to Risk Aversion and the Risk Premium

Utility theory has been developed to assist our understanding of how individuals will behave in the face of risk. It is possible to use utility theory to make inference about the risk premiums that investors with different risk profiles will require from different forms of risky investment.

Risk aversion was defined by Pratt (1964). As has been noted above, for a given utility function $U(X)$, it is necessary to require $U'(X) > 0$ for an individual to prefer more wealth to less and $U''(X) \leq 0$ to ensure that an individual is risk averse. However, neither $U'(X)$ nor $U''(X)$ indicate how risk aversion is changing. Pratt showed that we can see how risk aversion changes by looking at the risk premium an investor demands for an actuarially neutral investment. An actuarially neutral investment is defined as one that has an expected payoff of zero (for example, an investment that has a 50% chance of returning a profit of 10% of the investment and a 50% chance of returning a loss of 10%).

The individual can be said to be risk averse if X is a random variable and

$$E[U(X)] < U[E(X)]$$

That is, the individual prefers an investment with a known payout of $E[X]$ to one that has a random payout with the same expected value. The extent of this aversion to risk can be determined by looking at the risk premium the investor requires from a risky investment. One investor is said to have greater local risk aversion than another at all levels of wealth if, at all levels of wealth, the amount of money that an individual would pay in exchange for a risk is smaller.

Consider a decision maker's utility function $U(X)$. Let the person's starting level of wealth by W. The risk premium r is such that the investor would be indifferent between receiving a random payout or addition to wealth of Z and receiving the non-random or certain amount $E[Z] - r$.

For example, if the risky investment is an equity, then the risk premium from the equity is the difference between the expected return from the equity $E[Z]$ and the return an investor would be equally happy receiving with certainty, perhaps from a treasury bill, $E[Z] - r$. Thus, the equity provides an extra return of r to compensate the investor for risk. The risk premium that would be required by a particular investor would depend on the utility of wealth function, the starting level of wealth and the nature of the random variable Z.

More formally, we can define the risk premium as $r[W, Z]$: for a given utility function, the risk premium is a function of the starting wealth W and the distribution of the random additions to wealth from the risky investment (denoted by Z).

Therefore, we can say that the risk premium for a risky investment is such that

$$U[W + E[Z] - r(W, Z)] = E[U(W + Z)]$$

That is, $r(W, Z)$ is the deduction that can be made from the expected value of the return from the risky investment that leads the investor to be indifferent between the addition to wealth of $E[Z] - r(W, Z)$ with certainty and the addition to wealth from the risky investment, denoted by the random variable Z under conditions of uncertainty. $E[Z] - r(W, Z)$ can be regarded as the risk-free return equivalent to the risky investment and $r(W, Z)$ as the risk premium. It is the way in which the risk premium $r(W, Z)$ varies with the initial level of wealth Z that determines the pattern of changing risk aversion for an investor.

Pratt (1964) showed that the risk premium is proportional to

$$-\frac{U''(x)}{U'(x)}$$

at any particular level of wealth x.

Using Pratt's measure of risk aversion, the well-known result that the linear utility function has constant (zero) risk aversion is easy to confirm. It can also be shown that the quadratic utility function has increasing risk aversion with wealth, over the spectrum of wealth where diminishing marginal utility of wealth and increasing absolute utility of wealth applies. In other words, as individuals become wealthier they demand a higher risk premium from the same risky investment. This is not necessarily irrational, but it is counter intuitive. For this reason, the quadratic utility function is often regarded as an inappropriate utility function for representing the risk preferences of individuals. It can also be shown that an individual with an exponential utility function (for example, where $U(X) = -\exp[-cX]$ with $c > 0$) has the characteristic of constant risk aversion. This implies that, whatever their starting value of wealth, investors would take the same investment decision if a given amount of the initial wealth is invested.

For an individual with a log utility function, risk aversion and the required risk premium falls as wealth increases. In fact, the required risk premium is inversely proportional to income and investors will demand the same proportionate risk premium for given proportionate investments whatever their starting value of wealth. They will, therefore, take similar investment decisions when a given proportion (as opposed to a given absolute amount) of wealth is being invested, regardless of their current wealth. This seems intuitive.

Where utility functions are used in investment modelling, the analysis of Pratt (1964) would suggest that it is more appropriate to use exponential or log utility functions, rather than quadratic utility functions. As we shall see in Section 4.3, this has implications for the types of risk measure that should be used in investment analysis.

4.3 Summary Risk Measures

In Chapter 3 we discussed the various contexts in which risk could be measured. Risk could be related to the liabilities of the investor (so that the appropriate risk measure could be, for example, the standard deviation of the surplus of a pension scheme). Risk could be measured in absolute terms. Or risk could be measured against a fixed benchmark or index performance. All the risk measures that we will discuss below can be used in all these contexts. For the purposes of our discussions, we can define two categories of investment risk measure: symmetrical risk measures and downside risk measures. We can define four contexts within which risk can be measured: in absolute terms, relative to liabilities, relative to a fixed benchmark and relative to an index performance. Within the downside risk category, we can further define a number of categories: semi-variance, expected loss (or expected shortfall), and probability of loss (or probability of shortfall). Tail loss and value at risk will also be defined in our discussion and are very close in their characteristics to expected loss and probability of loss respectively.

There is more discussion of investment risk measures in a practical context, substantial numbers of references and references to Websites on risk measurement in Booth et al. (2003) and also in Dowd (2002).

4.3.1 Standard Deviation of Returns

Standard deviation of returns is computed using the whole probability distribution of returns. The standard deviation is calculated as follows:

$$\sqrt{\frac{\sum_{i=1}^{n} (x_i - \bar{x})^2}{n}}$$

If this relates to investment risk calculated in absolute terms, X is the random variable that represents the level of return. x_i would then represent the particular return, or value taken by the random variable, in period i. If the measure is being calculated for monitoring the risk of a pension fund, life fund, etc., then the random variable X could represent the surplus of the fund. If investment risk is being calculated relative to the returns from an investment index (that may form the target for an investment fund), then the mean of the distribution of returns is replaced by the index return in each observation. This will produce a measure known as "tracking error."

There are two features of standard deviation measures that are worthy of note. First, because it uses the whole probability distribution of returns (whether an empirical distribution or a theoretical distribution) no information is lost. Second, risk is represented by squaring the difference between each observation and the mean, so that proportionately greater weight is given to more extreme observations.

Standard deviation of returns as a risk measure is easy to compute, easy to apply, easy to understand and easy to work with at different levels of aggregation (e.g., the asset level, the asset class level, and the portfolio level). If investment returns are normally distributed, then all risk measures are directly related to the standard deviation of investment returns and, thus, no further risk measures are needed. On the other hand, standard deviation of returns can be shown to be an inappropriate measure of risk unless investors have quadratic utility functions or investment returns are normally distributed (or approximately so). It is generally thought implausible that investors have quadratic utility functions, as it can be shown that a quadratic utility function implies that the risk premium investors demand from risky investments increases as wealth increases (see Section 4.2.1). Whether investment returns are sufficiently close to being normally distributed for standard deviation of returns to be appropriate as a risk measure is an empirical matter. Certainly, in situations where extreme outcomes have to be closely monitored, it might well be dangerous to use standard deviation of returns as a measure of risk, rather than one of the downside measures. A review of whether the use of standard deviation of investment returns is a sufficiently good approximation to theoretically more appropriate measures can be found in Booth et al. (2003).

4.3.2 Downside/Shortfall Risk Measures

Symmetrical risk measures give equal weight to observations above and below the mean. Whilst this might be appropriate in some circumstances, the philosophy underlying downside risk measures is that it is possible to define a particular level of returns (a benchmark) above which adverse consequences are not serious for the investor. Therefore, risk measures should only take into account the likelihood of achieving a level of investment returns below this level.

The basic method for calculating downside risk is as follows:

- Identify the "benchmark" below which downside risk is to be measured (for example, a return of 3% p.a.).
- Identify where returns fall below the benchmark and calculate the shortfall (e.g., a return of −1% p.a. would represent a 4% shortfall).
- The shortfall would then be raised to a power (e.g., it would be squared if the risk measure were shortfall semi-variance; its absolute value would be taken if the risk measure were expected shortfall; etc.).
- This quantity would then be multiplied by the probability of occurrence.
- These quantities would then be summed to calculate the risk measure.
- If the probability of loss is the required risk measure, then the probability of the fund underperforming the benchmark would be calculated
- Often the square root of the shortfall semi-variance is used, thus creating a risk measure that is often described as semi-standard deviation.

Formally, the shortfall measure can be described as follows:

$$\frac{\sum_{x_i < r} (r - x_i)^z}{n}$$

where r is the chosen benchmark. The sum is calculated over all values of x where x_i, the return in period i, is less than r. If shortfall semi-variance is being calculated, $z = 2$; if expected loss or expected shortfall is being calculated, $z = 1$; if probability of loss or probability of shortfall is being calculated, $z = 0$. Probability of shortfall is simply the probability that the investment return (or the objective variable being measured) falls below the benchmark. Sometimes, the risk measure is calculated so that the divisor n for the downside risk measure only relates to the number of occasions where the return falls short of the benchmark (i.e., where $x_i < r$); this would lead to the calculation of a "conditional partial moment." For most purposes, it is better to divide by the whole sample because, whilst occasions of outperformance do not contribute to the risk measure, they are relevant when we are interested in calculating the probability of, or a measure of, underperformance.

Downside measures of risk are used in portfolio risk management and risk management more generally. A concept similar to "expected loss," for example, underlies the risk-based capital method of regulating insurers in the U.S. The family of shortfall risk measures has an underlying logic. The measures are often known as "partial moments," because they relate to summary statistics such as the standard deviation and expected value of the distribution but are calculated using only part of the distribution. Thus, shortfall semi-standard deviation is the "downside equivalent" of standard deviation as a risk measure. These risk measures can be justified as being theoretically rational, as long as certain assumptions are made about either the properties of the probability distribution of returns or the utility function of the investor. Many of these risk measures and the corresponding investor utility functions that make them rational were introduced by Markowitz and are discussed, for example, in Markowitz (1991).[1] There are weaknesses of downside risk measures. In many contexts, an extreme upside event can be an indicator of risk (for example, if an investment fund outperforms the benchmark index by 10% it is an indication that the portfolio is significantly different from the index and, in other years, an extreme negative outcome may be possible). This is an aspect of a wider problem that downside risk measures limit the information they use. Also, downside risk measures are not mathematically tractable when individual asset classes are aggregated into portfolios: simulation techniques often have to be used to obtain meaningful results.

All shortfall measures can be used in a variety of contexts. The objective variable could be, for example, the surplus of a pension or life fund (or the return on surplus). Alternatively, the measure could be redefined with reference to a fund-specific benchmark so that x is related to the underperformance relative to an index rather than to the absolute value of returns. Haberman et al. (2003a,b) use shortfall risk measures in the context of risk measurement in defined-benefit pension schemes.

4.3.3 Value at Risk

Value at risk (VaR) is often used as a measure of risk in the banking industry. It can also be used in portfolio management, and by insurance companies and pension funds. VaR is closely related to the family of shortfall risk measures. The purpose of VaR is to measure the predicted "worst loss" from a portfolio with a given confidence level. In terms of mathematical notation, VaR can be defined with reference to the following quantity:

$$\int_{-\infty}^{V_1} f(v)\,dv = 1 - c = P(v \leq V_1)$$

where $f(v)$ is the density function of investment returns (or other objective variable, such as the surplus of a life or pension fund), c is the VaR confidence level, and V_1 is the value at risk. V_1 is defined such that the probability of obtaining a return less than V_1 is equal to one minus

---

[1]These aspects of Markowitz's work are not generally appreciated. It is often suggested that Markowitz only supported using symmetrical risk measures.

the desired confidence level. Thus, if the desired confidence level is 95%, VaR is defined such that the probability of obtaining a return less than V_1 is 5%. It is easy to see how VaR relates to probability of loss or probability of shortfall: the probability of loss below the VaR (V_1) is 5%. VaR simply focuses on the amount of loss that could be incurred with a given probability, whereas probability of loss or probability of shortfall focuses on the probability of a outcome worse than a given outcome.

VaR is used in a number of practical circumstances, in banking (for example, to set capital), in pension fund asset allocation studies and in portfolio risk management. Conceptually, it is a very straightforward risk measure to use. It is easy to explain to management. For example, if the 10% VaR is −5%, in answer to the question, "How much could we lose?" then the answer is conceptually clear: "On only one in ten occasions will you make a return of less than −5%." VaR is also useful for identifying the level of "serious" losses from a portfolio.

There are weaknesses of VaR, however. VaR ignores the magnitude of the losses beyond the level of the value at risk. For example, if the 90% value at risk is −5%, then 10% of occasions are expected to lead to a return of less than −5% but we have no information as to how much worse than −5% these losses will be. With the exception of probability of shortfall, the other risk measures that we have considered all provide us with more information because they take into account not just the likelihood of an adverse outcome but also its magnitude. Also, only a relatively small amount of data is often used to determine the VaR, as it concentrates on the tail of the distribution: statistical assumptions used to calculate it can, therefore, be quite strong or confidence intervals around VaR can be wide. Overall, if the VaR is of interest to management then it is perfectly appropriate to calculate it, although it would be helpful if a confidence level around the VaR were also presented and if VaR was presented alongside other measures of investment risk.

4.3.4 Practical Issues when Calculating VaR

There are two main methods of calculating VaR. One is sometimes described as the "delta-normal" method, where it is assumed that investment returns follow a normal distribution (other distributions could be assumed, but the practical implementation will be more difficult). The other method uses historical simulation. There is a danger that either or both of these methods can understate value at risk for particular portfolios. In the delta-normal method, this can happen if probability distributions of returns are not as assumed (e.g., if they are fat tailed). In both methods, problems can arise if there are subjective factors that lead us to believe that particular portfolios have a higher risk level than is indicated by historical data. If risk-adjusted performance targets, risk limits, and so on were only based on the calculated value-at-risk figure, portfolio managers could take much bigger risks than was intended. Because the manager would be rewarded for these risks by higher expected returns, he might have an incentive to take such risks. This issue is formally analysed by Ju and Pearson (1999). This is not a reason for not using VaR approaches. However, quantitative risk measures should just be one aspect of a risk management process.

In fact, there is a paradox with regard to the practical usefulness of VaR as a risk measure given these theoretical problems. If we assume a standard distribution of returns, such as the normal distribution, when calculating VaR, then all risk measures (including VaR) will rank portfolios in exactly the same way. On the other hand, if we cannot assume such standard distributions, then there may be differences between the rankings provided by different risk measures, but VaR could then suffer from the practical shortcomings that we have identified above. The shortcomings of value at risk are analysed formally in the literature on coherent risk measures (see Section 4.5).

4.3.5 Tail Loss

There is a series of closely related risk measures developed from the VaR concept. These measures have various names (for example, tail loss, T-VaR, or accumulate VaR). The measures overcome some of the problems of VaR described above, as they measure the extent of possible losses below the VaR. However, they still suffer from the difficulty that they are calculated using a relatively small part of the distribution of returns and, therefore, do not use all the available data. The tail-loss measures are specific practical examples of downside or shortfall risk measures, discussed in Section 4.3.2.

In general, if the data are available to calculate VaR, then it will be available to calculate tail loss. There is no disadvantage of using tail loss, other than that it is more difficult to present to decision-makers who do not have a quantitative background.

Mathematically, this class of risk measures is quite complex. However, in principle, they are easy to explain. We can choose a particular confidence level for VaR, we calculate the VaR and then examine all the possible outcomes below that VaR, assess their probability, and calculate the expected loss below the VaR.

4.4 Combining Risk and Return Measures

In practice, when asset allocation decisions are taken there will be a range of decision criteria (see, for example, Section 4.6), and quantitative as well as non-quantitative issues will be taken into account. As a result, it is not always necessary to develop a formal mechanism for trading risk and return in order to determine an "optimal" portfolio. However, in any practical application, it is useful to have a theoretical framework in order add rigour to the decision-making process. Decision criteria and formal portfolio selection frameworks will be discussed in greater detail in Chapter 5. At this stage, we will just mention the simplest decision criterion that could be used.

In determining the optimal portfolio, the investor could maximise the following function:

$$F(R_t) = E(R_t) - a\sigma(R_t) \qquad \text{where } a \geq 0 \qquad (4.3)$$

R_t is the objective variable (which could be investment returns, value of surplus, etc.) so that $E(R_t)$ and $\sigma(R_t)$ represent the objective variable's expected value and standard deviation respectively. Clearly, the function could be generalised further. Any risk measure could be used instead of standard deviation, and the function to be maximised could also include expected returns and a risk measure defined relative to the benchmark (formal risk-adjusted performance measures will often be used in these circumstances; see Adams et al. (2003)). a is a risk-aversion parameter. As $a \to 0$, the investor will have no aversion to risk and the investor will simply try to maximise the expected value of the objective variable.

4.5   Coherent Risk Measures

A body of theoretical literature has been built up that helps us to understand the practical problems when applying risk measures in different situations. Part of the theoretical literature relates to the origins of risk measures as derived from utility functions (as discussed in Section 4.2). Another aspect relates to the "coherence" of risk measures. Without necessarily relating a risk measure to an underlying utility function, it is possible to determine whether a particular risk measure is "coherent" with reference to four properties defined by Artzner (1999) (these issues are discussed and developed in more detail in, for example, Dowd (2002)). Artzner demonstrated that a risk measure RM(V) should have the following properties:

1. *Monotonicity.* If $V_1 \leq V_2$, where V_1 is the out-turn for portfolio 1 and V_2 is the out-turn for portfolio 2, in all states of the world then RM(V_1) $\geq$ RM(V_2). In other words, if the outcome for portfolio 2 is always not worse than that for portfolio one, in all states of the world, then portfolio 2 cannot be more risky and this should be demonstrable through the risk measure.

2. *Translation invariance.* $RM(V + k) = RM(V) - k$. That is, if we add a constant to the value of the portfolio (for example, by adding cash) then it reduces the risk by that constant.
3. *Homogeneity.* $RM(bV) = bRM(V)$. That is, scaling a portfolio by a factor b simply scales its risk.
4. *Sub-additivity.* $RM(V_1 + V_2) \leq RM(V_1) + RM(V_2)$. That is, if we add together two portfolios, then the risk of the combined portfolio cannot be greater than the sum of the risks of the two separate portfolios. Assume, for example, that VaR fulfilled this condition. If one portfolio has a VaR of 100 and the other a VaR of 150, then, if this condition holds, the combined portfolio should not have a VaR of more than 250. (In fact, VaR does not fulfill this last condition; see below).

It is quite clear that VaR fulfills the first three properties (as do the other risk measures we have considered). However, it does not *necessarily* fulfill the last property (although in most practical situations it will). As well as VaR, probability of shortfall is also not a coherent risk measure. This is not surprising given the relationship between the two measures. We have already explained, in conceptual terms, the problems of using VaR. The conceptual weakness that we identified (the fact that VaR does not take account of the extent to which outcomes fall short of the VaR) underlie its theoretical weaknesses as identified in the literature on coherent risk measures. It should be noted that this does not mean that VaR is of no practical value (see Section 4.3.3). Downside risk measures other than the probability of shortfall, such as expected shortfall and downside semi-variance, do fulfill all the conditions for coherent risk measures. They also use more information from the distribution of returns.

4.6 The Use of Shortfall Constraints

The idea of a shortfall constraint was introduced in Chapter 3. Investors may be able to represent the trade off between risk and return by maximising a function of risk and return, as described in Section 4.4. However, such a linear function might not represent the investor's pattern of risk aversion appropriately because there may be an aversion to particular outcomes (for example, outcomes that lead to insolvency). In most ranges of outcomes, risk may be a linear function of the value of the outcome or of a standard risk measure, but at particular levels of outcome there may be a significant loss of utility. A number of methods can be used to represent the risk of not achieving a particular, pre-defined, level of investment return or other objective variable. Using utility theory, a discontinuity can be put in the utility function at the point at which particularly adverse consequences are assumed to exist. A portfolio can then be found which maximises expected utility. This method has the advantage of being completely general and can be used to take into account the degree of risk aversion not just at the point of the discontinuity, but also at all possible levels of investment outcome. Using utility functions in this way has a number of practical difficulties. First, it may be difficult to measure the appropriate length of the discontinuity. Also, it may be difficult to rationalise the way in which risk is being measured. Finally, it may be difficult to perform the modelling exercise in practice.

An alternative method of dealing with aversion to a particular outcome is by using a "shortfall constraint" or "probability constraint." Like the use of shortfall probability or value at risk as a risk measure, the use of a shortfall constraint does not give rise to consistent results within an expected utility maximisation framework. An example of a shortfall constraint would be if the investor were to reject the portfolio if

$$\Pr(R \leq m) \geq q$$

where R is the investment return or the level of surplus, m is a critical level of surplus or return, and q is a probability (such as 0.05). If $m = 0$ then q could be regarded as the probability of ruin.

A probability constraint could be combined with other measures of risk so that the portfolio chosen could be the utility-maximising portfolio or the most appropriate portfolio on an efficient frontier (see Chapter 5), subject to it fulfilling the probability constraint. The shortfall constraint could then be an implicit way of representing a discontinuity in the utility function. Alternatively, the approach could be to take asset allocation decisions on the basis of a function of the mean and standard deviation of surplus, but where portfolios are rejected if the value at risk (or other shortfall risk measure) were higher than a given level.

4.7 Conclusion

In this chapter we have developed a number of different approaches to investment risk measurement that can be applied along with the portfolio selection and asset allocation techniques to be discussed in Chapter 5. It is worth considering the application of these measures as part of the concluding comments of this chapter. The first technique that we will consider in Chapter 5 is that of immunisation. This is an investment risk minimisation technique and, therefore, the risk measures that we have considered here are not relevant. Immunisation is intended to eliminate the risk of not meeting liabilities. The next technique we will consider in Chapter 5 is efficient frontier analysis. In theory this can be used to identify investment portfolios that have the lowest level of investment risk for a given level of investment return, using any of the measures of investment risk that we have defined. However, in general, whilst efficient frontier analysis has been carried out using shortfall semi-variance (see Ferguson and Rom (1994) and Sing and Ong (2000)), practical computational requirements tend to lead users of efficient frontier analysis to use standard deviation of returns as the measure of investment risk. The same comment applies for the development of efficient frontier analysis where institutional liabilities are incorporated so that "return on surplus" is the objective variable and standard deviation of return on surplus is used as the risk measure. Asset modelling and asset liability modelling using simulation are more flexible techniques. These enable the user to employ a greater range of risk measures and decision criteria. Value at risk and other shortfall risk measures tend to be used as well as standard deviation of investment returns when stochastic simulation techniques are used in practice. Haberman et al. (2003b) give examples of efficient frontiers calculated using downside risk measures in the context of defined-benefit pension schemes. Thus, simulation, using stochastic investment models, is sufficiently flexible to allow the use of whichever measure of investment risk is appropriate to a particular investor.

References

Adams, A. T., Booth, P. M., Bowie, D., and Freeth, D. S. (2003). *Investment Mathematics*. Wiley, London, U.K.

Artzner, P. (1999). Application of coherent risk measures to capital requirements in insurance. *North American Actuarial Journal* 3(2), 11–25.

Booth, P. M., Matysiak, G. M., and Ormerod, P. (2003). *Property Portfolio Risk*. Investment Property Forum, London, UK.

Dowd, K. (2002). *An Introduction to Market Risk Measurement*. Wiley, U.K.

Ferguson, K. W. and Rom, B. M. (1994). Post modern portfolio theory comes of age. In: *Proceedings of the 4th AFIR Colloquium*, Vol. 1, Society of Actuaries.

Haberman, S., Day, C., Fogarty, D., Khorasanee, M. Z., Nash, N., Ngwira, B., Wright, I. D., and Yakoubov, Y. (2003a). A stochastic approach to risk management and decision making in defined benefit pension schemes. *British Actuarial Journal* 9(3), 493–586.

Haberman, S., Khorasanee, M. Z., Ngwira, B., and Wright, I. D. (2003b). Risk measurement and management of defined benefit pension schemes. *IMA Journal of Management Mathematics* 14(2), 111–128.

Ju, X. and Pearson, N. D. (1999). Using value at risk to control risk taking: how wrong can you be? *Journal of Risk* 1(2), 5–36.

Markowitz, H. M. (1991). *Portfolio Selection*. Basil Blackwell, Oxford.

Pratt, J. W. (1964). Risk aversion in the small and in the large. *Econometrica* 32, 122–136.

Sing, T. F. and Ong, S. E. (2000). Asset allocation in a downside risk framework. *Journal of Real Estate Portfolio Management* 6(3), 213–224.

Chapter 5

Portfolio Selection Techniques and Investment Modelling

5.1 Introduction

In this chapter we look at various techniques that can help us in the process of portfolio selection. The first technique, that of immunisation, is a risk minimisation technique that can be used in particular circumstances. The second set of techniques is based on the portfolio selection models of Markowitz (1952, 1991). They indicate how we can trade risk and return in a coherent framework. Portfolio selection techniques are then developed and generalised, e.g., to trade risk and return defined in terms of the risk of not meeting actuarial liabilities. In order to solve some of the more complex problems of portfolio selection and use a range of risk measures that may be more complex to manipulate than standard deviation of investment returns, it is sometimes necessary to use simulation, based on stochastic modelling. In the final parts of this chapter we introduce stochastic investment models and show how they can be applied in simple situations. Stochastic investment modelling has many of the characteristics of value-at-risk (VaR) modelling in banking risk management, although, technically, VaR refers to the computation of a particular point in the probability distribution of outcomes rather than the modelling technique itself. See Chapter 1 for a discussion of, and some further references to, the similarities between VaR modelling and stochastic investment modelling in the non-bank area and see Chapter 4 for discussion of the concept of VaR.

5.2 Immunisation

5.2.1 Derivation of Conditions

The theory of immunisation is most easily applied to investment entities that have a series of fixed liabilities due at particular times. Thus, it may be applied within life insurance companies to non-profit liabilities, to pension funds in respect of pensions in payment and certain types of investment or annuity funds. It is more difficult to apply the theory to non-life insurance companies because of the degree of uncertainty of the times of payments made by non-life companies. Nevertheless, the broad conclusions of the theory — the importance of matching the timings of expected asset and liability cash flows — can be applied. We will discuss the theory of immunisation in the context of a life insurance company or defined-benefit pension fund.

Every so often a life insurance company or pension fund will perform a valuation. The valuation will be performed for a number of purposes. One purpose will be to ensure that, at realistic expectations of future interest rates, the assets are sufficient to meet future liabilities. The theory of immunisation gives us guidance as to how we should invest in a portfolio of fixed-interest securities, to give us protection from unexpected changes in future interest rates.

We will assume that the rate of interest that the investing institution can earn on its assets is compatible with that which it uses to value its liabilities. Further, to begin with, we assume that the assets are exactly sufficient to meet future liabilities (in reality, any extra assets are assumed to be invested separately). Finally, we assume that all the assets are invested in fixed-interest securities and that they are held to meet liabilities fixed in a specific currency.

Consider, first, what would happen if the institution had liabilities that consisted of a number of payments due in 20 years time and all assets were held in cash. Assume that the liability and asset values were exactly equal, at a rate of interest of 6%, the rate expected to be achieved on long-term bonds (appropriate to the liabilities). If interest rates were to fall to 4%, then the assets would be insufficient to meet future liabilities in 20 years time. A similar problem would arise if assets were longer than liabilities and if interest rates were to rise. The theory of immunisation indicates how assets should be invested to provide protection against changes in interest rates.

Define $V_A(\delta)$ as the present value of the assets at force of interest δ and $V_L(\delta)$ as the present value of the liabilities at force of interest δ. We have assumed that the present value of the assets equals that of the liabilities, i.e., $V_A(\delta) = V_L(\delta)$. Now consider a small change in the general level of interest rates to $(\delta + h)$. The Taylor expansion would give

$$V_A(\delta + h) = V_A(\delta) + hV'_A(\delta) + \frac{h^2}{2}V''_A(\delta) \cdots \qquad (5.1)$$

$$V_L(\delta + h) = V_L(\delta) + hV'_L(\delta) + \frac{h^2}{2}V''_L(\delta) \cdots \qquad (5.2)$$

where $V'_A(\delta)$ and $V'_L(\delta)$ represent the first differential of the present value of the assets and liabilities respectively, taken with respect to (δ), and $V''_A(\delta)$ and $V''_L(\delta)$ represent the second differential of the present value of the assets and liabilities respectively, taken with respect to δ. The excess of the value of the assets over that of the liabilities is

$$V_A(\delta + h) - V_L(\delta + h) = V_A(\delta) - V_L(\delta) + hV'_A(\delta) - hV'_L(\delta)$$
$$+ \frac{h^2}{2}V''_A(\delta) - \frac{h^2}{2}V''_L(\delta) \cdots$$

We began by assuming that $V_A(\delta) = V_L(\delta)$. Let us also assume that $V'_A(\delta) = V'_L(\delta)$. Thus, $V_A(\delta + h) - V_L(\delta + h) = (h^2/2)(V''_A(\delta) - V''_L(\delta))$, with higher order terms being ignored, as h is assumed to be small. Whatever the sign of h (the change in the force of interest), $h^2/2$ will be positive. Therefore, if there is a small movement either up or down, in the force of interest, the value of the assets will be slightly greater than the value of the liabilities, at the new force of interest $(\delta + h)$, assuming the following conditions hold:

$$V_A(\delta) = V_L(\delta) \tag{5.3}$$

(i.e., the present values of assets and liabilities are equal at the starting interest rate) which was our starting assumption:

$$V'_A(\delta) = V'_L(\delta) \tag{5.4}$$

and

$$V''_A(\delta) > V''_L(\delta) \tag{5.5}$$

Combining Equation 5.3 and Equation 5.4, we obtain

$$\frac{V'_A(\delta)}{V_A(\delta)} = \frac{V'_L(\delta)}{V_L(\delta)}$$

which is our definition of volatility (see Chapter 1), applied to the present values of assets and liabilities, so that the condition in Equation 5.4 requires that the discounted mean terms or volatilities of assets and liabilities are equal. It can be shown (see Adams et al. (2003)) that the condition in Equation 5.5 requires that the spread of the terms of the asset cash flows around the discounted mean term is greater than the spread of the terms of the liability cash flows around the discounted mean term.

Thus, if we start from the position that the present value of the assets is equal to the present value of the liabilities and the volatility (or discounted mean term) of the assets is equal to the volatility (or discounted mean term) of the liabilities, and if the spread of the asset terms around the discounted mean term is greater than that of the liability terms, then a small uniform change in the force of interest will lead to the institution making a small profit. An investor in this position is said to be "immunised" from changes in the rate of interest. The theory of immunisation can be attributed to Redington (1952).

A special case of immunisation occurs when every liability outgo (the outgo at time t, to be denoted by L_t) is matched by an asset proceed (the asset proceed at time t, to be denoted by A_t). For the purposes of this discussion, an asset proceed is a known fixed coupon or capital payment from a fixed-interest security. In this case, no change in interest rates can affect the financial position of the fund. Every liability outgo is exactly matched by an asset proceed. The present value of the assets will equal the present value of the liabilities at any rate of interest (in the Taylor expansion, used to demonstrate immunisation, all the derivatives are equal and $V_A(\delta) = V_L(\delta)$, leaving $V_A(\delta + h) - V(\delta + h) = 0$, whatever the value of h). The theory of absolute matching was demonstrated by Haynes and Kirton (1952).

5.2.2 Observations on the Theory of Immunisation

The theory of immunisation shows the effect on an investor if there is a small, uniform change in the force of interest. If interest rates change by a large amount (so that later terms in the Taylor expansion are significant) then it is possible that the investor may be detrimentally affected by such a change. It should be said, in this connection, however, that it is really the equality of volatilities or durations of the assets and liabilities that is most important. If the volatilities of the assets and liabilities of the investor are equal, then it will be quite effectively protected from changes in interest rates — even if the changes are substantial.

The theory of immunisation tells us little about the result of a non-parallel shift in the yield curve: for example, a downward movement in short-term interest rates combined with an upward movement in long-term interest rates. More sophisticated risk-management techniques are needed to ascertain the effects of changes of this sort. A problem related to this is that (except in the special case of absolute matching) an investor could be immunised but face difficulties because the cash inflows in any period are insufficient to meet cash outflows. The institution should, therefore, use cash-flow-modelling techniques to identify the years in which this is likely to be the case.

The theory of immunisation has been developed to deal with the situation where fixed-interest assets are used to match fixed liabilities, such as non-profit life insurance liabilities. It is not easily adjusted to allow for with-profit liabilities or to deal with the various risks inherent in equity investment. There are also a number of practical difficulties inherent in the implementation of the theory, even when the assets and liabilities are of the appropriate type. For example, normally, when a non-profit life assurance contract is written, there will be a negative expected cash outflow in the early years (for an annual premium endowment assurance, in the early years, the expected death benefit will normally be less than the premium). Even when contracts of different remaining terms to maturity are combined together in a fund, the discounted mean term of a fund's liabilities can often be much longer than the discounted mean term of the longest available bonds. The formula for the discounted mean term of an annual coupon perpetuity, for example, is $1/i$, where i is the redemption yield. At a 5% rate of interest the discounted mean term is 20 years. This is the longest possible discounted mean term from a bond where the coupon rate is less than the redemption yield.

The conditions for immunisation change continuously over time; clearly, it would not be practical to manage a fund where the asset portfolio was continuously changed, so that the discounted mean term always matched that of the liability portfolio. Essentially, the theory, in its simple form, does not allow for transaction costs.

If the conditions for immunisation cannot be met in practice, then, clearly, the conditions for the more restrictive special case of absolute matching are unlikely to be achieved in practice. It is tempting, therefore, to discard the theory of immunisation as being of no practical use due to the above difficulties. This would be premature. It is true that more modern methods (to be described later in the book) should be used in the financial management of investing institutions. However, the theory of immunisation still gives useful guidance.

5.2.3 The Usefulness of Immunisation in Practice

Even though, because of the difficulties described above, the theory of immunisation cannot provide us with precise answers as to how assets in the investment portfolio should be selected, it still gives useful guidance. With regard to its non-profit business, a life office should always be aware of the approximate term of that business, and invest in bonds of an appropriate length. If a life office does not invest in bonds of an appropriate length to match its non-profit business, then it should be aware of the risks and be able to quantify those risks.

As far as with-profit business is concerned, part of the benefits paid under with-profit policies will be guaranteed (the guaranteed sums assured plus any attaching reversionary bonuses already declared). A life office can attempt to immunise this part of the liability in the same way as it would immunise non-profit business. The office has greater investment freedom with the remainder of the assets backing with-profit business and can invest in assets that will provide the balance of risk and expected return that the office seeks.

If the insurance company has considerable free assets, then it may deviate from the most secure possible investment policy without endangering solvency. Again, the theory of immunisation will help the office to quantify the risks of such mismatching.

The theory of immunisation would probably not be used to a great extent in the management of a defined-benefit pension fund. An exception may be in the case of a closed fund or a mature fund where it is desired to manage the risk of pensions in payment and other fixed liabilities. However, the theory does provide us with a theoretical justification for pursuing the sort of investment policy that may, in any case, seem intuitively sensible. For example, that part of a pension fund that is being invested to meet fixed pensions in payment should, if the mismatching risk is to be minimised, be invested in medium- to long-term bonds.

Similarly, a non-life company may not use immunisation formally. Again, investment policy in practice would, however, follow that which we would expect intuitively from the formal application of immunisation. A non-life company would prefer shorter term bonds, reflecting the shorter term of its liabilities.

Thus, the theory of immunisation is a model that has a number of shortcomings. Its use should never preclude the use of more sophisticated techniques. Nevertheless, it has provided and still does provide useful insights into the type of investment policy that an institutional investor should follow — particularly in the presence of fixed liabilities.

5.3 Modern Portfolio Theory

Having discussed a model that gives us guidance as to the particular term of fixed-interest investments that should be selected in a portfolio of fixed-interest securities, we will now move on to discuss modern portfolio theory, which can give guidance on the risks inherent in different types of asset portfolio and on the split of assets between risky and risk-free investments.

5.3.1 Portfolio Diversification

A risk-averse investor will wish to obtain the highest expected return for a given level of risk or, for a given expected return, bear the lowest level of risk. Most modern portfolio theory models have, in terms of their applications, concentrated on defining risk in terms of standard deviation of short-term monetary returns. We considered the limitations of this risk measure in Chapter 4. It is difficult to apply efficient frontier models, such as those that we develop below, using other measures of risk because of the complexity of the calculations, although simulation techniques can be used.

If we consider a portfolio of n investments, where the expected return from the ith investment is $E(R_i)$ and the proportion of the portfolio invested in the ith investment is x_i, then the expected return from the portfolio $E(R_p)$ is

$$E(R_p) = \sum_{i=1}^{i=n} x_i E(R_i)$$

The standard deviation of the return from the portfolio σ_p^2 is

$$\sigma_p^2 = \sum_{i=1}^{i=n} x_i^2 \sigma_i^2 + \sum_{i=1}^{i=n} \sum_{j=1}^{j=n} x_i x_j \sigma_{ij}$$
$$i \neq j$$

where σ_i^2 is the variance of return of the ith investment and σ_{ij} is the covariance between the returns from the ith and the jth investments. We may assume that all $x_i \geq 0$ and $\sum_{i=1}^{i=n} x_i = 1$. The first of these two assumptions could be relaxed, particularly if there is access to derivative markets, so that short selling is possible.

Even with this limited development of portfolio theory, we can still obtain some useful insights. Those insights are based on the assumption that a risk-averse investor will try to choose portfolios that minimise the standard deviation of return for a given expected return. *All other things being equal*, an investor will prefer an investment that has a lower standard deviation to one that has a higher standard deviation. Second, lower covariances between the returns from investments will reduce the risk of the portfolio. Third, the addition of a further investment, to a portfolio, is likely to reduce the overall variance of the returns from the portfolio; however, this need not be the case if the variance of the new investment is high relative to that of the existing investments and the positive correlation of returns with other investments is also high. More generally, an important insight of modern portfolio theory is that the risk of an investment should not be considered in isolation from an analysis of the portfolio of which it will form part. In a portfolio context, the risk of a particular asset is the marginal contribution of that asset to the risk of the portfolio. This will depend not just on the standard deviation of returns (or other risk measures) of the investment itself, but also on the relationship between the returns on the investment and the returns on other investments in the portfolio.

These ideas provide a theoretical justification for diversifying a portfolio, a procedure that has considerable intuitive appeal. In particular, diversification into international investments, as has been discussed in Chapter 3, can considerably reduce the risk of an investment portfolio, even if the international investments are volatile. It is worthwhile showing the effects of diversification by way of a simple example.

Example 5.1

An investment fund has a choice of three investments, a, b, and c. Investments a and b both have an expected return of 10% per annum and a standard deviation of return of 10% per annum. Investment c has an expected return of 12% per annum and a standard deviation of return of 11% per annum. The correlation coefficient between all investments is 0.5. Show that by spreading its assets equally between all three investments the insurance company can build a portfolio with a higher expected return and lower variance of return than if assets are split equally between investments a and b.

Let E_i be the expected return from investment i, let σ_i^2 be the variance of return from investment i, and let σ_{ij} be the covariance of returns between investments i and j.

$$E_a = 0.1; \quad E_b = 0.1; \quad E_c = 0.12$$

$$\sigma_a^2 = 0.01; \quad \sigma_b^2 = 0.01; \quad \sigma_c^2 = 0.0121$$

$$\sigma_{ab} = 0.1 \times 0.1 \times 0.5 = 0.005; \quad \sigma_{ac} = 0.1 \times 0.11 \times 0.5 = 0.005; \quad \sigma_{bc} = 0.0055$$

A portfolio with 50% of its assets in investment a and 50% of its assets in investment b has an expected return of

$$0.5 \times 0.1 + 0.5 \times 0.1 = 0.1$$

and a variance of return of

$$0.5^2 \times 0.01 + 0.5^2 \times 0.01 + 2 \times 0.5^2 \times 0.005 = 0.0075$$

A portfolio with one-third of its assets in each of a, b, and c has an expected return of

$$1/3 \times (0.1 + 0.1 + 0.12) = 0.10667$$

and a variance of return of

$$(1/3) \times [0.01 + 0.01 + 0.0121] + 2[(1/3)^2 \times 0.005 + (1/3)^2 \times 0.0055 \times 2] = 0.00712$$

Thus, the addition of a further investment, with a higher variance, reduces the overall portfolio variance whilst increasing the expected return.

We will now consider some of the ways in which it has been suggested that mean–variance analysis can be used to construct a portfolio.

5.3.2   Efficient Portfolios

If we construct a portfolio made up of many investments, each with known expected return, variance of returns and covariance of returns with other investments, then we can calculate the expected return and variance of returns of the portfolio. If we do this for all possible portfolios, then we will find that the expected return and standard deviation of returns from the various portfolios all lie within an area that will take the form of the shaded area shown in Figure 5.1. The line AB is known as the efficient frontier. If we assume that investors are risk-averse and measure risk by considering variance of returns, then investors will only choose portfolios that lie on AB. If they chose any other portfolio, then they could increase the expected return or reduce the variance of return by moving to AB.

Whilst it may be self-evident that the investor will wish to choose one of the points on the efficient frontier, we still need to determine which point should be chosen. A solution to this problem can be found using utility theory or indifference curve analysis. In the context of the modern portfolio theory model, the ideas are more clearly illustrated by means of indifference curves.

The indifference curves $U1$, $U2$, and $U3$, superimposed on Figure 5.1 to produce Figure 5.2, each represent combinations of risk and return that provide investors with a given level of satisfaction. Curves upwards and to the left offer a higher return for a given risk level and provide a higher level of satisfaction than curves below and to the right. Therefore, the investor will choose the combination of risk and return that leads him to the highest indifference curve (point D). Investors with different indifference curves will choose different portfolios, but all investors choose "efficient" portfolios (in the mean–variance sense) if they regard variance of returns as the measure of investment risk.

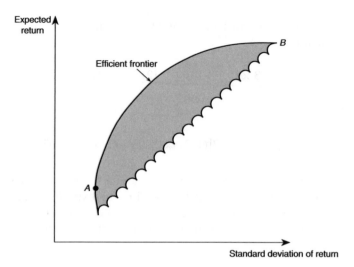

FIGURE 5.1 Mean–variance efficient frontier

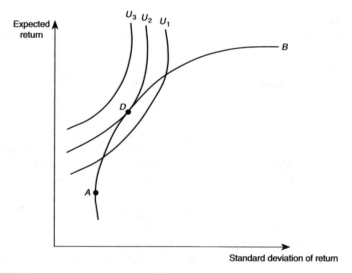

FIGURE 5.2 An investor's optimum portfolio

5.3.3 Capital Market Line

We can now introduce the concept of a risk-free asset that investors can include in their portfolio. If the model is concentrating on short-term nominal returns, then the risk-free asset could be 3-month Treasury bills. The risk-free asset has a certain return over the given time horizon.

If a portfolio is made up just of the risk-free asset, then the expected return will be R_f, where R_f is the known return from the risk-free asset.

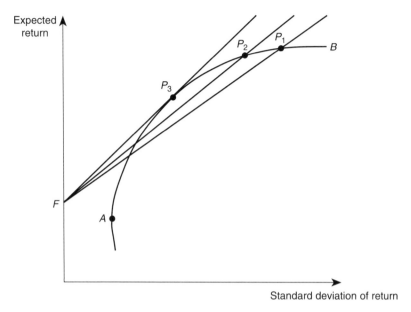

FIGURE 5.3 Available portfolios when risk-free asset is held

Such a portfolio would have no risk and the risk/return position from such a portfolio is denoted by point F in Figure 5.3.

Now consider a portfolio combining the risk-free asset and risky securities. If the portfolio of risky securities has an expected return of $E(R_m)$ and variance of return σ_m^2, then the expected return $E(R_p)$ and variance of return of a portfolio σ_p^2, where proportion a is invested in risky securities and $(1-a)$ in the risk-free asset are

$$E(R_p) = aE(R_m) + (1-a)R_f \qquad (5.6)$$

and

$$\sigma_p^2 = a^2\sigma_m^2 + (1-a)^2 \times 0 + 2a(1-a) \times 0 = a^2\sigma_m^2 \qquad (5.7)$$

The combination of any risky portfolio with the risk-free asset will give rise to a new set of portfolios which fall on the straight line joining F with the risky portfolio as shown in Figure 5.3. P_1, P_2, and P_3 are efficient portfolios when the risk-free asset is not considered. However, we can go beyond P_1, P_2, and P_3 by borrowing at the risk-free rate of return and investing the borrowed capital in the risky portfolio. Portfolios on the curve AB and on all the lines emanating from F that touch the curve AB are among the possible portfolios. The expected return and variance of

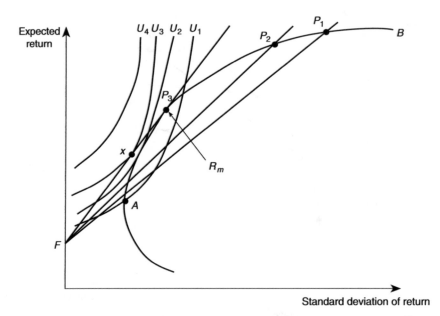

FIGURE 5.4 The optimal combination of risky and risk-free assets

return from these portfolios can be calculated using Equation 5.6 and Equation 5.7.

Which portfolio maximises the investor's utility or takes him to the highest possible indifference curve? The highest possible indifference curve is attained if the investor takes a combination of the portfolio implied by point P_3 (denoted by R_m on Figure 5.4) and the risk-free security. The optimal point x is also shown in Figure 5.4. Clearly, there is always a point on line FP_3 that will be superior to any point on line AP_3, FP_1, or FP_2.

If all investors have the same risk-free asset and have the same view of expected risk and return for risky securities, so that they compute the same efficient frontiers, then this theoretical development has interesting implications for investment decision taking. All investors will choose risky portfolios that are identical and mix them in different proportions with the risk-free security to produce the optimal degree of risk and return. Taken together with the efficient markets hypothesis, this provides a justification for indexing portfolios. This is an equilibrium investment market theory that relies on fairly stringent conditions being met. We have already suggested that different investors will have different risk-free securities, depending on their liabilities, investment time horizons, investment objectives and so on. It is not clear that such equilibrium theories of asset pricing and investment market behaviour are a useful approximation to real-world investment markets.

Nevertheless, such abstractions are always useful in providing insights. Even if investors have different risk-free assets and construct different efficient frontiers, then the theoretical development of this section suggests that, in certain circumstances, investors may be able to separate the decision of which risky portfolio to hold from the decision of how much of the risk-free asset to hold.[1] Once the efficient frontier has been constructed and the risk-free security selected, a decision to take more or less risk is executed only be moving up or down the line FP_3 not by changing the composition of the risky portfolio. However, even this insight may not have direct practical applicability to all investors. In many cases there will not be a risk-free asset that is exactly appropriate for the investor. In such cases, the investor would simply select the optimal portfolio from the efficient frontier.

5.3.4 The Capital Asset Pricing Model

For our purposes, it is not necessary for us to take the ideas behind modern portfolio theory any further. The interested reader can consult a standard textbook such as Adams et al. (2003) and papers that developed the original ideas, such as that of Sharpe (1964). However, it is useful to mention how these theories have been developed further into the capital asset pricing model (CAPM). The CAPM confirms the result that all investors will choose the same "risky" portfolio of assets and mix it with risk-free assets. It also proposes a relationship between risk and return for a security based on the β measure of risk that is related to the covariance of the return from the security with the return from the market. In effect, the theory proposes an equilibrium-pricing model for securities. The CAPM reinforces the idea that risk that can be diversified away does not matter; furthermore, an asset with a high covariance of returns can contribute more to the risk of a portfolio than an asset with a higher variance of returns but that also has a low covariance of returns with other assets in the portfolio. The CAPM then distinguishes between the systematic risk of a security (that cannot be diversified away) and the specific risk of a security that can be eliminated in a diversified portfolio.

5.3.5 Modern Portfolio Theory: Insights and Limitations

Whilst modern portfolio theory and equilibrium pricing models, such as CAPM, provide useful insights, they have a number of difficulties, which

---

[1]See also the concept of "separation theorem," e.g., in Adams et al. (2003) and references cited therein.

requires that actuaries use different models. The main insights of modern portfolio theory are:

- Investors want to "trade off" risk and return.
- Risk can be measured in much the same way as returns can be.
- Some portfolios of investments can be ruled out as a result of having both higher risk and higher return.
- Diversification is important in reducing risk.
- The risk of a security is determined by its contribution to the risk of the portfolio.
- The risk of a security can be divided into diversifiable and non-diversifiable risk.
- The market portfolio of assets has a special place in investment strategy.

The restrictive assumptions underlying much of modern portfolio theory may be appropriate in certain investment modelling situations; however, some of them are incompatible with actuarial realities. The following features limit the applicability of traditional portfolio theory models:

- It is not necessarily reasonable to measure risk purely in terms of the standard deviation of portfolio returns (see Chapter 4), and other risk measures may be difficult to use in computational models.
- All investors do not have the same investment decision-making time horizon or view risk in the same way (which is necessary to define a unique risk-free security). Investors will have different time horizons (depending on the length of their liabilities) and view risk differently (some investors considering real returns and some considering returns relative to salary increases, for example).
- Modern portfolio theory, in its standard form, does not approach problems where investors are investing to meet liabilities, the value of which may be correlated with returns from assets.
- Models such as CAPM do not deal with the reality of different investors facing different tax rates and with foreign currency investments.

In addition to these difficulties in applying modern portfolio theory models in actuarial work, there is much discussion as to whether abstract models, such as CAPM, adequately describe investment markets (see Adams et al. (2003)). Some of the difficulties mentioned above are not encountered if the investor simply relies on the insights from modern portfolio theory discussed in Section 5.3.1 and Section 5.3.2 and does not rely on the results of abstract models such as CAPM. Certain other difficulties can be dealt with by extending the scope of the theory: this will be covered below. The first difficulty that was identified above could be dealt with either by extending the scope of the theory or by taking a different approach altogether, such as stochastic asset modelling using simulation.

5.4 Extension of Portfolio Theory to Include Actuarial Liabilities

5.4.1 Portfolio Optimisation in the Presence of Liabilities

Portfolio selection techniques for institutions with liabilities were developed by Wise (1984), further developed by Wilkie (1985) and then generalised by Sherris (1992). Contributions have also been made on the subject in the financial economics literature, notably by Sharpe and Tinte (1990) and Leibowitz et al. (1994). In this chapter, we will look at the subject of portfolio optimisation in the presence of liabilities from three perspectives. We will begin by looking at a simplified version of the Wilkie (1985) exposition. We will then generalise this, based on insights from Sharpe and Tinte (1990), Leibowitz et al. (1994) and Booth (2002). Then we will look at the framework set out by Sherris (1992). This subject area can be regarded as a formal mathematical introduction to the subject of asset/liability modelling. We will broaden the discussion to incorporate stochastic investment modelling in Section 5.5, Section 5.6 and the Appendix.

Wilkie (1985) considers the problem of portfolio selection in the presence of liabilities as an extension of the mean–variance portfolio selection framework discussed in Section 5.3. One additional decision variable is brought in related to the "price of the portfolio" or the initial level of assets invested. In a simplified version of the Wilkie framework, we can ignore this complication.

The starting point for analysis is that a pension scheme or life fund invests in two possible assets or asset classes to meet a particular liability at some point in the future. The "ultimate surplus" is the difference between the value of the assets and liabilities at that future time. It is easiest to conceptualise this in terms of "running off" a portfolio of assets and liabilities until all liabilities have been met so that we can consider the financial position of the institution at the time the final liability has to be met. In effect, there is a single-period time setting, a decision is made about asset allocation at the beginning of the period and the allocation cannot be revised in the light of further information.

The ultimate surplus S can be defined as

$$S = A(x_1 R_1 + x_2 R_2) - L \tag{5.8}$$

where A is the total amount of assets, x_i is the proportion of assets invested in asset i, R_i is the accumulation factor (one plus the rate of return) from asset i over the relevant time period and L is the amount of the liability at the "run off" time. From multiplying the expression out, it is clear that the amount of the surplus is simply equal to the amount of the assets less the amount of the liability at the time at which the liability becomes due. The stochastic variables are the accumulation factors in relation to the assets and the amount of the liability and the decision variables are the proportions invested in the two asset classes.

It is of interest to look at expressions for the expected value and variance of the surplus. The expected value E of the ultimate surplus is

$$E = Ax_1E_1 + Ax_2E_2 - E_l \tag{5.9}$$

where E_i is the expected value of the accumulation factor of the ith investment and E_l is the expected value of the liability l. As would be expected, the expected surplus depends on the accumulation factor or rate of return on the assets and the expected value or rate of growth (sometimes described as rate of return) of the liability. There are no new insights from this observation. Just as high expected return assets lead to high expected return portfolios in the asset-only framework in Section 5.3, high expected return assets lead to high expected surplus portfolios in the asset/liability framework that we have introduced here.

The variance V of the ultimate surplus can be defined as

$$V = A^2x_1^2V_1 + A^2x_2^2V_2 + V_l + 2A^2x_1x_2C_{12} - 2Ax_1C_{1l} - 2Ax_2C_{2l} \tag{5.10}$$

where V_i is the variance of return from asset i, C_{12} is the covariance between the values of the two assets, and C_{il} is the covariance between asset i and the liability value. This result is the straightforward application of statistical theory to determine the variance of a linear combination of random variables. It holds regardless of the distribution of the asset values and the liabilities.

The most notable result of this work is that it can be seen that an asset that has its returns positively correlated with the liability can reduce the variance of the ultimate surplus, even if the asset return itself has a high variance. Thus, for example, this work can provide a scientific justification for the inclusion of equities in a pension fund portfolio if they have a high covariance with pension fund, salary-related liabilities; whether this is the case, of course, is an empirical matter. It is also important to note that the greater the variance of the liability, other things equal, the greater will be the variance of the ultimate surplus.

In the same way as we can produce efficient frontiers for asset portfolios in the Markowitz framework, we can produce efficient frontiers for institutions that indicate asset portfolios that minimise variance of ultimate surplus for a given level of expected surplus.

5.4.2   Connections Between Redington and the Wise–Wilkie Approach

Before generalising this model, it is useful to look at the relationship between the Redington theory (of immunisation) and the approach to portfolio selection in the context of meeting actuarial liabilities that we have discussed above.

Redington, in developing the theory of immunisation, looked at the interaction between assets and liabilities. In particular, he found the asset mix that, under certain conditions, would lead to the minimum variance of surplus (or, strictly speaking, a small positive increase in surplus). However, Redington did not suggest any way in which the investor should trade variability of surplus for increased expected surplus. There was also no suggestion as to how surplus funds (i.e., funds not necessary to provide for liabilities) should be invested.

The Wise–Wilkie approach allows for more general fluctuations in financial conditions, whilst still including interactions between assets and liabilities in the model. In addition, some attempt is made to "trade off" variability of ultimate surplus for expected ultimate surplus. However, in order that tractable mathematical solutions could be found, the analysis was strictly within a mean–variance framework. Wilkie uses and develops further the Markowitz concept of an "efficient frontier": first, by including liabilities and, second, by using a three-dimensional approach where one of the variables could be the price of the portfolio (equivalent to including the initial level of surplus). In our exposition, we have not included this additional variable.

5.4.3   Generalisation of Portfolio Optimisation
in the Presence of Liabilities

The discussion in Section 5.4.1 provides an intuitive introduction to the asset/liability modelling framework. It is helpful to extend it further in two ways. We will discuss these extensions in the context of a defined-benefit pension plan. The first issue relates to the definition of the objective variable, defined as surplus S in Section 5.4.1. Various different objective variables and objective functions have been proposed for pension plan asset/liability modelling. Wilkie (1985) and Sherris (1992) effectively use absolute levels of surplus, as we have used above. Focusing on the absolute level of surplus of a pension scheme is helpful when performing asset/liability modelling for particular schemes. However, it is less useful for more general theoretical and empirical work, because different pension schemes will have different starting values of surplus (although Wilkie dealt with this by introducing the "price of the portfolio," equivalent to the starting value of the assets, as an additional variable).

In order to standardise comparisons between institutions, an alternative would be to use "surplus return" as the objective variable, in the same way as return on assets can be used as an objective variable in "asset only" modelling. Surplus return could be defined in a similar way to the return on an asset portfolio (i.e., $SR_t = (S_t/S_{t-1}) - 1$, where S_t is the surplus at time t and SR_t is the surplus return in year t). However, whereas the return on an asset portfolio gives rise to the same rate of increase in invested wealth, whatever the starting value of wealth, so that it is possible to derive tractable portfolio selection results using utility theory, this is not the case with "surplus return." For example, a 10% increase in surplus or 10% surplus return will give rise to a different increase in pension fund sponsor and fund member wealth if the starting surplus were £100 million, compared with if it were £1 million. Surplus return can also produce infinite or undefined values when the starting or finishing level of surplus is non-positive.

An alternative has been proposed by Leibowitz et al. (1994), who use "funding ratio return" (effectively the rate of increase of the funding ratio or in the ratio of assets to liabilities). As this is the difference between two ratios divided by a ratio, it can give rise to results that are difficult to interpret. For example, if there is a 10% increase in the funding ratio when it is at a level of 1.1, this will imply a smaller increase in wealth than if the initial funding ratio were (say) 2. Leibowitz et al. recognise this and suggest approaches to adjust for these problems. However, it is not necessary to complicate the approach in the way that they suggest.

In an approach proposed in the real-estate finance literature, Chun et al. (2000) use what they also describe as "surplus return" as the decision variable but in fact, define it rather differently from Leibowitz et al. If we define surplus return as SR_t above, Chun et al. use $SR_t(S_{t-1}/A_{t-1}) = (S_t - S_{t-1})/A_{t-1}$. Thus, they use surplus return standardised by the starting level of assets in the scheme. In other words, they focus on the increase in surplus expressed per unit initial assets. Such a measure, or a similar measure standardised by the starting level of liabilities, would seem an appropriate objective function for a pension fund, as it indicates the proportionate increase in member's benefits (plan liabilities) that is possible or the proportionate return of assets to the employer either directly or by reducing future contributions. It is reasonable, therefore, to focus on the amount of any increase in surplus expressed per unit initial assets (or liabilities). In developing this approach to obtain more general equations, we will use the following additional notation:

SR'_t is the surplus return, standardised for initial asset values, in year t
A_{t-1} are the assets invested in the scheme at time $t-1$
r_i is the rate of return on asset i (N.B. not the accumulation factor)

r_l is the rate of return (rate of increase) in the plan's liabilities
L_{t-1} is the scheme liabilities at time $t - 1$

$$SR'_t = \sum_{i=1}^{i=n} x_i r_i - r_l \frac{L_{t-1}}{A_{t-1}}$$

The standardised surplus return variable that is our objective variable can be regarded as the rate of return on the asset portfolio (found by multiplying the proportions invested in different asset classes by the rate of return from those asset classes) less the adjusted rate of return on the liabilities. The adjustment to the rate of return on the liabilities is the ratio of liabilities to assets at the beginning of the period. It can easily be shown that this is equivalent to the increase in surplus per unit initial assets. Once again, the control variables are the proportions invested in different asset classes (although it may also be possible to control r_l to some extent if there is any discretion with regard to benefits paid by the pension scheme).

The expected value of the standardised surplus return objective variable is:

$$E(SR'_t) = \sum_{i=1}^{i=n} x_i E(r_i) - E(r_l) \frac{L_{t-1}}{A_{t-1}} \tag{5.11}$$

where $E(r_i)$ is the expected rate of return from asset i, $E(r_l)$ is the expected rate of growth (rate of return) of the liability, and the variance of the standardised surplus return objective variable is

$$\sigma(SR'_t)^2 = \sum_{i=1}^{i=n} x_i^2 \sigma_i^2 + \sum_{i=1}^{1=n} \sum_{j=1}^{j=n} \sigma_{ij} x_i x_j \frac{L_{t-1}}{A_{t-1}} + \sigma_l^2 \frac{L_{t-1}^2}{A_{t-1}^2} - \sum_{i=1}^{1=n} \sigma_{il} x_i \frac{L_{t-1}}{A_{t-1}} \tag{5.12}$$

where σ_i is the standard deviation of return from asset i, σ_{ij} is the covariance of return between asset i and asset j, σ_l is the standard deviation of return of the liability l, and σ_{il} is the covariance of return between asset i and the liability.

Thus, as might be anticipated from the earlier analysis of the two-asset case, the variance of the redefined objective variable depends on the variance of the assets (weighted by the proportions invested in the different assets), the covariance between the assets, the variance of the liabilities, and the covariance between the assets and the liabilities. Indeed, if Equation 5.11 is multiplied through by A_{t-1} and Equation 5.12 multiplied through by A_{t-1}^2, then the equations look very much like Equation 5.9

and Equation 5.10 respectively, except that it has been generalised to n assets and the notation is more formal.

Once again, the institution can choose investment portfolios that minimise variance of the surplus return variable for a given value of the expected value of standardised surplus return. Booth (2002) shows the results from an actual modelling exercise that develops individual points on a mean–variance efficient frontier. The calculation of the efficient frontier can be undertaken on a spreadsheet program such as *Microsoft Excel*. There is a very good explanation of how to use this package efficiently for the purpose of calculating mean–variance efficient frontiers in Adams et al. (2003). This can easily be adopted to incorporate liabilities in the *Excel* optimisation process.

It is worth taking this analysis one step further. So far, we have considered how we might determine the mean and variance of surplus or standardised surplus return for an institution. This allows us to calculate a mean–variance efficient frontier and, therefore, choose portfolios from that frontier. If variance or standard deviation is our chosen risk measure (see Chapter 4 for a discussion of the different types of risk measure that can be used in different contexts) then we can propose a way of choosing an "optimal" portfolio from a range of "efficient" portfolios. The objective of the institution should be to choose appropriate values for the asset proportions (and possibly the liability structure, if that can be controlled) to maximise an objective function relating to SR'_t that weights the standard deviation of standardised surplus return negatively and its expected value positively.

Thus, we can suggest that the investor should choose the investment portfolio to maximise an objective function such as

$$F(SR'_t) = E(SR'_t) - a\sigma(SR'_t)$$

where a can be regarded as a risk aversion parameter. This method of combining risk and return in a single function was also discussed in Chapter 4. The higher a is, the greater will be the investor's aversion to risk. If $a = 0$ then the investor will be risk neutral and will try to maximise expected surplus. As $a \to \infty$, the investor will choose the portfolio that will minimise the variance of the plan surplus.

5.4.4 Portfolio Selection in an Asset/Liability Framework Using a Generalised Approach to Risk

In Chapter 4 we have discussed utility theory as the basis for developing risk measures that could be used by investors in practice. In Section 5.4.1 to Section 5.4.3 we have considered how an investor might look at the asset allocation decision in a technical framework, but using only variance of the objective variable (related to the surplus of the institution) as the measure of risk. Sherris (1992) has suggested that the institution should invest to maximise the expected utility of surplus. The approach is more general, therefore, because the utility function used can reflect any risk profile relevant to the investor.

Returning to the simplified example in Section 5.4.1, with two assets for convenience, consider the ultimate surplus given by Equation 5.8. We note that the model can easily be generalised to n assets. In general, the investor has the objective of maximising

$$E[U(S)]$$

i.e., maximising the expected value of the utility of surplus.

Interesting results can be found if it is assumed that the investor has an exponential utility function. Sherris assumed that investors had an exponential utility function, so that the objective of the investor is to *maximise*

$$E[-\exp(-S/r)]$$

where r is a parameter reflecting the risk tolerance of the investor. This is equivalent to *minimising*

$$E[\exp(St)]$$

where $t = -1/r$. Thus, the investor is choosing the asset proportions to minimise the moment-generating function of the probability distribution of the ultimate surplus.

If the ultimate surplus is assumed to follow a normal distribution, then it is possible to find analytically the portfolio mix that maximises expected utility. It is the portfolio that minimises

$$E[\exp(St)] = \exp(Et + Vt^2/2)$$

where E is the expected value of the surplus, given by Equation 5.9, and V is the variance of the surplus, given by Equation 5.10.

This, of course, takes us back to a mean–variance framework. We are required to minimise a function of the mean and variance of surplus. This arises as a result of explicitly assuming a normal distribution of the ultimate surplus. The general approach involves maximising a function that represents the utility of surplus. It has been shown that, if the investor has an exponential utility function, this involves minimising the moment-generating function of the probability distribution. In particular cases (for example, where we assume a normal distribution of returns), analytical solutions can be found. In practice, the utility function used should reflect the investor's own view of risk. The probability distribution of investment returns need not necessarily follow a normal distribution — it may well be skewed. In this more general framework, it is not necessarily possible to find a mathematical solution to the asset allocation problem. If this is the case, then we may be able to obtain a better understanding if we use stochastic investment models and stochastic models of the liabilities to simulate probability distributions for the surplus and for the asset accumulations. More general measures of risk can then be employed in determining optimal portfolios.

It should also be added that Sherris has generalised the portfolio selection problem further. As in the Wilkie framework, it was recognised that the initial level of surplus was an important decision variable. The institution, where surplus is distributed to a group such as shareholders or with-profit policyholders, needs to determine the most desirable distribution policy. Discussion of this problem takes us beyond the scope of this particular chapter, but the reader should be aware of the issue.

A further issue of great importance is how the optimal asset allocation changes over time. This takes us from what is effectively a single-period portfolio selection model to a multi-period model with periodic reallocation of assets. This issue is beyond the scope of the chapter.

5.5 Stochastic Investment Models

This short section cannot do full justice to the field of stochastic investment modelling. In this section, we will introduce briefly the subject of time series modelling in general, as well as the development of investment models. A fuller presentation of the theory of time series modelling can be found in time series texts such as that by Mills (1999). Applications of econometric techniques to times series modelling more generally are presented by McAleer and Oxley (2002). More detail on the general principles of the development of stochastic investment models can be found in Adams et al. (2003). Investment models or critiques of models are presented in papers by Wilkie (1986, 1995), Geoghegan et al. (1992), Huber (1995, 1997a), Smith (1996), Yakoubov et al. (1999) and Huber and Verrall (1999). This chapter is not intended to give a full and up-to-date account of stochastic investment modelling, but to give a guide to some of the principles. The model to be described was one of the first detailed models to be developed. It is still frequently used and its properties are amenable to analysis. Other categories of model, such as that proposed by Smith (1996), take a different approach and are less easy to analyse in a strightforward manner. Yakoubov et al. propose an extension of the Wilkie approach. Some involved in the modelling process may feel that random-walk models are most appropriate for investment modelling; these can be developed as special cases of the Wilkie model.

As Huber (1997); and Huber and Verrall, (1999) discuss, stochastic investment models fall into two basic categories. Those that are intrinsically based on fundamental economic relationships could be described as econometric models. Those that are intrinsically based on the analysis of past data could be described as statistical models. Some models (such as the Wilkie model) do not fall clearly into either category. Whilst it is convenient to categorise stochastic investment models in this way, it should be borne in mind that a statistical model should not be completely inconsistent with financial and economic theory. Similarly, an econometric model should not use relationships that are incompatible with the data. One can also distinguish between the short-term and long-term properties of a model. A particular model may have poor short-term forecasting capability and yet be capable of producing long-term distributions for financial and investment variables that represent economic "common sense."

5.5.1 Economic Investment Models

To illustrate the fundamental structure of an econometric model, we will discuss some of the economic and financial relationships that could be found in such a model. The discussion is not intended to identify all the possible relationships between economic and investment variables, but simply to illustrate the form such a model might take. We will consider the investment categories of cash, conventional government bonds, index-linked government bonds and equities; and the economic variables considered are inflation, money supply, government borrowing, the savings ratio and the rate of return on capital. Returns from cash are determined to a large extent by inflation and monetary policy (short-term interest rates being the main instrument of monetary policy). We could, therefore, have an equation of the form

$$i_t = f_1(r_t, r_{t-1}, m_t, {}_1e_t)$$

where i_t is the cash return at time t, r_t and r_{t-1} are the rates of inflation at times t and $t-1$, m_t is the rate of increase in the money supply at time t, and ${}_1e_t$ is a random error component. The lags and functional forms of the relationships may be complex; however, it is not necessary to develop specific relationships to understand the principles behind an economic-based model.

The long-term government bond yield could be regarded as being a function of inflationary expectations, short-term interest rates, the savings ratio and government borrowing thus:

$$y_t = f_2(r_t, r_{t-1}, r_{t-2}, \text{PSBR}_t, s_t, i_t, {}_2e_t)$$

where y_t is the long-term gilt yield, r_t, r_{t-1}, and r_{t-2} are recent inflation rates (which people may use to form expectations about long-term inflation), PSBR_t is government borrowing, s_t is the savings ratio, and ${}_2e_t$ is a random error component.

The prospective real yield from index-linked government bonds could be modelled as a function of the variables that determine the yield on conventional government bonds, without the inflation element, thus:

$$j_t = f_3(\text{PSBR}_t, s_t, i_t, {}_3e_t)$$

where j_t is the real yield from index-linked government bonds.

Prospective equity returns could be modelled as a function of index-linked government bond yields and the rate of return on capital:

$$g_t = f_4\big(j_t, \text{ROC}_t, {}_4e_t\big)$$

where g_t is the prospective equity return at time t, ROC_t is the rate of return on capital at time t, and ${}_4e_t$ is a random error component. It should be noted that all the error components would be different and have different structures in each relationship. It should also be noted that it might be desirable to model functions of the variables i_t, y_t, j_t and g_t rather than their absolute values.

A number of points need to be made about this framework which has been drawn up to illustrate possible links between variables in an economic model. First, for long-term modelling purposes a "generator" would be needed for the economic variables; this would have to be developed from some form of statistical time series model. This technique is different from techniques used in econometrics, where predictions are based, in the short term, on already revealed values of explanatory variables. Equations then have to be postulated representing links between different economic variables. Second, some form of autoregressive component would probably be necessary; thus, current cash rates, for example, could also depend on the difference between the cash rate at an earlier time and its predicted value at that time.

There are a number of problems with economic-based investment models. Perhaps the two major problems are the difficulties of finding a consensus on the appropriate economic relationships and finding appropriate time series models for the economic variables themselves.

It is possible that economic and financial relationships are not generally amenable to statistical modelling; at the very least, we cannot expect econometric relationships to remain stable over time.

5.5.2 Statistical Investment Models

A purely data-driven stochastic investment model would attempt to model investment return patterns from the past data, as a time series, without looking for any fundamental economic model or including exogenous economic variables. Relationships between variables being modelled can, however, be included. It is generally regarded as desirable to produce time series models of stationary data, i.e., a time series that has no systematic change or trend in the mean. The data would often be differenced to make it stationary. First, it is worth discussing some of the time series processes that could be modelled by a data-driven stochastic investment model. To illustrate the point, we will choose variables that relate to the level of, or changes in the level of, the retail price index.

Defining the level of the retail price index at time t as Q_t, Q_t could be said to follow a random walk if

$$Q_t = Q_{t-1} + Z_t$$

where Z_t is a purely random process with mean zero and variance σ_z^2. The expected change in the price index each year is zero. A random-walk model is sometimes used for share prices. It would generally be regarded as inappropriate for the retail price index. First, there is an assumption of zero average inflation; second, we might expect changes in the retail price index to be correlated with previous changes. A drift parameter could be inserted to represent a non-zero average increase in the price index.

An autoregressive process has a random element and an element that depends on previous values taken by the variable in question. The parameters in the model must satisfy certain constraints, or the variable in question will not revert to its mean levels. The mean level of the variable could be zero or more generally some other level μ. If inflation is stationary (as opposed to the price level being stationary) then the difference between the log of successive values of the price index will be stationary. A first-order autoregressive process for the change in the log of the price index with mean zero could be of the form

$$\nabla \ln Q_t = \alpha \nabla \ln Q_{t-1} + Z_t$$

where we require the constraint $\alpha < 1$, where $\nabla \ln Q_t = \ln Q_t - \ln Q_{t-1}$ and where Z_t is a random variable with mean zero and variance σ_z^2.

If the mean level of inflation is not zero, then we could model the deviation from the mean as an autoregressive process thus:

$$\nabla \ln Q_t - \mu = \alpha(\nabla \ln Q_{t-1} - \mu) + Z_t$$

where μ is the mean force of inflation and $\alpha < 1$.

The autoregressive process would be of order p if it could be modelled as follows:

$$\nabla \ln Q_t = \alpha_1 \nabla \ln Q_{t-1} + \cdots + \alpha_p \nabla \ln Q_{t-p} + Z_t$$

where restrictions have to be imposed on the parameters α_1 to α_p to ensure stationarity.

Alternatively, deviations from a constant non-zero mean rate of inflation can be modelled thus:

$$\nabla \ln Q_t - \mu = \alpha_1(\nabla \ln Q_{t-1} - \mu) + \cdots + \alpha_p(\nabla \ln Q_{t-p} - \mu) + Z_t$$

The random-walk model may be regarded as desirable for the price index if a central bank were following a zero inflation target and if increases in the price level, when they occurred, did so as a result of sudden random shocks to the economy, with no lags in the economic system. The autoregressive model is useful where the variable being modelled depends on past values and on a random component. This might be relevant if a central bank were following a price-level target.

As has been shown, a constant mean level for inflation can also be included so that the departure from the mean level of inflation is modelled as an autoregressive process and the inflation level gradually moves back to the mean.

A moving-average process could be represented by

$$\nabla \ln Q_t = Z_t + \beta_1 Z_{t-1} + \cdots + \beta_q Z_{t-q}$$

where Z_t is a random process that may have zero mean and variance σ_z^2. Again, the parameters have to satisfy certain constraints. For example, if $q = 1$, $|\beta_1| < 1$.

A moving-average process has a random element and elements that depend on parameters and values taken by the random element in previous periods. It is therefore a weighted average of the previous random elements. If there are lags in the economic system, the moving average process may be a good way of modelling changes in $\ln Q_t$.

If non-zero average inflation were assumed, an appropriate moving average model could be:

$$\nabla \ln Q_t - \mu = Z_t + \beta_1 Z_{t-1} + \cdots + \beta_q Z_{t-q}$$

The main feature of the statistical models described here is the emphasis on modelling the links between current and past values of the variable in question and looking for appropriate random processes that will produce appropriate conditional probability distributions for the variable being modelled. The danger of statistical models is that no particular underlying economic model is being used to explain movements in financial variables that are heavily affected by economic factors. However, it should be mentioned that only the simplest examples of how a time series model could be developed have been described here. Statistical models can be expanded to include interactions between variables within the model; e.g., see Chan (2002), who develops a vector autoregressive moving-average time series model.

A combination of autoregressive and moving-average processes can be included in so-called autoregressive integrated moving-average (ARIMA) models; see Chatfield (1996). The particular data set being modelled may not be stationary. A good example of this is the price level. As has been mentioned, if inflation is a stationary or mean-reverting process, the difference between the log of successive values of the price index will be stationary, but the price index will not be stationary. To obtain the value of the variable under consideration, the values of the variable being modelled need to be summed; hence, the use of the term "integrated" model.

An ARIMA model for the price index combines the features of the autoregressive and moving-average processes to produce a process of the form

$$\nabla \ln Q_t = \alpha_1 \nabla \ln Q_{t-1} + \cdots + \alpha_p \nabla \ln Q_{t-p} + Z_t + \beta_1 Z_{t-1} + \cdots + \beta_q Z_{t-q}$$

where all variables have been defined previously. An ARIMA model such as the one specified above would be described as an ARIMA$(p, 1, q)$ model. The numbers in the parentheses refer to the order of the autoregressive process, the order of differencing of the time series, and the order of the moving-average process respectively. This ARIMA model for the change in the log of the retail price index is an autoregressive moving-average model for the force of inflation. The ARIMA model corresponding to the autoregressive and moving-average models where non-zero inflation is assumed would be of the form

$$\nabla \ln Q_t - \mu = \alpha_1 (\nabla \ln Q_{t-1} - \mu) + \cdots + \alpha_p (\nabla \ln Q_{t-p} + \mu)$$
$$+ Z_t + \beta_1 Z_{t-1} + \cdots + \beta_q Z_{t-q}$$

Statistical models can simultaneously model a number of time series variables. For example, if the time series of the retail price index and gilt yields are being modelled, then these variables can also be modelled as functions of each other using transfer functions. Thus, statistical models do not totally preclude economic interactions. However, the form of interaction that can be modelled is limited.

5.5.3 Statistical Models with Shocks

The statistical models described above assume that it is appropriate to model a time series as a linear combination of other time series variables, with the addition of white-noise components. This may not reflect economic reality. Clarkson (see discussion in Wilkie (1986)) and Booth (see discussion in Wilkie (1995)) have pointed out that inflation has tended to have different mean values during different epochs. It would not be possible to difference data to produce a stationary time series, if this were the case. There seems to be considerable evidence for this characteristic of inflation data. In the U.K., for example, different time periods, during which different economic policies were followed, clearly had an impact on the pattern of inflation. Over the last century, these time periods have included: the adoption of the gold standard (here, there was a tendency for the price level to be stationary); times of war; the post-World War II period, during which time policymakers largely ignored the main causes of inflation when determining and using policy instruments; the period of monetary targeting; and, at the current time, a policy of central bank independence and inflation targeting. One would expect each of these regimes to produce a different form of statistical process for inflation.

Criticisms of standard statistical models can be addressed by accepting that any models used should be specific to the economic environment. If it is believed that a process is generally stationary but subject to occasional "shocks," then a model can include a random shock element. For example, Clarkson (1989) suggests that an inflation series could be modelled as

$$\nabla \ln Q_t = \nabla \ln Q_{t-1} - \theta(\nabla \ln Q_{t-1} - \mu) + \alpha(t) + \beta(t)$$

where μ is an average rate of inflation, $\alpha(t)$ is a random component, and $\beta(t)$ is a parameter that represents the movement of the level of inflation to different mean levels in different epochs, of random length.

Such a model could be regarded as attractive for a number of reasons. Neo-classical economists and monetarist economists generally believe that inflation is the result of the failure of political institutions or central banks. In this context the $\beta(t)$ component could be regarded as modelling the possibility of changes in institutional arrangements which may change the overall average rate of inflation for a significant but random period.

The use of a model including a random shock component suffers from one major disadvantage: it is very difficult to produce reliable parameter estimates for the shock component. Shocks, by definition, are not sufficiently numerous and regular for there to be adequate data to model their frequency, average magnitude and variability. For this reason, random shock models have not tended to be used in practice.

If a linear statistical model is used when a random shock model is appropriate, then various difficulties could be encountered. The model will not reflect the underlying statistical process. In a mean-reverting model, the mean will be some kind of average of all the means experienced over several epochs. That mean may be different from the mean in any particular epoch. In addition, the variance of the model will be completely misspecified. In any epoch, there may be a greater degree of certainty about the variable being modelled than the model indicates. Furthermore, the effect of shocks causing outliers will be to increase the overall variance of the residuals in the model whilst the distribution of the residuals is inherently misspecified.

There are reservations about all published stochastic investment models and genuine debate about the methods that should be used to fit models. Some of the more desirable model forms, from a theoretical point of view, are difficult to fit in practice. We now move on to discuss the form of the most widely documented model used in the U.K. actuarial field, i.e., the Wilkie model.

5.5.4   The Wilkie Stochastic Investment Model

5.5.4.1   *Form of the Model*

The Wilkie model was presented in Wilkie (1986) and was expanded and updated in Wilkie (1995). Different model forms were then considered and models fitted to more financial variables. Parameters were also re-estimated.The principles of the Wilkie model are well illustrated and the explanations are more detailed in the 1986 version. However, readers interested in the subject, and particularly in other model forms, such as ARCH models, should also refer to Wilkie (1995). We will only consider the narrower model here. However, a detailed case study of the Wilkie real-estate model is presented in the Appendix to the chapter. The case study shows some of the benefits and problems of using the Wilkie model approach and provides an opportunity for an in-depth critique. The case study is adapted from Booth and Marcato (2004).

The narrower Wilkie model considers four investment variables: the retail price index, share dividend yields, share dividends and conventional irredeemable government bond yields. In this description, Wilkie's notation will be used throughout to aid comparison with the original source. The driving force for all variables is the level of inflation; for this reason, the model has often been described as having a cascade structure. The model for the retail price index at time t, $Q(t)$ is

$$\nabla \ln Q(t) = QMU + QA(\nabla \ln Q(t-1) - QMU) + QSD \times QZ(t) \qquad (5.13)$$

where $\nabla X(t) = X(t) - X(t-1)$.

QMU represents a fixed mean annual force of inflation. An appropriate value of 5% has been suggested. However, this point illustrates well the problem of using this kind of linear model. It is arguable that the "mean" rate of inflation has never been 5% in the U.K., but that it has varied between values lower than 5% (from 1700 to 1939) and values significantly above 5% (from 1955 to 1990). At the current time, an inflation-targeting regime is in place that targets inflation of 2.5%.[2] It may be reasonable to use that value for QMU. $\nabla \ln Q(t-1) - QMU$ is the difference between the last period force of inflation and the fixed mean (QA being a constant). The final term is a random component. Inflation is therefore modelled as a first-order autoregressive process. There is no econometric input into the equation.

The discussion in the annex is directly relevant to the discussion of the dividend yield and dividend models. The dividend yield is modelled as an autoregressive process and as a function of inflation thus:

$$\ln Y(t) = YW \nabla \ln Q(t) + YN(t) \qquad (5.14)$$

where $\ln Y(t)$ is the log of the dividend yield; YW is a constant, and $YN(t)$ represents a constant, plus a function of deviations of past values of $YN(t)$ from that constant, plus a random term. Thus, share yields effectively depend on inflation and a first-order autoregressive process. The constant in the $YN(t)$ term is suggested to be $\ln(0.04)$.

A high rate of inflation implies a high share yield in this model: one economic justification for this would be if, in times of high inflation, capital values did not respond quickly to increases in money dividends. The term may also reflect the possibility of money illusion, whereby investors confuse increases in nominal with increases in real interest rates in times of high inflation, thus leading to lower equity values. Also, inflation generally has an adverse impact on an economy and may lead to lower share prices (higher dividend yields) at least in the short run.

---

[2]In fact, the U.K. inflation targeting regime is based on a 2% target using the harmonized index of consumer prices. This is broadly equivalent, in the long term, to a target for the retail price index measure of inflation of 2.5%, which was, in fact, the target prior to 2004.

Wilkie models the level of dividend yields and the level of dividends. A combination of dividend yields and the level of dividends allows the user to obtain increases in equity capital values and total returns. The level of dividends depends on inflation—both immediately and with a time-lag effect. In addition, the index of dividends is also affected by the residual of the previous period from the dividend yield model. There is a constant element (which effectively represents mean real dividend growth). Finally, there is a random term that has an immediate and a lagged effect.

Specifically, the model can be written as follows:

$$\nabla \ln D(t) = DW \left[\frac{DD}{1 - (1 - DD)B} \right] \nabla \ln Q(t) + DX \nabla \ln Q(t)$$
$$+ DMU + DY \times YE(t-1) + DE(t) + DB \times DE(t-1)$$

(5.15)

$D(t)$ is the index of dividends at time t. A value, chosen by Wilkie, of $DX = 0.2$ indicates that 20% of any increase in the general level of prices will be reflected in dividend increases in that year. The DD term in brackets represents a series of coefficients, the effect of which is to represent the effect of lagged inflation on dividends, with the B term being a lag operator. The sum of the coefficients is one. The lag term is multiplied by DW, for which Wilkie suggests a value of 0.8. This, combined with the value of $DX = 0.2$, ensures that an increase in the general level of prices feeds through completely into an increase in dividends, albeit with a time lag. Wilkie suggests a value of zero for DMU, implying zero long-term real dividend growth. There is further discussion of the issue of assuming unit gain[3] in the Appendix.

The last three terms are essentially random components; however, it is worth commenting further on the $DY \times YE(t-1)$ term. Given a negative value for the constant DY (Wilkie suggests -0.2), a positive residual from the dividend yield model would imply a reduction in dividends (from this influence) in the following year. The rationale for this is that an increase in capital values over and above that predicted by the model may occur in one year as a result of investors changing anticipated future real dividend growth. Changes in anticipated real dividend growth do not appear directly in the dividend yield model and, therefore, will be reflected in the residual. The residual will be positive if anticipated real dividend growth decreases so that the actual dividend yield rises above that predicted by the model (and vice versa, if anticipated dividend growth

---

[3]This is the property whereby a change in one variable by 1% (e.g., the price level) is ultimately reflected by a change in a dependent variable (e.g., the level of dividends) even if there is a long and complex series of lags before the change is reflected.

falls). This change in dividend yield will then be followed by the anticipated change in dividend levels.

The above approach adds an extra dimension to the model, but it does have some weaknesses. First, the real dividend growth anticipated by investors may take place several years into the future, and the model does not allow for this possibility. Second, the residual component of the dividend yield model may arise as a result of investors changing the real rate of interest that they are willing to accept from equities. This also is not reflected in the model.

Wilkie suggests a value of zero for the DMU term. This is a constant and represents the long-term mean rate of real dividend growth. Users of the model can apply a different value for DMU.

The model for the conventional irredeemable government bond yield consists of a third-order autoregressive model for real consol yields and an allowance for expected future inflation. Expected future inflation is determined by a weighted average of current and past levels of inflation. Specifically, the model is as follows:

$$C(t) = CW\left[\frac{CD}{1-(1-CD)B}\right]\nabla \ln Q(t) + CN(t) \qquad (5.16)$$

where

$$\begin{aligned} \ln CN(t) = \ln CMU + (CA1 \times B + CA2 \times B^2 + CA2 \times B^3) \\ \times (\ln CT(t) - \ln CMU) + CY \times YE + CSD \times CZ(t) \end{aligned} \qquad (5.17)$$

The first term on the right-hand side of Equation 5.17 represents the level of inflationary expectations as measured by a weighted average of past inflation rates. Therefore, it uses an adaptive expectations model, rather than a rational expectations model. The parameter CW is set to unity, indicating that inflationary expectations are fully incorporated in nominal bond yields; this parameter could, of course, be varied to represent different degrees of money illusion. The CMU term, within the CN(t) term, represents the mean real rate of return, which is set at 3.5%. Wilkie suggests that good long-term forecasts can be found by setting CA2, CA3, and CY to zero. This would leave a first-order autoregressive process and a random component. The suggested parameter value for CA1 is 0.91, so that the real yield moves slowly back to the mean.

5.5.4.2 *Use of the Model*

In this section we look at how the Wilkie model can be used. We will also bring out further the fundamental characteristics of the model. The purpose of using a stochastic investment model is to enable more complex problems to be considered using simulation than can be solved analytically: the stochastic investment model can be used to simulate probability distributions of future investment returns, where it is not possible to find an analytical distribution. Therefore, it is possible to obtain measures of variability using stochastic models, which is not possible using a deterministic approach. In using the model, it is necessary to choose the parameter values appropriately. Wilkie has estimated parameter values, which are given in the previous section. However, the user can choose different parameter values. It is also necessary to provide starting values for the various variables that are modelled autoregressively. In principle, two forms of starting values are reasonable. First, the long-term average values of the variables could be used (Wilkie describes these as "neutral values"). Second, current values could be used (combined with recent past values where, for example, a series of lagged values of inflation comes into the model). The choice of starting values will depend on the purpose for which the model is used.

The neutral starting value (or long-term mean value) for the annual force of inflation was suggested as 5%. We have commented on that above, and 2.5% might be a more sensible value in current conditions; the starting value for the dividend index is of no consequence (it is merely an index number). The neutral starting value for inflation in the term for the lagged effect of inflation on the dividend yield would also be 2.5% if the starting value for inflation were 2.5%. A starting value is needed for the random residual from the dividend yield model (as this impacts on the level of dividends): the neutral value is zero. The conventional irredeemable government-bond yield model requires a starting value for the real bond yield and for inflation and the "carry through" from past inflation. The neutral value for these last two variables is 2.5% in each case if the neutral value for inflation is 2.5%; the real yield is 3.5%. A minimum value for the bond yield was set equal to 0.5% by Wilkie, to avoid the possibility of negative long-term bond yields when inflation is very negative over a sustained period. Users may wish to update estimates of parameters and use their own starting values. At the time at which the model was calibrated by Wilkie there was virtually no current academic or practical discussion of the possibility of deflation and the impact of deflation on short- and long-term interest rates. There is more serious discussion on that subject now (e.g., see Yates (2003) and www.bankofengland.co.uk for speeches given by the Bank of England's staff and Monetary Policy Committee members on the subject). However, it should be noted that deflation had occurred over the period Wilkie fitted the model and Wilkie did discuss the implications of deflation on the model parameters.

As well as calculating the four direct series discussed so far, three derived series can be calculated. The first is the share price $P(t)$, which is straightforward to calculate from the dividend yield and dividend index.

Also, a "rolled-up" share index, with dividends rolled up net or gross of tax, can be calculated.[4] Finally, a rolled-up government-bond investment index can be calculated, where income is reinvested net or gross of tax.

Wilkie performed a number of tests of the model, and further testing has been performed. Notable published studies of the Wilkie model are those by Geohegan et al. (1992), which concentrated on the earlier model, Huber (1995, 1997a) and Huber and Verrall (1999). The tests by Geohegan et al. were carried out by performing a large number of simulations and examining returns, standard deviations of returns and correlations between different asset classes at different time horizons. Some of the more interesting results are presented here.

The mean rate of inflation is around 5%, as would be expected. The standard deviation starts at 5% for a term of 1 year and decreases for the compound average rate of inflation over the longer term. The mean real return on shares is around 5% over the medium and longer terms. As might be expected, the standard deviation of returns is very high over short terms and then drops very quickly. There is negative short-term correlation between returns from equities and inflation (due to the YW term), but a positive long-term correlation, as an increase in the price level eventually feeds through into dividends. The real return on shares has less uncertainty attached to it than the money return. The mean return on conventional government bonds was between 8 and 9%, reflecting a 3.5% real yield over longer terms; with different, perhaps lower, values for mean inflation, the values for the mean return on conventional government bonds would be correspondingly lower. The standard deviation of the return from conventional government bonds is high over 1 year (reflecting their high volatility), drops steadily until term 20 (reflecting their long-term guarantee) and then increases slightly (reflecting the uncertainty surrounding the interest rate at which coupons are reinvested). There is a strong negative correlation between inflation and short-term consol returns, as an increase in inflation will lead, quite rapidly, to a rise in consol yields and a fall in capital values. There is a positive correlation (of 0.43) between the short-term real rate of return on conventional government bonds and that on shares; however, this correlation falls to almost zero at longer terms. The short-term correlation is not dissimilar to that which one finds between annual bond and equity returns in practice.

The user can change the parameter values representing the mean values of the different series. In particular, as has already been mentioned, the user may wish to change the long-term mean inflation and real dividend growth rates. This will not, in general, change the structure and cross-correlations that arise from the model.

---

[4]In the U.K., tax paid on dividends cannot now be reclaimed by non-taxpayers. This situation is quite common in tax systems internationally.

The above results are reasonably intuitive. It is also worthwhile considering some of the difficulties. First, the model is a linear model and, therefore, does not allow for the possibility of sustained changes in the mean level of variables; this may be particularly important with regard to inflation. Kitts (1990) suggests that the residuals obtained for the retail price index model are not independent, perhaps reflecting the reality of sustained bursts of high and low (or negative) inflation not captured by the model. This is a point we have already noted in our discussion of models that incorporate shocks.

Where it is thought that a constant variance of the variable being modelled does not exist, well-established ARCH techniques could be used: e.g., see Adams et al. (2003). However, Geohegan et al. (1992) question the value of ARCH models for longer term modelling.

A periodic change in the mean level of variables could be captured by a model incorporating random shocks (see Section 5.5.2). Huber (1995, 1997a), too, finds both the Wilkie model as a whole and the parameter estimates to be unstable over time. Ludvik (1993) comments on the correlations between the residuals of the various parts of the model and the cross-correlations between the returns from the various asset categories. Ludvik suggests modelling the equity/gilt yield ratio directly. He also suggests imposing a structure on the residuals so that there would be correlations between the residuals from various series. In limited cases, in fact, Wilkie does include such terms in his model, reflecting the relationship between the residuals from one series and the values generated for other series (see above).

A further difficulty with the Wilkie model has been identified. The distribution of the residuals, empirically, would not appear to be normal, as is assumed in the fitting techniques. The data would appear to suggest that a fatter tailed distribution of residuals should be used. The model can, therefore, be said to be less reliable in indicating the probability of extreme outcomes. This may be an important shortcoming if the model is to be used for forecasting extreme events, such as insolvency. Some practitioners believe that the probability of extreme negative values of inflation, obtained from the model, is too high. However, these comments have been made before the recent changes to the U.K.'s monetary policy regime. Once again, they indicate the problem of combining together data from a number of different economic regimes or epochs. In the period from 1945 to 1997, it could indeed be argued that the probability of deflation was very low. In other periods (from 1650 to 1945 and, possibly, since 1997) this may not be so.

As has been noted, the Wilkie model has been extended and tested further in Wilkie (1995). Other fitting techniques have also been considered. The extended model included wages, index-linked government bond yields, property yields and income, and foreign currency investments. Some of the criticisms discussed here were addressed in that paper.

The Wilkie stochastic investment model is just one of many approaches to investment modelling, all of which have different shortcomings. Unlike many alternative models, the Wilkie model has been fully published and, therefore, exposed to criticism. It may be the case that the short-comings of alternative models are greater than those of the Wilkie model. In many cases, there will be insufficient data to model certain features of investment markets. Therefore, it is important that model users are cautious when interpreting results and using results for practical applica-tions. We will now move on to describe some of the uses of stochastic investment models.

5.6   Asset Liability Modelling

Earlier in this chapter we discussed methods of determining investment portfolios that could reduce the probability of insolvency (absolute matching and immunisation) and methods that would indicate the risk/return trade-off of investment strategies in the face of liabilities (Wilkie's extension of modern portfolio theory). Simulation techniques allow more general approaches to be taken to portfolio selection.

Investment models can be used to simulate investment returns from different portfolios. The stochastic behaviour of different assets can be modelled and probability distributions of asset accumulations determined (these probability distributions reflecting the characteristics of the invest-ment model). The correlation between the assets can also be modelled and probability distributions of portfolios of assets determined.

Booth (1995) simulated the accumulation of a large number of investment portfolios. This is not the most recent or sophisticated study of its type. However, the results are intuitive and presented in a way that enables them to be interpreted in the context of the earlier treatment of risk measures in Chapter 4.

Five hundred different asset portfolios are used, consisting of different combinations of U.K. equities, conventional irredeemable government bonds, index-linked government bonds and cash. The Wilkie model is used, although it was adjusted for some asset classes to allow for the modeller's subjective view of the behaviour of different asset classes. The accumulation consists of the investment of 35 annual lump sums that were equal in constant purchasing power terms. Therefore, it takes the form of a defined-contribution pension scheme investment. The simulation exercise produces the probability distribution for the accumulation of the portfolio. All accumulations are measured in constant purchasing power terms.

With long-term stochastic investment modelling, mean–variance efficient frontier analysis can be performed in a similar way to that which is used

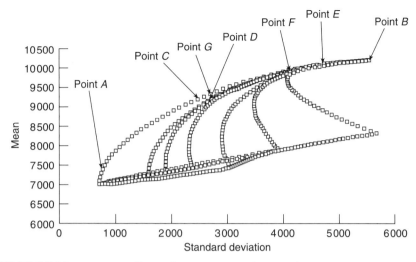

FIGURE 5.5 Mean–variance efficient frontier: 35 annual accumulations

in modern portfolio theory. However, more general approaches to risk can also be used. It is of interest to look at some of the points on the mean–variance efficient frontier, shown in Figure 5.5.

The expected utility-maximising portfolio, using a log utility function, is point E. This consists of 76% equities, 12% conventional government bonds and 12% index-linked government bonds. Adding a constraint that a real return of at least 1.5% per annum should be achieved over the 35-year period, with probability of at least 0.97 (so that the highest expected utility portfolio was chosen of those which met the constraint), takes us to a 68% equities and 32% index-linked government bonds portfolio (point F). Tightening the constraint, so that the probability of achieving a real return of 1.5% had to be at least 0.99 for the portfolio to be considered, would lead us to choose a portfolio consisting of 54% equities and 46% index-linked government bonds (point G).

Points A, B, C and D all indicate other portfolios with different characteristics. Both A and B are efficient. A is low return and low risk (90% index-linked government bonds and 10% equities). B is high return and higher risk (100% equities). Point C is an efficient portfolio with medium risk (46% equities and 54% index-linked government bonds). Point D is away from the efficient frontier. It has 40% equities, 30% index-linked government bonds and 30% conventional government bonds. Because all calculations are carried out in constant purchasing power terms, adding more conventional government bonds can increase risk without increasing the expected return.

Different investors have different liabilities and, therefore, do not always trade risk and return in the same way. Conventional government bonds can have a lower expected return than equities because they are risk free for some investors. For the investor who measures all returns in real terms, conventional government bonds may also be more risky.

We can also model the liabilities of an institution in conjunction with the assets. The correlation between the assets and liabilities will become evident, as the financial factors that affect asset values will often affect liability values in the same way. Thus, inflation, which may drive the stochastic investment model, may also affect the liabilities; this would be particularly evident for a pension scheme. A particular objective variable, such as the surplus of an institution when all liabilities are run off, needs to be identified and simulations carried out. Asset portfolios can then be chosen that lead to a satisfactory probability distribution for the objective variable.

It is easiest to think in terms of a closed defined-benefit pension fund or a life fund and consider the ultimate surplus after all liabilities have been met. As far as the assets are concerned, the aim would be to generate a probability distribution for A_T, where T is the point at which all liabilities have been repaid. A_T would depend on the split of assets between investment categories and on the reinvestment of investment income. The return from the assets would be simulated over the T time periods. If, for example, at the beginning of the year t the assets had value A_{t-1}, then the value at the end of the year t would be

$$A_{t-1}\left(x_1 R_1 + x_2 R_2 + \cdots + x_j R_j\right)$$

where x_i is the proportion invested in the ith asset type and R_i is the accumulation factor in respect of the ith asset type. R_i is a random variable which, in the Wilkie model, will depend mainly on previous values of the variable and inflation, together with a random component.

The complexity of the correlation relationships between successive values of R_i and the accumulation factor for different assets would make it impractical to derive a probability distribution for A_T other than by simulation. Simulation could be used to generate a probability distribution for the value of the liabilities L_T. In a pension scheme, in particular, L_T would also be dependent on inflation (and on real wage inflation, which could also be modelled).

Each simulation should model the assets and the liabilities together. A simulation that had higher than average inflation, for example, may lead to higher asset and liability values but may result in little change in the pattern of ultimate surplus, depending on how well matched the assets and liabilities were. The objective variable, surplus S_T, would be modelled as

$$S_T = A_T - L_T$$

Investment portfolios can then be chosen to meet particular decision criteria. For example, the user could derive mean–variance efficient frontiers. Portfolios could be selected which maximise the utility of ultimate surplus or which maximise expected surplus subject to a shortfall constraint. If a mean–variance efficient frontier were derived, then a trade-off function between mean and variance would be necessary to determine the desired asset allocation. This trade-off function would be different for different investors.

This closed-fund methodology can be adapted to model a continuing defined-benefit pension fund which has a particular contribution rate set by the actuary. The assets and liabilities can be modelled to find the distribution of ultimate surplus, given the contribution rate. This can be done for different investment portfolios. Various approaches to investment risk can then be used to assist in the portfolio selection decision; the appropriate trade-off between expected ultimate surplus and standard deviation of surplus can be found. Various shortfall constraints can also be used to assist in portfolio selection. For example, portfolios that lead to a particular probability of having insufficient assets, or which had a VaR above a certain level,[5] could be rejected; it should be borne in mind that, in this case, the practical reality of having insufficient assets would be to require an increase in the contribution rate.

A further probability constraint could be a requirement to appear solvent, on a particular deterministic basis, at regular interim valuations, with a particular probability. The application of international accounting standards, for example, will require schemes to show pension fund surpluses and deficits, determined on a particular basis, in company accounts. Companies may wish to avoid deficits of a particular level calculated on this basis. Stochastic modelling can be used to determine the investment strategy that, for a given contribution rate, will lead to a sufficiently low probability of failing a short-term funding test.

It should be pointed out that where there are two control variables (long-term surplus and short-term funding position) there may be a conflict between the investment strategies implied by the different control variables.[6] It is up to the sponsor of the fund to resolve this conflict, bearing in mind the priorities of the fund. Haberman et al. (2003a,b) provide an example of using stochastic asset/liability modelling for a continuing defined-benefit pension scheme, employing a range of one-sided risk measures as well as efficient frontier-type analysis to resolve such conflicts. There is further discussion of portfolio selection using simulation methods in Daykin et al. (1994).

---

[5]See Chapter 4 for a discussion of VaR.

[6]Although acceptance of this statement has implications for acceptance of the efficient markets hypothesis, this is an issue that should be considered explicitly.

There are theoretical and practical difficulties of applying stochastic models. First, financial variables are difficult to model. Complex models may well hide inadequacies that cannot easily be understood, whereas simpler models may not represent the behaviour of variables well. It is difficult to test long-term stochastic models against the data; indeed, even if they "pass" such tests the financial series may well begin to behave differently in the future. It is the nature of economic and financial variables that capturing their behaviour effectively for modelling purposes is difficult or, arguably, impossible. We have noted specific examples of this problem in our discussion of the Wilkie model. Sometimes, academics working in stochastic modelling distinguish between model risk and parameter risk. Parameter risk is the risk that parameters such as expected return are incorrectly specified or estimated. Indeed, parameters may be unstable. Model risk — the risk of choosing the wrong model — is particularly difficult to quantify when modelling financial variables. A consideration of some of the difficulties of modelling financial time series, from the philosophical point of view, can be found in Booth (1997), Huber and Verrall (1999) and Huber (1997b).

5.7 Conclusion

Mathematical models that assist portfolio selection and asset allocation decisions can be augmented by simulation techniques. These techniques allow the user to employ less-restrictive assumptions and, thus, solve a wider range of practical problems. The development of stochastic investment models is a basic requirement for long-term simulation work. These can be econometric or statistical, or a combination of the two. The Wilkie model, which is commonly used, is a statistical model incorporating some elements of prior economic theory. When simulation techniques are used, more appropriate ways of measuring investment risk can be incorporated in portfolio selection models. Stochastic asset liability modelling can then be used to select optimal portfolios to meet different decision criteria.

Appendix A. The Wilkie Model: A Case Study

A.1 Introduction

In this appendix, a review of the Wilkie model applied to the real-estate sector is presented. Parameters are calculated for different model forms and the results of the fitting process discussed in detail. The analysis highlights some of the advantages and disadvantages of the Wilkie model structure. It should be noted that real-estate performance data are sometimes regarded as having a different structure than that from other asset classes, in that they are more likely to exhibit cyclical behaviour. This may make the data more amenable to modelling using the Wilkie approach. It is also worth noting that the issue of whether there is "unit gain" from inflation to real-estate rents and capital values is a controversial one in the real-estate literature, and this issue needs to be examined in the fitting process. In the development of the real-estate model below, other relationships and parameters are added where it is felt appropriate. The issues discussed in this appendix are relevant to the other asset classes to which the Wilkie model is fitted, particularly to the equity markets. The material in this appendix is also presented in Booth and Marcato (2004) and reproduced here with permission of the Institute of Actuaries. The subject of the dynamics and performance characteristics of real-estate markets, as well as the relationship between real-estate returns and inflation, is discussed by Brown and Matysiak (2000). Yield models are discussed in Section A.2 to Section A.4; income or rent models are discussed in Section A.5.

A.2 Modelling Real-Estate Yields

A.2.1 The Wilkie Real-Estate Yield Model

The Wilkie model is described in detail in Wilkie (1995). It models real-estate yields and rents separately. The data used are the Jones Lang Wootton Indices from 1967 to 1994.

The yield model can be described as follows:

$$P(t) = \exp[\text{PI} \times I(t) + \text{PN}(t)]$$

or

$$\ln P(t) = \text{PI} \times I(t) + \text{PN}(t)$$

with

$$\text{PN}(t) = \text{AR1}[\ln(\text{PMU}), \text{PA}, \text{PSD}]$$

where $P(t)$ is the rental yield on the real estate index and $I(t)$ is the rate of inflation at time t; the other terms are as defined below and have an obvious relationship with equivalent terms in the equity and bond market models.

Thus the property yield is described as a function of a mean value (PMU), inflation, a first-order autoregressive parameter (PA) and a random term PSD $\times N(0,1)$. More explicitly, we could write the model as

$$\ln P(t) = \text{PI} \times I(t) + \ln(\text{PMU}) + \text{PA}[\ln P(t-1) - \ln(\text{PMU})] + \text{PSD} \times N(0,1)$$

$$(A.1)$$

There are various possible explanations for including the parameter representing the relationship between inflation and property yields. If there were no money illusion or lags in the economy, then a rise in the general price level would feed directly and immediately through into rents and capital values and inflation should not affect yields. However, if for example, inflation leads to long-run increases in rents but not in the rents receivable from a real-estate portfolio (because of the periodic nature of rent reviews), then capital values may increase to anticipate future increases in rents but current yields fall because rents respond with a lag. There are also aspects of money illusion that could lead to a relationship between rental yields and inflation. The same relationship is incorporated in the Wilkie U.K. equity model, described in Section 5.5.3. Nevertheless, in fitting the general model, Wilkie found the inflation parameter insignificant and, therefore, proposed a simple AR(1) model.

Wilkie found the following parameter values for that AR(1) model:

PMU = 7.41%
PA = 0.9115
PSD = 0.1177

A.3 Further Development of a Stochastic Real-Estate Yield Model

Prior theory would suggest that a relationship between real-estate yields and equity and bond yields might exist. It would normally be thought that an increase (decrease) in bond yields would be associated with an increase (decrease) in real-estate yields. There are a number of reasons for supposing that such a relationship might exist. The most obvious reason is that equity, bond and real-estate values can be determined in a discounted income model. If there were an increase in risk-free rates of return with all other variables remaining the same, then we would expect to see an increase in yields in all markets. The problem of valuation smoothing,[7] however, may complicate any such relationship. Indeed, if there were cyclical patterns in bond and equity yields, then valuation smoothing may produce a negative relationship between bond and/or equity yields and real-estate yields. There is also a prior case for testing the relationship between short-term interest rates and real-estate yields. It is

---

[7]Real-estate data are accumulated from surveyors' valuations of properties. It is generally believed that a degree of what is termed "valuation smoothing" comes into the performance data as a result of this. This feature of real-estate data weakens the relationship with returns from other asset classes. As a result, analysts often adjust real-estate performance data so that it should reflect more closely the underlying transactions. There is a considered analysis of this problem in Booth and Marcato (2004) and the references cited therein.

feasible that increases in short-term interest rates, signaling a tightening of monetary policy, may take effect through investment and credit channels, adversely affecting real-estate capital values (e.g., see Pepper (1994)); we might, therefore, expect a positive relationship between short-term interest rates and real-estate yields.

Therefore, we generalise Wilkie's model and add the following parameters:

$D(t)$ is the dividend yield from equities
$C(t)$ is the redemption/running yield from consols
$T(t)$ is the return from Treasury bills (as a proxy for short-term interest rates)

The generalised model therefore becomes

$$\ln P(t) = \text{PI} \times I(t) + \text{PD} \ln D(t) + \text{PC} \times \ln C(t) + \ln(\text{PMU})$$
$$+ \text{PA}[\ln P(t-1) - \ln(\text{PMU})] + \text{PT} \times T(t) + \text{PSD} \times N(0,1) \qquad \text{(A.2)}$$

There is no clear economic rationale for expressing the model in log form, rather than using untransformed variables, although the former does give better diagnostic statistics. It can also be said that there is no clear economic case for not using log variables, so it seems reasonable to be guided by the data in this case. After fitting the model using this basic structure, further autoregressive relationships were then investigated. In the next section, a generalised form of the Wilkie model using different forms of real-estate data is estimated.

A.4   Results of Fitting the Stochastic Model Using the Investment Property Databank Indices

The same approach as that used by Wilkie is used here. The yield data series from the Investment Property Databank (IPD) annual index from 1971 to 2000 (inclusive) is used to estimate the models; this is a different data set from that used by Wilkie. We found no model forms with any relationship between yields and inflation. Thus, the next step is to focus on the relationship between real-estate yields, bond yields, equity yields and Treasury bill returns, together with the autoregressive parameters. The fact that inflation did not have a significant impact on real-estate yields is not of great concern. A prior case can be made for including inflation in such models (see Section 5.5); however, the case is not an overwhelming one.

Autoregressive parameters of order one are significant, as are bond yield, equity yield and Treasury bill parameters. The coefficient on the bond yield parameter (PT) is negative. This should not be of great concern. Whilst prior theory might indicate some relationship between bond yields and real-estate yields (through the risk-free rate of return), it should be noted that most studies find a very low, zero or negative correlation between U.K. bond returns and U.K. real-estate returns from valuation-based indices: the correlation between returns is higher when de-smoothed real-estate returns are used (see Booth and Matysiak (2004)).[8] Nevertheless, whilst these findings from modelling real-estate yields do not contradict other studies, the result is counterintuitive, and further research into the various relationships between securities markets and the real-estate market would be worthwhile. The coefficient on equity yields (PD) is positive; there is generally a positive correlation between U.K. equity returns and real-estate returns even where the latter are computed from valuation-based indices, so this result is not surprising. From the class of AR(1) models, one of the following models for real-estate yields, as computed from valuation-based indices, could be used.

Simple Model 1

$$\ln P(t) = \ln(\text{PMU}) + \text{PA}[\ln P(t-1) - \ln(\text{PMU})] + \text{PSD} \times N(0,1) \qquad \text{(A.3)}$$

where

PMU $= 6.1\%$ (s.e. $= 1.23$)
PA $= 0.83$ (s.e. $= 0.10$)
PSD $= 0.10$
$\bar{R}^2 = 0.72$

The Lagrange multiplier (LM) statistic was used to test for serial correlation in the residuals for all the models. In this case the LM statistic was 1.17, which indicates that there is no evidence of serial correlation in the residuals. The results are very close to those found by Wilkie and the model structure is identical, although without a parameter here to reflect the impact of inflation on yields. The fact that different data providers and different data sets have been used for both investigations gives some confidence in the results.

Simple Model 2—In simple model 2, a second-order autoregressive parameter was included. The following results were found:

PMU $= 6.1\%$ (1.23)
PA $= 1.1$ (s.e. $= 0.18$)
PA(2) $= -0.30$ (s.e. $= 0.18$)
PSD $= 0.10$
$\bar{R}^2 = 0.72$

---

[8]De-smoothing data involves removing the impact on data of valuation smoothing.

where PA(2) is the second-order autoregressive parameter. Again, there was no evidence of serial correlation in the residuals.

Complex model 1—In the complex model, parameters for the bond yield, equity yield and Treasury bill returns, but not for inflation, were included (i.e., Equation A.2). The parameter values are then

PMU $=6.1\%$ (1.23)
PA $=0.90$ (s.e. $=0.08$)
PD $=0.35$ (s.e. $=0.10$)
PC $=-0.33$ (s.e. $=0.09$)
PT $=0.11$ (s.e. $=0.06$)
PSD $=0.08$
$\bar{R}^2 = 0.82$

There is some evidence of serial correlation in the residuals, although the hypothesis of "no serial correlation" would not be rejected at the 95% level. It should be noted that the sign on the bond yield parameter PC is negative, the implications of which are discussed above.

Complex Model 2—Significantly better diagnostic statistics are achieved by including a second order auto-regressive parameter. When this is included, all the parameters that are significant in complex model 1 remain significant. The results when a second order autoregressive term is included in the complex model are:

PMU $=6.1\%$ (s.e. $=1.23$)
PA $=1.19$ (s.e. $=0.15$)
PA(2) $=-0.31$ (s.e. $=0.14$)
PD $=0.40$ (s.e. $=0.10$)
PC $=-0.40$ (s.e. $=0.10$)
PT $=0.14$ (s.e. $=0.06$)
PSD $=0.07$
$\bar{R}^2 = 0.83$

There is no evidence of serial correlation in the residuals.
For practical simulation purposes, the simple models are clearly significantly easier to use (particularly if real estate is the only category being modelled). However, it will miss any relationships between equity, bond and real-estate yields. A positive autoregressive parameter of order one, as has been used above, can be explained by the process of valuation smoothing; in theory, the same could be said for positive autoregressive parameters with greater lags. However, there is extremely strong evidence of *negative* second-order serial correlation. Furthermore, including a second-order autoregressive parameter improved the

diagnostic statistics of the models. There is no obvious economic or financial reason for negative second-order serial correlation, either in the literature on efficient markets or in the literature on the influence of valuation smoothing on real-estate indices. However, the evidence for negative second-order serial correlation is so strong that its inclusion should be considered in long-term forecasting models. It should be added that there is also strong evidence of second-order serial correlation in the capital return indices: it is not just a function of the yield data. It is possible that the combination of both valuation smoothing and some form of cycle in the yield could create complex autoregressive structures with alternating positive and negative signs. It is worth noting that the second-order serial correlation does not exist in all U.K. commercial real-estate data sets over all time periods. But, this is an important "uninvestigated feature" of real-estate data in the U.K. real-estate finance literature.

In Booth and Marcato (2004) there is further analysis involving fitting the model to longer data series and also to property company data. These forms of investigation are important because they enable the modeller to attain a greater understanding of the long-term properties of the data and also any structural instability. It is worth noting that, when real-estate data going back to 1921 are used, the second-order autoregressive parameter is insignificant. It is possible that this parameter appears to be significant in the shorter data series because of particular features of the returns during the shorter period.

A.5   The Wilkie Real-Estate Income Model

A.5.1   Form of the Wilkie Rent Model

The Wilkie real estate income model is based on the Wilkie equity income model. Adapting his notation, Wilkie's model is as follows:

$$R(t) = R(t-1) \exp[\text{RW} \times \text{RM}(t) \times I(t) + \text{RMU} + \text{RSD} \times N(0,1)] \quad \text{(A.4)}$$

where $R(t)$ is the level of rents in year t, $\text{RM}(t)$ is a term (defined in more detail below) that captures the impact of past values of inflation on current rental growth, $I(t)$ is current inflation, RMU is the mean level of real rental growth, and RSD is the standard deviation of the residuals. RW is a parameter indicating the influence of past inflation on rents and RX the impact of current inflation on rents. Simplifying, this becomes

$$\ln[R(t)/R(t-1)] = \text{RW} \times \text{RM}(t) + \text{RX} \times I(t) + \text{RMU} + \text{RSD} \times N(0,1) \quad \text{(A.5)}$$

This provides the equation for the force of rental growth (i.e., the annualised rental growth expressed in continuous time). The force of rental growth, therefore, depends on the impact of earlier inflation and current inflation (both also expressed in continuous time), a mean growth term and a random term. Key issues for modelling rents are whether and how quickly inflation feeds through into rents. Because the model is being developed for use in asset/liability modelling and, therefore, relates to a group of standing properties, the rental index used is the rent that would be expected to be received from a portfolio, not the rents that would be received from properties that have just been reviewed. Therefore, passing and not market rents are modelled.

A.5.2   Unit Gain from Inflation to Rents

The empirical evidence on the relationship between increases in the general price level and commercial property returns is mixed. Prior theory would suggest that a rise in the general price level would lead to a rise in rents. Almost certainly there would be leads and lags, depending on the transmission mechanism of monetary policy and inflation through the economic system and also depending on rent review periods. As is noted earlier in this chapter, one of the problems of long-term actuarial modelling is how to make judgements about whether to include parameters for which there is an economic logic even if the data set from which parameters are estimated does not provide evidence of an empirical link. This problem is exacerbated by the fact that modelling has to be undertaken over a long time horizon, even where the data period used for estimation of the model is short.

In modelling investment returns, we should make the distinction between expected and unexpected inflation. Some assets (e.g., long-term bonds) should reflect expected but not unexpected inflation in returns. Index-linked bonds, on the other hand, should reflect not just expected inflation in current prices, but also, in their long-run returns, should reflect unexpected inflation (as the cash flows increase due to increases in the price level). The situation is more complex for real investments (such as real estate and equities). One might expect these investments to reflect both expected and unexpected inflation in their returns; however, their returns also depend on a number of other factors in the real economy. For example, the transmission mechanism of inflation may lead to costs to businesses that impair real performance. As has been mentioned the rent review structure of real estate leases may also lead to a lag before unexpected inflation is reflected in rents.

Brown and Matysiak (2000) review the evidence for whether expected and unexpected inflation are reflected in real-estate returns across a number of countries. There is considerable evidence that both are reflected to some extent, and some evidence that both are fully reflected. It seems reasonable, as an a priori assumption, to assume unit gain from inflation to real-estate rents, but with a lag. There are institutional factors (the rent review structure) that would prevent unexpected inflation being reflected in rental growth immediately, even in an economy with no friction and no money illusion.

At first, the assumption is made that there would be unit gain from the general price level to the level of rents, but where the increase in rents occurs with a lag structure. Wilkie also makes this assumption. The significance of the parameters is tested and alternatives investigated where necessary. The model is estimated assuming unit gain and then with the parameters RW and RX unconstrained. A result that $RW + RX > 1$, as Wilkie found, would not seem consistent with prior theory is it would indicate that a 1% rise in the price level would lead to more than a 1% rise in rents in the long term. However, a result that $RW + RX < 1$, or that the parameters are not significant, might be of interest, as it would confirm earlier work that the empirical relationship between inflation and rents is, at best, ambiguous.

In terms of the notation, the inflation feed-through mechanism in the Wilkie structure can be explained as follows. The parameter on current inflation (RX) and the parameter on lagged inflation (RW) sum to one (if we wish to require long-run unit gain). The influence of lagged inflation (through RW) on real-estate rents is assumed to arise through the following process:

$$RM(t) = RD \times I(t) + (1 - RD)RM(t - 1)$$

i.e., the lagged inflation feed-through for year t is equal to a parameter (RD) times inflation in year t plus $(1 - RD)$ times the lagged inflation input in the previous year. The lagged inflation input in the previous year depends on the lagged inflation input the year before, and so on, so that all previous values of inflation are included with declining weights. For example, if $RD = 0.2$, then $(1 - RD) = 0.8$. As $RM(t - 1)$ would have been equal to $0.2I(t - 1) + 0.8RM(t - 2)$, so $RM(t) = 0.2I(t) + [0.8 \times 0.2I(t - 1) + 0.8 \times 0.8RM(t - 1)]$. Continuing this process, the parameters on all previous values of inflation can be found and the weights on current and all previous values of inflation from this aspect of the process will sum to unity. This is then multiplied by the parameter RW, which distributes the influence of inflation between current inflation and past value of inflation. The influence of current inflation is $RW \times RD + RX$. So, if $RX = 0.4$, then the influence of a 1% increase in the price level on rents in the same year would be $100(0.6 \times 0.2 + 0.4)\%$. The total influence of inflation is $RX + RW$. Wilkie estimates that $RD + RX$ is greater than unity, and so he constrains $RX = 0$.

Wilkie estimates the following parameter values:

$RW = 1$
$RD = 0.11$ (s.e. $= 0.07$)

RMU = 0.0006 (s.e. = 0.0152)
RSD = 0.0661 (s.e. = 0.009)

Thus, mean real rental growth is negligible and the influence of inflation is mainly through lagged inflation.

A.5.3 Estimation of the Rent Model

This form of the Wilkie model for real-estate rents was re-estimated using the IPD rental series from 1973 to 2000. It is notable that no significant coefficients could be found of the form RW or RX, despite a correlation coefficient of 0.64 between current inflation and current rents. In order to investigate the lag structure of inflation further, a slightly more generalised form of the model of the impact of lagged inflation was then used. However, to avoid over parameterisation, only 4 years lagged inflation were used. The model form was then as follows:

$$\ln[R(t)/R(t-1)] = RX(0) \times I(t) + RX(1) \times I(t-1) + RX(2) \times I(t-2) + RX(3)$$
$$\times I(t-3) + RX(4) \times I(t-4) + RMU + RSD \times N(0,1)$$

$$\text{(A.6)}$$

Only RX(0) is significant, although the other inflation parameters are all positive. As these forms were therefore unsatisfactory, the following model form was fitted:

$$\ln[R(t)/R(t-1)] = RX \times I(t) + RY \times RI(t-1) + RMU + RSD \times N(0,1)$$

$$\text{(A.7)}$$

where RI is the force of rental growth in period t. The rationale for this model is that the upward-only rent review structure should lead to the deferral of the impact of market rental growth on passing rents (see Chapter 2), regardless of whether this increase in rents arises from inflation or is a "real" change in rents. Thus increases in rents in any one period will be related to increases in previous periods. Long-run unit gain would not necessarily be expected in this model as the autoregressive parameter will capture, in part, the impact of past values of inflation on current rents, parameters for which have been excluded as insignificant. The RMU parameter represents long-term mean real rental growth. This is −0.8%, so that long-term rental growth is slightly less than the rate of inflation.

The parameters estimated for this model are:

RMU = − 0.0083 (s.e. = 0.06)
RX = 0.31 (s.e. = 0.13)
RY = 0.67 (s.e. = 0.14)
RSD = 2.94
$\bar{R}^2 = 0.66$

There is no evidence of serial correlation in the residuals, and other diagnostic statistics were satisfactory; so this would seem to be an appropriate model form for long-term actuarial modelling of real estate rents.

Thus, for the rental model, Booth and Marcato (2004) suggest that the force of rental growth should be modelled as a function of inflation, a first-order autoregressive component, a mean growth rate, and a random term. This is different in detail from the model estimated by Wilkie, but there are similar characteristics.

References

Adams, A. T., Booth, P. M., Bowie, D., and Freeth, D. S. (2003). *Investment Mathematics*. Wiley, UK.

Booth, P. M. (1995). The management of investment risk for defined contribution pension schemes. In: *Transactions of the XXV International Congress of Actuaries*, vol. 3.

Booth, P. M. (1997). Knowledge, wisdom or understanding? *The Actuary* 7(8), 24–25.

Booth, P. M. (2002). Real estate investment in an asset/liability modelling context. *Journal of Real Estate Portfolio Management* 8(3), 183–198.

Booth, P. M. and Marcato, G. (2004). The measurement and modelling of commercial real estate performance. *British Actuarial Journal* in press.

Booth, P. M. and Matysiak, G. A. (2004). How unsmoothing should affect pension plan asset allocation. *Journal of Property Investment and Finance*. 22(2), 147–161.

Brown, G. R. and Matysiak, G. A. (2000). *Real Estate Investment: A Capital Market Approach*. Financial Times Prentice Hall, Harlow, UK.

Chan, W. S. (2002). Stochastic investment modelling: a multiple time series approach. *British Actuarial Journal* 8(Pt. 3), 545–591.

Chun, G. H., Ciochetti, B. A., and Shilling, J. D. (2000). Pension plan real estate investment in an asset/liability framework. *Real Estate Economics* 28(3), 467–492.

Chatfield, C. (1996). *The Analysis of Time Series: An Introduction*. Chapman and Hall, London.

Daykin, C.D., Pentikainen, T., and Pesonen, M. (1994). *Practical Risk Theory for Actuaries*. Chapman and Hall, London.

Geoghegan, T. J., Clarkson, R. S., Feldman, K. S., Green, S. J., Kitts, A., Lavecky, J. P., Ross, F. J. M., Smith, W. J., and Toutoundsi, A. (1992). Report on the Wilkie stochastic investment model. *Journal of the Institute of Actuaries* 119(Pt. 2), 173–228.

Haberman, S., Day, C., Fogarty, D., Khorasanee, M. Z., Nash, N., Ngwira, B., Wright, I. D., and Yakoubov, Y. H. (2003a). A stochastic approach to risk management and decision making in defined benefit pension schemes. *British Actuarial Journal*. 9(3), 493–586.

Haberman, S., Khorasanee, M. Z., Ngwira, B., and Wright, I. D. (2003b). Risk measurement and management of defined benefit pension schemes. *IMA Journal of Management Mathematics*, 14(2), 111–128.

Haynes, A. T. and Kirton, R. J. (1952). The financial structure of a life office. *Transactions of Faculty of Actuaries* 21, 141–197.

Huber, P. P. (1995). A review of Wilkie's stochastic investment model. Actuarial Research Paper No. 70, City University, London.

Huber, P. P. (1997a). A review of Wilkie's stochastic investment model. *British Actuarial Journal* 3(Pt. 1), 181–210.

Huber, P. P. (1997b). A study of the fundamentals of actuarial economic models. Dissertation for the award of Ph.D., City University, London.

Huber, P. P. and Verrall, R. J. (1999). The need for theory in actuarial models. *British Actuarial Journal* 5(Pt. 2), 377–396.

Kitts, A. (1990). Comments on a model of retail price inflation. *Journal of the Institute of Actuaries* 117, 407–414.

Leibowitz, M. L., Kogelman, S., and Bader, L. N. (1994). Funding ratio return. *Journal of Portfolio Management* 21, 39–47.

Ludvik, P. M. (1993). The Wilkie model revisited. In: *Proceedings of the 2nd AFIR Colloquium*, vol. 2, Instituto Italiano Degli Attuari.

Markowitz, H. M. (1952). Portfolio selection. *Journal of Finance* 7, 77–91.

Markowitz, H. M. (1991). *Portfolio Selection*. Basil Blackwell, Oxford.

McAleer, M. and Oxley, L. (2002). *Contributions to Financial Econometrics*. Blackwell Publishing, Oxford, UK.

Mills, T. C. (1999). *The Econometric Modelling of Financial Time Series*. Cambridge University Press, Cambridge.

Pepper G. T. (1994). *Money, Credit and Asset Prices*. Macmillan, London.

Redington, F. M. (1952). Review of the principles of life office valuations. *Journal of the Institute of Actuaries* 78, 286–315.

Sharpe, W. F. (1964). Capital asset prices: a model of market equilibrium under conditions of risk. *Journal of Finance* 19, 425–442.

Sharpe, W. F. and Tinte, L. G. (1990). Liabilities — a new approach. *The Journal of Portfolio Management* (Winter), 5–10.

Sherris, M. (1992). Portfolio selection and matching: a synthesis. *Journal of the Institute of Actuaries* 119(Pt. 1), 87–106.

Smith, A. D. (1996). How actuaries can use financial economics. *British Actuarial Journal* 2(Pt. 5), 1057–1174.

Wilkie, A. D. (1985). Portfolio selection in the presence of fixed liabilities: a comment on the matching of assets to liabilities. *Journal of the Institute of Actuaries* 112(Pt. 2), 229–278.

Wilkie, A. D. (1986). A stochastic investment model for actuarial use. *Transactions of the Faculty of Actuaries* 39, 341–403.

Wilkie, A. D. (1995). More on a stochastic investment model for actuarial use. *British Actuarial Journal* 1(Pt. 5), 777–945.

Wise, A. J. (1984). The matching of assets to liabilities. *Journal of the Institute of Actuaries* 111(Pt. 3), 445–485.

Yakoubov, Y. H., Teeger, M. H., and Duval, D. B. (1999). The TY model. Paper presented to the Staple Inn Actuarial Society, Staple Inn Actuarial Society.

Yates, T. (2003). Monetary policy and the zero bound to nominal interest rates. *Bank of England Quarterly Bulletin* 43(2), 27–37.

Part II

Life Insurance

Chapter 6

Fundamental Features of Life Insurance

6.1 Introduction

Life insurance forms part of the long-term insurance business written by life insurance companies (synonymously referred to as **life offices**). Life insurance companies provide a vital financial service to individuals and to firms who wish to insure themselves against financial losses that might be incurred as a result of any of the following:

death
survival to a particular time or over a particular period
sickness or disability

Life offices issue contracts of insurance, contingent on any of the above events, which are long term or permanent (that is, the policyholder does not need to reapply for renewal of his/her contract after each year of cover, but is guaranteed renewal over the duration (*term*) of the contract, as agreed when the policy is issued). We cover the first two headings in this part of the book: sickness and disability contracts are covered in Part V.

Life insurance is an example of financial intermediation, as we described in Chapter 1. The policyholder's financial asset can be changed dramatically through life insurance—e.g., by converting a regular investment (or *premium*) into a term assurance benefit; see Section 6.2. Other contracts make a less material alteration to the policyholder's asset: for example, unit-linked policyholders accumulate a share in a specified pool of assets (Section 7.2) in much the same way as the policyholder could accumulate funds directly, without the intermediation of the insurance company.

In this part of the book we describe the operation and management of long-term insurance from an actuarial viewpoint. Unless stated otherwise, a U.K. or U.S. context is assumed, although many of the principles described are transferable to the operation of long-term insurance in other countries. U.K. examples are usually used where a numerical illustration is required.

In this chapter we give a brief description of the types of long-term insurance contract available, and the commercial, taxation and supervisory framework within which life offices have to operate. Chapter 7 and Chapter 8 describe the technical operation of long-term insurance, including the estimation of reserves and the recognition of profit; the impact of supervisory regulation is considered in Chapter 9.

Chapter 10 focuses on the nature and management of the various types of life insurance company risk. The activities undertaken to control each source of risk are described in this context. In Chapter 11 we describe the actuarial role in the risk management of life insurance companies.

6.2   Basic Contract Structures

The main types of life insurance contract are briefly described below. (For more detail see Diacon and Carter (1992), for U.K. contracts, and Black and Skipper (2000) or Atkinson and Dallas (2000) for U.S. contracts. The nature of an insurance contract is described in more detail in Section 12.3.)

6.2.1   Term, Term-Life or Temporary Assurance

This contract pays a lump-sum benefit upon the death of the insured person (the *life assured*) within a specified period. There are a number of variants of this contract available, particularly in the U.S. Normally, level premiums are paid for a level benefit during the fixed term of the policy. However, some contracts offer guaranteed renewal of the policy up to some limiting age, such as 65 or 70. Under this arrangement policyholders can effect a new term assurance at the expiry of their original contract, for the appropriate level of premium applicable to a new policyholder at that age, without producing further evidence of health. The term of the renewed contract would be the same as that of the contract just expired. Where the contract is 1-year renewable, then policyholders essentially pay an increasing premium each year for the level death benefit, the insurance expiring at the contractual limiting age for renewal.

6.2.2   Whole Life Assurance

A lump sum benefit is paid on the death of the life assured, whenever it should occur. This type of contract, in various forms, makes up the bulk of life insurance issued in the U.S., but forms a much smaller (though still significant) proportion of U.K. business. A particularly important form is the *universal life* contract, described in more detail in Chapter 7.

6.2.3 Endowment Assurance

A lump-sum benefit is paid on survival of the assured life up to the end of a specified period. In theory, the contract can be issued without any benefit being paid on earlier death (in which case it would be a **pure endowment assurance**). More often a death benefit would be paid: this could be a return of premiums (with or without interest), an amount related to the policy's value (similar to its discontinuance value — see Section 6.7.4), or a lump sum equal in size to the survival benefit sum assured. The importance of this type of contract in the life insurance market varies dramatically by country, being hardly ever issued nowadays in the U.S., whereas it remains popular in the U.K. and elsewhere in Europe, where it is used essentially as a savings vehicle (e.g., for saving towards retirement), or as a means to repaying a loan.

6.2.4 Annuity

An annuity is a contract that pays regular payments to the policy-holder (the *annuitant*), in return for an initial premium (often called a *consideration*). The payments continue for as long as the annuitant survives, although many annuities include minimum guaranteed periods during which payments will continue to be made, even if the annuitant has died beforehand. Annuities may be issued to more than one life, in which case payments would usually be payable until the second death, with the amount of payment reducing after the first death. Amounts of payment can be level, or they may be linked to some inflation or investment index.

Annuity contracts may begin from the moment the consideration is paid, in which case it is an *immediate annuity*, or it may commence at some agreed date after the consideration is received, in which case it is a *deferred annuity*. Proceeds from endowment contracts are often used to purchase annuities, so that the combination of the two contracts in sequence is similar in effect to a single deferred annuity policy. Nevertheless, the two types of arrangement can have very different consequences for policyholders and insurance companies alike, especially when guaranteed terms are involved. The important difference is that the deferred annuity contract will define any guaranteed benefit in terms of the regular income it promises, whilst the endowment will define any guarantees in terms of the lump sum promised on survival (which becomes available at the start of the income payment period).

A special case of the endowment-plus-annuity combination deserves special mention: this is where the conversion of the lump sum endowment proceeds to income at retirement are subject to guaranteed minimum terms, promised from the time the policy was sold. It is this kind of guarantee that led to the effective downfall of Equitable Life in the U.K. in 2000. (See also Section 6.7.4.8 and Section 10.5.6.)

The purpose of any deferred annuity type of arrangement is to provide a replacement source of income after a person has retired from their lifelong employment.

6.3   Life Insurance Products

Many of these policy types are often issued together as single contracts, producing combinations of benefits and other features which are desired to meet the public financial need. Contracts also differ according to their benefit and charging structures. These include:

Nonparticipating (or *nonprofit*) contracts, which may be:
   conventional
   organized as a linked accumulating fund
   organized as a nonlinked accumulating fund
Participating (*or with-profits*) contracts, which may be:
   conventional
   organized as a nonlinked accumulating fund

These are described in detail in Chapter 7 and Chapter 8. Each combined contract (or *product*) which is issued by a life office is defined by a particular set of terms and conditions (the *policy conditions*). Policies may also be issued either to individuals separately, or as a *group* arrangement (or *scheme*) where a single contract covers a group of people who are associated in some way — for example, the employees of a single employer or members of a commercial trade association. The products that are available at any particular time reflect the current demands for them by the public, which are determined by a combination of:

need
sales pressure
popular opinion and media pressure
legislation

Personal taxation legislation has particularly significant influences on the types of product that are available. For example, some products are sold with a view to mitigating personal taxation liabilities (such as may occur on the disposal of a person's estate), whilst other aspects of legislation affect the tax treatment of the products themselves. For example, the investment returns on the assets backing certain types of product might be exempted from taxation in the hands of the insurance company, whilst other products might not be so generously treated. In order to qualify for this favorable treatment, the products may have to satisfy certain design criteria; in this way tax legislation can directly influence the types of product available.

Competition between offices for business can lead to considerable variation in the products offered by different companies, even where ostensibly they are designed to meet similar needs. (Thorough discussions of the main financial needs and the products which can be used to meet them are to be found in Diacon and Carter (1992) and in Black and Skipper (2000).) For the ordinary member of the public, this means that there is a large and confusing array of products, many of which are complex and difficult to understand; therefore, it is generally necessary for these products to be sold through professional intermediaries. One of the key roles of these intermediaries is to advise the public as to which products (and, in the case of independent advisers, which companies' products) would be best suited to meet their financial needs.

6.3.1 Product Distribution

There are essentially four modes of distribution for life insurance products:

1. independent intermediaries (or brokers)
2. employed sales forces
3. tied agents
4. direct sales

6.3.1.1 Independent Intermediaries

These are generally firms from whom members of the public seek advice regarding their personal insurance arrangements. These arrangements may include other financial services, such as direct investments and other types of financial intermediation, as well as insurance and annuity products. Independent intermediaries act on behalf of their clients, to select both the insurance company and the insurance (or other) product which best meets their clients' needs in each case. In other words, their advice must be completely free from bias towards any product or product provider due to some other interest, such as remuneration for any completed contract.

Independent intermediaries are generally remunerated by the payment of commission from the insurance company, usually as a large one-off, or *initial*, commission paid at the completion of a sale, followed by regular much smaller amounts (the *renewal* commission) payable with any future premium installments. The initial commission can be very large, sometimes greater than a whole year's premium for the sale of a very long-term annual premium contract; renewal commission would usually be in the range of 0 to 5 percent of each premium. This mode of remuneration produces an obvious conflict of interest against true independence, and generally requires regulation to control it.

6.3.1.2 *Employed Sales Forces*

An employed sales force forms part of the staff of an insurance company, from which all their remuneration arises, whether as commission or salary. As employees they will sell to the public the products provided by their employer. They are therefore acting clearly as agents of the company, rather than of the client. Nevertheless, a professional approach to selling will usually be fostered, so as to enhance the company's long-term reputation and hence not endanger its future business prospects. Legislation may also exist to control selling practices in this situation.

6.3.1.3 *Tied Agents*

These include individuals or firms who, incidentally to their main business, sell insurance products to the public from one or a number of particular providers, remuneration being entirely by commission. Examples would be a bank or other financial institution that would advise its customers about their various insurance needs, and recommend appropriate products from the insurers to which they are tied. This may be taken one stage further, by the parent bank or institution setting up its own subsidiary life insurance company and tying itself to this, so that the profits generated can be retained within the whole organization. (Insurance sold this way by banks is sometimes referred to as **bancassurance**.) The distribution of the insurance business is much aided by the vast databases of (often loyal) customers that the banks have, giving bancassurance companies, in particular, a considerable marketing advantage over companies using other distribution channels.

6.3.1.4 *Direct Sales*

All modes of distribution that exclude a professional intermediary are referred to as *direct sales*. Methods include selling via newspaper or magazine advertisements (so-called *off-the-page* selling), sending advertisements directly to people's homes (often packaged into personalized letters), telephone sales, and sales channeled through the Internet or via interactive television. A disadvantage of direct selling is that it is harder to provide advice, often because the company can only gain a limited amount of information about the customer, and this may restrict the types of contract that can be sold by this means. Nevertheless, insurers are often keen to use this route where possible, as it can be a very cheap and efficient means of selling their products.

6.3.2   Who Initiates the Sale?

Traditionally, life insurance was always considered to be sold, rather than bought. This is because people tend to treat death and its consequences as taboo subjects and, therefore, avoid the issue of life insurance until it is too late to do anything about it. So in order to "sell" life insurance, the companies (or their intermediaries) have to go out to the customers and remind them of their (real) life insurance needs. But it is not true to suggest that the adage holds universally. For example, a significant proportion of business sold by independent intermediaries is client-initiated, and many types of direct sales require a significant initiative from the prospective applicant (for example, responding to advertisements or searching the Internet for information).

In any case, it must ultimately be true that customers will only buy (or at least retain) contracts of life insurance if they have a real need for the protection or financial service it provides them. We will talk about this more in Section 6.7.

6.4   Taxation of Long-Term Business

Life insurance companies are trading enterprises which, as with all such bodies, are subject to the appropriate taxation of their home nation. Taxation may be based on any number of measures of company earnings, including investment income, profits, and premium income. Where tax treatment differs between product types, as we described in Section 6.3, the company has to keep separate accounting records of the different funds in order to have its tax liabilities properly assessed. These funds will be referred to as *nontaxed* or *taxed* funds as appropriate.

6.5   Supervision and Regulation

There are two principal reasons for regulation of the life insurance industry (Black and Skipper, 2000). First, the long-term nature of many contracts means that the ultimate value of the product will depend upon the insurer still being in business at the date of claim, which could be many years into the future. Even if most customers could reasonably assess the current financial soundness of an insurer, which is itself unlikely, there would be no safeguard to protect a policyholder from a company that becomes insolvent before the contractual claim event occurs.

Second, customers have imperfect knowledge of the products available: many do not understand the benefits provided by and charges associated with the products being offered for sale, and in many cases are unable to assess a product's value for money.

Regulation has therefore arisen, for long-term (and short-term) insurance, in two main areas:

1. Solvency assessment and control, to ensure that claims will be paid.
2. Selling practices.

With the market for insurance becoming increasingly global, it is becoming more and more important that there are internationally agreed standards of insurance supervision. To this end, insurance supervisors across the world have come together to form the International Association of Insurance Supervisors (IAIS), in order to develop and promote standards of insurance supervision. The principles of the IAIS (IAIS, 2003) cover:

Insurance core principles (details of the structures and procedures that have to be in place for any insurance supervision system to work).
Supervision of international insurers with particular reference to their cross-border operations.
Conduct of insurance business (i.e., selling practices, as above).
Supervision of insurance activities on the Internet.
Capital adequacy and solvency (i.e., solvency assessment and control, as above).
Supervision of reinsurers.

The interested reader is referred to IAIS (2003) for more details.

6.5.1 Solvency Regulation

Insurers generally have to obtain some form of authorization or license in order to trade in long-term insurance. Continuation of authorization is then dependent on companies meeting certain obligations, which include the continuing demonstration of adequate solvency in accordance with prescribed methods of valuation of assets and liabilities (see Chapter 9). Failure to meet these obligations will result in intervention by the supervisor in the company's affairs. Meeting the continuing requirements of the regulations is, therefore, of paramount importance to every life office.

6.5.2 Regulation of Selling Practices

Regulation will typically include some or all of the following elements.

1. Requirement for intermediaries to be properly trained, including explicit licensing.

2. Restricting the activities of intermediaries. For example, regulation may prevent certain types of intermediary from selling policies of more than one insurance company.
3. Regulation of advertising, including the calculation of illustrated benefits.
4. Disclosure of information to the client, e.g., the status of the intermediary, amount of commission, projected surrender values (see Section 10.5.2).
5. Requirement that intermediaries should provide best advice for the customer on the basis of full information about the customer's needs and circumstances.

The aim of the legislation, again, is to protect the consumer, by ensuring that ethical practices are adopted by insurers and intermediaries when promoting their products for sale, and that, as far as possible, the consumer is properly advised about the nature of the product being sold.

6.5.3 Other Areas of Regulation

Other common areas of regulation include the following:

Underwriting — for example, restricting the extent to which certain evidence can be taken into account when making individual risk assessments, such as the results of genetic tests (see Section 10.4.1 for more about underwriting).

Policy terms and conditions — examples include the specification of minimum amounts of benefit payable on policy discontinuance (see Section 10.5.2 for more about discontinuance).

Premium rates — regulations may impose restrictions on the premium rates or charges that may be levied for any contract. This might involve the rates themselves (e.g., maximum and/or minimum rates), or particular elements of the premium basis (e.g., specify the rates of interest and mortality to be used but allow the company flexibility to choose "appropriate" expense assumptions).

6.5.4 The Influence of Professional and Ethical Practice

From a business point of view, professional and ethical practice is to do with gaining and maintaining consumer trust and confidence. From a moral point of view, professional and ethical practice should follow from an obligation to society to do what is right. Whilst one can argue as to which is the real motive, the reality is that insurance companies' activities are influenced by professional and ethical considerations.

One of the main features of a professional approach to life office management is a full consideration of the interests of the public, and of policyholders in particular. This can affect almost every decision that an insurance company might make: for example, premium rates, profit distribution, selling practices, reserving, underwriting, investment policy, discontinuance terms (the list goes on).

Not only should the company have the *will* to do what is right, it also needs the *competence* to carry it out. (It is not much consolation to policyholders of an insolvent insurance company for the company to say "Well, we didn't *intend* to become insolvent.")

In the same way as policyholders need to have confidence in their insurance companies, so the companies need to have confidence in their advisors and decision makers. Actuaries often play key roles in advising life insurance companies, and professional actuarial associations exist to foster appropriate practice amongst their membership. The professional bodies often publish guidelines to their members about how they should behave in a whole range of circumstances. In a sense, therefore, these professional guidelines (and those of other professionals involved in managing insurance companies) provide a tangible set of limits within which the insurance company would be bound to operate if it were to achieve the standards of professionalism to which it aspires.

Clearly, actuaries (and others) have to be competent in the advice that they give, but other factors (such as the development of a trusting relationship with the company) are important if the advice is to have effect. The individual must, of course, put aside his or her own personal interests when going about his or her own work.

6.5.5   Provision of Compensation

In some countries, should insurance companies fail despite the existing regulatory protection, some compensation for policyholders may be provided for. For example, in the U.K. a proportion of the fair value of policyholders' contracts will be met (if necessary) from a centrally administered fund which is supported by contributions from all the U.K. insurance companies.

In addition to specific insurance legislation, life insurance companies are also subject to corporate accounting legislation, which, in particular, usually requires the publication of annual accounts. The main purpose of this legislation is to enable shareholders and other providers of capital to assess the profitability of the business, so as to make judgements about the value of their holdings and of a company's prospects.

6.6   Life Office Ownership

Life offices are either *mutual* or *proprietary* (proprietary companies are called *stock* companies in the U.S.). A proprietary office is owned by its shareholders, who provide the office with capital and share in its profits through the payment of dividends. The shareholders are the ultimate controllers of a proprietary company, having the right to vote at general meetings and to elect the office's directors. Shareholders' liability is limited to the value of their individual holdings. Proprietary offices can increase capital either by retaining profits or by issuing new shares.

A mutual office has no shareholders, but is owned by certain of its policyholders, who pay a higher premium for this privilege. These policy-holders, therefore, usually have the right to vote at general meetings, although they rarely do so, and to share in all the profits of the office, usually by receiving additional benefits (called *bonuses* or *dividends*) under their policies. Policyholders' liability is, therefore, essentially limited to the premiums paid. Mutual offices can only increase their capital by retaining profits, or by purchasing certain kinds of financing reinsurance arrangements (see Section 10.5.3), although the sale of certain products can also increase capital (see Section 8.2.2, Section 9.3.2 and Section 10.5.3).

The types of contract described above for mutuals (so-called participating contracts — see Chapter 8) can also be sold by proprietaries. Whilst the policyholders would normally not have any ownership rights, they would be entitled to some share of the profit distribution. The allocation of profits between the two parties is an important conflict of interest for these companies to resolve.

6.7   Long-Term Insurance Risk Management

6.7.1   Personal Risk Management

The death of a person can cause financial hardship to other people. For example, the death of a man or woman who shares his or her income with other members of the family (partner and/or children) will leave the survivors of the family with less income on which to survive thereafter. A financial loss is therefore incurred by the other members of the family on the event of the death of this individual. The event (death) consti-tutes a *risk* which, without insurance, is fully borne by the dependants of the person who dies.

If the family has considerable savings, or replacement sources of income are available after the person's death, then the loss of this income on death might be bearable by the dependants; that is, the consequences of the event are not financially crippling. Hence, the family may consider retaining the risk; that is, making a (conscious or unconscious) decision not to take out insurance against this particular risk.

On the other hand, the event may have dire financial consequences for the survivors. They may be unable to meet their outgoings, e.g., due to the cost of a continuing mortgage or high level of rent, the costs of providing basic necessities such as food and clothing, or the cost of looking after the children while the surviving partner goes out to work. In this case, the family would need to take out insurance to cover their risk.

Even if the loss of income on death is not financially crippling, it could cause a significant deterioration in the standard of living of the survivors, and hence in their quality of life. Life insurance can, therefore, be seen as a means of protecting the standard of living of those who are dependent upon a person's continuing income (or wealth).

A major risk to the standard of living is due to retirement (or, rather, the risk of living beyond retirement age), whereupon normal income from employment will end. This, again, is a risk to both individual and dependants. Deferred annuities or similar arrangements can be used to provide replacement income.

A related "survival" risk is that which is protected by an annuity. Some individuals may have accumulated a significant amount of wealth by the time they retire. Their risk management decision is thus whether (and how much) of this to convert into an annuity. The insurance provided by an annuity is that it will keep being paid for as long as the person survives, whereas under the alternative of drawing the money from personal savings the money could run out before the person dies. (See also Chapter 21 for further discussion.)

Thus, personal risk management involves the decision to retain or to insure against each source of risk to which the individual is exposed. In this regard, nonlife (or general) insurance and sickness and disability insurance contingencies are equally important; see Part V.

6.7.2 Insurance Company Risk

The business of a life insurer is to manage the collective risks passed to it by its policyholders. At this point it is useful to define what is meant by risk from the point of view of a life insurance company: a future event will constitute a risk to the company if the event can have an adverse financial effect on the company, and where the occurrence of the event and/or the intensity of its financial effect are uncertain at the present time.

The primary risks that are managed by life insurance companies are those due to the uncertainties caused by the demographic determinants of life insurance policy claims; that is, due to the incidence of death and survival among its policyholders. These risks will be referred to collectively as the company's *insurance risk*.

6.7.3 Managing Insurance Risk: Risk Pooling

The following description of risk pooling applies equally to general insurance; see Section 12.1.

Let us assume that an insurer issues n identical insurance contracts at time t_0, which terminate at time t_1, to n independent risks (lives). Premium income of p_n will be received, assumed payable at time t_0. Shareholders will also provide capital of $u_0 = wp_n$; i.e., capital provision is assumed to be a proportion w of the premium income p_n. We will ignore investment income and expenses throughout. Now, the capital remaining at the expiry of the insurance contract, at time t_1, will be

$$U_1 = (1 + w)p_n - C_n$$

where C_n is the total amount of claims incurred over the period from the n policies.

The insurer can be described as technically solvent at time t_1 if it has nonnegative assets at that time, i.e., if $U_1 \geq 0$. At time t_0 we can, therefore, consider the office to have a certain *probability of insolvency* α:

$$\alpha = \Pr(U_1 < 0) = \Pr(C_n > (1 + w)p_n)$$

which, with w and p_n determined at outset, will depend entirely upon the distribution of C_n.

Now insolvency has dire consequences for both the policyholders and the shareholders of the company: all claims subsequent to insolvency will not be honored (that is, policyholders receive less than the contractual amounts under their policies); shareholders will suffer the total loss of their capital.

Furthermore, the insurance service provided by the company to the public will cease. Thus, it is in the best interests of all parties to keep the probability of insolvency to a minimum. However, there are other aspects of the insurance process of interest to each party.

1. The shareholders are keen to increase the returns on their capital. The return on capital can be identified as

$$R = \frac{p_n - C_n}{u_0} = \frac{1}{w}\left(1 - \frac{C_n}{p_n}\right)$$

with expected value

$$E(R) = \frac{p_n - E(C_n)}{u_0} = \frac{1}{w}\left(1 - \frac{E(C_n)}{p_n}\right)$$

2. The policyholders are keen to minimize the cost of their insurance. The average cost of the insurance, per unit of premium paid, can be defined as

$$L = \frac{p_n - C_n}{p_n} = 1 - \frac{C_n}{p_n}$$

with expected value

$$E(L) = 1 - \frac{E(C_n)}{p_n}$$

Hence, we can also see that

$$R = \frac{L}{w} \quad \text{and} \quad E(R) = \frac{E(L)}{w}$$

Now the probability of insolvency can be reduced by increasing the number of independent risks insured n. This is due to the "law of large numbers," which implies that

$$\lim_{n \to \infty} \left\{ \Pr \left[\left| \frac{C_n}{n} - \frac{E(C_n)}{n} \right| < \varepsilon \right] \right\} = 1$$

where ε is an arbitrarily small constant.

The average claim cost per policy will, therefore, tend towards the expected claim cost, as n increases. This is the *pooling of risk* principle, which underlies all insurance. Hence, we can envisage, for a given value of w and for a certain premium rate, a value of n which will produce an acceptably low probability of insolvency (Cummins, 1991). It should be noted, however, that where there is heterogeneity amongst the risks, the law of large numbers may not be entirely valid (Daykin et al., 1994).

Once α is reduced to an acceptable level, therefore, further increases in n can be translated into reductions in capital, premium rate, or both. Thus, we can reduce L by reducing the premium rate, $p_n/E(C_n)$, and we can increase R by reducing w by a greater proportion than L is reduced. Hence, with sufficient lives insured, the relative amount of capital provided can be small enough to produce an adequate rate of return to the shareholders, while the cost of insurance to the policyholders can also be kept to an acceptable level. The quantity wp_n would generally be referred to as the *risk-based capital*, i.e., the amount of capital that an insurer needs to hold to leave the company with a particular level of overall risk.

Example

An insurer issues 10 000 identical policies at time t_0 to 10 000 identical and statistically independent lives, having the following characteristics:

Sum assured payable on death before t_1: £10 000
Expected proportion dying before time t_1: 0.05
Single premium: £550

The shareholders provide initial capital of £2.5 million (45% of total premium income). Expenses and investment income will be ignored.

We assume that the number of claims is approximately normally distributed, with

$$\text{mean} = 10\,000 \times 0.05 = 500$$

$$\text{variance} = 10\,000 \times 0.05 \times 0.95 = 475$$

The total premium income plus capital (£000) $= 10 \times 550 + 2500 = 8000$, so insolvency will occur if the total claim amount exceeds £8 million, that is, if there are more than 800 claims. Hence we can calculate the probability of insolvency as

$$\alpha = \Pr\left(Z > \frac{800 - 500}{\sqrt{475}}\right) = \Pr(Z > 13.765)$$

where Z is a unit normal random variable. Hence α is infinitesimal. The expected cost of insurance is:

$$1 - \frac{5000}{5500} = 0.091$$

and the expected return on capital is:

$$\frac{0.091}{0.45} = 0.202$$

The actual cost of insurance, and the return on capital, will have the probability distribution shown in Table 6.1.

Hence, even with only a modest number of lives, all three requirements of the insurance process (that is, of solvency, return on capital, and cost of insurance) appear to be quite easily satisfied. The risk pooling process is further considered in Section 12.3 in relation to general insurance.

Table 6.1 Probability distribution of the return on capital and the cost of insurance for an insurer with 10 000 identical risks

Return on capital (%)	Cost of insurance (%)	Probability
<3	<1	0.02
3–11	1–5	0.14
11–20	5–9	0.34
20–29	9–13	0.34
29–37	13–17	0.14
>37	>17	0.02

6.7.4 The Influence of Modern Product Designs on Life Insurance Company Risk

Modern life insurance has many complexities in addition to the situation so far assumed. The principal additional factors are that contracts are long term with conditions fixed at outset, sums assured differ between policies, and there are different types of contract. Furthermore outcomes, such as solvency and returns, depend additionally upon the returns from investment, the expense outgo, and the effects of voluntary discontinuance of policies.

6.7.4.1 Long-Term Renewability

Since the Equitable Society was formed in the U.K. in 1762, life insurance contracts have been sold which guarantee continuation of cover year after year, at the same level of premium and benefit and with no further evidence of medical health (following the scientific principles laid down by James Dodson). The duration of this guarantee can be for as long as the whole of life (for example, as in a whole of life policy or an annuity). As the only opportunity for controlling the terms of the contract is at the very outset of the policy, prior to acceptance, the initial underwriting of the policy is of considerable importance in controlling the risk. Also, it becomes harder to predict claim costs the further into the future the prediction is made (often referred to as the increasing funnel of doubt), so that there is an increasing chance that the average claim experience will differ from that allowed for in the premium rates over the duration of a long-term policy. Hence, there is a greater risk of making an overall loss (and hence an increased risk of insolvency) with long-term rather than short-term contracts of this nature.

Long-term renewability causes additional complications. Consider a portfolio of policies under which fixed and level annual premiums are payable. While the average premium received will be the same each year,

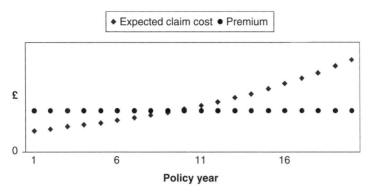

FIGURE 6.1 Annual expected claim cost and premium income under a long-term annual premium life assurance contract

the average annual claim cost under such policies will not. Contracts of assurance (for example, whole of life or temporary assurances) will have a gradually increasing average annual claim cost with increasing policy duration, as the policyholder's age and the probabilities of claiming rise accordingly. This is illustrated in Figure 6.1.

The excess of premium income over claim costs in the early years of the contract cannot be taken as profit to be distributed to shareholders. This excess income has to be retained by the office until such time as it needs to be drawn upon to pay the deficit of income that arises later on, when average claim outgo exceeds premium income. The amounts of such income retained at any time are called the *reserves*: they reflect the amount of assets that the office has to hold at that time in order that the claim costs that are expected to arise under the existing policies in the future can be paid (after allowing for future premium income).

The determination of appropriate levels of reserve for the office to retain is a matter of vital importance, and involves considerable difficulty. The difficulties arise from the unknown amount of liability represented by each contract; for example, it is not known in advance when (or even if, for some contracts) a claim will be made under a particular policy.

6.7.4.2 Distribution Risk

The uncertainty of the cost of claims before the policies have terminated means that it is only possible to assess the actual profit (or loss) earned on any portfolio once the contracts cease to be in force. Unless a life office ceases to sell policies in the future, it will never be in a position where all its contracts have terminated. Nevertheless, shareholders and policyholders reasonably expect profits to be distributed at regular intervals, usually annually. Assessment of profits, therefore, has to be made on the basis of the difference between the office's assets and the reserves which it has set aside in order to meet future liabilities. If profits are distributed on the basis of this assessment, and in the event the reserves allowed for in the assessment turn out to have been insufficient, then the office will have paid out too much to its shareholders and/or policyholders, and insolvency could occur as a result. Thus, the office's distribution policy runs a risk of *overdistribution*, which we could describe as *distribution risk*.

6.7.4.3   Capital Strain

The need to prevent overdistribution leads naturally to companies holding prudent (i.e., large) reserves. This requires capital, as the premiums received at the start of a policy will not necessarily be large enough to set up the reserves required at that time (see Chapter 9). Holding adequate reserves is usually a supervisory requirement (see Section 6.5.1), so that failure to have enough reserves can lead to supervisory intervention and/or subsequent failure of the business. The most likely single cause of this would be if the company's new business volumes were growing too rapidly. The initial capital input needed at the start of many policies is often referred to as *new business strain*.

6.7.4.4   Discontinuance Risk

Long-term contracts, under which more than one premium is to be paid, provide the opportunity for policyholders to discontinue their contracts; that is, to stop paying premiums at some point during the policy term and thereby cause the contract to terminate, or at least continue in a somewhat reduced form. Many policy types may have built up significant reserves at the time of any discontinuance, so that it is only fair that some compensation is paid to the discontinuing policyholder. The fair (or *equitable*) amount to be paid is, however, very difficult to determine, bearing in mind the unknown amount of the reserve itself. There is, therefore, a possible risk of overpayment on the discontinuance of any contract in this way — this constitutes what might be referred to as a *discontinuance risk*.

6.7.4.5   Variable Sums Assured

The presence of policies with very large amounts of benefit (*sums assured*) leads to significant increases in the variability of returns, and to the risk of insolvency (see Section 10.4). This leads to the need for *reinsurance*, in which the insurance is shared with another company, or with several other companies.

6.7.4.6 Different Contract Structures

The events that constitute a claim differ between contracts, as described in Section 6.2. Different benefit and charging structures can exist (such as in linked and nonlinked contracts, and with-profit and nonprofit policies; see Chapter 7 and Chapter 8). All these differences can lead to different risk profiles, often reflecting the very different levels of guarantees provided, and hence to the need for different approaches to their management.

6.7.4.7 Investment Risk

The delay between the receipt of premiums and the payment of claims means that the life office often holds the reserves for its policyholders over considerable periods of time. This money can be invested in order to obtain a return that can be used for the benefit of policyholders (by reducing the premium the policyholders have to pay or by increasing their benefits), and/or for the benefit of the shareholders by producing higher profits. However, this also leads to *investment risk*; e.g., where the return on the office's investments is lower than that assumed in the calculation of the office's premium rates. Reserves can also be discounted to allow for future investment returns, but again there is a risk of understating the reserves if the rate of discount assumed is too high. Investment returns are a major source of profit under many contracts and the office's investment policy (that is, the decisions the management take to hold, buy, or sell particular assets) is a major element of the office's overall management strategy.

6.7.4.8 Guaranteed Policyholder Options

Some contracts grant policyholders the right to additional benefits, to be taken at their choice, on terms that are guaranteed in advance. Examples include the following.

Lump survival benefits under endowment policies are often used to purchase annuities and hence provide retirement income. Some of these policies guarantee that a minimum rate of conversion will be applied at the conversion date if the company's normal conversion rates are less favorable at that time (we will refer to these as *guaranteed annuity options*).

Policies often provide rights to policyholders to take out new additional policies in the future, at the company's then standard rates of premium; this enables policyholders to take out life insurance in future (at normal prices) without having to provide the usual evidence of health (see also Section 6.2.1).

The cost of the first option is likely to be zero if there is no significant change to economic and investment conditions between the time the option is purchased and the time that it is exercised. On the other hand, a significant change in conditions over this period could lead to a very significant cost to the company. Because all such policies will be similarly affected, the cumulative cost of meeting these guarantees could become financially crippling to the company, if not properly managed. This is precisely what happened to the Equitable Life Assurance company in the U.K., in 2000: inadequate reserves were held towards the cost of these guarantees, allowing overdistribution of profits to occur (in pursuit of competing management aims to maintain high benefit levels under with-profit policies over the period). We discuss the management of these risks in Section 10.5.6.

The health option is likely to be more easily dealt with, as the expected additional cost of the death claims can be quite accurately predicted and the cost shared between all policyholders who choose to have the option under their contract.

6.7.4.9 Expense Risk

In carrying out its various activities a life office incurs certain expenses of management. The following list summarizes these activities.

Policy administration:
receiving premiums
investing surplus income (asset management)
administering reinsurance
paying claims
providing information
selling policies
    giving advice
    giving information
    providing policy documents
    underwriting
    marketing

Actuarial management:
setting premium rates
product design
risk assessment
experience monitoring
setting reserves
profit distribution
investment policy
reinsurance policy

Supervisory obligations:
preparing reports and accounts
taxation

Expenses are paid for by charges to the policyholder, either by an addition to the premium rate or by a reduction from the benefits. This leads to an *expense risk*: that the total charges made for expenses will not be as large as the actual expenses incurred. Life insurance companies can be particularly vulnerable to expense risk due to the fact that level premiums may be receivable over a large number of years, while inflation over the period could be materially higher than was assumed in the premium basis. The company would be most seriously affected were this to be accompanied by falling new business sales, so that the so-called *expense overrun* could not be covered from surplus generated by new business issued at more appropriate prices.

The company's risks can, therefore, be divided into three main sources: insurance risk, investment risk, and business risk. The latter refers to the risks that arise from the company's business activities, which therefore include the expense, discontinuance, and distribution risks described above. The nature of these risks and their management are described in more detail in Chapter 10.

6.7.5 Managing the office

A life office is not a passive body. It has a management whose job it is to look after the needs of the people who have interests in the business; that is, to manage the risks to best effect for its policyholders and shareholders. The decisions and actions of the management (which we might call management policy, or management strategy) can make significant differences to the profile of risks and returns faced by the office.

This, therefore, leads to a further risk to the insurance company: that which is due to management incompetence or, even, due to unprofessional conduct (see Section 6.5.4). A natural tendency to satisfy shareholders' short-term demands at the risk of policyholders' long-term security is probably the greatest threat in this regard.

Before we can consider the full nature of the risks incurred by life offices and the way in which they can be managed, we need to describe in more detail the financial nature of long-term insurance business, which arises from the types of contract that the companies issue and through the supervisory regulation of the business. This is described in Chapter 7 to Chapter 9.

References

Atkinson, D. B. and Dallas, J. W. (2000). *Life Insurance Products and Finance*. The Society of Actuaries.

Black, K. and Skipper, H. D. (2000). *Life Insurance*, 13th edition. Prentice-Hall.

Cummins, J. D. (1991). Statistical and financial models of insurance pricing and the insurance firm. *Journal of Risk and Insurance* 58, 261–302.

Daykin, C. D., Pentikainen, T., and Pesonen, M. (1994). *Practical Risk Theory for Actuaries*. Chapman & Hall.

Diacon, S. R. and Carter, R. L. (1992). *Success in Insurance*. John Murray.

IAIS. (2003). *IAIS Principles, Standards and Guidance*. www.iaisweb.org (under link "principles and standards").

Chapter 7

Nonparticipating Life Insurance

As referred to in Section 6.3, nonparticipating (or nonprofit) life insurance falls into three main categories:

1. conventional
2. organized as a linked accumulating fund
3. organized as a nonlinked accumulating fund

It should be mentioned that not all of these types of contract are necessarily found together. For example, while they are all sold in the U.S., the nonlinked accumulating fund is only found in the U.K. in a participating form (see Chapter 8).

7.1 Conventional Nonparticipating Contracts

7.1.1 Operation of the Fund

In a traditional *nonprofit* (or *nonparticipating*) contract the contractual terms are agreed between company and policyholder at the outset of the policy. These terms cannot be varied thereafter. These contracts are the simplest of all in their operation: the policyholder agrees to pay a premium, which may be single or annual, in return for a guaranteed benefit which will be payable on the occurrence of the insured event(s) within a specified period of time. This "specified period of time" may be for the whole of life, or the insurance may be limited to a particular number of years (the *term* of the contract).

We have already described in Section 6.2 how term-assurance contracts can incorporate guaranteed renewability options, particularly prevalent in the U.S. Another U.S. product feature is that referred to as *re-entry* (see Black and Skipper (2000) or Atkinson and Dallas (2000)). Such products allow policyholders, at specified durations of their policies, to subject themselves to a repeat underwriting test (see Section 10.4.1). Should they pass the test then they can pay a reduced premium for a number of years thereafter, reflecting the fact that they have become *select* lives once more (see Bowers et al. (1986)). Should they fail the test at the re-entry date, thereafter they become subject to a higher premium rate than other policyholders of the same age.

A second feature of U.S. nonparticipating contracts is where *indeterminate* premiums are allowed. Here, the premium to be charged can be reassessed at any stage by the company, in order to reflect changes to expected future conditions (particularly relating to future mortality, expenses, and investment returns). Premiums can be increased or reduced when they are reviewed, but are subject to a contractual maximum level. Premiums are usually lower than the maximum. This feature essentially represents a form of risk sharing between the company and its policy-holders.

The operation of a nonprofit policy within a life office is shown in Figure 7.1. Policyholders pay premiums that accumulate in a fund of assets. The expenses of the office are paid out of the fund, and the policyholders' benefits are paid as claims occur. The profits being earned by the fund are distributed to the providers of capital periodically from the fund.

Example

We will use the following simple hypothetical example to illustrate how a nonprofit fund operates. Note: a basic knowledge of actuarial

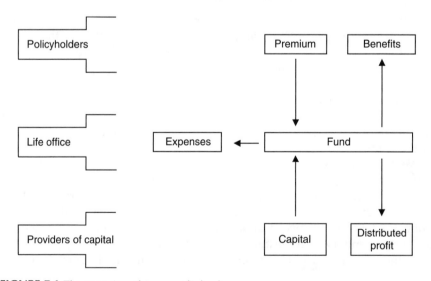

FIGURE 7.1 The operation of a nonprofit fund

mathematics is assumed, but see Bowers et al. (1986) or Atkinson and Dallas (2000).

Contract
Nonprofit endowment
Term 3 years
Sum assured £1000

Premium basis
Expected mortality rate at age x (q_x)   5% each year (for all x)
Expected return on assets   10% per annum
Expected expenses
   year 1   50% of premium*
   year 2   5% of premium*
   year 3   6% of premium*
Profit margin/contingency loading: 7.5% of premium*

*premium excluding profit margin (= "best estimate" premium)

    All expenses are assumed to occur at the beginning of each respective policy year, while all claims are assumed to be paid at the end of each year. (Note that, in reality, mortality rates will rise with increasing age over most age ranges, and most actuarial modeling exercises would include such an assumption in practice. Mortality assumptions are most often linked to some standard mortality table deemed to be appropriate for the country and class of life concerned. However, it can be shown that, at least over a short period of years, an assumption of level mortality makes very little difference to the projected outcomes of the fund. Level mortality will, therefore, be assumed for the numerical examples shown here, for reasons of simplicity.)

    The premium excluding profit margin (the "best estimate" premium) is given by

$$\frac{1000(q_x v + p_x q_{x+1} v^2 + {}_2 p_x v^3)}{0.5 + 0.95 v p_x + 0.94 v^2 {}_2 p_x} = £377.3$$

where x refers to the entry age of the policyholder

$$p_x = 1 - q_x$$

and

$$v = 1/1.1$$

Hence the total premium is £377.3 × 1.075 = £405.6.

There are two main reasons why we should include a margin, such as the 7.5% loading used here. First, given that the above premium basis assumptions are "best estimate," there is broadly a 50% chance that actual experience will turn out to be less favorable than assumed. There is, therefore, a significant risk of the company losing money on the business if no margin is included. Second, the lack of any expected profit will make the business completely unattractive to providers of capital, and capital would undoubtedly be necessary to meet supervisory reserving requirements (see Chapter 9). Therefore, the margin is required: (a) to reduce the financial impact on the company of adverse experience; (b) to provide a suitable return to the providers of capital (i.e., to service the capital).

We will assume that 10 000 identical policies are issued at time $t = 0$, and that (for a start) the actual experience is exactly the same as that expected according to the premium basis. Voluntary discontinuance will be ignored.

The accumulation of the fund is described by the following formula:

$$F_{t+1} = F_t + P_t + I_t - E_t - C_t$$

where F_t are the assets (fund) accumulated by the beginning of year t, and P_t, I_t, E_t, and C_t are respectively the total premium income, investment income (including all capital gains,) expenses, and claim outgo incurred during year t.

The accumulation of our example nonprofit fund is shown in Table 7.1. Items of outgo are shown in parentheses.

It is informative to consider each individual policy's contribution to the accumulation shown in Table 7.1. In this example, because the expected contribution from each policy is the same, this simply involves dividing all the items shown in Table 7.1 by the number of policies in force at the beginning of each year. The result is shown in Table 7.2.

The individual policy's share of the fund at any point in time is known as its *asset share*. It has no value at time 3 because there are no policies left to share in the remaining assets, although there are assets left in the fund (the profit of £983 000). It should also be noted that part of the asset

Table 7.1 Operation of nonprofits fund by policy year (£000)

Policy year	Premium	Expenses	Interest	Claims	Fund
1	4056	(1887)	217	(500)	1886
2	3853	(179)	556	(475)	5641
3	3661	(204)	910	(9025)	983

Table 7.2 Individual in-force contribution to nonprofits fund (£)

Policy year	Premium	Expenses	Interest	Claims	Asset share
1	405.6	(188.7)	21.7	(50)	198.5
2	405.6	(18.9)	58.5	(50)	625.0
3	405.6	(22.6)	100.8	(1000)	–

shares shown in Table 7.2 at times 1 and 2 represents the accumulations of the profit margins in the premium up to that point, and should perhaps be excluded from the policyholders' asset shares. This raises the question of allocating assets between the policyholders and the shareholders (or other capital providers), the former representing the *policyholders' reserves*, the latter representing the accumulated, or retained, profit.

7.1.2 Reserving and Profit Recognition: Realistic Basis

The concept of a reserve was introduced in Section 6.7.4. The reserve is the amount of assets that the life office needs to hold at a particular point in time (the *valuation date*) such that, allowing for future premium income and for the anticipated returns from investment, the liabilities of the life office will be met as they fall due. The liabilities of the life office can be defined as:

1. The obligations that the company has under the contracts currently in force to pay benefits according to the terms of those contracts.
2. The expenses that the life office will incur in the future in order to administer the contracts of insurance it holds.
3. The obligation to provide a return to those with a stake in the company for the capital that they have provided; that is, to the with-profits policyholders and/or shareholders, as appropriate.
4. Taxation.

It is not meaningful to try and place a value on the future distribution of profit to shareholders, as by definition this distribution depends on the future profits made and is essentially a balancing item.

The reserve at time t can therefore be calculated as

$$\text{Reserve} = (\text{Value of future benefits}) + (\text{Value of future expenses})$$
$$- (\text{Value of future premiums}) \tag{7.1}$$

The tax liability is allowed for in this expression by taking values of future interest and expenses net of the appropriate rate(s) of tax. To be consistent with the treatment of future shareholder's profits (which have been excluded), future tax payable on shareholders' profits would also be ignored in this calculation.

The values of the items in Equation 7.1 are all unknown as at the valuation date, as they are dependent upon the outcome of a number of uncertain events, such as future mortality and investment returns. The life office must, therefore, make an estimate of the required amount of reserve, by making appropriate assumptions about future mortality, investment return, expenses, and so on. What constitutes an appropriate basis for the assumptions depends very much upon the purpose for which the reserve estimates are required. If the reserve is required in order to identify the amount of profit that the company could safely distribute without incurring too great a risk of insolvency, then the assumptions would naturally represent a cautious view of future events. This would, in effect, produce a level of reserve which would almost certainly (that is, with a high probability) be sufficient to meet the existing liabilities as they fall due, and in turn this would help prevent overdistribution of profit. This is essentially the rationale behind the supervisory valuation; see Chapter 9.

On the other hand, should the reserve be required in order to calculate a best, or most realistic, estimate of the profit retained in the fund, then the assumptions made for the reserve calculation would be the best estimates of the outcomes for the variables concerned. In effect this would produce a reserve estimate that would have an approximately 50:50 chance of being higher or lower than the true level of reserve, as the estimated value would be (approximately) the expected value of the true level of reserve (which can be considered to be a random variable). Such an approach would be appropriate for the production of internal management accounts, whose purpose would be to keep the management informed of the realistic profitability of the office.

7.1.2.1 *True and Fair Accounting*

A realistic approach would also be desired to meet the corporate accounting regulations, in order to show a "true and fair" view of the company's financial condition.

At the time of writing, a project is being progressed by the International Accounting Standards Board to develop, ultimately, an International Accounting Standard for Insurance (see Hairs et al. (2002)). The key principle proposed is that both assets and liabilities should be reported in insurance company balance sheets at their "fair value," where fair value is defined as "the amount for which an asset could be exchanged or a liability settled between knowledgeable, willing parties in an arm's length transaction" (Hairs et al., 2002, p. 204).

For assets that are frequently and easily traded on the stock markets (in a "deep liquid" market (Hairs et al., 2002, p. 204)), then fair value would be equivalent to market value. For other assets, and nearly all liabilities, no deep liquid market exists, and hence some proxy should be taken. Valuing infrequently traded assets requires the kinds of direct approach described in Chapter 1. For liabilities, the idea is to value them using assumptions that a typical market trader (in liabilities!) would make when determining the amount he or she would charge to take then on. These assumptions would not have a "best estimate" character in the way that we have described above, but would include some margin related to the uncertainties inherent in the experience to which the assumptions relate. In essence, therefore, finding a fair value of a liability involves trying to ascertain its (theoretical) market value; in practice, this will be assessed by valuing using assumptions that can be described as "best estimate plus market value margins" (Hairs et al., 2002, p. 217).

7.1.2.2 Numerical Examples

Let us now introduce realistic reserves into our numerical example. In Case 1 we will assume a purely best estimate for the reserve calculation, while in Case 2 we will introduce a possible market value margin in the reserving basis, so as to mimic the fair valuation basis.

Case 1

For Case 1 we will assume the best estimate basis to be the mortality, interest, and expense assumptions used for calculating the premiums, but without any additional loading for future adverse contingencies. The profit that is recognized by adopting this reserving basis is shown in Table 7.3.

Note that this example suggests that the company only has to hold reserves at the end of each year. In practice, of course, companies have to hold appropriate reserves for their policies in force at all times. Hence, the appropriate reserve would need to be set up immediately after the first premium had been received, and would continue to be held right up to the maturity date of the policy (at which point, of course, the reserve would become exactly equal to the maturity benefit, in this example). The profit

Table 7.3 Profit recognition on a best estimate basis (Case 1): accumulation to the end of each year

Time t	Total assets (£000)	Reserves per policy (£)	Number in force	Total reserves (£000)	Retained profit (£000)
1	1886	113.1	9500	1075	812
2	5641	526.1	9025	4748	893
3	983	0	0	0	983

Table 7.4 Profit recognition on a best estimate basis (Case 1): annual profit per policy in force at start of each year

Policy year	Total retained profit at start of year (£000)	Interest on existing profit (£000)	Total profit at end of year (£000)	Profit earned during the year[a] (£000)	Profit per policy (£000)
1	0	0	812	812	81.2
2	812	81.2	893	0	0
3	893	89.3	983	0	0

[a]Excluding interest on existing profit at start of year.

that is earned each year, with respect to each policy in force at the beginning of each year, is shown in Table 7.4.

Hence, according to this reserving basis, we make all our profit in the first year and no (new) profit thereafter. To put this another way, the total profit retained by the company when we reserve on our best estimate basis is the current present value of all the profits we expect to make from our policy during its existence.

The retained profit at time t (when reserving on best estimate assumptions) can also be considered as

Retained profit = (Accumulated value of past profit to time t)

+ (Present value of future expected profits from time t)

Given that the profit (in this example) comes solely from the profit margin of £28.30 in each premium, you can check the above retained profit figures by accumulating (and discounting) past (and future) payments of profit margin as at time t.

Case 2

The Case 2 reserving basis includes a 10% margin in all assumptions. Therefore, the basis is

Expected mortality rate at age x, (q_x) 5.5% each year (for all x)
Expected return on assets   9% per annum
Expected expenses
        year 1   55% of premium*
        year 2   5.5% of premium*
        year 3   6.6% of premium*

*premium excluding profit margin

The profit recognized by reserving on this basis is shown in Table 7.5. The profit earned each year is then worked out in Table 7.6.

Table 7.5 Profit recognition on a fair value basis (Case 2): accumulation to the end of each year

Time t	Total assets (£000)	Reserves per policy (£)	Number in force	Total reserves (£000)	Retained profit (£000)
1	1886	131.0	9500	1244	642
2	5641	536.7	9025	4844	797
3	983	0	0	0	983

Table 7.6 Profit recognition on a fair value basis (Case 2): annual profit per policy in force at start of each year

Policy year	Total retained profit at start of year (£000)	Interest on existing profit (£000)	Total profit at end of year (£000)	Profit earned during the year[a] (£000)	Profit per policy (£000)
1	0	0	642	642	64.2
2	642	64.2	797	91	9.6
3	797	79.7	983	105	11.7

[a]Excluding interest on existing profit at start of year.

It can be seen that making the reserving basis more conservative (Case 2 compared with Case 1) has reduced the profit in the first year and increased profits in subsequent years; that is, increasing reserves delays the emergence of profit.

The annual profit per policy is normally derived from a standard cash-flow calculation. Hence, the cash flow in year t is

$$cf_t = pm_t + i_t - e_t - c_t$$

where pm_t, i_t, e_t, and c_t are respectively the amounts of premium, investment income, expenses, and claims incurred during year t.

The profit during year t is

$$pro_t = cf_t + ir_t - \Delta r_t$$

where ir_t is the interest during the year on reserves held at start of year t and

$$\Delta r_t = \text{Increase in reserves during year } t$$
$$= (\text{Reserves held at end of year for survivors})$$
$$- (\text{Reserves held at start of year})$$

These items are all expressed as average amounts per policy in force at the start of each year, which will be denoted by lower case symbols,

Table 7.7 Annual average cash flows per policy in force at start of each year

Policy year	Premium	Expenses	Interest	Claims	Cash flow
1	405.6	(188.7)	21.7	(50)	188.6
2	405.6	(18.9)	38.7	(50)	375.4
3	405.6	(22.6)	38.3	(1000)	(578.7)

Table 7.8 Profit recognition on a realistic basis: annual cash flows and profit (£)

Policy year	Cash flow	Reserve	Increase in reserve	Interest on reserve	Profit
Case 1 (best estimate basis)					
1	188.6	113.1	107.4	0	81.2
2	375.4	526.1	386.7	11.3	0
3	(578.7)	0	(526.1)	52.6	0
Case 2 (fair value basis)					
1	188.6	131.0	124.4	0	64.2
2	375.4	536.7	378.9	13.1	9.6
3	(578.7)	0	(536.7)	53.7	11.7

to separate them from total amounts received, for which upper case symbols will be used. The figures for our examples are shown in Table 7.7 and Table 7.8.

Finally, the *profit signature* can be calculated either by taking the annual in force profit from Table 7.8 (last columns), and multiplying it by

$$\frac{\text{Number of lives at start of year } t}{\text{Number of lives at start of year 1}}$$

or by taking the total profit earned during the year (penultimate column of Table 7.6) and dividing it by the number of policies in force at the start of year 1. The profit signature is therefore the average profit earned in each year for a single policy taken out at time 0. The profit signatures for this contract (given all the various assumptions we have made) are shown in Table 7.9 for both Cases 1 and 2. The key elements of interest so far from the above are shown in Table 7.10.

Table 7.9 Profit signatures of a single nonprofits policy

Policy year	Case 1	Case 2
1	81.2	64.2
2	0	9.1
3	0	10.5

Table 7.10 Summary: nonprofit fund accumulation to end of each year (£000)

Policy year	Case 1			Case 2			
	Profit for year	Total profit	Total reserves	Profit for year	Total profit	Total reserves	Total assets
1	812	812	1075	642	642	1244	1886
2	0	893	4748	91	797	4844	5641
3	0	983	0	105	983	0	983

Note that the profit for the year is essentially the sum of the profit signatures for all policies issued at time 0. The total reserves at the end of the year can also be considered as the assets (notionally) allocated to this tranche of policyholders, while the retained profit represents the assets (notionally) allocated to the shareholders (and/or with-profits policyholders).

A number of important observations should be made.

1. The accumulated profit recognized at any point during the term is notional and not certain. For example, the experience of the fund could be such that negative profits (that is, losses) are incurred in years 2 and 3, to the extent that the accumulated profit at the end of the term would be less than at time 1. This possibility is more likely under the (less prudent) Case 1. Should the first year's recognized profit have been distributed to the shareholders at time 1, then the fund would have had a shortfall by the end of the term despite being apparently solvent at time 1. This is an example of overdistribution, which is discussed further under the heading of "distribution risk" in Chapter 10.

2. A reserve is a point estimate of a quantity that is determined by the outcome of a number of random variables, including future mortality and investment returns. Hence, where, as in Case 1, a reserve is a best estimate of the assets required to cover the fund's liabilities, based as it is on the most realistic assessment of future conditions, then the true (but unknown) value of the reserve will be higher or lower than the estimated value with approximately equal probability.

3. The amount of profit recognized at any point during the policy term is entirely dependent upon the reserving basis. Making a reserving basis more conservative (as in Case 2 compared with Case 1) delays the emergence of profit over the policy term.

7.1.3 Reserving on the Premium Basis

If we calculate reserves for a particular contract on exactly the same basis as the premiums were calculated, then the profit (or loss) which emerges will reflect the extent to which actual experience has been better (or worse) than that initially expected when the contracts were originally sold. This is a useful standard, because it indicates immediately the ongoing adequacy of the premium rates that are in force.

The main disadvantage with this approach is that contracts that are still in force may have been issued so long ago that the premium basis assumptions are very different from current "best estimates" of the future. Hence, the reserves may not be considered "realistic"; that is, they cannot really be considered as the amount of assets needed to be held in order to meet the expected cost of the future liabilities under the contracts currently in force.

As soon as we vary the reserving (or *valuation*) basis, however, we immediately anticipate all of the profit (or loss) that we would expect to achieve in the future on account of the difference between the valuation basis and the premium basis. In other words, part of the future profit (or loss) in relation to the premium basis is capitalized at the valuation date, as a "one-off" profit (or loss), due to the change in valuation basis. Thereafter, until the valuation basis is changed again, the emerging profit reflects the difference between the actual experience and that assumed in the valuation basis. Hence, if we are attempting to show profits realistically and fairly, then any such one-off *valuation profit* should be separately shown in the accounts and an explanation given as to its nature.

The above leads to an alternative standard of reserving, i.e., to calculate reserves on the basis of the *current* premium basis for each class of business. Hence, the profit shown each year represents the extent to which actual earnings differ from the assumptions implicit in the policies currently being sold. The profit emerging is then of direct relevance to considerations of the ongoing adequacy of the office's premium rates.

There is a second, quite distinct method of accounting for profit, particularly popular for internal management accounts, which is called the *embedded value* approach. We will return to this later in Section 9.3.1.

7.1.4 Cash Values

As we described in Section 6.7.4, policyholders who voluntarily discontinue their life insurance policies are usually entitled to some compensation. Should a company no longer have any liability under a contract due to its discontinuance, then the reserve that it would need to hold would immediately fall to zero. Therefore, considerations of equity would suggest that the reserve so released should become immediately available to compensate the discontinuing policyholder.

Of course, it would be inappropriate to pay the current level of *supervisory* reserve as a surrender value, as this reserve includes a considerable amount of capital that shareholders (or with profit policyholders) have provided to maintain supervisory solvency. The process of determining surrender values is not simple, and we will return to this subject again in Section 10.5.2. For the present, we note that such cash-equivalent values (or just cash values) for life insurance contracts form an important part of the overall profile of benefits that they provide, and reserving calculations are often important in helping to determine appropriate values.

7.2 Linked Accumulating Nonparticipating Contracts

In the U.K. this kind of business is usually referred to as *unit-linked* or just *linked* business. In the U.S. it is known as *variable*, or *dynamic*, life insurance. Hereafter, we will refer to it by its U.K. name.

There are significant differences between unit-linked and the conventional nonprofit products described in the previous section. Most importantly, premiums less deductions accumulate in an explicitly designated fund of assets. The policyholder's fair share of this asset pool at any time represents the policy's current cash value which, net of any surrender penalty, would be payable to the policyholder on discontinuance. Because of the direct link between the assets and the policyholder's cash value (hereafter referred to as the *unit fund*), the value of this unit fund will, in theory, rise or fall exactly in parallel with increases or decreases in the values of the underlying assets; hence the origin of the description "variable" life insurance. This implies that the policyholders share in all of the aggregate investment experience of the asset pool, and that none of the investment risk is borne by the life office itself.

Another major feature of this type of contract is that the various components of the product are *unbundled*. This refers to the fact that, each year, specific amounts of money are deducted from the policyholder's assets to pay for the company's expenses, and to pay for the insurance benefit provided under the contract. This allows for a very important aspect of product design to be included: to give policyholders choice over how much premium they wish to pay each year, and how much benefit they wish to have, under their contracts. This premium and benefit flexibility has led to the name *universal life* for this product, as it is commonly known in the U.S. These products are all whole-life policies, but in the U.K. they are often marketed as endowment contracts. This is done on the basis that the pure endowment benefit at maturity is simply the value of the policyholder's unit fund at that time, although (unlike a conventional pure endowment) the insurance company does not incur any survival risk.

It should be noted that some unit-linked contracts in the U.S. (and historically in the U.K.) have fixed benefits and premiums. These are simply referred to as *variable whole life insurance* in the U.S., and are not unbundled contracts in the strict sense, though the accumulation of the unit fund follows the same principles as for universal life.

It should be noted that universal life products are also available without a direct link between the policy's unit fund and the value of the designated asset pool. These *nonlinked* versions are described in Section 7.3 and Section 8.1.4.

The rest of this section describes how unit-linked contracts specifically operate in the U.K. Unit-linked contracts elsewhere will be organized along similar principles but will differ in detail. A description of the U.S. approach can be found in Black and Skipper (2000).

7.2.1 Operation of the Fund

The operation of a portfolio of unit-linked policies within a life fund is shown in Figure 7.2. A proportion of the premium paid by the policyholder will be allocated to the unit fund. This allocation rate may be less than, equal to, or more than 100%. Any unallocated premium, plus explicit charges that may be deducted from the unit fund, accumulate in a nonunit fund. The "nonunit fund" is simply the company's cash, and is often colloquially referred to in the U.K. as the "Sterling" fund. The nonunit fund is essentially analogous to the nonprofit fund under conventional nonprofit contracts. The costs of insurance benefits and of expenses are deducted from this fund, the balance accumulating as profit (or loss) for the shareholders and/or the with-profits policyholders. Distributions of profit would be made from time to time from the nonunit fund.

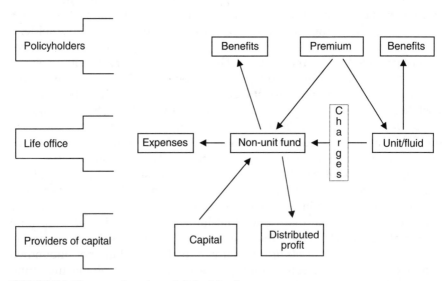

FIGURE 7.2 The operation of a unit-linked fund

The unit fund effectively belongs to the policyholders, who have contractual rights to their shares of this fund under any claim. In total, the unit fund can be considered as the sum of the contractual shares under all of the policies currently in force.

The policyholders' contractual shares are determined by a system of *unit allocation*. Each policyholder is allocated a number of units of the fund whenever a premium is paid. Each unit is of equal value at any particular date, but the value (or price) of a unit will vary over time, reflecting the value of the underlying assets at any particular time. The units will also have two prices on any particular date: the price at which they can be purchased (the *offer price*) and the price at which they can be sold (the *bid price*). A policyholder will, therefore, be allocated the following number of units on the payment of a premium:

$$\frac{\text{Amount of premium allocated}}{\text{Offer price of one unit}}$$

which will be added to the total of his or her existing units. At any future time (say t years after the start of the policy), the policyholder's contractual share will equal

$$U_t = \text{(Total number of units allocated to policyholder)}$$
$$\times \text{(Bid price of one unit)}$$

The difference between the bid and offer prices of a unit reflects the costs of purchasing and selling assets on behalf of the money received or paid out. Hence, whilst £P of the policyholder's money may be allocated to the purchase of units, part of that £P is lost by the office in the transaction costs involved in buying assets in the fund with that money. Further transaction costs are incurred in selling assets to pay benefits. Hence, the bid price of units is always lower than the offer price, with a difference typically of the order of 5% (that is, with the bid price being 95% of the offer price); this difference is referred to as the *bid–offer spread*.

Assuming the unit fund is expanding (in which case assets are being purchased), the unit price itself is determined from

$$\text{Offer price} = \frac{\text{Total purchase price of assets underlying the unit fund}}{\text{Total number of units in issue}}$$

where the purchase price includes the buying costs that would be involved (i.e., with the buying costs added to the market values of the assets). The bid price is then 95% of the offer price (assuming a 5% bid–offer spread). This basis of unit pricing is called the appropriation pricing basis.

In order to avoid cash flow problems, companies usually have more units in issue than are currently allocated to policyholders. So, whilst the general long-term trend might be for the unit fund to expand, short periods may occur during which more units are cashed in by policyholders than are bought by them. Rather than sell assets, the company can just hold the spare units itself (i.e., increasing the number of unallocated units in issue). This will mean that the unit prices can remain on an appropriation (purchasing) pricing basis during short periods of net de-allocation of units.

If the unit fund is contracting (in which case assets are being sold), then the unit price would be determined from

$$\text{Bid price} = \frac{\text{Total selling price of assets underlying the unit fund}}{\text{Total number of units in issue}}$$

where the selling price is net of the selling costs that would be incurred (i.e., with the selling costs deducted from the market values of the assets). The offer price is then the bid price divided by 0.95. This pricing basis is referred to as the expropriation pricing basis.

In a similar manner to the above, if there is a net allocation of units during a general trend for de-allocation, then the company can allocate some of its spare units to policyholders and thereby avoid the physical need to buy assets, thereby keeping the units on an expropriation basis.

The company will only change from one basis to another when it is expected that the new basis will remain appropriate for a significant period of time. Changing from one basis to another may cause a discontinuity in the returns passed on to policyholders, where the per-unit value of the sum of the market buying and selling costs differ from the bid–offer spread.

The policyholder's contractual share of the unit fund is more briefly referred to as the *face value* (or just *value*) of the policyholder's units. Sometimes, the face value of the units may be equal to the contractual benefit payable (for example, on maturity of a unit-linked endowment policy), or it may affect the value of other types of claim. For example, the benefit on death may be the greater of the value of the units and some guaranteed amount, or it may equal the face value plus some additional amount (see below for more details). On voluntary discontinuance the policyholder would probably be entitled to an amount which cannot exceed, but may be less than, the face value, depending on the design of the contract. We discuss this further in Section 7.2.5.

Modern unit-linked contracts rarely provide guaranteed benefits on discontinuance or maturity, which means that the policyholders' benefits are determined entirely by the performance of the underlying portfolio of assets to which the policies are linked. This has led to some very high policy returns on some occasions; it has also led to low returns or even losses on others. In other words, under the contract between policyholder and life office, the policyholder takes on the whole of the investment risk under his or her policy, in return for the potential to participate fully in the investment profits of the fund.

Linking policyholders' contractual values to a specific asset portfolio lends itself to some attractive product design features. The office can offer a wide variety of portfolios to which policyholders could choose their policies to be linked; for example, corporate stocks (equities; which could also be differentiated by industrial sector and degree of risk), fixed interest securities (bonds), index-linked securities, property (real estate), and so on. Many funds are linked to the performance of overseas investments. The choice of fund-link will reflect the policyholder's individual preference for risk and return, as described in Section 3.3.1. The discussions of investment strategies for unitized funds in Section 3.6.1 and Section 3.8.1 are also relevant here. Policyholders should benefit from the fact that their investment experience is pooled with that of many other policyholders, through the much greater diversity of assets which can be held on behalf of a large number of people than by single individuals. Policyholders are also expected to benefit from the management of the various portfolios by the life office's investment managers. Most offices also offer mixed funds, each containing a variety of asset types in proportions managed by the office, as a means of increasing diversification further and thereby reducing policyholders' individual risks of incurring large losses.

In modern unit-linked contracts, the charges, including any unallocated premium, are closely related to the expected costs that are likely to be incurred by the insurance company on behalf of each policy. For example, if the office expects to incur expenses per policy of £20 each year, then the policyholder may be charged £20 each year (plus any margin for error and/or profit). As expenses tend to rise with inflation, then policyholders' charges may even allow for this; for example, they may be increased annually in line with some inflation index, or they might be reviewed periodically by the office (for example, every 5 years). This is an example of the operation of an indeterminate premium structure, as introduced in Section 7.1.1.

As well as administration costs, including commission, policyholders will be charged for the cost of the policy's insurance benefits and, of course, to produce a contribution to profit. Again, the charge for the benefits is closely linked to the expected cost, and this, in turn, depends upon the terms of the contract. For example, a contract may specify a minimum death benefit of £S say, such that the benefit paid on the policyholder's death would be the value of the policyholder's units (see above), or £S, whichever was the higher. If £U_t is the value of the units under a policy after t years, then the expected cost of the death benefit for year t would be

$$\begin{cases} q(S - U_t) & S > U_t \\ 0 & S \leq U_t \end{cases}$$

where q is the relevant probability of dying over the year. Hence, again the charge made to the policyholder each year would be closely linked to this.

Any charge made from the unit fund will be effected by the cancellation of the number of units that, at bid price, will be sufficient to pay for the charge. Hence, a charge of £X will be made by deducting the following number of units from the unit fund:

$$\frac{X}{\text{Bid price of one unit}}$$

A further element of the charging structure is the charge made to cover the costs of the investment management of the policyholders' unit fund (see above). This is usually made by deducting a fixed proportion of the value of the unit fund each year, and a value in the range 0.25% to 1.00% per annum would be typical. This means that policyholders are charged for this service (that is, for the benefit of expert investment management) in proportion to the size of their investment, which would seem to be fair.

The amounts of each charge, or at least the method by which they are calculated and the circumstances under which they might be varied, are all set out in the policy conditions and form part of the contract between the life office and policyholder. Unit-linked policies, therefore, have a much more complex array of policy conditions than a conventional nonprofit contract, but the policyholder can see clearly where his or her money is being spent — much more clearly, in fact, than under a conventional policy. On the other hand, the policyholder loses at least some of the level of guarantee provided by a conventional nonprofit contract.

Unit-linked policies afford the life office a great deal of flexibility in terms of product design, which of course includes the specification of the charging structure. One of the key elements in such a product design is in charging appropriately for the very high expenses incurred in setting up the contract, as opposed to the much more modest annual expenses incurred thereafter. Initial expenses, including commission, can amount to 100% or more of one year's premium under a long-duration contract. There are essentially three main ways in which modern policies charge for initial expenses:

using a nil allocation period
using the annual fund management charges
using level annual charges

The use of a nil allocation period is the simplest to understand and also the most transparent (that is, the easiest to see) from the policyholder's viewpoint. This simply represents a period of time, usually expressed in months, during which *all* of the premium paid by the policyholder is paid into the nonunit fund. This period can vary between just a few months to well over a year in length, depending on the contract (and life office) concerned. Judicious choice of the nil allocation period by the life office can also allow a fair proportion of profit to emerge at very early durations.

The second method involves using a much higher rate of annual fund management charge than would otherwise be necessary to cover the ongoing costs. The additional level of charge would be designed to be sufficient to repay the initial expenses over the duration of the contract. Sometimes, the higher fund management charge would only be applicable for a limited period; alternatively the higher fund management charge would only apply to part of the unit fund (for example, to the units purchased by the first 2 years' premiums). The level of fund management charge needed to cover the initial expenses would obviously differ according to which structure was used.

The third method involves the deduction of an additional fixed-level charge each year, usually by making a level reduction to the proportion of premium allocated to the unit fund each year. The period over which the charge is deducted may be shorter than the policy term. As with the second method, the fixed annual charge would be designed to compensate the office for its initial expenses and to provide an appropriate margin for profit.

Example

We will again use a simple example, to illustrate the operation of a unit-linked fund.

Contract
Unit-linked endowment

Term	5 years
Death benefit	Higher of £10 000 or bid-value of units
Survival benefit	Bid-value of units
Annual premium	£300

Charges

Bid-offer spread	5%

For mortality — the following percentages of sum at risk, deducted at the start of each year:

Year 1	0.00277
Year 2	0.00310
Year 3	0.00347

Year 4	0.00391
Year 5	0.00440

For expenses — the following deducted at the start of each year:

Year 1	£220
Year 2	£22.9
Year 3	£23.8
Year 4	£24.8
Year 5	£25.7

Fund management charge: 0.5% of the bid value of the units, deducted at the end of each year.

Experience assumptions

Mortality (q-type rates):

Year 1	0.00251
Year 2	0.00281
Year 3	0.00315
Year 4	0.00354
Year 5	0.00398

Maintenance expenses — the following assumed incurred at the start of each year:

Year 1	£200
Year 2	£20.8
Year 3	£21.6
Year 4	£22.5
Year 5	£23.4

Investment management expenses: 0.35% of the bid-value of units incurred at the end of each year.

Rates of return on assets:

Unit fund:	10% p.a.
Nonunit fund:	6% p.a.

Note that the different rates of investment return assumed for the unit and nonunit funds reflect the probable different natures of the underlying investments in each case.

Claims are assumed to be payable at the end of each year.

The operation of the funds is in accordance with the following formulae (amounts per policy):

Unit fund

$$u_t = [u_{t-1} + a\, p_t(1-b) - z_t - m_t](1 + i_t^u) - fmc_t$$

where u_t is the bid-value of unit fund at end of year t, a and b are the allocation rate (100%) and bid–offer spread respectively, p_t, z_t, m_t, and fmc_t are the amounts of premium, expense charge, mortality charge and fund management charge respectively paid in year t, and i_t^u is the rate of investment return earned by the unit fund during year t. Items are self-explanatory except for fmc_t and m_t, which are:

$$fmc_t = c[u_{t-1} + a\, p_t(1 - b) - z_t - m_t](1 + i_t^u)$$

where c is the fund management charging rate (0.5%);

$$m_t \begin{cases} 0 & \text{for } [u_{t-1} + a\, p_t(1 - b) - z_t](1 - c) > S \\ q_t\{S - [u_{t-1} + a\, p_t(1 - b) - z_t](1 - c)\} & \text{otherwise} \end{cases}$$

where q_t is the mortality charging rate for year t (as defined in the charging structure) and S is the minimum sum assured under the policy (£10 000). The amount deducted from the sum assured in this calculation is the value of the unit fund at the end of the year assuming no growth in value (that is, assuming zero investment return). This will generally produce a conservative estimate of the sum at risk at the end of the year (which is the amount in the curly brackets). Alternative assumptions could, of course, be used.

Nonunit cash flow

The nonunit cash flow is calculated as:

$$cf_t = \{p_t[1 - a(1 - b)] + z_t - e_t^z + m_t\}(1 + i_t^n) + fmc_t - e_t^{fm} - c_t$$

where e_t^z and e_t^{fm} are the amounts of administration and fund management expenses incurred during year t, i_t^n is the rate of investment return earned by the nonunit fund during year t, and c_t is the expected (average) cost of claims from the nonunit fund during year t.

For any death claim during the year the claim cost to the nonunit fund is equal to the sum at risk (that is, the sum assured less the unit fund at end of year) if positive, or zero otherwise. Multiplying this amount by the assumed experienced mortality rate gives the expected claim cost (assumed incurred at the end of the year).

The per-policy unit fund calculations are shown in Table 7.11 and the per-policy nonunit cash flows are shown in Table 7.12. Table 7.13 plots the progress of the company's unit and nonunit funds, assuming that 10 000 (identical and independent) policies are issued. And a summary of the profit and accumulation under a unit-linked fund is given in Table 7.14.

Table 7.11 Per-policy calculation of the unit-fund

Year	Bid value of units at start of year after allocation	Admin expense charge	Mortality charge	Unit fund at end of year before fund charge	Fund management charge	Unit fund at end of year after fund charge
1	285.0	220.0	27.5	41.2	0.2	41.0
2	326.0	22.9	30.1	300.4	1.5	298.9
3	583.9	23.8	32.8	580.0	2.9	577.0
4	862.1	24.8	35.8	881.7	4.4	877.3
5	1162.3	25.7	39.0	1207.4	6.0	1201.3

Table 7.12 Per-policy nonunit cash flows, per policy in force at the start of each year

Year	Balance of premium	Expense and mortality charges	Admin expenses	Interest	Fund management charge	Fund management expenses	Expected death cost	Cash flow
1	15	247.5	200	3.8	0.2	0.1	25.0	41.3
2	15	53.0	20.8	2.8	1.5	1.0	27.3	23.2
3	15	56.6	21.6	3.0	2.9	2.0	29.7	24.2
4	15	60.6	22.5	3.2	4.4	3.1	32.3	25.3
5	15	·64.7	23.4	3.4	6.0	4.2	35.0	26.5

7.2.2 Reserving for Unit-Linked Business

One of the attractive features of unit-linked business is the fact that the policyholders' reserves are (mostly) explicitly defined; that is, the amount of assets needed to be held to meet the policyholders' liabilities is simply the sum of all the unit funds at any particular time. This is because policyholders' benefits are defined in terms of the value of a portfolio of assets, so that the portfolio of assets itself becomes the reserve required to meet the liabilities. It does not matter what happens to the value of that portfolio of assets: the value of the liabilities will always be equal to it, so that no possibility of insolvency can arise. It should be noted, however, that reality will differ from this theoretical ideal to some extent; for example, purchases and sales of assets in the unit fund can never be exactly matched to the cash income and outgo generated by the contracts in force. Hence, the value of the assets underlying the unit fund at any time will not be precisely equal to the value of the unit liabilities. Hence, the reserves for the unit fund may need to be increased above the total face value of the units in force, in order to allow for this mismatch. Companies generally endeavor to keep the extent of mismatch to a minimum.

The situation is different for the nonunit fund, however, which operates identically to a conventional nonprofit fund, except for one very significant difference: annual costs can be closely related (or *matched*) to the annual

Table 7.13 Operation of a unit-linked fund by policy year: totals for 10 000 policies (£000)

Unit fund

Year	Bid value of allocated premiums	Expense charges	Mortality charges	Interest	Fund management charges	Death claims	Unit funds
1	2850	2200	275	37	2	1	41
2	2843	228	300	272	15	8	2973
3	2835	237	326	524	29	18	5723
4	2826	246	355	795	44	32	8668
5	2816	254	385	1084	60	47	11822

Nonunit fund

Year	Charges at start of year	Admin expenses	Interest	Fund management charge	Fund management expenses	Death claims	Nonunit assets (funds)
1	2625	2000	38	2	1	250	413
2	678	207	53	15	10	272	669
3	712	215	70	29	20	295	950
4	749	223	89	44	31	320	1258
5	787	231	109	60	42	346	1595

Table 7.14 Summary: unit-linked fund accumulation (£000)

Policy year	Profit for year	Total profit at end of year	Unit funds at end of year	Total assets at end of year
1	413	413	409	823
2	231	669	2923	3642
3	240	950	5723	6673
4	251	1258	8668	9925
5	261	1595	0	1595

contributions (charges) such that, provided the charges being made are adequate, the outgo in every year will be expected to be met from the same year's income. However, owing to the random elements that contribute to it, the amount of annual cash flow is unpredictable and may be negative, even where the office has attempted to match closely its charges to its expected costs.

As far as recognizing profit is concerned, provided that each policy year's cash flow can be regarded as an independent individual insurance transaction, then, once the year is complete, the profit or loss revealed at the end of the year will be real. Profit is simply the difference between income and outgo during the year, and directly reflects the differences between the experience and the current charging rates under these policies. However, there are certain situations in which this approach to profit recognition may not be adequate. For example, in some unit-linked contracts the amounts or rates of charge may be guaranteed not to vary for part or all of the policy term, and there may be other reasons why the income and outgo each year may not be as closely matched to each other as in the above example. This may cause the expected profits in some future years of a policy's life to be negative, such that a large portfolio of such policies would be expected to be in shortfall over these years. Where these losses cannot (or will not) be made good by other measures, such as by increasing the charges to these policyholders when required, then prudent offices would take care to retain profits from earlier policy years that can then be released to pay for the excess outgo in the loss-making years. In other words, the office would need to hold nonunit reserves for those unit-linked policies that are expected to produce losses (that is, negative cash flows) at some stage in the future.

Table 7.15 shows the per-policy cash flows that would arise in our numerical example, if future experience were all to be 50% worse than previously assumed (i.e., expenses and mortality 50% higher, interest 50% lower). The one exception is that the expenses at the very start of the policy remain at the original level of £200. We now see a series of negative cash flows. If we anticipate this change in experience in advance, then we can avoid these future losses in the following way.

Table 7.15 Unit-linked per-policy cash flows, with 50% worse experience (see text)

Year	Balance of premium	Expense and mortality charges	Admin expenses	Interest	Fund management charge	Fund management expenses	Expected death cost	Cash flow
1	15	247.5	200	1.9	0.2	0.2	37.5	26.9
2	15	53.0	31.2	1.1	1.4	1.6	41.0	−3.2
3	15	56.6	32.4	1.2	2.7	3.0	44.7	−4.7
4	15	60.8	33.8	1.3	4.0	4.6	48.9	−6.3
5	15	65.1	35.1	1.4	5.3	6.3	53.4	−8.1

Table 7.16 Effect on profit of holding nonunit reserve at end of year 4 (see text)

Policy year	Cash flow	Nonunit reserve at start of year	Interest on reserve	Increase in reserve at end of year	Profit
1	26.9	0	0	0	26.9
2	−3.2	0	0	0	−3.2
3	−4.7	0	0	0	−4.7
4	−6.3	0	0	7.8	−14.1
5	−8.1	7.8	0.2	−7.8	0

Table 7.17 Effect on profit of holding nonunit reserves

Policy year	Cash flow	Nonunit reserve at start of year	Interest on reserve	Increase in reserve at end of year	Profit
1	26.9	0	0	20.2	6.7
2	−3.2	20.3	0.6	−2.6	0
3	−4.7	17.7	0.5	−4.1	0
4	−6.3	13.6	0.4	−5.8	0
5	−8.1	7.8	0.2	−7.8	0

Looking first at the final year's expected loss of £8.1 (a total expected loss of £79,830 from all 9880 policies in force at the start of year 5), we could prevent this occurring by holding a nonunit reserve at the start of year 5 that was at least equal to £79,830/1.03 (anticipating interest of 3% over year 3), or £7.8 per policy in force. The resulting profit is shown in Table 7.16. The loss in year 5 has now been removed, while in year 4 there is an increased loss of £14.1. Therefore, we need to repeat the above process for each preceding year, until all the future anticipated negative cash flows have been eliminated. The result is shown in Table 7.17. So the effect (and indeed the purpose) of introducing nonunit reserves is to anticipate future losses and to charge them to earlier profits.

In the operation of unit-linked business, two types of reserve may therefore be required: a unit reserve and, sometimes, a nonunit reserve.

The latter is an estimate, as the amounts upon which it is based are subject to uncertainty, and so the reserve may prove to be too high or too low, in exactly the same way as for conventional nonprofits business. However, it helps to quantify the amount of profit that needs to be retained from time to time, in order to prevent overdistribution and possible future insolvency.

7.2.3   Actuarial Funding

It is possible that even a prudent projection of future cash flows could result in a positive future profit stream. This situation can particularly arise when alternative charging structures have been used for recovering initial expenses. Where this involves high regular fund management charges (see Section 7.2.1), then a reserving adjustment called actuarial funding may be employable.

Example

We will use the same numerical example as in Section 7.2.1, but with the following differences:

the expense charge in year 1 is reduced from 220 to 22 (consistent with the level of regular expense charge in subsequent years)
the annual fund management charge is increased from 0.5% to 7.5% of the end-year unit fund value

The resulting progress of the unit fund and of the nonunit cash flows is shown in Table 7.18 and Table 7.19 respectively (compare with Table 7.11 and Table 7.12).

This new product design has the advantageous marketing structure of broadly level charges, which gives the policyholder a much larger unit fund value earlier in the policy term. (The maturity value of the unit fund at the end of the term is little changed, however, as the increased fund management charges essentially compensate for the loss of the high initial expense charges of the original design.)

The nonunit cash flows have changed dramatically, with a highly negative value in the first year, followed by significantly higher values in subsequent years. The company, therefore, requires a significant input of capital every time one of these policies is issued.

The amount of this capital can be reduced by actuarial funding. With actuarial funding, the unit reserve purchased by each allocated premium is reduced by part or all of the future fund management charges that will ultimately be earned on this money. In our example, the bid value of

Table 7.18 Per-policy calculation of the unit-fund, high fund management charge structure

Year	Bid value of units at start of year after allocation (1)	Admin expense charge (2)	Mortality charge (3)	Unit fund at end of year before fund charge (4)	Fund management charge (5)	Unit fund at end of year after fund charge (6)
1	285.0	22.0	27.5	259.6	19.5	240.1
2	525.1	22.9	29.6	519.9	39.0	480.9
3	765.9	23.8	32.3	780.8	58.6	722.2
4	1007.2	24.8	35.5	1041.6	78.1	963.4
5	1248.4	25.7	39.0	1302.1	97.7	1204.4

Table 7.19 Nonunit cash flows, per policy in force at the start of each year, high fund management charge structure

Year (1)	Balance of premium (1)	Expense and mortality charges (2)	Admin expenses (3)	Interest (4)	Fund management charge (5)	Fund management expenses (6)	Expected death cost (7)	Cash flow (8)
1	15	49.0	200	−8.2	19.5	0.9	24.5	−150.1
2	15	52.6	20.8	2.8	39.0	1.8	26.7	59.9
3	15	56.1	21.6	3.0	58.6	2.7	29.2	79.1
4	15	60.3	22.5	3.2	78.1	3.6	32.0	98.5
5	15	64.7	23.4	3.4	97.7	4.6	35.0	117.8

the premium allocated at the start of the first year, after deducting the expense and mortality charges, is £236. Now at the end of the first year, we expect to receive a charge of amount

$$236 \times (1 + u_1) \times 0.075$$

where u_t is the unit fund growth rate over year t. In terms of the assets currently backing the unit fund, this charge has (present) value

$$236 \times \frac{1 + u_1}{1 + u_1} \times 0.075 = 236 \times 0.075$$

Similarly, the charge expected in the second year has present value

$$236 \times (1 - 0.075) \times 0.075 \times {}_1 p_x$$

remembering that the charge for that year will only be received from those policies that survive to the start of the year (and assuming the policyholder is aged x at entry). Continuing in this way, we find that the present value of the future charges from this allocation is

$$236 \times 0.075\ddot{a}_{x:\overline{5}|}$$

where $\ddot{a}$ is calculated at rate of *discount* 0.075. This is the maximum amount by which we can reduce our unit reserve at time 0. So the reserve that has to be *held* is

$$236[1 - 0.075\ddot{a}_{x:\overline{5}|}] = 236[1 - d\,\ddot{a}_{x:\overline{5}|}] = 236A_{x:\overline{5}|}$$

(see Bowers et al. (1986)).

Using the given mortality assumption, and rate of discount $d = 0.075$, $A_{x:\overline{5}|} = 0.67886$. So the *actuarially funded reserve* is £160.2. The company then capitalizes an immediate additional profit of $236.0 - 160.2 = £75.8$ (which is the same as the value of $236 \times 0.075\ddot{a}_{x:\overline{5}|}$ obtained earlier).

Now consider the remaining profit earned on the unit fund, remembering that we have already deducted the expense and mortality charges. In 1 year's time, the reserve required to be held for these units is

$$236 \times 1.1 \times [1 - 0.075\ddot{a}_{x+1:\overline{4}|}] = 240.1A_{x+1:\overline{4}|} = 240.1 \times 0.73323 = 176.0$$

The company also has to cover the cost of paying the normal (unfunded) reserve for those who die during the year, which has expected cost:

$$q_x(240.2 - 176.0) = 0.00251 \times 64.2 = 0.2$$

The available money to pay for all this is the initial asset of £160.2 plus interest (at 10%). So our net profit for the year is

$$160.2 \times 1.1 - 176.0 - 0.2 = 0$$

In other words, the assurance funding factor valued at $d = 0.075$ (the total fund management charge) is the maximum amount of fund reduction possible in order for the unit fund to remain at least self-financing in future. Applying this (maximum) level of actuarial funding to our example gives the figures shown in Table 7.20 and Table 7.21.

We can see that actuarial funding has achieved a significant reduction in initial strain. By making the product more capital efficient, it will increase the return on capital compared with the case without funding.

If we look carefully at Table 7.21, we can see that we have actually introduced a degree of mismatching of cash flows: we have outgoes (the investment management expenses — column (7)) occurring at the end of the

Table 7.20 Calculation of profit arising from the unit fund when actuarial funding is used

Year	(1)	(2)	(3)	(4)	(5)	(6)	(7)	(8)
1	236.0	0.67886	75.8	160.2	16.0	176.0	0.2	0
2	232.5	0.73323	62.0	346.5	34.7	380.9	0.3	0
3	228.9	0.79210	47.6	562.2	56.2	618.1	0.3	0
4	224.7	0.85587	32.4	810.4	81.0	891.2	0.2	0
5	220.3	0.92500	16.5	1094.9	109.5	1204.4	0	0

Note:
(1) Value of new units bought at start of year, net of expense and mortality charge.
(2) Actuarial funding factor.
(3) Profit made from funding of allocated money $= [1 - (2)] \times (1)$.
(4) Total funded value of unit fund at start of year, after expense and mortality charges.
(5) Interest on unit fund $= (4) \times 0.1$.
(6) Total funded value of unit reserve needed to be held at end of year.
(7) Additional death cost expected at end of year.
(8) Residual profit $= (4) + (5) - (6) - (7)$.

Table 7.21 Calculation of total profit arising from the nonunit fund when actuarial funding is used

Year	(1)	(2)	(3)	(4)	(5)	(6)	(7)	(8)	(9)
1	15	75.8	49.0	200	-3.6	0	0.9	24.5	-89.2
2	15	62.0	52.6	20.8	6.5	0	1.8	26.7	86.8
3	15	47.6	56.1	21.6	5.8	0	2.7	29.2	71.0
4	15	32.4	60.3	22.5	5.1	0	3.6	32.0	54.7
5	15	16.5	64.7	23.4	4.4	0	4.6	35.0	37.6

Note:
(1) Balance of premium = Table 7.19 (1).
(2) Profit made from funding of allocated money = Table 7.20 (3).
(3) Expense and mortality charge = Table 7.19 (2).
(4) Administration expenses = Table 7.19 (3).
(5) Interest $= [(1) + (2) + (3) - (4)] \times 0.06$.
(6) Residual annual profit from unit fund = Table 7.20 (8).
(7) Fund management expenses = Table 7.19 (6).
(8) Expected death cost = Table 7.19 (7).
(9) Profit $= (1) + (2) + (3) - (4) + (5) + (6) - (7) - (8)$.

years that are not covered by any direct positive cash flows at that time. To rectify this, we should anticipate an amount slightly less than the full 7.5% fund management charge in our actuarial funding factors (a logical amount to assume might be 7% per annum, as this leaves 0.5% per annum to emerge annually, the same as emerged annually under the original product design). The effect would be to decrease slightly the profit emerging at the time of premium allocation (column (2)), and to increase (from zero) the residual profit emerging from the unit fund every year (column (6)). Provided we choose our funding factors appropriately, therefore, we should leave ourselves with enough profit emerging at the end of each year to cover the investment management expenses, as required. (Actuarial funding of amounts other than the full fund management charge is actually more complex than described here; more details can found in Puzey (1990).)

This illustrates a general principle: that the amount of actuarial funding should not be so extreme as to lead to negative cash flows emerging at any time in the future; as we explained in Section 7.2.2, all policies should remain self-financing overall, once they have been issued.

There are two other points in relation to actuarial funding to be borne in mind:

1. Profits or losses can arise over time due to mortality differing from that expected in the actuarial funding; therefore, it may be necessary to hold an additional reserve in the fund to cover such possible future losses.
2. The value of the units payable on the discontinuance of a policy must not be greater than the actuarially funded unit reserve.

We cover this second point further in Section 7.2.5.

7.2.4 Unit-Linked Business with Level Initial Expense Charges

The third method of charging for initial expenses, stated in Section 7.2.1, is where a level percentage deduction is made from each premium before it is allocated to units.

The effect on cash flows will be similar to that when high fund management charges are used, as shown in Section 7.2.3. However, because the charges are level in monetary terms, rather than increasing with the growth of the unit fund, the capital strain in the first year should be less severe with this method.

Nevertheless, it will still be the case that a large loss in the first year will be followed by larger (than normal) positive cash flows in subsequent years, and it is desirable to rebalance these profit flows in order to reduce the initial strain, if possible. This can be achieved by holding *negative nonunit reserves*.

The underlying principle is similar to the idea behind actuarial funding, in that we wish to take credit at the start of the policy for the charges made for the initial expenses, i.e., the future nonallocated premiums. If the amount of each nonallocated premium is X, then the maximum credit we can take for these future charges at time 0 must be

$$Xa_{x:\overline{n-1}|}$$

assuming the charges of X are levied for n years, at the beginning of each year, the life is aged x at entry, and the first year's charge has just been received.

What rate of interest should we use to calculate $a_{x:\,\overline{n-1}|}$? To answer this we have to consider what is going to happen to the company's reserves. By holding a negative nonunit reserve of $Xa_{x:\,\overline{n-1}|}$ our net reserve for the policy after the allocation of the first premium will be:

$$V_{0+} = U_{0+} - Xa_{x:\,\overline{n-1}|}$$

where U_{0+} is the value of the unit reserve after the allocation of the first premium (i.e. at time 0+). Now it would be inappropriate to make a deduction from *unit*-fund assets at time 0+, because we would then be relying on the future receipts of X each year to make good the unit fund shortfall. The problem here is that the unit fund growth rate could be higher than the rate of interest used to calculate $a_{x:\,\overline{n-1}|}$, so that the payments of X would not be sufficient.

So, as the name suggests, $Xa_{x:\,\overline{n-1}|}$ needs to be a negative *non*unit reserve. The answer is to make this reduction from assets *somewhere else* in the company, leaving the unit fund itself fully invested in its matching assets. The negative nonunit reserve, therefore, operates like a loan, borrowing from the assets backing *other* policies on the company's books, in exchange for future repayments of X per annum from the unit-linked policies' future cash flows. The negative nonunit reserve then takes the role of the backing asset for some other (nonlinked) liability, and so will demand a fair rate of interest on the investment. The rate of interest for calculating $a_{x:\,\overline{n-1}|}$, therefore, needs to be a *high enough* rate to meet the investment requirements for the policies that are effectively making the loan.

In summary:

The amount of negative nonunit reserve held at the start of any year t will be $Xa_{x+t-1:\,\overline{n-1}|}$, based on a suitably high interest rate.

The deduction of the negative nonunit reserve from the total reserves at the start of the policy will make a large positive contribution to profit on day 1, thereby reducing or removing the initial strain.

Each subsequent year the annual receipts of X from each premium, for each surviving policyholder, compensate for the reduction in negative nonunit reserves each year. So, after the first year, all future years' profits are reduced by X compared with what they were originally.

We also need to note that:

1. V_{0+} must always be higher than the current surrender value. This implies a surrender penalty at least as large as the current level of negative nonunit reserve. This also means that the negative nonunit reserve can never exceed the unit reserve in size, as this would lead to a negative total reserve.

2. The cash flows for the policy net of the change in negative nonunit reserve each year must remain positive.
3. There must be sufficient *positive* nonunit reserves held elsewhere in the company in order to finance the desired negative nonunit reserves.
4. There is a risk from mortality being higher than anticipated in the calculation of the negative nonunit reserve, as this would leave the company with inadequate repayments for the loan. Assuming a high mortality rate in the calculation of $a_{x+t-1:\,\overline{n-1}|}$ is therefore prudent.

7.2.5 Discontinuance Terms for Unit-Linked Policies

A normal starting point for the discontinuance value (cash value) of a unit-linked policy would be the unit fund, less any surrender penalty. The need for a surrender penalty equal to the amount of any actuarial funding or of any negative nonunit reserve has already been mentioned in the preceding two sections. These are necessary in order to recoup all the initial expenses incurred by the policy, where the charges up to the date of surrender have not yet been sufficient to achieve this. Looking at this in a slightly different way, if we perform these reserve-reduction techniques in relation only to those future charges that are designed to recover the initial expenses, then the reductions achieved by the (appropriate) method will give us precisely the surrender penalty required.

A further deduction might be desired by the company in order to recoup some or all of the future *profits* that would be lost following discontinuance. This helps to stabilize the company's profits from these policies, making them less sensitive to changes in discontinuance rates. However, the extent to which these additional penalties are possible will depend on other factors, particularly marketing and competitive considerations, as well as issues of fairness.

Surrender values for the three different charging structures described in the preceding three sections are now summarized.

(1) Nil-allocation or high initial expense charge:

$$SV_t^{(1)} = U_t - EFP_t$$

where SV_t is the surrender value, U_t is the face value of the unit fund, and EFP_t is a calculation representing some or all of the expected present value, as at time t, of the future profits from the policy that would have been earned after time t.

(2) High fund management charges with actuarial funding:

$$SV_t^{(2)} = FU_t - EFP_t$$

where FU_t is the actuarially funded value of the unit fund at time t.

Note that $SV_t^{(2)}$ would be calculated this way, i.e., using an actuarially funded unit reserve value, even if the company was not *actually* using actuarial funding in its reserves; the future fund management charges still need to be collected at surrender in order to recoup the initial expenses fully.

(3) Level part allocation with negative nonunit reserves:

$$SV_t^{(3)} = U_t - NNUR_t - EFP_t$$

where $NNUR_t$ is the amount of negative nonunit reserve at time t.

Similar to (2), on surrender the amount of would-be negative nonunit reserve has to be deducted in the surrender value calculation, even if the company was not using negative nonunit reserves in practice.

In all the above cases EFP_t would be a desirable deduction, though (as we have discussed) not necessarily a feasible one.

The extent of any surrender penalties would normally be explicitly stated in the policy conditions. So, unlike many nonlinked policies, the company may have no discretion regarding the calculation of the surrender value for any unit-linked policy *once it has come into force*, because the terms of surrender are effectively guaranteed in the contract.

7.3 Nonlinked Accumulating Nonparticipating Contracts

Contracts of this form are issued in the U.S., under the generic name *current assumption whole life* (abbreviated as CAWL) contracts. Under these contracts a whole-of-life death benefit is payable as usual, which is fixed throughout the policy term. However, the cash value of the contract (per policy in force) accumulates according to the relation

$$CV_t = (CV_{t-1} + P_t - CH_t)(1 + j_t)$$

where CV_t is the cash value of policy at end of year t, P_t is the premium allocated to policy in year t, CH_t is the charges deducted in year t, and j_t is the interest credited to the contract for year t, assuming that charges and premiums occur at the start of each policy year, for simplicity. The form of j_t represents the fundamental difference between this type of contract and both unit-linked contracts and accumulating with-profits (AWP) contracts (see Section 8.1.4). j_t is related to the new money rate of interest over the year in question. It would normally be equal to the actual yield, less some margin by way of an interest rate charge. There is also a contractual minimum value of j_t, typically of the order of 4 to 5% per annum.

For comparative purposes we could write the per-policy accumulation of a unit-linked fund in the same way, namely:

$$U_t = (U_{t-1} + P_t - CH_t)(1 + i_t)$$

where U_t is the value of the unit fund at the end of year t and i_t is the actual overall rate of return on the assets supporting the unit fund over year t less some small margin (the fund management charge).

The differences between i_t and j_t are therefore that:

1. j_t is related to reinvestment yields while i_t is determined by returns from the whole portfolio of assets.
2. j_t is subject to a minimum guarantee, whereas i_t is not (indeed i_t can be negative).

As a result of these differences, the accumulating cash value may well differ appreciably from the policy's asset share from time to time, which (at least where the cash value exceeds the asset share) would require a penalty to be applied to the cash value should the policyholder choose to discontinue his or her contract.

If we wish to find a realistic value of the contract, then the presence of the interest rate guarantee leads to option-like features that need to be modeled carefully, especially if fair valuation techniques (Section 7.1.2) are being used. (This also applies to AWP; see Haberman et al. (2003).) More details about this benefit type can be found in Black and Skipper (2000).

References

Atkinson, D. B. and Dallas, J. W. (2000). *Life Insurance Products and Finance*. The Society of Actuaries.

Black, K. and Skipper, H. D. (2000). *Life Insurance*, 13th edition. Prentice-Hall.

Bowers, N. L., Gerber, H. U., Hickman, J. C., Jones, D. A., and Nesbitt, C. J. (1986). *Actuarial Mathematics*. The Society of Actuaries.

Haberman, S., Ballotta, L., and Wang, N. (2003). Modelling and valuation of guarantees in with-profit and unitised with-profit life insurance contracts. Under review.

Hairs, C. J., Belsham, D. J., Bryson, N. M., George, C. M., Hare, D. J. P., Smith, D. A., and Thompson, S. (2002). Fair valuation of liabilities, *British Actuarial Journal* 8(P II), 203–340.

Laker, R. J. and Squires, R. J. (1985). Unit pricing and provision for tax on capital gains in linked assurance business, *Journal of the Institute of Actuaries* 112, 117–161.

Neill, A. (1977). *Life Contingencies*. W. Heinemann, London.

Puzey, A. S. (1990). Actuarial funding in unit linked life assurance — derivation of the actuarial funding factors. Actuarial Research Paper No. 20, City University, London.

Chapter 8

Participating Life Insurance

As referred to in Section 6.3, participating (or with-profits) life insurance falls into two main categories: either conventional, or organized as a non-linked accumulating fund.

Note that, by definition, all participating life insurance business must be non-linked. Participating contracts, in some form or other, are widespread throughout the developed world.

8.1 Different Distribution Methods

Under a conventional participating, or with-profits, contract, policyholders are guaranteed a certain level of benefit (the sum assured) in return for the payment of fixed single or annual premiums. To this extent it resembles a non-profit contract, and the mode of accumulation of premiums in the fund is also essentially similar to a non-profit fund. However, there is one very significant difference: under the with-profits contract the life office increases the return to the policyholder by means of additional periodic distributions whose amounts are not determined in advance, but are determined from time to time by the office to reflect the individual policyholders' shares of the office's profits. This is why these policies are known as *participating* contracts: they participate in the profits earned by the fund.

With-profits policyholders earn their rights to share in the office's profits by contributing to the office's capital, and thereby also bearing their share of the office's losses. (Indeed, as we shall see, it is difficult to identify at what stage the with-profits policyholders are sharing in losses as opposed to profits; it is better simply to think of these policyholders as sharing in the office's experience, whether this be good or bad.) This is achieved by with-profits policies providing a much lower level of guaranteed sum assured for a given amount of premium than would be provided for the same premium by a non-profit contract. Hence, at least at early policy durations, the benefit payable (on death, say) will be lower than that which would be payable under the non-profit contract. Indeed it is possible for the overall return to the holder of a with-profits contract to be lower than that of the equivalent non-profit contract if, for example, the office's experience turns out to be less favorable than that assumed in the premium basis of the non-profit policies. On the other hand, should the experience turn out to be *more* favorable (as would usually be expected), then the with-profits policyholder could be considerably better off.

8.1.1 Contribution Method

Away from Europe, profits are usually distributed to policyholders in the form of an annual **dividend**. This dividend is calculated for each policy to be in proportion to the different sources of surplus generated by the policy over the last year, that is from the excess of expense loadings over actual expenses, the excess of expected mortality costs over actual costs, and from the excess of the actual investment return compared with the anticipated return. The dividend payable for the particular year is calculated using a formula of the following type:

$$D = K\left[(V_0 + P - e')(i' - i) - (q' - q)(S - V_1) - (e' - e)(1 + i)\right]$$

where V_0 and V_1 are respectively the policy reserves at the beginning and the end of the year, S is the sum payable on death, P is the premium paid, i, e, and q are respectively the rates of interest, expenses, and mortality assumed in the valuation basis (i.e., in the calculation of V_0 and V_1 for the year in question); i', e', and q' are respectively the actual experienced rate of interest, amount of expenses, and average mortality rate over the year, and K is some constant.

For simplicity the formula above assumes expense and premium cashflows occur at the start of the year, while claims are payable at the end of the year.

Should $K = 1$ then this implies that all of the profit earned in relation to the valuation basis during the year (see Section 7.1.3) will be distributed at the end of the year. Usually $K < 1$, so that not all of the total surplus earned during the year is distributed, which will result in an increase in the company's free assets, or capital. The position where $K > 1$ could also apply, especially if the with-profits fund were in decline and excess surplus could be distributed back to the policyholders. Furthermore, an element of smoothing of dividends from year to year is normally practiced by life offices, so that the policyholders' annual returns are subject to less volatility than the returns experienced by the company itself. (The smoothing process is described further, in the U.K. context, below.) However, in every case, the use of the dividend formula ensures that all policyholders' dividends are proportionate to their contribution to the company's profits, and hence this is an extremely equitable method of profit distribution.

8.1.1.1 Types of Dividend Payment

Policyholders can usually elect to have their dividends paid in any one of the following ways:

1. As cash.
2. As a reduction to the premium payable during the next year (the reduction would be equal to the amount of cash that would have been payable in (1)).
3. To be used as a single premium to purchase additional contractual benefit, i.e., to increase the overall sum assured payable on a contractual claim. This form of payment is exactly equivalent to the *reversionary bonus* of the U.K. system (see below).
4. The dividend is left in the fund to accumulate on deposit. The accumulated amount would be payable on a policy claim, whether contractual or through voluntary discontinuance.
5. To be used as a single premium to purchase additional death benefit for the next year only, i.e., to purchase a 1-year term assurance.

A *terminal* dividend is also usually paid on a claim. This is a further addition to the total claim amount, which, of course, is never guaranteed in advance of the claim. It is equivalent to the terminal bonus in the U.K., which is described below.

8.1.2 Uniform Reversionary Bonus Method

This is the method that has been in use in the U.K. ever since the first distributions were made by the Equitable Society in the late 18th century. As such, it probably represents a historical accident, as the contribution method was found at that time to be too cumbersome for the highly labor-intensive administrative systems of the period, even though that method is almost certainly superior in concept and clearly more equitable than the uniform reversionary bonus system. Had with-profits life insurance started later in the U.K., as in the U.S., then it too would almost certainly have been able to utilize the contribution method.

In the U.K., the reversionary bonus for a particular year is expressed as some proportion of the contractual sum assured, existing reversionary bonuses, or both. These proportions, known as *rates* of bonus, are described as uniform because the same rates are applied to all contracts in force on a particular date, regardless of any other factor such as age, policy term, or expired policy duration as at the distribution date. Once a reversionary bonus has been added to a policy's contractual benefit, it cannot be taken away; i.e., it forms part of the policy's guaranteed benefit from the moment at which it is added. Hence, the guaranteed benefit under a U.K.

with-profits policy follows a rising stepwise progression with increasing duration, in which the size of each step cannot be predicted in advance, other than knowing that future steps can never descend.

An office's reversionary bonus structure usually takes one of the following forms:

Simple bonus. The amount of bonus declared under each policy at a particular date is a particular proportion of the guaranteed sum assured. Hence if b_t is the bonus rate declared for year t, and S is the guaranteed sum assured under a policy, then the new bonus for year t (NB_t) for that policy will be

$$NB_t = b_t S$$

Compound bonus. The amount of bonus is declared as a particular proportion of the total of the sum assured and existing bonus at time t. Hence if B_t is the total reversionary bonus attaching to a policy immediately preceding the declaration for year t, then

$$NB_t = b_t(S + B_t)$$

Super-compound bonus. The super-compound bonus is similar to compound, except that different rates (i.e., proportions) of bonus are applied to the sum assured and to the existing bonuses. Hence if $b(s)_t$ and $b(b)_t$ are respectively the rates of bonus on the sum assured and the existing bonus, then

$$NB_t = b(s)_t S + b(b)_t B_t$$

A terminal bonus is paid at the claim date of a with-profits policy. Its amount is never guaranteed in advance (although it cannot be negative), so that the life office can in theory pay whatever level of terminal bonus that it feels is appropriate for a claim at any particular time.

8.1.3   Revalorization Methods

These methods are used in most European countries other than the U.K. With these methods, the profit each year is distributed in proportion to the size of the reserve held for each policy. So, if the proportion of reserves to be distributed at a particular time t is r_t, say, then the amount distributed to a policy with current reserve V_t would be $r_t V_t$.

There are two main methods of distributing (i.e., paying out) this amount to policyholders. The first is to increase both the policy benefits and the future premiums by the same proportion r_t. So, if the current benefit payable on a claim is S_{t-}, and the current level of annual premium is P_{t-}, then the levels of benefit and premium immediately after the distribution would be

$$S_{t+} = (1 + r_t)S_{t-}$$

$$P_{t+} = (1 + r_t)P_{t-}$$

The logic behind this is as follows. Assuming companies value their policies using a net premium valuation method, the reserve at time t, before distribution, would be (for a typical whole of life policy for age x at entry and with duration s years)

$$V_{t-} = (S_{t-})A_{x+s} - (NP_{t-})\ddot{a}_{x+s}$$

If the company now wishes to distribute profits to the value (cost) of $r_t V_{t-}$, then the total policy liability will increase to

$$V_{t+} = V_{t-}(1 + r_t)$$

Multiplying through we get

$$V_{t+} = (1 + r_t)(S_{t-})A_{x+s} - (1 + r_t)(NP_{t-})\ddot{a}_{x+s}$$
$$= (S_{t+})A_{x+s} - (NP_{t+})\ddot{a}_{x+s}$$

which is the net premium policy value for a sum assured of S_{t+}. So, increasing both the sum assured and the current premiums by the same proportion, r_t, distributes the required amount of profit to each policyholder.

Some policyholders find this kind of distribution structure meets their needs. Although r_t is not linked in any direct way to inflation, investment yields show some correlation with inflation. The overall effect is then a policy whose size (benefits and premiums) tends to reflect inflation increases, at least over the long term, and some policyholders find this a useful property of their savings vehicles.

On the other hand, some policyholders could find the premium increases unacceptable, in which case an alternative distribution, which increases only the sum assured, is usually available. In this case, the amount distributed for a particular policy in year t would be b_t, found by solving for b_t in

$$(1 + r_t)V_{t-} = (1 + b_t)(S_{t-})A_{x+s} - (NP_{t-})\ddot{a}_{x+s}.$$

8.1.4 Accumulating with Profits

This method of profit distribution is also a variant of the accumulating fund benefit structure, referred to at the start of the chapter. It is thought to be unique to the U.K., where it was originally known as *unitized with-profits*. It is defined by the following fundamental relation; although actual practice may not be quite as simple as implied by this expression, it includes all the essential features of the contract:

$$U_t = (U_{t-1} + P_t - CH_t)(1 + g)(1 + b_t)$$

where U_t is the face value of the policyholder's fund (the policy fund) at the end of year t, P_t is the premium allocated to the policy in year t, CH_t are the charges deducted in year t, g is the guaranteed annual rate of growth of the policy fund, and b_t is the bonus interest rate declared for year t.

It is assumed that all premiums and charges are incurred at the beginning of each year; adjustments to the relation would have to be made where this is not the case. The charges can have a variety of forms for a variety of purposes, similar to those of the other accumulating fund contracts (see Section 7.2.1). Explicit fund management charges may be made, but these are often applied implicitly by making reductions to the rate of bonus declared each year.

Similar to the other distribution methods, the regular bonus is declared each year for the year in question. This may be done retrospectively, as with conventional with-profits, but more commonly the bonus rate is declared at the start of the year to which the rate will apply. This is because policy fund values are usually updated more frequently than annually; i.e., monthly, weekly or often daily. The guaranteed and bonus growth rates are therefore converted into the equivalent rate of accumulation for the frequency in question. Hence, an office which augments its policy funds monthly would accumulate them during year t at the following rate every month:

$$\left[(1 + g)(1 + b_t)\right]^{1/12}$$

The guaranteed rate g is usually fixed at a fairly low rate. In the early days of this product (the mid 1980s), guaranteed rates of up to 4% per annum were commonly found. Nowadays, in recognition of the prevalent low interest rate environment, guaranteed rates other than zero are rarely offered. Note that a guaranteed rate of 0% must not be considered synonymous with no guarantee; the face value of the policy fund is guaranteed not to fall in the future.

The total contractual benefit payable at any time t is equal to the policy fund value plus a terminal bonus. We might write this as

$$B_t = U_t + TB_t = U_t(1 + r_t)$$

where B_t is the total contractual claim benefit at end of year t, TB_t is the amount of terminal bonus payable at end of year t, and r_t is the current rate of terminal bonus at end of year t.

If the contract also provides an additional insurance benefit, such as to pay at least £X on death within a specified period, then the contractual benefit over that period will be the maximum of B_t and X at the date of claim.

8.2 Profit Distribution Strategies

8.2.1 History

Historically, under the U.K. system, reversionary bonuses were deemed to represent profit originating from revenue in the with-profits fund, i.e., physical income (including investment income and realized capital gains) less physical outgo. Terminal bonuses were developed in the second half of the 20th century in the U.K. in order to allow life insurers to invest more heavily in equity and property investments without threatening the company's solvency. This allowed with-profits policy returns to benefit from the increasing real returns achieved on such investments, compared with the returns from fixed-interest assets which were becoming inadequate in real terms as inflation became more significant over the period. This led to a significant amount of equity and property investment of U.K. life insurers' with-profits funds, with more than 50% invested in these assets for a mature fund. This proportion can, however, vary quite dramatically over relatively short time periods in conditions of high stock market volatility. A sustained period of falling stock markets in the first years of the current century caused significant reductions to the proportion of equities in life insurer's backing assets. This was due firstly to the fall in the value of the assets themselves, and secondly to companies switching into fixed-interest assets in order to prevent heavily reduced solvency margins from reducing further.

Even though most U.K. business is now of the accumulating with profits (AWP) type, the new structure has, in the main, carried through the same overall philosophy as its predecessor. Hence, the aim is to have a significant terminal bonus component, allowing investment in higher yielding assets and so generating higher returns for policyholders. The jury is still out as to whether this will be successfully achieved; in practice, it may turn out to be more similar to the contribution method in the U.S., at least in terms of its investment backing. For the contribution method itself, there is much less investment in real assets by U.S. life insurers compared with the U.K., reflecting the fact that annual dividends tend to have higher cost than reversionary bonuses, leaving much less profit retained for terminal distribution (if any). For the revalorization methods, as there is no terminal distribution at all, the majority of contracts are backed by more-or-less 100% fixed-interest portfolios. (See Section 3.6 for more discussion of the investment of with-profits life insurance funds.)

Note that although AWP may appear to be very similar to a unit-linked contract, its nature can be very different. This is particularly because of the guarantees and consequent backing assets. An AWP policy with high terminal bonus will have some similar characteristics to a unit-linked policy backed by equities, in that a fall in the value of the backing assets will produce a reduction in benefits payable in both cases (although AWP benefits are also smoothed over time; see Section 8.2.3). However, the value of the policy fund U_t will be invariant in the face of market value falls (reflecting the fact that the fund growth rate g cannot be negative), whereas the whole of the unit fund value of a unit-linked policy is at risk.

8.2.2 Profit Deferral

In any system in which there are both regular and terminal distributions, life offices usually have, as part of their bonus philosophy, a strategy of maintaining a balance between regular and terminal profit distribution that is broadly in keeping with the mix of assets in which the fund is invested. A terminal bonus or dividend essentially represents a distribution of *deferred* profit: profit that was earned over the duration of the policy but was not distributed as it was earned each year. The profits retained during the policy term build up into a pool of (temporarily) "free" assets, increasing the company's solvency until such time as the accumulated pool is paid out as terminal bonus.

Companies prefer to increase the amount of profit deferral because of the flexibility afforded by the terminal bonus structure: it allows them to invest more in assets with volatile market values, because terminal bonus rates can be *reduced* over (relatively) short time scales. The increased ability to invest in equities and property is expected to feed through to higher overall payouts for policyholders, though this means some having to suffer low returns on those (hopefully rare) occasions when equity and property returns show prolonged poor performance (e.g., as occurred to policies maturing shortly after the turn of the millennium).

8.2.3 Profit Smoothing

In the case of the U.K. reversionary bonus and AWP methods, offices appear to prefer to declare regular bonus rates that vary only slightly from year to year, if at all, while the balance of accumulated profit is made up by a more variable terminal bonus. However, in practice, even terminal bonuses vary less violently than, say, the market value of the fund's assets. As a result of this practice, not all of the fluctuations in the fund's profitability are passed on to the with-profits policyholders; that is, the policyholders effectively receive a *smoothed* share of the accumulated profits of the fund.

The idea of paying a smoothed share of accumulated profits is common to all methods of distribution, but to differing extents. Annual cash dividends under the contribution method are smoothed from year to year, as we said earlier, but (because of the more exact nature of their calculation) they will not be expected to be the *same* each year, for any individual policy. Even the revalorization methods, which are basically formula driven, will present a fairly smooth pattern of profit distributions over time, as changes to the market values of the underlying assets are largely ignored in the calculation of r_t. (r_t is instead calculated as a smoothed book value return — based on investment income and *realized* gains earned in relation to the book values of the underlying assets. Even the gains (or losses) made in relation to the book value price are spread over the lifetime of each asset.) The revalorization methods, therefore, adopt a passive approach to smoothing; the company has very little discretion over what the value of r_t is in any one year. All other methods involve greater or lesser extents of discretion in what is paid out, as we now describe.

For the three other methods (contribution, reversionary bonus, AWP), let us consider the policyholder's "smoothed share." A logical starting point for considering the policyholder's overall entitlement is the policy's *accumulated asset share* (see Section 7.1.1), less, in the case of a proprietary office, any contribution to shareholders' profits. The with-profits policyholders may also share in the profits (and losses) from the nonprofit business, which will increase (or decrease) each policy's entitlement accordingly. Furthermore, as with the case of the contribution method of distribution described above, the office may consider it prudent to increase or decrease the company's long-term capital base (i.e., its estate) depending on the long-term solvency needs of the office. This would result in policy entitlements being accordingly lower or higher than their current asset shares. It should be noted that policyholders will not be explicitly led to any formal expectation of receiving a return which equates to the policy asset share. Their expectations are purely that they will share, equitably, in the profits (or losses) of the company and that this share will be a smoothed version of what the company actually earns over time.

However, it is generally convenient to consider these with-profits policies as paying benefits whose values are equal to their asset shares, plus a smoothing adjustment which may be positive or negative according to conditions, subject also to a guaranteed minimum benefit level. Hence, when asset share values are temporarily relatively high (e.g., due to unusually high stock market values) the levels of benefit paid may be of lower value than the asset share. When asset share values are temporarily depressed (e.g., due to unusually low stock market values) then the reverse may be the case. The extent to which with-profits policyholders' benefits follow these fluctuations in stock market values, and in other sources of profit, reflects the second aspect of an office's bonus philosophy, i.e., the extent of smoothing which is practised by the office. This can vary considerably between different offices and between different distribution methods. Clearly, on the occasions where payouts exceed asset share values, then there has to be a contribution from the office's capital. The contributions are in the opposite direction when asset shares exceed benefit levels. Should asset shares fall below the value of the current guaranteed benefits, then again a contribution from capital would be required. This would represent a contribution to pay for the guarantees under the contract, and should be considered separately from the general smoothing process, which implies a zero net contribution to or from capital over the long run. To pay for this guarantee, a deduction would be made from the policyholders' asset shares each year (i.e., a "capital charge"; see also Needleman and Roff (1996)).

8.3 Discontinuance Terms for With-Profits Policies

When with-profits policies are discontinued, a surrender value will (normally) be payable. Broadly speaking, most companies would aim to pay a surrender value that was close to the asset share. This is unlikely to cause many public relations problems, except where surrender occurs early on in the policy term when asset shares tend to be very low because of the high initial expenses. Companies will pay higher surrender values than justified in order to keep the public happy (though payouts for the nonsurrendering policies must suffer, as ultimately these early surrender losses have to be paid for by them).

For AWP there is more of a public relations issue with surrender because of the existence of the (published) value of the policy fund U_t. Recall that the *contractual* benefit B_t payable at time t is

$$B_t = U_t + TB_t$$

Because B_t is essentially the *smoothed* asset share at time t, it would not always be appropriate to pay B_t (or even U_t) as a surrender value, as policyholders will tend to choose to surrender at times when B_t (or U_t) exceeds the asset share (causing loss to the company — an example of adverse selection). Instead, the policyholders would generally be credited with an amount at time t calculated according to the following formula:

$$DV_t = B_t - MVA_t$$

where DV_t is the "discontinuance value" of policy at time t and MVA_t is the "market value adjuster" (MVA) at time t.

The MVA allows the office to pay less than the current benefit level B_t on discontinuance where asset shares are lower than this. This can remove the risk of loss on discontinuance and, hence, reduce the adverse selection risk.

Further general considerations regarding discontinuance values are discussed in Chapter 10.

8.4 Overall Operation of a With-Profits Fund

We will use the U.K. with profits methods to illustrate the overall operation of with-profits funds. This is summarized in Figure 8.1 (α is the proportion of distributed profit which is attributed to the shareholders. For mutual offices $\alpha = 0$.)

8.4.1 Numerical Example

We will use the following hypothetical example to illustrate how a conventional with-profits fund operates, in a mutual U.K. life office; the differences for a proprietary office will be considered in Section 8.7.

Contract
With-profits endowment

Term	10 years
Sum assured	£1000

Premium basis

Expected mortality rate	2.5% each year
Expected return on assets	10% p.a.
Expected expenses	
year 1	50% of office premium
years 2–10	5% of the office premium, inflating at 5% p.a.
Reversionary bonus rate	4% p.a. compound
Terminal bonus rate	25% of all other benefits at the date of claim

According to these assumptions, the annual office premium is £142.95.

We will first of all assume, as usual, that our experience is exactly equal to that expected according to the premium basis and, furthermore, that the office declares exactly the rates of bonus assumed in the premium basis. For simplicity, it will also be assumed that no charge is deducted from the accumulating asset shares to pay for the cost of guarantees, although in practice some charge would be made, as discussed above.

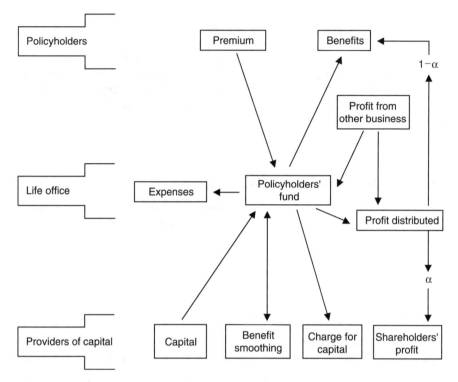

FIGURE 8.1 The operation of a with-profits fund

Under these assumptions the benefit paid on death during policy year t will be equal to

$$1000 \times 1.04^{t} \times 1.25$$

which, as usual, is assumed to be paid at the end of the year of death. For policyholders surviving the 10-year term, the maturity benefit would be

$$1000 \times 1.04^{10} \times 1.25 = 1850$$

Assuming the cost of claims to follow this progression, the accumulation for a tranche of 10 000 identical policies issued at the same time is as shown in Table 8.1.

The final two rows of Table 8.1 need further explanation. Row 10 shows the accumulation of the fund to the end of the tenth policy year immediately before the maturity claims are paid to the survivors. Row 10(M) shows the cost of the maturity benefits, which are then deducted from the fund.

The per-policy asset share is the fund at the end of the year, divided by the number of policies in force at the end of the year.

Table 8.1 Operation of a with-profits fund by policy year (£000)

Year	Premium	Expenses	Interest	Claims	Fund	Per-policy asset share
1	1430	(715)	71	(325)	461	47.3
2	1394	(73)	178	(330)	1630	171.5
3	1359	(75)	291	(334)	2872	309.8
⋮	⋮	⋮	⋮	⋮	⋮	⋮
8	1197	(84)	984	(358)	10470	1282.1
9	1167	(86)	1155	(363)	12343	1550.2
10	1138	(88)	1339	(368)	14364	1850.3
10(M)	–	–	–	(14364)	0	0

As expected, under the assumption that the experience is equal to the premium basis, the accumulated fund at the expiry of the policy term is exactly zero, i.e., no profit or loss has accumulated to the life office. In a mutual life office, this is as it should be: the with-profits policyholders, who are the legitimate owners of the funds, are simply paid back through the benefit structure the amount that the office has earned from the investment of their premiums, after deducting their share of expenses.

It should be noted that asset shares will "chase" the increasing claim amount under a with-profits policy as policy duration increases, becoming equal at maturity (or, in the case of a whole-life policy, where the last (oldest) policyholder of a tranche dies). This relationship will be broadly maintained even where experience causes asset shares to differ from their expected course, because benefit levels will be varied in response to these changes by varying the bonus rates. On the other hand, under a nonprofit policy, assuming no profit is distributed during the life of the contract, the asset share value at maturity could lie below or (more likely) substantially above the benefit level, reflecting the fact that benefit levels are completely independent of the emerging experience of the fund. This difference between the operation of nonprofit and with-profits funds is of fundamental significance to the management of these funds.

8.5 Determining Dividends or Bonus Rates

8.5.1 Contribution Method

Determining dividends involves the determination of the experience items for the dividend formula quoted in Section 8.1.1, i.e., the average mortality rate, investment return, and expense for the policy in question. These items are obtained by looking at the experience of homogeneous groups of policies, grouped together by common characteristics of, e.g., age, sex, duration, smoker status, and policy size. The value of the factor K has already been discussed.

The terminal dividend is found as the (smoothed) residual amount of the asset share at maturity, after deducting the amount of the sum assured and any other benefit then payable.

8.5.2 Revalorization Methods

These methods are almost entirely formula driven, as described in Section 8.1.3.

8.5.3 Reversionary Bonus Method

This is one of the most difficult methods to manage in practice, and so we will discuss this in some detail. We will also use this to illustrate the types of problem that actuaries encounter, e.g., in balancing the requirements of smoothness, fairnes, and affordability.

Consider the fund accumulation shown in Table 8.1, with experience exactly equal to the premium basis. If we were to calculate the policy values of the contracts at the end of each year, adopting the same assumptions (i.e., that reversionary bonuses will continue to be added in the future at the rate of 4% per annum, and that a terminal bonus of 25% will continue to be paid on all claims), then the total policy values at the end of each year would be exactly equal to the total fund values shown in Table 8.1. As the fund is essentially the sum of all the policy asset shares, then this would indicate that the current experience (the asset shares) is exactly adequate to meet the expected future payouts under the contracts, according to the current level of benefit (i.e., assuming current rates of bonus continue to be declared into the future). This method of calculating the value of the benefits is referred to in the U.K. as a *bonus reserve valuation*, for obvious reasons, although essentially it can be thought of as simply a standard gross premium valuation in a with-profits context.

Should the fund value (the total asset shares) exceed the bonus reserve value, then assuming the expected experience materializes, the fund is in surplus and continuation of current bonus rates would lead to a residual fund after all the policies have terminated, i.e., it would make a contribution to the office's capital. The reverse position would lead to a deficit, and hence to a contribution *from* capital. In such situations, policyholders would be paid benefit levels that are lower or higher than their asset shares could support. Over the long run, such an imbalance would need to be rectified, i.e., by varying reversionary and/or terminal bonus levels in the future to an appropriate extent.

However, in reality, asset shares are extremely volatile. Any change in asset shares may well not be permanent—indeed values may rise and fall from year to year in an essentially random way. The life office may, therefore, respond (e.g., to a higher than expected rise in asset shares at the end of a particular year) by increasing bonuses, but not to the full extent supportable by the asset shares at that time, thereby waiting to see whether asset values in the future would justify making further increases thereafter. Hence, from time to time, benefit levels (represented by these policy values) may be higher or lower than asset shares as the office "smoothes out" the excessive variations in the experience. This is precisely the way in which with-profits benefit levels equate to smoothed asset share values, as we described earlier.

In order to illustrate the process, we will assume that, during the first 4 years, asset values in our example vary according to the following pattern:

Year 1: as expected
Year 2: +5% to the expected result
Year 3: −20% from the expected result
Year 4: +10% to the expected result

In this example, were the office not to vary its bonus rates at all over the period, then the ultimate surplus at the termination of the contracts would be close to zero. Hence, the office could adopt a policy of keeping bonus rates constant over the period, representing the greatest degree of smoothing that the office might adopt in the circumstances. The result of this strategy is shown in Table 8.2.

(Note that the contribution to the capital in year t is essentially the new profit earned during the year, and is equal to

$$CP_t = (F_t - V_t) - (F_{t-1} - V_{t-1})(1 + i_t)$$

where i_t is the return on assets during year t. The accumulated contribution at the end of year t is simply the accumulation of all previous

Table 8.2 Fully smoothed benefits under with-profits contracts with fluctuating asset shares[a]

Policy year	F_t (£000)	V_t (£000)	CP_t (£000)	$F_t - V_t$ (£000)	V_t/F_t	TB rate (%)
1	461	462	0	0	1.00	25
2	1875	1630	245	245	0.87	25
3	2513	2872	(628)	(359)	1.14	25
4	4178	4193	380	(15)	1.00	25
⋮	⋮	⋮	⋮	⋮	⋮	⋮

[a]F_t: fund at end of year t; V_t: total policy values at end of year t; CP_t: contribution to capital; $F_t - V_t$: accumulated contribution; TB: terminal bonus.

years' contributions at the earned average rates of investment return. By definition, this is also equal to $F_t - V_t$.

While the balance between asset shares and policy values is restored at the end of year 4, in year 2 the policy benefit levels were 87% of asset share values, while in year 3 the benefit levels exceeded asset shares by 14%. Does this matter? It does not really matter for death claims, where the actual benefits *paid* exceed the asset shares by a considerable extent in any case. However, in practice, a life office will have many tranches of policies on its books in any particular calendar year, and there will of course be maturities as well as death claims during the year. Hence, the paying of benefit levels at 87% of asset shares may represent a serious shortfall in the returns obtained under maturing policies and may, therefore, be possibly considered as inequitable to the policyholders concerned. In other words, this fully smoothed bonus philosophy might represent too much smoothing.

It should also be noted that maturity is the only claim event under these contracts for which the claim amount is directly comparable to the asset share, i.e., without any dependence upon assumptions of future experience.

Therefore, life offices normally make a priority of ensuring that appropriate benefit smoothing is achieved under all its currently *maturing* policies, and this is essentially the way that companies determine their terminal bonus rates.

Table 8.3 shows the effect in our numerical example of a partially smoothed benefit structure, as well as that of adopting no smoothing at all. Individual fund and policy values are omitted for brevity. This can be compared with the fully smoothed position shown in Table 8.2. Note that neither of these possibilities have involved any change to the regular bonus rates; these have remained constant (i.e., fully smoothed) in all cases.

The "no-smoothing" regime passes on all the fluctuations in asset values to the policyholders, a practice that would be contrary to the

Table 8.3 Unsmoothed and partly smoothed benefits under with-profits contracts with fluctuating asset shares[a]

Policy year	Partly smoothed				No smoothing	
	CP_t (£000)	$F_t - V_t$ (£000)	V_t/F_t	TB rate (%)	V_t/F_t	TB rate (%)
1	0	0	1.00	25	1.00	25.0
2	108	108	0.94	27	1.00	28.6
3	(336)	(218)	1.09	23	1.00	20.0
4	241	1	1.00	25	1.00	24.9
⋮	⋮	⋮	⋮	⋮	⋮	⋮

[a]For key: see Table 8.2.
Note: where there is no smoothing, the values of CP_t and $F_t - V_t$ are zero for all t.

generally accepted philosophy of profit distribution under these contracts. To operate under such a philosophy would require daily valuations and daily changes in terminal bonus rates, which would be totally impracticable and would mean, in effect, that these policies would offer no more protection from fluctuations in investment experience than a unit-linked contract. The partly smoothed option might, therefore, represent the practical alternative: benefit levels vary with asset shares, but not to the full extent, e.g., staying broadly within 10% above or below the underlying asset values.

The precise compromise between no smoothing and full smoothing adopted by any office in practice depends largely upon policyholders' expectations, which in turn will be influenced at least in part by the way the office has operated in the past, unless this has been explicitly contradicted by information given to the policyholders when the contracts were issued. Clearly, some smoothing would be expected, otherwise the policyholders would have taken out unit-linked rather than with-profits policies. Offices will then tend to adopt a smoothing policy that is consistent with its approach in the past, as it is reasonable to assume that this is what their policyholders would expect them to do (or at least any changes in policy would be gradual and/or would be introduced after giving the policy-holders proper notification). As a result of this, every office's smoothing strategy is unique. There is no single "correct" approach; the differences simply mean that the policyholders of different offices are effectively owners of different products, receiving a different profile of benefits between one office and another.

Under what circumstance might reversionary bonus rates be changed? This will only tend to occur when there has been such a sea change in the *expected* future experience that to continue the existing reversionary bonus rates indefinitely into the future would be contrary to the company's overall strategy for the mix of reversionary and terminal bonus, or even might be unaffordable. (In such a case, the bonus reserve value of all the company's with-profit liabilities *excluding* all the terminal bonus payments would exceed the current asset shares under those policies.) The bonus reserve valuation is then a way of testing the "bonus-earning capacity" of the fund; indeed, numerical solutions can be found (in terms of rates of bonus) that represent that capacity, i.e., those rates of bonus that make the current bonus reserve value exactly equal to assets shares. (See Section 11.5.2 for further discussion of this topic.)

8.5.3.1 *Profit Distributions with Overlapping Tranches*

We noted in section 8.1.2 that the uniform reversionary bonus system was considerably less equitable than the contribution method used in other countries, such as in the U.S. The revalorization methods are also reasonably equitable, at least with respect to the distribution of investment surplus. A fully equitable approach under the U.K. system would be one in which benefit levels for all policies were closely matched to their underlying asset shares. The use of the same reversionary bonus rates for all policies, regardless of their expired and unexpired durations, does not naturally lead to this effect. However, by making suitable adjustments to terminal bonus rates (e.g., by making them a function of policy duration) a broadly equitable result can be obtained.

8.5.4   Accumulating With-Profits Method

As with conventional with-profits, the total payout B_t is designed to represent a smoothed version of the policy asset share, subject to a minimum level equal to the face value of the policy fund U_t. Note that under AWP contracts the benefit level itself is equated to the smoothed asset share, whereas under conventional with-profits a realistic bonus reserve valuation of the benefits is required in order to make this comparison, except for policies at maturity. This direct relationship between benefit levels and asset shares under AWP allows the management to assess the suitability of the current benefit levels under all its contracts without any reference to future bonus rates or future returns, which is a considerable advantage over the conventional contract.

8.5.4.1   *Numerical Example (Mutual Office)*

Contract
Accumulating with-profits endowment
Term	More than 3 years
Death benefit	Higher of £1000 or value of policy fund
Annual premium	£200

Charges
Year 1	75% of first annual premium
	(equivalent to a 9-month nil-allocation period)
Year 2 onwards	
Mortality	4% of sum at risk per annum
Expenses	2.5% of annual premium plus £25
	per annum per policy

Note: future charges are not guaranteed

Bonus structure

The policy fund increases at a guaranteed rate of 4% per annum. Bonus interest (which cannot be negative) is declared at the end of each year, as a proportion of the policy fund. For the sake of this example, the declared rate will be assumed to be the same each year at 4% per annum.

Terminal bonus is payable, declared as a proportion of the total policy fund. Its rate will be varied each year so as to produce benefit levels that equate with smoothed asset shares.

Interim bonuses will be assumed to operate each year, at the rates of bonus declared at the end of the previous year.

Experience assumptions

	Year 1	Year 2	Year 3
Return on assets (%)	10	8	12
Mortality rate (%)	2.5	5	2
Expenses per policy (£)	125	20	35

Incidence assumptions

All charges are assumed to be deducted at the beginning of each year; expenses are also assumed to be incurred at the beginning of each year, and claims at the end of each year. The annual premium is paid at the beginning of each year.

Calculating the mortality charge

If M_t is the mortality charge payable at the beginning of year t, then we will assume that, with a minimum of zero:

$$M_t = q[SA - (U_{t-1} + P_t - Z_t - M_t)(1 + g)(1 + mb_t)(1 + mr_t)]$$

where q is the mortality rate assumed for charging purposes, SA is the sum assured, Z_t is the charge for administrative expenses during year t, mb_t is the interim reversionary bonus rate for year t, mr_t is the interim terminal bonus rate for year t, and $[\,]$ is the sum at risk.

The deduction from the sum assured represents the estimated total contractual benefit payable on claim at the end of year t. This leads to

$$M_t = \frac{q[SA - (U_{t-1} + P_t - Z_t)(1 + mi_t)]}{1 - q(1 + mi_t)}$$

where

$$1 + mi_t = (1 + g)(1 + mb_t)(1 + mr_t)$$

Fund development

The development of an AWP fund, assuming 10 000 identical policies issued at time 0 and assuming the experience given above, is shown in

Table 8.4 Development of an accumulating with-profits fund

Year t	Premium £000	Expenses £000	Interest £000	Claims £000	Fund £000	N_t	AS_t
1	2000	(1250)	75	(250)	575	9750	58.97
2	1950	(195)	186	(488)	2029	9262	219.04
3	1852	(324)	427	(185)	3799	9077	418.49
⋮	⋮	⋮	⋮	⋮	⋮	⋮	⋮

Note: F_t = Fund (total assets) at end of year t
N_t = Number of survivors at end of year t
AS_t = Per-policy asset share at end of year t

Table 8.5 Accumulation of the policy fund and calculation of benefit levels for an accumulating with-profits contract (£)

Year t	P_t	CH_t	U_t	sr_t	r_t	B_t	AS_t	B_t/AS_t
1	200	(150.00)	54.08	9.0%	8.0%	58.41	58.97	0.99
2	200	(60.98)	208.85	4.9%	7.0%	223.47	219.04	1.02
3	200	(53.55)	384.29	8.9%	8.0%	415.03	418.49	0.99
⋮	⋮	⋮	⋮	⋮	⋮	⋮	⋮

Table 8.4, which is constructed in exactly the same way as for a conventional with-profits fund, such as shown in Table 8.1.

A possible progression of the policy fund and benefit levels over the first 3 years of our example contract is shown in Table 8.5.

While r_t in Table 8.5 is the rate of terminal bonus that is assumed to be payable on claims at time t, sr_t is the supportable terminal bonus rate, such that

$$U_t(1 + sr_t) = AS_t$$

Hence, sr_t would be the terminal bonus rate which would be declared if no smoothing of benefits occurred, while a comparison between B_t and AS_t shows the extent of smoothing implied by the actual terminal rates chosen.

This example serves to show how easy it is to compare asset shares with benefit levels under the AWP design and, hence, to judge the appropriate levels of bonus to declare. Provided expense and other charges are well matched to actual costs, then the policy fund value will closely follow asset shares with duration, which implies more equitable benefit levels for the policyholders. Variations in bonus rates will then largely reflect variations in asset returns.

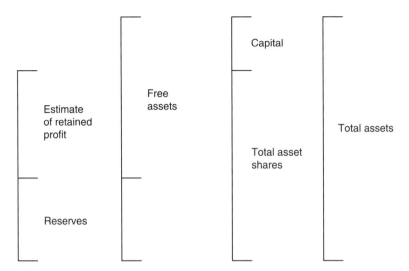

FIGURE 8.2 Subdivision of assets in a with-profits fund

8.6 Reserving for With-Profits: Earned Profit Reserves

It is helpful to consider the assets of a with-profits office as subdivided in the way shown in Figure 8.2. We will refer to the "fund" as being synonymous with the total asset shares. Hence, the total assets of the office consist of the fund plus capital. In a mutual office, capital is usually referred to as the company's estate; a proprietary office will also have an estate, but in addition may have substantial shareholders' capital. The distinction between the two may be hard to define.

The estimate of the retained profit is the part of the current asset shares that are considered to represent profits earned for the policyholders by the valuation date, but which have not yet been distributed to them; hence, we will also refer to this as the *earned profits*. In general, the difference between the fund value and the reserves is called a *surplus*, whose value can be seen to be dependent upon the assumptions used to value the assets and the liabilities. With the fund value defined by the asset share calculation, then the amount of the earned profit is mostly dependent upon the assumptions and method of calculating the reserves. So, by calculating the reserves in an appropriate manner, we should be able to estimate the earned profits emerging each year. We shall refer to this as the "earned profit approach" to reserving.

8.6.1 Example: Reversionary Bonus Method

The key principle of the earned profit approach is to show, as retained profit, the profit that has been earned by the valuation date but has not yet been distributed to the policyholders. This means firstly that reversionary bonuses that have been declared to date must be included in the reserve calculation, because these are profits that have already been distributed. Second it has important implications for the valuation of the future premiums. Recall from Section 7.1.2 the numerical example of Case 1 (Table 7.3 and Table 7.4), where using a realistic reserve capitalizes all the *future* profits and includes them in the retained profit calculation at any time. We clearly need to avoid doing this if we want to exclude future (not yet earned) profit from our retained profit calculation. We can achieve this for our with-profits business by valuing realistically, but *excluding* the future profit loadings from the value of the future premiums.

This should become clear by looking at our numerical example (first introduced in Section 8.4). We can see that, at the outset of the policy, the total value of the future profit loadings payable in the premium is equal to the value of all the reversionary and terminal bonuses expected to be paid on claims according to the premium basis (this is often called the *bonus loading* in the premium). Hence, assuming that profits are to emerge uniformly over the policy term, the annual profit loading for the contract in the numerical example is £52.68 per policy, which means that the future premium that we will value in our earned profit reserve calculation is £142.95 − £52.68 = £90.27 per annum. The earned profit reserve at policy duration t is then

$$V_t = (\text{ Value of sum assured})$$
$$+ (\text{ Value of reversionary bonuses declared up}$$
$$\text{to the end of year } t - 1)$$
$$+ (\text{ Value of future expenses})$$
$$- [(\text{Value of future premiums}) - (\text{Value of future bonus loadings})]$$

all calculated on the original premium basis.

(A close relative of this approach is the *net premium valuation method*. The net premium method can be used for calculating earned profit reserves *whether or not* we are using premium basis assumptions; we will return to this in the next chapter.)

Using these reserves, and assuming the experience is also equal to the premium basis, the surplus emerging using this reserving basis is shown in Table 8.6. In Table 8.6, the surplus for the year "before bonus" is the new surplus generated during the year, excluding the value of the bonus declared for the year in question. The latter is then shown as the "cost of new bonus," which is essentially the amount of assets newly allocated at this time to the policyholders. The cost of new bonus in year 10 refers to the cost of the reversionary bonus for that year; the cost given under year 10(M) refers to the cost of the terminal bonus at the maturity date. As none of the

Table 8.6 Emergence of surplus in a with-profits fund (£000)

Policy year	Surplus for year before bonus	Cost of new bonus	Net surplus for year	Accumulated surplus	Total reserves	Total assets
1	461	(183)	278	278	183	461
2	537	(200)	337	643	987	1630
3	474	(218)	256	963	1909	2872
⋮	⋮	⋮	⋮	⋮	⋮	⋮
8	403	(356)	47	2435	8035	10470
9	389	(396)	(7)	2671	9672	12343
10	376	(442)	(66)	2873	11491	14364
10(M)	0	(2873)	(2873)	0	0	0

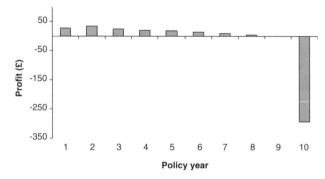

FIGURE 8.3 Profit signature for a with-profits fund

terminal bonus is distributed until the policies become claims, the emerging surplus gradually accumulates to the amount of terminal bonus payable at the maturity date.

The profit signature (as introduced in Section 7.1.2) for the policy in our example can be found by dividing the net surplus for the year in Table 8.6 by the number of policies originally issued (10 000). The result is illustrated in Figure 8.3.

The accumulated surplus can be considered as the with-profits policies' contribution to the office's capital. This contribution is expected to be temporary, because it would ultimately be expected to be paid back to the with-profits policyholders when their policies terminate, mostly as terminal bonus. We will refer to the difference between the office's total assets and its earned profit reserves as the company's free assets, these being the total of the accumulated surplus, the estate, and (for proprietary companies) the shareholders' capital. The free assets, therefore, represent an estimate of the office's current total amount of working capital and, therefore, are of considerable practical significance. The uses, including the investment, of the company's free assets have already been described in Section 3.6.3. The earned profit approach to reserving also forms the basis of the supervisory solvency assessment, which is described in Chapter 9.

8.6.1.1 *Interim Bonuses*

In the example shown in Table 8.6, the accumulated fund has already been reduced by claims during the year. These claims would include each terminating policy's fair share of the year's reversionary bonus and terminal bonus, despite the fact that the office would not have decided, at the time of claim, what its level of bonus should be for that year. This reflects life-office practice of paying an interim bonus on such claims at a rate that is separately announced by the office. While there is a risk that these claims will, as a result, receive slightly different payments from those to which they should have been entitled, it is much more convenient than paying the bonus only after it has been declared, which may be more than a year after the claim date. The financial risks (and inequities) caused by this practice are, therefore, felt to be slight, relative to the expense and inconvenience of the alternative.

8.6.2 Example: Accumulating With-Profits Method

At first sight, the earned profit reserve would seem to be obvious: it should equal U_t, the policy fund value at time t. This includes past additions of bonus interest but excludes the current terminal bonus (as it should, as this is past profit that has not yet been distributed to policyholders). Furthermore, no credit has been taken for any future profits in this value.

   In practice, the application may not be quite so straightforward. The policy has a guaranteed future growth rate of g per annum; if realistic returns i per annum are less than g, then the realistic value of paying the existing policy fund on future contractual claim would be (for entry age x and policy term n years)

$$U_t A'_{x+t:\,\overline{n-t}|}$$

valued at rate $i' = (i-g)/(1+g)$.

   If $i<g$, then i' will be negative, so that the assurance function will have a value greater than unity. If $i>g$ (as it will be in most circumstances), then our principle of not taking credit for future profits means that we should assume a "net" interest rate of $i'=0$, so that we end up with our usual reserve value of U_t in these conditions.

   There remains the issue of future expenses, the cost of any additional death benefits, and the future premiums. First, we can assume that the company's liability with regard to the *past* premiums, which is U_t, will not be dependent on the payment of future premiums. In other words, we can consider the future premiums as paying solely for future benefits, in which case they can be ignored. (There may be cases where this assumption will not hold; their consideration is beyond the scope of this book.)

The expenses and additional death costs will be covered, in theory, by future charges (note: whether or not future premiums are payable). If this *truly* is the case, then again these can be ignored. In this respect, the charges and costs can be considered as analogous to the nonunit cash flows of a unit-linked policy, so that we may need to hold an additional "nonunit" reserve at certain times in order to finance any negative cash flows expected in the future. The methods described in Section 7.2.2 can, therefore, be used to calculate this nonunit reserve.

In summary, the earned profit reserve for AWP is

$$V_t = U_t A'_{x+t: \overline{n-t}|} + NUR_t$$

where NUR_t is the nonunit reserve required (if any) at time t, and A' is calculated at rate

$$i' = \max\left(\frac{i-g}{i+g}, 0\right)$$

Returning now to our numerical example, we will assume for simplicity that the earned profit reserves are simply equal to the policy fund value U_t. Using the policy fund values from Table 8.5 and multiplying the relevant amounts by the number of policies in force, the progression of the reserves, free assets, and surplus are shown in Table 8.7. Note that in Table 8.7 the "free assets" are synonymous with the "accumulated surplus" of Table 8.6. The net new surplus and the cost of new bonus for conventional and accumulating with-profits are compared in Table 8.8, expressed as a proportion of the total assets at the end of each year. Values are taken from Table 8.6 and Table 8.7.

Examination of Table 8.6, Table 8.7 and Table 8.8 shows that the cost of new bonus at early policy durations is a much smaller proportion of

Table 8.7 Emergence of surplus in an accumulating with-profits fund (£000)

| Policy year | End of year | | | During year | |
	Total assets	Total reserves	Free assets	Cost of new bonus	Net surplus for year
1	575	527	48	(20)	48
2	2029	1934	95	(74)	43
3	3799	3488	311	(134)	205
⋮	⋮	⋮	⋮	⋮	⋮

Table 8.8 Net new surplus and cost of bonus for conventional and accumulating with-profits, as a proportion of total assets

Policy year	Net new surplus (%)		Cost of bonus (%)	
	Conventional	Accumulating	Conventional	Accumulating
1	69	8	40	3
2	21	2	12	4
3	9	5	8	4
⋮	⋮	⋮	⋮	⋮

the fund value under the AWP contract than it is for the conventional policy. On the other hand, the surplus emerging from an AWP fund is a much smaller proportion of the fund than for conventional with-profits. However, these items appear to be more stable over time in an AWP fund, and an important point is that we have not yet considered the effect of supervisory regulation on the reserves and, hence, on the free asset position; recall that the earned profit reserves have so far been calculated on "best estimate" assumptions of future experience. This will be pursued in Chapter 9.

8.7   Distribution of Shareholders' Profits

The only difference in the operation of a proprietary with-profits office compared with a mutual office is in terms of the transfer of profit to the shareholders' funds.

In the U.K., shareholders normally take a fixed proportion of the profit from the with-profits business (a typical proportion is 10%). The shareholders receive their share at the same time as the policyholders do, so they will get one-ninth of the value of all regular bonuses when they are declared and one-ninth of all terminal bonuses when they are paid out (i.e., at death or maturity).

One point needs to be mentioned about profits from conventional nonprofit and/or unit-linked business, where a proprietary company also has with-profits policyholders. Either the system can be as shown in Figure 8.1 (where shareholders receive 10% of *all* profits earned by the company, from all sources of business), or the profits from the other business are "ring fenced" so that they *all* go to the shareholders, while only the with-profits business is split in the 90 : 10 ratio.

Similar systems apply for other distribution methods. In the case of the revalorization methods, the proportion of profit distributed to the policyholders is stated in the policy conditions; the rest is paid to the shareholders.

8.8   A Comment on Meeting Consumer Needs

The nature of an accumulating fund contract, such as AWP, makes it ideally suited to two particular aspects of product design:

1. It is best suited to providing lump-sum survival benefits with minimum cost being expended *en route* in providing death benefits. This immediately makes it attractive as a vehicle for pension provision, for which contributions would be accumulated over the policyholder's working life to produce a lump sum on survival to the retirement date, which would then be exchanged for some sort of annuity to provide pension payments thereafter.
2. Under a conventional regular-premium with-profits or nonprofit contract, the benefits, including any declared reversionary bonuses, are only guaranteed on the condition that the level premium continues according to the terms of the contract. This makes it very difficult to vary, and particularly to reduce, contribution levels. Under accumulating fund contracts, current benefit levels are generally independent of future premiums and, hence, contributions can be varied from year to year with relative ease. This again makes the design extremely suitable as a pension policy, as contributions to a pension would be expected to vary from year to year due to their dependence upon personal income.

Thus, it is therefore not surprising that the majority of insured individual pension arrangements in the U.K. are now funded by unit-linked or AWP contracts, or even both. A deferred annuity arrangement (the *flexible premium deferred annuity*) with similar flexibility to these contracts is used in the U.S. for personal pension provision (see Black and Skipper (2000)). Conventional with-profits remains suited to the provision of endowment or whole life death benefits, with the unique feature of being able to provide a rising guaranteed benefit level which is substantially in excess of asset shares, at least at early policy durations. This is a feature that is not naturally provided by any accumulating fund route.

Further reading on the operation of with-profits in the U.K., and on the current issues surrounding the management of the business, can be found in Clay et al. (2001, including discussion).

References

Black, K. and Skipper, H. D. (2000). *Life Insurance*, 13th edition. Prentice-Hall.
Clay, G. D., Frankland, R., Horn, A. D., Hylands, J. F., Johnson, C. M., Kerry, R. A., Lister, J. R., and Loseby, L. R. (2001). Transparent with-profits — freedom with publicity. *British Actuarial Journal* 7, 365–465.
Needleman, P. D. and Roff, T. A. (1996). Asset shares and their use in the financial management of a with-profits fund. *British Actuarial Journal* 1, 603–688.

Chapter 9

The Regulation of Solvency and its Effect on the Emergence of Profit

The various aspects of the statutory regulation of long-term business insurers were introduced in Section 6.5. It is not appropriate in a book of this nature to describe current legislation in detail; here, we are interested in the general impact that the legislation has on the operation of the life insurers' business, and in particular its implications for reserving, profit recognition, and profit distribution. However, we do emphasize the actuarial role in this process, particularly with regard to reserving.

9.1 The Statutory Solvency Assessment

The supervisory authority should be concerned that insurers provide value for money, while not putting at risk the solvency of the office. This can be framed in terms of policyholders' reasonable expectations (PRE; see Shelley et al. (2002) for a recent discussion). Now policyholders would "reasonably expect" an insurer:

1. To take all necessary steps to ensure solvency and, hence, the honoring of its obligations to its policyholders.
2. To pursue, by ethical means, strategies that will improve the returns earned by the insurer on the premiums paid by the policyholders, where the quality of returns is a function both of their expected value and their variability.
3. To distribute its returns in an equitable way amongst its policyholders, by the setting of reasonable nonprofit premium rates or charges and/or through bonus distributions, and amongst its shareholders, as appropriate.

It can be seen, therefore, that policyholders' reasonable expectations are in respect of their office's behavior. Hence, these expectations can also be reasonably influenced by:

The office's behavior in the past.
How the office has stated that it intends to behave in the future.

While the main aim of statutory regulation should be to ensure that policyholders' reasonable expectations are met by life offices, regulation should strive also to affect companies in a completely unbiased way, without favor to particular firms, and also to incur companies in as little intrusion and cost as possible commensurate with fulfilling its purpose.

The aim of continued solvency is paramount. Without solvency, and moreover without public *confidence* that a company will remain solvent for a very long time, no substantial life insurance business would happen. Regulation to this effect can target a number of areas:

1. Controlling the premium rates charged, or specifying the bases that have to be used in pricing.
2. Restricting the company's investments, either by imposing a floor (minimum) on the proportions invested in secure matching assets, and/or by imposing a ceiling (maximum) on the proportions invested in high risk assets.
3. Make (minimum) restrictions on the amounts of assets that have to be held in reserve to meet the liabilities (i.e., a reserving requirement).

There are other sensible things that regulators can do in order to help to ensure solvency, for example:

Vet persons who take executive responsibility within insurance companies. Keep control of the market by enforcing authorization or licensing requirements on those who wish to do such business.

The methods followed in the E.U. and the U.S. are mixtures of methods (2) and (3). This approach is consistent with a liberated approach to regulatory control: essentially you (the company) can do what you want (within reason!) provided that you hold enough reserves to prevent your insolvency, and thereby protect your policyholders' financial interests. The argument then is that the market can be left to deal with the rest — value for money, equity, etc. This is, of course, something of a simplification, but represents the general idea.

9.1.1 Statutory Reserving Requirements

The general approach of the E.U. and U.S. methods is to make a conservative assessment of the company's free-asset position, with the requirement that the free assets so revealed are "adequate" to meet the nature of the risks that are borne by the office. This suggests that it is probably more appropriate to talk about the statutory test as a test of the office's "adequacy" rather than of its solvency, as such, with the assumption that solvency is assured if adequacy is satisfied; see MacDonald (1993). The principles behind the E.U. approach can be found in Groupe Consultatif des Associations d'Actuaires des Pays des Communautes Europeennes (1990), and for the U.S. approach in Black and Skipper (2000). Summaries of approaches adopted in other countries are given in Black and Skipper, and also by Scott et al. (1996). An international approach is currently under consideration by the International Association of Insurance Supervisors (IAIS, 2002), as described in Section 6.5.

The key principle for any solvency assessment is invariably one of *prudence*. For a company to pass the supervisor's solvency test it should mean that the company has a high probability of remaining solvent under all possible future conditions that can be considered to be *reasonably likely to happen*. We will take this to be our definition of prudence.

To ensure that prudence is achieved in this context, the regulations must place some controls both on the way that the assets are valued, and the way that the liabilities are valued, for the purpose of demonstrating solvency. So one approach, as in the E.U., is to require that the assets are valued prudently (i.e., to be of lower value than their best estimate) and/ or the liabilities are valued prudently (i.e., to be of higher value than their best estimate). If we call these (prudent) values A_t and V_t respectively (for a valuation as at time t) then we obtain a prudent measure of the free assets from $A_t - V_t$. (These *statutory free assets* are also synonymously referred to as the company's *solvency margin*.)

It is not enough that this value should be positive; as we said earlier, this solvency margin has to be *adequate*, i.e., sufficient to cover the risks borne by the office. For example, two companies with the same reserve value may have very different policies on their books, or may be adopting very different investment strategies, in which case they are going to need different amounts of free assets in order to leave themselves with equal probabilities of insolvency. So the regulations need to specify an adequacy test that this solvency margin has to satisfy.

This test is defined by requiring the statutory free assets to be at least as large as a specified amount. This amount will be a function of the risk characteristics of the insurer. In the E.U., the amount is called the *required minimum margin of solvency* and has the following general formula:

$$SM = f_1 V + f_2 (SA - V) \qquad (9.1)$$

where SM is the required minimum solvency margin, V is the statutory value of liabilities net of reinsurance, SA is the sum assured payable on death (net of reinsurance), and f_1 and f_2 are factors multiplied to the value of the liabilities and to the sum at risk respectively.

This formula is essentially calculated on a per-policy basis, and then summed over all policies to give the overall required margin for the company. It is a kind of *risk-based capital* assessment of the company's investment and insurance risks (see Section 6.7.3, where risk-based capital is introduced). The f_1 factor reflects the investment risk, the f_2 factor reflects the insurance risk, and the values of those factors vary by class of business, in accordance with the perceived contribution to the overall insolvency risk made by that class. The logic of applying these factors to the size of the reserve and the sum at risk on death respectively should be apparent. Typical values, for the highest risk classes, are $f_1 = 0.04$ and $f_2 = 0.003$.

The method is certainly a crude one, though this is not to say that it is necessarily inappropriate or inadequate, as the values of the factors can be chosen conservatively. However, it is not likely to be very accurate. On the other hand, because the free assets have already been calculated on prudent bases, which in turn should take into account much of the inherent riskiness under each contract, then this lack of inaccuracy in the required minimum margin should not be so much of an issue.

The other way of tackling the solvency test problem, in this framework, is illustrated by the U.S. method. Here, the assets and liability values are calculated much less prudently than under the E.U. regime, producing a larger free-asset value. However, the calculation of the minimum solvency margin (actually called risk-based capital in the U.S.) is much more detailed. It also produces a much larger value, like for like, compared with the E.U. equivalent. In particular, the different sources of risk, namely asset-default risk, insurance risk, asset-liability mismatching risk, and business risks, are explicitly taken into account.

9.1.2 Breaching the Supervisory Solvency Regulations

Companies whose free assets do not meet the required minimum solvency margin are subject to "intervention" by the supervisor. In the E.U. there are two levels of intervention: the first is triggered by a breach of the required minimum described above; the second (more intrusive) intervention occurs if the free assets fall below a more critical smaller level (usually one-third of the first level). Under the U.S. system, companies showing free assets greater than 200% of risk-based capital are not questioned. Companies with progressively lower amounts of free assets relative to risk-based capital are subject to increasing severity of intervention, with complete control of the company being assumed by the supervisor for companies holding less than 70% of the required risk-based capital level.

Clearly, any breach of the statutory reserving requirements may have dire consequences for a life insurance company. One such consequence might be to prevent the company from issuing any more new business. Even where this is not the case, the adverse publicity associated with any apparently failing company would normally be sufficient to ensure that it would fail to sell any more new business anyway. Therefore, it is imperative for life offices to strive to meet the regulatory requirements, and this forms an overriding priority for the management of companies.

9.1.3 Dynamic Solvency Testing

It is not sufficient for life offices just to meet the statutory solvency requirements at the present time: both the regulators and the company wish to have confidence about *continuing* to meet the regulatory requirements over at least the medium-term future, under a range of possible future scenarios.

In some territories, regulations now include a requirement to submit results of model office projections that show the company's future solvency position under a range of conditions. Whilst this requirement does not directly lead to additional reserves being held at the valuation date, it does place a constraint on any aspect of management strategy that could prejudice its *future* statutory solvency. We describe the methodology involved in Chapter 11.

9.2 The Valuation of the Liabilities

We have already discussed asset valuation in Part I of this book, to which you are referred. Here, we will concentrate on the liability valuation.

9.2.1 The Valuation Method

For conventional (nonprofit or with-profit) business, there are essentially two methods of valuation: the *gross premium* and *net premium* methods. We will look at both in turn. Bear in mind throughout that a key principle that any method should satisfy for the purpose of statutory supervision is that it must be prudent.

9.2.1.1 Gross Premium Method

For nonprofit policies, the formula is essentially that given in Section 7.1.2, which is

$$V_t = (\text{Value of future benefits}) + (\text{Value of future expenses})$$
$$- (\text{Value of future premiums}) \tag{9.2}$$

The premiums valued in this formula are the full premiums actually payable under the contract, referred to as the *office premiums* or the *gross premiums*.

The great advantage of this method is that all the items of the net liability outgo are explicitly included. The main potential difficulty is that all the future profit loadings are included in the value of the premiums, are hence valued as an asset and so count towards the company's retained supervisory profits as at the valuation date. This would not seem prudent, as it takes credit for profits that have not yet been earned (and may not *be* earned if, for example, the policy were discontinued in the future).

This may not be such of a problem in practice if sufficiently conservative assumptions are used for the valuation basis. So, if the valuation (reserving) assumptions are more conservative than the premium basis margins for adverse experience and profit put together, then the actual office premium will be loss-making on those valuation assumptions. Whilst prudence requires us not to capitalize future profits, it *must* require us to anticipate all future losses. The gross premium formula, including as it does the actual (loss-making) office premium, will therefore achieve this result.

Where the valuation basis still leaves the office premium as profit making, then it would be necessary to take future surrenders into account, as to do so may increase the value of the liabilities (depending on what the surrender values are). This clearly could become very complicated, particularly as surrender is a policy option and a feasible outcome for the company is that a very large proportion of its policyholders could all surrender over a very short time period. This is why many valuation regulations stipulate that the supervisory reserve must not be less than any current guaranteed surrender value (often extended to be the current surrender value whether guaranteed or not).

When we turn to conventional with-profit business, the problems of the gross premium method become worse. The office premium now includes much higher profit (i.e., bonus) loadings. The gross premium approach will compensate for this, in theory adequately, by including an explicit value of the future bonuses. So, the formula for the reserve would be

$V_t = $ (Value of sum assured and declared bonuses)

$\quad + $ (Value of future profit distributions) $ + $ (Value of future expenses)

$\quad - $ (Value of future office premiums) $\hspace{4cm}$ (9.3)

The problem arises when we consider the basis that companies might choose for their supervisory reserves: the future profit distributions are clearly unknown, and, whilst they *should* be included in accordance with policyholders' reasonable expectations, a supervisor will find it very difficult to assess whether the allowance is appropriate or not, because companies have different profit distribution structures and philosophies.

In summary, the gross premium approach is perfectly suited to conventional nonprofit business, provided that the valuation assumptions are stronger (more conservative) than the premium basis. (Note that this can be checked by recalculating the policy premium using the valuation basis — if this "valuation premium" exceeds the office premium, then the condition is satisfied.) This situation usually applies to most nonprofit business in the U.K., for example, where competitive pressures tend to limit the size of the margins that can be included in the office premiums, and the valuation assumptions are generally prudent. The method is also suitable for single-premium or paid-up policies, where there are no future premiums to be capitalized. In all other situations, however, the method is difficult to use, and is especially problematic for handling with-profits policies.

9.2.1.2 Net Premium Method

The net premium policy value is defined as

$$V_t = \text{(Value of future benefits)} - \text{(Value of future net premiums)} \quad (9.4)$$

where the net premium is the annual premium that would be charged, according to the valuation rates of interest and mortality, for the policy benefits; no allowance is included in the net premium for future expenses and profit distributions. Noting that

$$\text{(Implied profit and expense loading)} = \text{(Office premium)} - \text{(Net premium)} \quad (9.5)$$

then the net premium value of the liabilities makes the implicit assumption that, for a nonprofit policy:

$$\text{(Value of future expenses)} = \text{(Value of implied profit and expense loadings)} \quad (9.6)$$

and for a with-profits policy:

$$\text{(Value of future expenses)} + \text{(Value of future profit distributions)} = \text{(Value of implied profit and expense loadings)} \quad (9.7)$$

While this assumption may not necessarily be valid, the great advantage is that the method specifically excludes *all* margins from the premium being valued, *other than* margins included in the *valuation* assumptions of interest and mortality themselves (remembering that the net premium is calculated using the valuation assumptions). So, the problem of capitalizing future profit margins, for both nonprofit and with-profits policies, is neatly avoided. Readers should note that this method is a general case of the valuation of earned profit reserves, for with-profits business, that we described in Section 8.6.1.

A valid criticism of the "pure" net premium formula is that it overvalues policies that have high initial expenses. These expenses are paid for by a level additional loading in each annual premium, and by excluding these from the net premiums we are essentially treating them as a future liability. An adjustment to the net premium value (called the Zillmerized net premium value) is therefore often used, which explicitly includes the annual initial expense loading into a so-called Zillmerized net premium. So, for a whole-life policy with sum assured S, initial expenses (in excess of regular expenses) of I, aged x at entry, the reserve after t years would be

$$V_t = SA_{x+t} - ZNP\ddot{a}_{x+t} \tag{9.8}$$

where

$$ZNP = \frac{SA_x + I}{\ddot{a}_x} = NP + \frac{I}{\ddot{a}_x}$$

and *NP* is the valuation net premium.

What problems are there with the net premium valuation? The biggest practical problem is that the value of the implied loadings (i.e., of the difference between the future office premiums and the assumed Zillmerized net premiums) may not be as large as a prudent explicit value of the future expenses would be. This situation would be most likely when the valuation basis (of interest and mortality) is much more conservative than the premium basis (including all loadings). Even where the expense assumption is large enough in total, it will not be very realistic for each policy, as it assumes that all future expenses are, in effect, level, whereas in reality they will most likely be increasing over time. (This can be mitigated to a degree by using a reduced valuation rate of interest.) The net premium method will also be inappropriate where there are no contractual future premiums, i.e., for single-premium and paid-up policies. (See Fisher and Young (1965) for more details about the net premium method.)

So, in summary, it would seem that the net premium method is appropriate for just the circumstances under which the gross premium method is *least* appropriate. In particular, it would be appropriate for conventional regular premium nonprofit contracts where the valuation basis is not excessively prudent, and for regular premium with-profits contracts under almost all circumstances. When using the method, checks have to be made that the implied future expenses being valued are sufficient, and adjustments have to made (or additional reserves made) where this is not the case.

9.2.1.3 Unit-Linked and Accumulating With-Profits Contracts

For these types of contract, the cash-flow methods described in Section 7.2.2 and Section 8.6.2 will apply, but subject to using assumptions that are appropriate for their supervisory valuation.

9.2.2 The Valuation Assumptions

The assumptions made must be suitably prudent, as discussed in Section 9.1.1. Assumptions of future mortality would be made with reference to the company's past experience and also in relation to any published industry standard experience, allowing for all "reasonably likely" adverse future outcomes. The allowance for future expenses has to be similarly framed in relation to the company's past experience.

The valuation rate of interest should have prudent regard to the rate of return that is likely to be earned from the assets being used for the reserve. For this purpose, the company's admissible assets (see Section 3.6.5) would be divided between the various contract types in a way that most suitably reflects the nature of those liabilities (this is referred to as the *hypothecation* of the assets to the liabilities). For example, nonprofit annuity contracts would be allocated a range of the company's fixed-interest securities, of appropriate terms, in order to immunize those liabilities as effectively as possible (see Chapter 4). On the other hand, with-profits liabilities (assuming a significant amount of profit deferral) would be hypothecated a large proportion of real assets, reflecting the low solvency risk generally afforded by such contracts and the need to provide good real returns to the policyholders (i.e., real assets for real liabilities). The valuation rate of interest would then reflect, in some (cautious) way, the implied investment return obtainable on the assets hypothecated to the reserve for a particular contract type. Furthermore, allowance should be made for the returns expected on future investments to the extent implied by the contract, and these would also be subject to higher margins to reflect the greater uncertainty involved in estimating them. For example, assuming assets are included at their market values, a portfolio of immediate annuity contracts backed by closely matching fixed interest securities would require a valuation interest rate close to the average market redemption yield of the existing assets (as there should be little need for future investment). On the other hand, a portfolio of regular premium endowment assurances would be valued at a much lower rate of interest than the current redemption yield (even if backed by fixed-interest assets), in order to reflect the much greater future investment required.

9.3 The Effects of the Statutory Solvency Requirements on Life Office Operation

Where statutory requirements are imposed on the valuation of the assets and liabilities (see Section 9.1), the essential effects are to make the value of the assets smaller, and the value of the liabilities bigger, than their realistic values. For most (sensibly invested) companies, the main impact should be on the liabilities; we will assume this to be the case for the examples that follow.

9.3.1 Non-Profit Business

We will return to the hypothetical nonprofit fund used in Section 7.1. Table 9.1 shows the equivalent figures to Table 7.5, this time using a statutory reserve, calculated as follows.

Table 9.1 Statutory reserves and free assets in a nonprofit fund[a]

Time t	(1) £000	(2) £	(3)	(4) £000	(5) £000	(6) £000	(7) £000
0+	2170	252.9	10000	2529	(359)	124	(483)
1	1886	225.5	9500	2142	(256)	108	(364)
2	5641	590.5	9025	5329	312	224	88
3	983	0	0	0	983	0	983

[a]Time $t=0+$ shows the situation at the commencement of the contract immediately after the first premium has been paid.
Note:
(1) F_t: total assets.
(2) Statutory reserve per policy.
(3) Number of policies in force.
(4) ^SV_t: total statutory reserves.
(5) $^SFA_t^{(b)}$: statutory free assets before solvency margin $= F_t - {^SV_t}$.
(6) SM_t: total required minimum solvency margin.
(7) $^SFA_t^{(a)}$: Statutory free assets after solvency margin $= {^SFA_t^{(b)}} - SM_t$.

Reserve at time t:

$$V_t = 1000 \times A_{x+t:\,\overline{3-t}|} - ZNP\ddot{a}_{x+t:\,e\overline{3-t}|} \tag{9.9}$$

where

$$ZNP = \frac{1000 \times A_{x:\,\overline{3}|} + Z}{\ddot{a}_{x:\,\overline{3}|}} = \text{£}352.88$$

and Z is the Zillmer allowance for initial expenses. Here, we have assumed a (typically partial) allowance of $Z = 100$.

The valuation rate of interest is assumed to be 6% per annum, and the mortality basis to be the same as the fair-value reserving basis (see Section 7.1.2 for details).

The required minimum solvency margin is then calculated (for purpose of illustration) as

$$SM_t = 0.04V_t + 0.003(1000 - V_t) \tag{9.10}$$

using the rates of f_1 and f_2 mentioned in Section 9.1.1.

Hence, for this tranche of business, the office has to provide capital of £359,000 simply in order to set up the required statutory reserves for the business. A further £124,000 is required in order to provide for the required minimum margin of solvency, producing a total capital requirement at time 0+ of £483,000. By the end of the second policy year (in *this* example) the initial capital is no longer required, and by the end of the term, of course, no reserves are required to be held, so we are left with the same total profit as we had originally.

It is instructive to see the above in terms of the annual contribution to the free assets, which we will refer to as the *statutory profit*. We will calculate two types of statutory profit:

$$^SP_t^{(b)} = {}^SFA_t^{(b)} - (1 + i_t)\,{}^SFA_{t-1}^{(b)} \tag{9.11}$$

$$^SP_t^{(a)} = {}^SFA_t^{(a)} - (1 + i_t)\,{}^SFA_{t-1}^{(a)} \tag{9.12}$$

where i_t is the experience rate of interest (assumed to be 10% per annum here).

The values of these quantities are shown in Table 9.2 for ($t = 1$, 2, 3), alongside the realistic profit and fair value calculations for comparison. RP_t is the realistic profit taken from Table 7.4, and FVP_t is the fair value profit, taken from Table 7.6.

Hence, the requirement to meet the statutory regulations has produced statutory profits that are highly negative in the first year, with

Table 9.2 Emergence of statutory profit for a nonprofit fund (£000)

Time t	RP_t	FVP_t	$^SP_t^{(b)}$	$^SP_t^{(a)}$
1	812	642	(256)	(364)
2	0	91	594	488
3	0	105	640	886

positive profits in subsequent years that are higher than the realistic level. $^SP_t^{(a)}$ tends to be smaller than $^SP_t^{(b)}$ in all years except the last, at which point the required minimum margin of solvency falls to zero and is released as profit into the fund. Note that the accumulated amount to time 3 (at the actual experienced rate of return of 10% per annum) of all of these patterns of profit emergence is £983,000, illustrating how different reserving assumptions lead to changes in timing of profit emergence rather than to a change in its overall amount.

This example is, of course, rather atypical of most nonprofit funds, as the policy term is exceptionally short (this was chosen so that the numerical example would be easy to assimilate). In practice, policy terms are much longer than this, usually at least 10 years. In order to illustrate the more typical situation, therefore, we show in Figure 9.1 the values of the statutory profit signature after allowing for the statutory solvency margin $^SP_t^{(a)}$ for a "typical" 10-year nonprofit endowment.

Figure 9.1, therefore, shows the typical nonprofit pattern: a large negative profit (or strain) in year 1, small but gradually rising statutory profits thereafter, with a much larger profit in the year of maturity due to the release of the (by then) large statutory solvency margin.

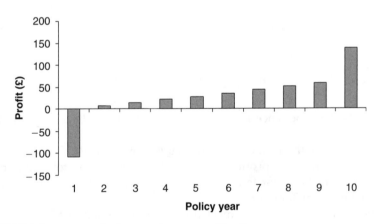

FIGURE 9.1 Statutory profit signature for a nonprofit policy

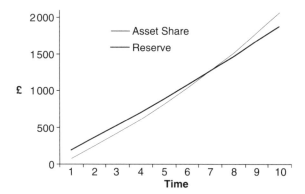

FIGURE 9.2 Comparison of asset share and supervisory reserve over time for a conventional nonprofit endowment assurance

In Figure 9.2, we show the progression of the asset share and the statutory reserve (including the required minimum solvency margin) with time, for the same policy as in Figure 9.1. It is clear that this policy requires capital support for about three-quarters of the policy term.

9.3.1.1 Shareholders' Profit

Because of the overriding necessity to meet the statutory solvency requirements, the amount of distributable profit cannot exceed the statutory free assets of the company at any time, and the statutory profit signature of a nonprofit fund represents the flow of capital to (or from) the shareholders' funds. The large strain in the first policy year (the "new business strain") produces significant additional capital requirements upon the issue of any new nonprofit policy.

One should not forget, however, that tranches of business are not issued in isolation. Were the fund stationary, for example, then the losses incurred from the policies just issued are more than compensated for by the profits from older policies (including those just maturing). The result would be an overall annual profit on the whole portfolio, which is very similar to the total realistic profit; however, the amount of reserves (and therefore assets held by the office) would be considerably higher than if only realistic reserves were held.

In inflationary conditions, however, funds are rarely even approximately stationary in monetary terms. If policy sizes broadly increase in line with inflation, then, even if constant numbers of policies are issued each year, the effect will be a fund that is more heavily weighted to the shorter durations. Total statutory profit will then be much lower than the total realistic profit, and if inflation is too high (or the fund is actually expanding in volume at too fast a rate) then there may be a net capital requirement each year rather than a contribution to shareholders' profit. This problem is, therefore, likely to be most acute for a new fund, or a rapidly expanding one.

9.3.1.2 Embedded Value Profits

The embedded value of a nonprofits fund at any time can be defined as

$$EV_t = {}^S FA_t^{(a)} + \text{(Present value of future statutory profits)} \qquad (9.13)$$

which is essentially the intrinsic value to the shareholders of the existing (in-force) business at that particular point in time. The present value of the future statutory profits would be the value of the expected profit signature of the whole fund discounted at an appropriate *risk discount rate*, which would be somewhat higher than the expected future rate of return on assets. The difference between the risk discount rate and the expected rate of return on assets reflects the additional return that the shareholders require for the use of their capital, which in turn will reflect the risk that the actual future profits will not be as high as anticipated in the expected profit signature. (See Section 11.2.2 for more discussion about the risk discount rate.)

The embedded value profit for any year is then the net increase to the embedded value during the year (Goford, 1985), defined by

$$P_t^{EV} = EV_t - (1 + i_t^\tau)EV_{t-1} \qquad (9.14)$$

where i_t^τ is the risk discount rate for year t.

We will again use our simple 3-year endowment policy as an example of this process. The relevant figures are shown in Table 9.3, where $PVFP_t$ is the expected value of future statutory profits at time t, calculated at a risk discount rate of 15% per annum.

This method is a useful alternative to the fair values method for demonstrating realistic profitability, with particular focus on the value of the business to the providers of capital. Comparison with Table 9.2 shows that (for the example considered here) both methods recognize a

Table 9.3 Embedded value profit for a nonprofit fund (£000)

Time t	${}^S FA_t^{(a)}$	$PVFP_t$	EV_t	P_t^{EV}
1	(364)	1094	730	730
2	88	770	858	18
3	983	0	983	(4)

significant proportion of total profit arising in the first year of a policy, i.e., when a product is sold.

9.3.2 Conventional With-Profits Business

The effect on the supervisory profits of holding statutory reserves is much the same for conventional with-profits as it is for nonprofit business. We will use the numerical examples from Section 8.4 and Section 8.6.1 to illustrate the effect. The following net premium reserve will be used, with value at time t (excluding the value of the reversionary bonus declared at time t) given by

$$^SV_t^{(e)} = (1000 + B_{t-1})A_{x+t:\,\overline{10-t}|} - ZNP\ddot{a}_{x+t:\,\overline{10-t}|} \tag{9.15}$$

where B_t is the declared bonuses up to and including time t and

$$ZNP = \frac{1000 \times A_{x:\,\overline{10}|} + 35}{\ddot{a}_{x:\,\overline{10}|}}$$

with 35 being the assumed Zillmer allowance for initial expenses, in this case.

The value of the newly declared reversionary bonus at time t is

$$0.04(1000 + B_{t-1})A_{x+t:\,\overline{10-t}|} \tag{9.16}$$

according to the stated assumption that the reversionary bonus is a constant 4% per annum compound, which is added to $^SV_t^{(e)}$ to give $^SV_t^{(i)}$: the statutory value of the liabilities at time t, including the cost of the new bonus. Finally, the statutory minimum margin of solvency is added to produce the total statutory reserve. The valuation rate of interest is 6% per annum, and the valuation mortality assumptions are taken to be the same as in the premium basis (see Section 8.4).

The relevant figures (which can be compared with those of Table 8.6) are shown in Table 9.4. The profit signature (the net surplus for the year per policy issued) is shown in Figure 9.3.

As expected, the reserves are higher and the profit lower than when realistic assumptions were used (Table 8.6). There is a relatively small "new business" strain in the first policy year, followed by significant and increasing statutory profits thereafter. Note how this contrasts with the realistic profit in which the annual profit gradually falls with increasing duration, as shown in Figure 8.3. At maturity (see Table 9.4, row 10(M)) the surplus released "before bonus" represents the release of the statutory minimum margin of solvency on the termination of the liabilities, which

Table 9.4 Emergence of statutory profit in a with-profits fund (£000)

Policy year	Surplus for year before bonus	Cost of new bonus	Net surplus for year	Accumulated surplus (free assets)	Total statutory reserves	Total assets (asset shares)
1	88	(255)	(167)	(167)	628	461
2	450	(270)	180	(4)	1634	1630
3	473	(287)	186	183	2689	2872
⋮	⋮	⋮	⋮	⋮	⋮	⋮
8	628	(399)	229	1568	8902	10470
9	668	(428)	240	1965	10379	12343
10	712	(460)	252	2413	11951	14364
10(M)	460	(2873)	(2413)	0	0	0

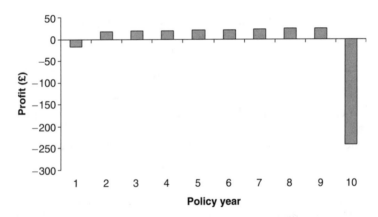

FIGURE 9.3 Statutory profit signature for a conventional with-profits policy

partly offsets the cost of the terminal bonus, as shown. Nevertheless, there remains a very significant net withdrawal of statutory free assets on the maturity of a with-profits policy, due to the cost of the terminal bonus.

The interesting thing to do, however, is to compare the pattern of supervisory profit emergence (Figure 9.3) with that for the conventional nonprofit policy for the same term (Figure 9.1). We can see that supervisory profit emerges much earlier for with-profits than it does for nonprofit. This is also clearly seen from a plot of asset shares and reserve values (Figure 9.4): the crossover point (at which time the policy changes from being a net user of capital to a net provider of capital) occurs *much* earlier on in the policy term, at around the end of year 2 (compare with Figure 9.2 where the cross-over occurs in year 7).

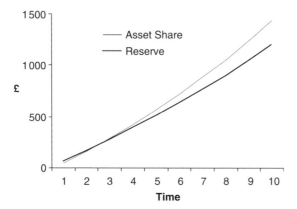

FIGURE 9.4 Comparison of asset share and supervisory reserve over time for a conventional with-profits endowment assurance

In summary, we can make the following observations:

1. The issue of with-profits business requires capital, though significantly less than for nonprofit business with the same level of premium income.
2. For all policy years except the first and last, a with-profits policy contributes significantly to the statutory free assets of the fund. Hence, except for extremely immature or vary rapidly expanding portfolios, the presence of a with-profits fund should provide substantial statutory free assets (i.e., capital) which can be used for other purposes (e.g., for financing the issue of nonprofit policies) as well as contributing to the statutory solvency of the whole office.
3. The build up of the statutory free assets under a with-profits fund is the deferred profit earned in previous policy years, ultimately being paid out as terminal bonus at policy maturity. This causes the large statutory loss seen at the end of the policy term.
4. The very different contributions to the statutory free assets made by with-profits and nonprofit policies should be appreciated (compare Figure 9.3 and Figure 9.4).

9.3.2.1 Shareholders' Profit in a Conventional With-Profits Fund

We can now examine the shareholders' profit from a proprietary conventional with-profits fund, according to the approach set out in Section 8.7. The relevant figures for our example are given in Table 9.5 and shown (per policy issued) in Figure 9.5. All assumptions are as previously, except that the annual premium has been increased to

Table 9.5 Emergence of free assets and shareholders' profits in a proprietary conventional with-profits fund (£000)

Policy year	Surplus for year before distribution	Cost of new bonus	Shareholders' profit	Net surplus for year	Accumulated surplus (free assets)
1	129	(255)	(37)	(162)	(162)
2	526	(270)	(38)	218	39
3	547	(287)	(40)	220	263
⋮	⋮	⋮	⋮	⋮	⋮
8	692	(399)	(53)	240	1827
9	730	(428)	(57)	246	2255
10	772	(460)	(61)	252	2732
10(M)	460	(2873)	(319)	(2732)	0

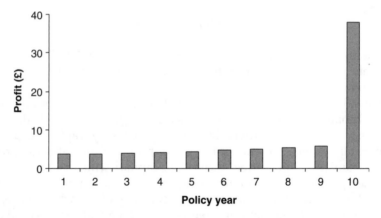

FIGURE 9.5 Shareholders profit from a conventional with-profits policy

£150.45 from £142.95 in order to cover the cost of paying to the shareholders one-ninth of the statutory value of the bonuses distributed to the policyholders each year.

Note that a positive contribution to shareholders' profit in the first policy year is only possible provided that free assets are available from elsewhere in the fund to make good the statutory shortfall. If not, then the bonus declaration should be reduced (down to zero, if necessary) and/or the shareholders will have to provide additional capital in order to increase the assets of the fund up to the level required for statutory solvency. This additional initial capital would be expected to be repaid during the duration of the policy through the appropriate choice of bonus rates (to which, of course, the shareholders' profits are inextricably linked).

A slight anomaly arises in this method of allocating profit to shareholders. If the actuary decides to reduce the valuation rate of interest (i.e., to take a more conservative view of the expected yield to be obtained from the assets) then the amount of profit (for the same rate of bonus) which is given to the shareholders will increase. It would also be in the shareholders' interests to increase reversionary bonus rates at the expense of reducing terminal bonus rates, as the shareholders would then receive their profits earlier. While this would benefit policyholders by increasing their level of guarantee, it will increase the risk of insolvency. These conflicting interests have to be considered carefully when determining bonus rates in proprietary offices, where the U.K. reversionary bonus system is used.

9.3.3   Other Types of With-Profits Business

The impact of the supervisory reserves on new business strain and profit emergence should be broadly similar for the other types of with-profits business described in Chapter 8. The main differences will largely relate to the different extents of profit deferral (i.e., the proportion of profits paid in terminal form) under each method; thus (for example), the lowest new business strain will be incurred for the U.K. reversionary bonus method and the highest will be under revalorization methods, with the contribution and AWP methods falling somewhere between the two.

For the contribution and revalorization methods, the net premium valuation would be the most appropriate valuation method, as for the U.K. system. For AWP, the basic methodology is as set out in Section 8.6.2; but see also the further discussion of the calculation of nonunit reserves for unit-linked contracts given in Section 9.3.4, where the arguments there apply equally to AWP.

Finally, for all with-profits contracts (except possibly the revalorization methods) consideration should be given to the rate of regular profit distribution that can be supported under the (conservative) valuation assumptions, given the smoothing requirements demanded to meet policyholders' reasonable expectations. For example, for an AWP policy currently paying 6% per annum bonus interest and 0% guaranteed, a valuation interest rate of (say) 4% per annum would only support a future bonus of not more than 4% per annum. In the valuation reserve the actuary should, therefore, assume future bonuses gradually reduce (in accordance with PRE) down to the affordable level over a period of a few years, so that the statutory reserves would need to be increased to cover these excess bonus payments, at least for those policies that have relatively short durations remaining.

9.3.4 Unit-Linked Business

The design of unit-linked contracts results in very different statutory solvency requirements compared with those of conventional nonlinked contracts. As explained in Section 7.2.2, if the policyholders' benefits are exactly equal to their unit reserves, then there is no risk of loss from variations in investment returns. Furthermore, if the annual charges are sufficiently closely linked to costs (and the future charging rates are not guaranteed), then cash flows may be expected to be positive throughout the lifetime of a policy and no nonunit reserve would need to be held. The statutory solvency requirements for unit-linked business, therefore, reflect this risk profile, and in essence require the following:

1. Unit reserves equal to the value of the policyholders' unit holdings (net of any allowable actuarial funding or negative nonunit reserves; see Section 7.2.3 and Section 7.2.4).
2. A reserve to cover any mismatch between the unit fund liabilities and the assets held in order to meet those liabilities (see Section 7.2.2).
3. Appropriate nonunit reserves to cover any anticipated future negative cash flows on existing business (see Section 7.2.2).
4. A statutory minimum solvency margin (or risk-based capital) requirement. This requirement can be relatively low for unit-linked contracts, as long as minimal guarantees are provided.

The methodology for calculating nonunit reserves (point (3)) has already been described in Section 7.2.2. However, the situation of a supervisory valuation is different from that explained in Chapter 7, which is (as always) that the reserves must be based on appropriately prudent assumptions of future experience. So while cash flows (income and outgo) could be well matched on a realistic, best estimate, basis, in a valuation we may have to assume significantly more adverse future experience, which may lead to future negative cash flows where none may be expected realistically. These must be reserved for, as a nonunit reserve, in the supervisory valuation. (These statutory reserves would also have to be allowed for when pricing the product — we will return to this in Chapter 11.)

Nevertheless, policies under which regular charging (rate) reviews[1] occur will not require *such* conservative valuation assumptions in order to be *sufficiently* prudent; should future experience become too adverse, the company has the right to increase its charges at the next review date if it needs to. The nonunit reserves would, therefore, not need to be so big

---

[1] Under most unit-linked policies, the company has the right to vary the policy charges and/or premiums periodically. This was described in Section 7.2.1; it is also very similar in nature to the indeterminate premium structure described in Section 7.1.1.

for such a policy compared with one where the future charges were all guaranteed.

A note of caution should be made here. Just because a policy's charges are *contractually* reviewable does not mean that the company will be *able* to increase charges in reality to whatever future level would be needed to cover costs. Policyholders will not like the price increases, and the result might be a large increase in surrenders (and no doubt involving healthy lives only, leaving the high-risk lives in the fund). The supervisory reserve should take full account of such possibilities.

Even with all these caveats, the unit-linked policy design does have a major advantage over its conventional nonlinked counterparts, at least potentially: a significant reduction or even avoidance of new business strain, which allows an office much more freedom to expand its business for example, without the risk of breaching the statutory solvency requirements.

It should be noted that it is always dangerous to apply general observations, such as those above, to the particular circumstances of a life office's actual unit-linked portfolio. It is also important for an office to hold statutory reserves in accordance with the actual risks involved, rather than to abide simply, and uncritically, by the letter of the regulations. The statutory valuation of a life office in general is, therefore, not a passive mechanical operation, but one requiring expert professional consideration and judgement, for which, of course, the role of the actuary is of paramount importance. This should always be borne in mind when considering the effect of the statutory regulations on the operation of any life office.

References

Black, K. and Skipper, H. D. (2000). *Life Insurance*, 13th edition. Prentice-Hall.

Fisher, H. F. and Young, J. (1965). *Actuarial Practice of Life Insurance*. Institute of Actuaries.

Goford, J. (1985). The control cycle. *Journal of the Institute of Actuaries Students Society* 28, 99–114.

Groupe Consultatif des Associations d'Actuaires des Pays des Communautes Europeennes. (1990). Report on the calculation of technical reserves for life insurance in the countries of the European Communities.

IAIS. (2002). Principles on capital adequacy and solvency. www.iaisweb.org.

MacDonald, A. S. (1993). What is the value of a valuation? In: *Proceedings of the 3rd Actuarial Approach to Financial Risks International Colloquium*, vol. 3, pp. 725–743.

Scott, P. G., Elliott, S. F., Gray, L. J., Hewitson, T. W., Lechmere, D. J., Lewis, D., and Needleman, P. D. (1996). An alternative to the net premium valuation method for statutory reporting. *British Actuarial Journal* 2(Part III), 527–621.

Shelley, M., Arnold, M., and Needleman, P. D. (2002). A review of policyholders' reasonable expectations. *British Actuarial Journal* 8(Part IV), 705–755.

Chapter 10

Life Office Risks and Risk Management

10.1 The Interested Parties

The general concept of life office risk was introduced in Chapter 6, where the risk was identified as arising from three main sources: insurance risk, investment risk, and business risk. It is the purpose of this chapter to look at these sources of risk in more detail, but it is first necessary to define the overall nature of the life office's risks.

A logical starting point is to consider the various interests of the three main parties with whom a life office is concerned, namely the policyholders, the shareholders (where appropriate), and the regulators.

10.1.1 Policyholders' Interests

In Section 3.6.4 we described how life insurance companies will seek to maximize returns, at least for their unit-linked and with-profits policyholders, subject to the constraint of taking sufficient precautions to maintain company solvency. This aim can be taken, perhaps, as a possible definition of success for the company; hence, life office risk can be considered as the risk that it will fail to meet these aims. The risks are therefore that:

1. The office will become insolvent, and hence unable to fulfill its contractual obligations.
2. In some sense, the returns (or "value for money") provided for its policyholders will be inadequate.

A policyholder's interpretation of (2) will, of course, depend upon the type of policyholder concerned. A holder of a nonprofit policy will have already considered his or her contract as providing acceptable value for money when agreeing to enter into the contract; hence, such a policyholder only remains concerned with risk (1). However, offices must continue to ensure that nonprofit policyholders are not over-charged for the risks they present, otherwise the public will not purchase these contracts and will essentially be denied this important financial service.

A holder of a unit-linked policy will wish to ensure that the charges levied under the contract are fair, and also that the investment managers of the office secure a "reasonable" return on the assets in which the policyholder's unit fund is invested. What is "reasonable" in this context will depend very much on the nature of the unit fund or funds chosen by the policyholder. This choice will have been made on the basis of information provided by the insurance company, whether from the past performance of the funds or from the stated policy regarding their future management. In either case, the policyholder will have chosen the fund or funds whose expected profile of returns matches his or her own risk preferences (see Section 3.8). Hence, the duties of the life office are simply to ensure that the distributions of returns from each fund are in accordance with its stated (or implied) investment policy, and, subject to this condition, to seek to maximize the returns obtained from each fund.

With-profits policyholders have the most complex requirements, as they essentially contribute to the life office's capital and hence demand a fair return on that capital (see Section 8.6.1). This interest is obviously shared with the shareholders where the office is proprietary. As described in Section 8.2.3, a with-profits policyholder's reasonable expectation of returns can be considered to be some kind of smoothed asset share, the extent depending on the method of profit distribution concerned. There are a number of component elements in this expectation, relating to:

1. the smoothing policy adopted
2. the charge made for the policy guarantees
3. the investment of the accumulating asset share

All three of these elements depend very much upon the office's past behavior and stated intentions of future behavior in these respects. Furthermore, the office will have to invest its policyholders' asset shares in a way that is consistent with the expected smoothing policy and the policy guarantees, i.e., in a way that would not be expected to threaten the continuation of (1) and (2). It should be noted that, in many offices, the with-profits policyholders' asset shares will include an appropriate share of the office's profits from its other (nonparticipating) business, consistent with the use of the with-profits' capital in financing these activities. There may also be contributions to asset shares from the returns earned from the estate, reflecting policyholders' expectations of a share in these returns (in the case of a mutual company, this would be a 100% share).

10.1.2 Shareholders' Interests

The shareholders' interests are similar to those of the with-profits policy-holders, except of course that they are concerned with their own share of the office's profits or losses in relation to the capital they have provided. Like the policyholders, they will also expect some return from the estate, depending upon how much of the estate can be considered as owned by the shareholders; see Smaller et al. (1996). In proprietary with-profits offices, the transfer of profit to the shareholders is closely linked to the distribution of policyholders' profits (see Section 8.7), so that meeting the interests of the with-profits policyholders should normally also satisfy the expectations of the shareholders.

10.1.3 Regulators' Interests

As we mentioned in Chapter 9, the aim of the regulators is to ensure that the policyholders' reasonable expectations are met. Hence, again satisfying the policyholders' interests should, at least in theory, also satisfy the regulators. However, the regulations ensure that this is not left entirely to chance: companies have to satisfy the specified regulations in order to continue in existence as a going concern.

10.2 The Nature of the Risks

Bearing in mind the foregoing, we can summarize the risks that life offices have to manage into three main elements:

1. The risk that the office will become insolvent ("actual" or "true" insolvency).
2. The risk that the office will not be able to meet the statutory solvency requirements ("statutory" insolvency).
3. The risk that the returns to the policyholders and/or the shareholders will be "inadequate."

Actual insolvency (risk 1) occurs when the actual amount of a company's committed (i.e., certain) liabilities exceed the amount of assets at any point in time. While trading would then cease immediately, it is unlikely that the existing policyholders would receive very much compensation from the dispersion of the company's assets relative to the expected value of their former contractual benefits. The shareholders would certainly lose all of their equity.

Hence, risk (1) essentially represents the most severe outcome of risk (3); the return to shareholders and policyholders could not be more inadequate than this.

However, it is important to identify true insolvency as a separate risk, because it marks a point at which something dramatic happens: the office ceases to honor its contractual liabilities to its policyholders. Prior to that point, full contractual benefits may still have been payable on claims (financed, for example, by cash flow and/or by depletion of assets). The returns to policyholders, therefore, fall dramatically at the point of insolvency; the returns to shareholders will show a similar effect but it may be a little less abrupt, depending upon whether or not dividends were being paid up to the bitter end.

The above discussion of true insolvency is, of course, almost completely hypothetical. This is because, in practice, true insolvency would never be allowed to happen; statutory insolvency would always happen first, as a result of which the fund would normally be wound up or transferred to another insurer. In any case, the policyholders and shareholders will receive some value for their respective entitlements to the assets (which will not be zero), depending on the terms of termination agreed with the regulators, although it might leave policyholders with an entitlement to smaller benefits than originally contracted under their policies. Hence, risk (2) marks another occasion in which a sudden fall in policyholders' returns occurs, i.e., it has a similar effect to actual insolvency but with a much reduced degree of severity. The shareholders, however, will almost certainly lose the whole of their equity on statutory insolvency, as all of the available assets will be used to provide compensation to the policyholders.

Management needs to concentrate on managing the risks in the order (1) to (3). A completely solvent office would, therefore, be concerned with managing risk (3) (although not taking any action which would unduly threaten risk (2)), while an office with a small statutory solvency margin would be mostly concerned with managing risk (2). Should statutory solvency be irrevocably breached, then the concern would focus, immediately, upon risk (1).

Risk (1) will be prevented if, overall, the income to the office generated by each tranche of business is at least equal to the outgo incurred, by the time that the tranche is terminated. The true insolvency risk, therefore, depends upon the constituents of the office's income and outgo, i.e.:

for income—capital provision, premiums, investment income, and asset
    appreciation
for outgo—claims, expenses, tax, and distributions of profit

Risk (2) depends additionally on the generation of statutory valuation strain during the policy term. For example, one tranche of nonprofit policies would produce statutory insolvency at duration 0+ (i.e., immediately after the first premium has been paid) if there was insufficient capital provided; see Section 9.3.1. The most important additional intrinsic factor affecting risk (2) is product design, because of its importance in determining the statutory profit signature.

Risk (3) is the risk of not meeting policyholders' (and shareholders') reasonable expectations of making returns. The aim of life office management is, therefore, that these reasonable expectations are at least met, and in most cases this will manifest itself in attempts to increase profitability and to ensure a fair distribution of those profits. Attempts to maintain or increase policyholders' and shareholders' returns by overdistribution (i.e., by distributing consistently more profit than earned) will ultimately increase risks (2) and (1). On the other hand, persistent retention of profit (i.e., over and above what would be considered a reasonable charge for capital or for prudent management of the estate) is unfair to the policyholders and shareholders and contrary to expectation.

Of course, part of this management process is to ensure that the expectations of its policyholders are, in fact, reasonable. Hence, expectations should not be influenced by:

figures of past policy payouts in isolation of the conditions which led to those payouts

monetary projections of future policy payouts, unless properly qualified by the assumptions underlying the projections

Policyholders' expectations are likely to be significantly influenced by the advice given to them during the sale of their policies; the behavior of the sales intermediary is of paramount importance in shaping policyholders' expectations, reasonable or otherwise. Hence, an important element in controlling the risk is the control of the behavior of the intermediaries.

10.3 Risk Control

We are now going to consider the various contributing elements to the life office's overall risk and the ways in which these risks can be controlled. In order to do this, it is necessary, or at least very helpful, to have a benchmark against which experience can be compared. For conventional nonprofit contracts, the obvious benchmark is the office's premium basis: the assumptions that were made in setting the prices at which the policies were, and are being, sold. Should all these assumptions be exactly met in practice, then the office will earn the levels of profit it expected when the policies were issued and, provided the policies were appropriately priced, the office should meet all three of its risk objectives. Should the experience be worse, in aggregate, than that allowed for in the premium basis, then the amount of profit will be lower than anticipated. This will, ultimately, translate into lower than expected returns to the participating policyholders and/or the shareholders (depending upon which parties share in the profits from the nonparticipating business). The risk must, therefore, not only be considered in terms of whether the experience is worse than the chosen benchmark, but also by how much it is worse. Hence, the significance of the risk depends upon both its incidence and its intensity.

All other types of contract provide benefits that depend in some way upon future investment returns. Hence, the appropriate benchmark for the investment risk under these contracts will relate directly to what the corresponding policyholders consider to be adequate in the prevailing conditions: this was discussed in Section 3.6 and Section 3.8. The other sources of risk for these contracts, i.e., the insurance and business risk elements, can refer to the pricing basis benchmark described above. For example, under a unit-linked contract the company runs the (investment) risk that the returns it achieves on its linked assets will be inadequate for its policyholders; it also runs the risks that its expenses and mortality experience will be worse than that assumed in calculating its charges.

The control of life office risk should, therefore, be considered with respect to each of its three main sources described in Chapter 6, namely investment risk, insurance risk, and business risk. Life office investment risk was described in Chapter 3; we consider insurance risk and business risk below.

10.4   Insurance Risk

This is the risk that the cost of contractual claims will be higher than that assumed in the premium basis. This risk has two elements: the amount of claim and the incidence of claim. The incidence of claim is determined by the occurrence of contingencies specified in the policy conditions, such as death, survival, and sickness. The premiums are usually calculated on the basis of the expected behavior of these random events, according to some statistical model. The incidence of claim can, therefore, differ from that assumed due to:

error in the parameterization of the model, e.g., that the assumed expected mortality rates in an assurance premium basis are lower than the actual expected mortality rates
random variation around the actual expected incidence of claim
error in the choice of model structure

These elements have been discussed by Daykin et al. (1994), Cairns (1995) and by Hooker et al. (1996). They are sometimes referred to as *parameter error, process error,* and *specification error* respectively and are not confined to life insurance risk: they apply equally to the use of any model in actuarial management (see Section 11.7 and Section 14.3 for examples).

Most long-term policy claim amounts are determined by precise, fully specified rules. This contrasts with the majority of general insurance (short-term business) policies, which are contracts of indemnity (see Chapter 12), under which the benefit is determined by the event itself, i.e., by the amount of loss incurred by the event. Claim amounts under long-term insurance business are, therefore, fundamentally predictable, and this feature greatly assists in restricting the extent of the insurance risk.

Nevertheless, claim amounts still contribute significantly to the insurance risk of life offices, due to the potential concentration of large sums assured payable on the death of single individuals. If we have n_x independent policyholders aged x, each with probability q_x of dying before age $x + 1$, and for which $s_{x,i}$ is the sum assured payable on death to the ith such policyholder, then it can be shown that the variance of the random amount of total claim C_x under the n_x policies over the year of age is

$$\text{Var}(C_x) = q_x(1 - q_x) \sum_{i=1}^{n_x} (s_{x,i})^2$$

Hence, for the same total sum assured at risk and for the same number of policies insured, the variance of C_x will be greater the more concentrated the risk is in individual lives, even though the expected death cost will be the same. This is equivalent to saying that the greater the variability in the sum assured between individual policies, the greater will be the variability in claim cost and, hence, the higher the probability of insolvency (all else being equal).

10.4.1 Underwriting and Risk Classification

The main way in which life offices control their claim incidence rates (or "frequencies") under their life and sickness insurance contracts is by initial *underwriting*. Underwriting is the means by which life offices judge whether an applicant should be accepted for insurance, and if so to establish the terms of acceptance. The aim is to classify the applicants into the following general risk groups:

1. insurable at the office's standard rate of premium for the contract applied for
2. insurable, but with special conditions attached
3. possibly insurable at a later date
4. uninsurable

For marketing reasons, the premium rate assumptions should be such that the majority of applicants would be acceptable at the standard rate of premium. These standard rates will be specific to the age of the policyholder (and term of the contract, if appropriate), and in some cases may be dependent upon other risk characteristics, e.g., upon the smoking habits of the applicant. (See also Section 10.4.13.)

A minority of applicants should fall into the second category, which is referred to as the "impaired lives" category. The special terms may include a debt on the policy (a deduction from the normal level of benefit) which might reduce over the term of the policy, charging of a higher premium, or exclusion of certain causes of death from the terms of the insurance. The terms are decided on an individual basis by the life office's underwriters in order to reflect their best assessment of the particular mortality risk presented by the applicant. Examples of impairments leading to special terms include known health conditions which could affect future mortality (e.g., heart disease, diabetes, obesity), various "risky" aspects of lifestyle (e.g., smoking, alcohol consumption, drug use, sexual promiscuity, dangerous sports) and family history (e.g., a high incidence of certain diseases among close blood relatives can sometimes affect the terms of acceptance). There is now also the possibility of genetic test results providing information about mortality risk; see Section 10.4.1.1.

A small number of applicants (group (3) above) may be currently uninsurable because of uncertainty regarding the level of risk involved. Examples would include applicants who are recovering from an operation or currently receiving treatment for some serious condition. The life office may defer consideration of these applicants until some time in the future when the prognosis becomes clearer.

There will be a final group of applicants whose condition is such that their expected claim experience cannot be predicted with sufficient accuracy to enable them to be insured. This would include people with certain congenital diseases, and most people with terminal illnesses, such as cancer and AIDS. This group, which should form a very small minority of applicants, would be declined insurance.

The overall aim of the underwriting process is for the average mortality experience under each group of applicants to be reflected by the premiums or charges paid. In particular, underwriting seeks to avoid adverse selection, e.g., to prevent individuals who present a significantly higher than average mortality risk from being charged the standard (and hence inadequate) rate of premium, or from securing an exceptionally high level of insurance cover (see Section 10.4.2).

10.4.1.1   Genetic Testing

A developing concern in the area of risk classification and underwriting is the development of genetic testing. This is where individuals can be screened for the presence or absence of specific errors in their genetic constitution, which can (so far only in rare cases) be a relevant predictor of mortality risk. The most quoted example is the case of Huntington's chorea, where an early (pre-senescent) death can be accurately predicted by the presence of a particular genetic error, which is identifiable by a genetic test (see Barnaby (1997), Wilkie (1997), and Le Grys (1997)). However, the possibility of using genetic test results routinely to make accurate risk assessments of major causes of death, such as heart disease and cancer, still seem a long way off, if not impossible to achieve, especially when the significant influence of environmental factors is taken into account.

Despite its current rarity, there is considerable debate concerning the ethics of using genetic test results for insurance risk classification purposes (see Barnaby (1997) for a brief overview). Practice in different countries is certainly not uniform; in some territories, known test results may have to be disclosed in insurance applications, whereas in others the use of any genetic test information in underwriting is not allowed. To our knowledge, nowhere is it permissible practice for insurance companies to *require* applicants to undergo a genetic test.

10.4.1.2 The Underwriting Process

The process of underwriting is described by Diacon and Carter (1992), Luffrum (1989), and by Black and Skipper (2000). A brief outline will be given here. Further reading on underwriting can also be found in Fisher and Young (1965), and in Leigh (1990).

Proposers for long-term insurance usually have to complete a kind of medical questionnaire, normally on the proposal form. On the strength of their responses to the questions, the insurer may seek to obtain further information from the proposer's medical practitioner, and/or commission an independent medical examination. While each additional stage in this sequence provides the underwriters with an increasingly reliable means of assessment of the risk presented by the proposer, they necessarily incur additional expense. Hence, these additional stages (medical report and medical examination) are only followed when the risk appears to be potentially higher than usual. This would occur when:

The proposer indicates some unfavorable or potentially unfavorable aspect of his or her health or lifestyle in the initial proposal.
The level of benefit applied for is higher than some preset limit for the age of the applicant (reflecting the importance of higher than average benefit levels on the variability of the claim cost, as described earlier).

In this way, it is hoped to employ the more detailed underwriting procedures in the most cost-effective way, i.e., only in those areas in which the more accurate classification of the risk is most likely to lead to significant savings in claim costs in the future.

An office must, therefore, decide upon its desired compromise between underwriting cost and the accuracy of its risk classification, both of which will also affect the premium rate that it can charge. There may be marketing as well as risk control considerations which will affect this decision, as, for example, it would be clearly very expensive and inefficient to have a very competitive premium basis if the company's market consists of relatively high risks, such that the majority of its applicants might have to be accepted on "special terms."

10.4.1.3 Preferred Lives

In some territories (particularly the U.S.) life insurance premium rates differ according to a much wider range of rating factors, such as income level, occupation, geographical region, family history, and the height-to-weight ratio (Le Grys, 1997). The variation in premium rates is broadly $\pm 50\%$ to standard rates. The advantages are that good risks get cheaper rates (and insurance companies can attract more low-risk business), but the disadvantages are that bad risks can feel unfairly discriminated against, particularly where the cause of their higher risk is largely or entirely outside of their control (e.g., due to social class background or to family history). In the U.K., for example, insurers have made little headway in this direction, other than charging different rates for males and females, and for smokers and nonsmokers.

10.4.1.4 Underwriting Annuity Risks

Underwriting is not necessary for applicants for annuities, as they are "self-selected": people who perceive themselves to be in poor health are unlikely to purchase annuities, so that the average initial mortality of annuitants would be expected to be lower than that of the general population. The office must, therefore, seek to ensure that the mortality rates assumed in its annuity premium rates are at least as low as the average mortality that will be experienced by the annuitants, in order to avoid losses. It should be noted that population mortality rates in most developed countries have generally improved (i.e., reduced) over recent years, so that it is particularly important to incorporate anticipated future reductions in mortality rates in the annuity premium basis.

10.4.1.5 Impaired Life Annuities

A recent and increasingly popular development is the issuing of impaired-life annuities. Here, applicants for annuities who have poor health or have a predictably raised mortality risk (such as smokers) may be offered more favorable annuity conversion rates, reflecting their shorter life expectancy. This has proved more popular in the U.K., for example, than its assurance equivalent (the preferred lives concept described above). Nonetheless, some important ethical issues are involved, e.g., if smoking is used as a rating factor then it could provide people with a financial incentive to smoke. (For further information on impaired life annuities, see Ainslie (2000).)

10.4.2 Financial Underwriting

There is some evidence, particularly from North America, that higher mortality experience tends to be associated with policies having higher sums assured. It would be rational for policyholders who perceive themselves to be a high risk to be more likely to choose a high level of benefit; this is an example of adverse selection.

There are potential "moral hazards" in life insurance. There may be a temptation to induce early claims, e.g., by suicide. Particularly suspect cases are proposals for life insurance on the life of a different person from the one who will be paying the premiums, referred to as "life of another" cases. There is also a risk that claims may be fabricated, usually by claiming deaths to have occurred in some overseas country where identification and registration of deaths are not rigidly enforced.

Financial underwriting is designed to identify these potential risks, if possible, in the following ways:

1. In order to ensure that the proposer has a valid *insurable interest* in the life of the policyholder that is at least as large as the amount to be insured. An insurable interest can be considered to exist if the policyholder would incur financial loss by the insured event, i.e., on the death of the life assured. While all individuals are assumed to have unlimited insurable interest in their own or their spouse's lives, in any other "life of another" case the insurable interest has to be demonstrated. Valid "life of another" cases do exist, of course; examples include a person who may have an insurable interest in his or her parents; a firm, who may have an insurable interest in particular individual "key" employees, whose death in employment would be expected to incur a reduction in profit to the firm (known as *keyman* assurance). (See Diacon and Carter (1992) for more information on insurable interest.)

2. In order to ensure that the need for the insurance (and at the proposed level) is present, and that the prospect has adequate financial means to pay for the anticipated level of cover throughout the term of the policy. This should uncover most cases of moral hazard, such as intended suicide, and would also tend to reveal prospects who, however altruistically, perceive their mortality risk to be exceptionally high.

Financial underwriting risks can extend over different life offices; proposers could secure a very high total level of death benefit by taking out modest-sized policies with a number of different firms. Therefore, it is customary to ask proposers to state whether they have applied for policies from other life companies, although compliance is difficult to verify.

10.4.3   Reinsurance

Reinsurance is essential for reducing the risk caused by the concentration of sums assured under individual policyholders. The essential principle of reinsurance is that it is a means by which a single risk, accepted by one life office, is shared with another life office, or offices.

The whole insurance and reinsurance process can be summarized as follows (from Spedding (1989)).

1. A life office (the *direct-writing* office) issues a contract of assurance with a member of the public for a sum assured of £A.
2. The direct writing office (the *ceding* company) may effect a contract of reinsurance for £B ($B \leq A$) with a reinsurance company, with benefit payment contingent on the original risk (e.g., on the death of the policyholder). It may alternatively effect several contracts, each with different reinsurance companies, with total amounts reinsured equal to £B.
3. The reinsurance company may effect a contract (or contracts) of reinsurance with other insurers (the *retrocessionaires*) for amounts totalling £C ($C \leq B$).
4. Retrocession could continue until the original risk of £A is shared out among a number of insurance providers, at levels that individually do not constitute an excessive risk to each office. Clearly, the larger A is, the greater the number of offices that will generally be involved in collectively covering the risk.

Thus, reinsurance allows each office to keep the total benefits insured at any one time for any individual life below a certain maximum level, thereby restricting total claim amount variability and as a result helping to reduce the probability of insolvency. If the life office also acts as a reinsurer, then it will gain a greater diversity of risk in its portfolio of policyholders, which will also help to reduce variability, e.g., by reducing the incidence of nonindependent deaths among the lives assured. The reinsurance process also allows individual offices to accept much larger risks than they could do otherwise, in the knowledge that the office itself will only need to retain a certain amount of the total loss under its own account. This obviously helps the office to sell business, and the public benefits by being able to effect contracts for much higher sums assured than would otherwise be insurable by the industry.

10.4.3.1 *Proportionate Reinsurance*

Descriptions of the process of reinsurance can be found in Diacon and Carter (1992), and in more detail in Carter (1983), Spedding (1989) and Black and Skipper (2000). Almost all life reinsurance is arranged on a *proportionate* basis, in which reinsurance is set up in relation to individual policies. The amount payable by the reinsurer under a particular policy is fixed as a certain proportion of the sum assured under the original policy, or as a proportion of the sum at risk (see below).

There are essentially two methods of proportionate reinsurance: *original terms* and *risk premium*. These are known as *coinsurance* and *yearly renewable term reinsurance* in the U.S. Original-terms reinsurance involves paying a proportion of the full office premium to the reinsurer, in return for which the reinsurer will pay the same proportion of the total claim amount should a claim arise under the policy. The ceding office, therefore, builds up a smaller fund, for which correspondingly smaller reserves need to be held, under each reinsured policy. As the policy reserve does not form part of the office's insurance risk (it is required whether the life dies or survives over any particular year), then this transfer of a proportion of the policy reserves to the reinsurer would normally be considered unnecessary and would in most cases lead to a loss in potential investment profit to the ceding office from the assets thereby given up. For these reasons, original-terms reinsurance is usually restricted to term assurance contracts, in which reserves are relatively small and, hence, the loss of investment profit is relatively minor.

Original terms is particularly inappropriate for the reinsurance of with-profits contracts, as the reinsurer would have to pay its proportion of all future bonuses declared by the direct writer, which would lead to very high investment mismatching risks for the reinsurer. Here, the risk premium method would often be used (see below) but, as an alternative, original terms reinsurance with a deposit back arrangement is also possible.

Under this method fixed proportions of total premiums and claims (including bonuses) are paid, as usual under original terms. Additionally, the reinsurer pays back to the insurer a "return" premium (called the *deposit back* premium) which is calculated to be enough to enable the insurer to set up full reserves for the policy. In the event of a claim the reinsurer is then only liable to pay its share of the sum assured and bonuses *less* the reserve value at the claim date. The reinsurer's claim amount, therefore, reduces in size with increasing policy duration, and in the case of an endowment reaches zero for a claim at policy maturity, thereby relieving the reinsurer of its investment risk.

Table 10.1 Example of reinsurance under a surplus proportionate reinsurance arrangement with an office retention limit of £250 000

Policy no.	Total sum assured (£)	Amount reinsured (£)	Proportion reinsured (%)
1	1,000,000	750,000	75.0
2	600,0000	350,000	58.3
3	200,000	0	0

Original-terms reinsurance is usually arranged under a *quota share* or *surplus* basis. Quota share involves a fixed proportion of all the ceding office's new business, however big or small, being transferred to the reinsurer. This is an inefficient method of reducing insurance risk, as it reduces the premiums of small policies, which constitute little to the claim variability, while it might still allow significantly large retention of risk from the very large policies. It is, however, useful for financing purposes, to which we will return in Section 10.5.3.

Under the surplus method the excess of the sum assured under any policy over a certain agreed maximum *retention limit* will be reinsured. Examples are shown in Table 10.1.

This is a much more efficient method of reducing the office's insurance risk, as it targets the large policies only. It also has the advantage of definitely restricting individual retained policy sizes to a known maximum level, which is not achieved under quota share.

The *risk premium method* differs from original terms in that the amount reinsured is a proportion of the sum at risk rather than of the total sum assured. The reinsured sum assured for year t would, therefore, be calculated as

$$S_t^R = R(S - V_t)$$

where R is the the proportion reinsured, S is the total sum assured under the original policy, and V_t the reserve under the policy at the end of year t.

The premium, referred to as the *risk premium*, covers only the reinsured sum at risk, and as a result varies each year. The risk premium for year t, $(RP)_t$, would be given by

$$(RP)_t = Q_t S_t^R$$

where Q_t is the reinsurer's "risk premium rate," a large part of which would consist of the expected mortality rate, the remainder allowing for the expenses and a contribution to the profits of the reinsurer. Risk premium reinsurance can operate as a surplus arrangement, with the ceding office retaining a maximum sum at risk equal to its retention limit each year, or it can operate as quota share, in which a fixed proportion of the sum at risk is reinsured each year. These two methods are sometimes alternatively referred to as the *constant retention* and *reducing retention* methods respectively.

Risk premium reinsurance is the most efficient method for reducing the office's insurance risk, as it not only targets just the larger policies, it also allows the office to retain its reserves in full under all its policies, thereby minimizing the transfer of potential investment profit to the reinsurer. This method is thus well suited to "permanent" assurance contracts, such as whole-life policies or endowments, and also to the reinsurance of the additional insurance benefits under unit-linked or accumulating with-profits contracts. It is, of course, a little more complex to administer than original terms, as both risk premium and benefit may vary each year under each policy.

10.4.3.2 Administration of Reinsurance

The vast majority of life office reinsurance is arranged under a treaty between ceding office and reinsurer. For example, a "surplus" treaty might require all new term assurance policies taken on by the ceding office to be reinsured above the agreed retention limit, provided the total sum assured does not exceed a certain maximum level (this would be the limit to automatic cover, stated in the treaty). Reinsurance of larger amounts would have to be negotiated on an individual (or *facultative*) basis between ceding office and reinsurer. The premium rates, and any commission rates paid by the reinsurer, would also be stated in the treaty. Treaties are reviewed from time to time by mutual agreement between both parties.

The reinsurance of impaired lives would normally be done on a facultative basis. However, as reinsurers tend to obtain greater experience (through greater volumes) of impaired assured lives than most individual direct-writing offices, they usually provide their treaty offices with underwriting manuals to assist in the underwriting of substandard cases. This benefits both parties as it allows the ceding office to underwrite on the strength of a larger experience (and, therefore, with more confidence) than it could do alone, and the reinsurer benefits by knowing that the terms by which the ceding office has accepted the substandard risk are compatible with its own expected experience, so that it should be possible to reinsure the risk, if required, at equitable rates to both parties.

10.4.3.3 Determining Reinsurance Policy

In transferring part of the insured risk to a reinsurer, a life office must also be transferring part of its potential profit to the other company. When determining an appropriate reinsurance policy, therefore, an office has to compromise between reducing insurance risk on the one hand, and reducing potential profitability on the other: reinsurance is a means by which the office trades profit for stability.

Reinsurance policy determines both the methods and terms under which reinsurance is arranged. To control insurance risk, surplus original terms or risk premium methods would clearly be advocated, depending upon the type of policy, as discussed above. The key decision remaining is the choice of retention level. Lowering the retention level will increase stability but will reduce the potential profit. The retention level could, therefore, be placed at too high a level, in which case there may be too great a risk of incurring large net claims; or at too low a level, in which case the profit from the office's business is so reduced as to make the company nonviable, producing an inadequate return on capital. It is also possible that profit could be so restricted by low retention levels that the office would expect to make an overall loss on its business, which of course would also result in insolvency.

This question can only be fully addressed by a proper assessment of the contribution to the "three risks" of the office that the future claims experience is expected to make, as described at the start of Section 10.2. The optimum retention level should be such that any lower level would produce a reduction in risk that is not as valuable to the shareholders as the reduction in expected profit levels caused by the additional reinsurance. This is a complex matter, requiring considerable judgement, and detailed consideration is beyond the scope of this book.

A maximum retention level may in theory be determined from consideration of the following probability:

$$\Pr\{D(r) - E[D(r)] > x\}$$

where $D(r)$ represents the (random) total death strain over a period of time net of reinsurance according to a given retention level r for new policies; x is a certain maximum acceptable excess death strain over the period. The retention level should then be chosen such that this probability is not higher than some acceptable maximum small amount (ε, say). The quantity x would be chosen by reference to the expected contribution that the death claims will make to, say, the overall solvency of the office.

It should also be noted that we have so far only considered reinsurance policy from the point of view of controlling the insurance risk: consideration of the other risks may require other reinsurance solutions, as described below.

10.4.3.4 Nonproportionate Reinsurance

The main type of nonproportionate reinsurance for long-term business is catastrophe excess of loss insurance, under which the reinsurer would agree to pay the excess above a specified minimum of the total cost of claims (net of all other reinsurance) due to a single event (i.e., due to some catastrophe). The reinsurance claim amount would also be limited to some specified maximum.

This type of reinsurance is most frequently applied to group life insurance risks, where contracts are issued to employers which will pay death benefits to each of the firm's employees. The additional risk faced by the insurer under this kind of policy is where multiple claims arise from a work-based catastrophe (the World Trade Center disaster, in the U.S. in 2001 was an extreme example of this). Catastrophe reinsurance, therefore, protects the insurer from the risk of *nonindependent* claims, which will be most important for the group life case but can also operate with respect to an insurer's whole portfolio. So, while claim frequency is likely to be extremely low under such reinsurance arrangements, the insurance protection provided can be very significant. War risks, epidemics, and nuclear risks would normally be excluded.

Reinsurance is a much more important feature of risk control in general insurance, as described in Section 12.4, due to the much greater contribution that insurance risk makes to the overall risk of a general insurance company. Also, while similar in principle, the approaches and methods of reinsurance for life and general insurance are quite different, due to the different natures of the risks involved. The main features in this respect are that:

Life insurance is long-term, so that individual proportionate reinsurance will be arranged at the inception of a contract and then remain in force for the whole policy duration.

As the claim amounts under life insurance contracts are not dependent upon the amount of loss suffered at the claim event (i.e., they are not contracts of indemnity), then for most life insurance contracts reinsurance is subject to considerably less uncertainty than in general insurance.

You are referred to Chapters 12 and 14 for more details.

10.5 Business Risk

Business risk can be considered as the contribution to the life office's risks caused by the trading activities (i.e., the business) of the insurer. These can be further subdivided into the following elements: expense risk, discontinuance risk, new business risks (from capital strain and other sources), distribution risk, risks from guarantees and options, data risks, and taxation risk. These are described in the remainder of this chapter.

There is also an overriding management risk (sometimes called *operational* risk), i.e., that the company, unwittingly or otherwise, pursues strategies or makes decisions that turn out to be sub-optimal — poor management will contribute to poor business results, increased insolvency risk, or both. We describe the key areas of this type of risk in Chapter 11; see also Section 6.7.5.

10.5.1 Expense Risk

The operation of the business of a life insurer necessarily incurs expenses of management. These were summarized at the end of Section 6.7.4.

The expense risk is essentially the risk that the total expenses incurred by the office are more than the total expense loadings received by the office in its premium income (again using the premium basis as our benchmark). Now, ignoring interest, total expense loadings received by the office in any one year from a portfolio of n policies (including new policies issued during the year) can be written as

$$L = \sum_{i=1}^{n} L_i$$

where L_i is the expense loading received from the ith policy during the year.

The office's total expense outgo during the year can be written as

$$E = \sum_{i=1}^{n} E_i(c) + \sum_{i=1}^{n} E_i(p) + E(a)$$

where $E_i(c)$ are the expenses under policy i which are a known proportion of the premiums received (e.g., commission), $E_i(p)$ are the expenses incurred by the explicit existence of policy i, but which are not a predetermined proportion of the premium received, and $E(a)$ is the total of all other expenses incurred by the office.

The division between $E_i(p)$ and $E(a)$ can be appreciated by considering how much of the office's expenses would not have been incurred if policy i did not exist. Examples of $E_i(p)$ would include the cost of medical examination for a new policy; postage, stationery, and premium collection costs associated with maintaining the policy; and costs incurred with administering the claim settlement for a terminating policy. However, $E(a)$ expenses can be considered to be those incurred whether any particular policy existed or not. This includes the vast majority of staff costs, costs of maintaining office property, marketing, actuarial, and accounting costs.

Hence, $E_i(c)$ and $E_i(p)$ expenses can be considered as explicitly volume related, while $E(a)$ expenses are independent of volume (in the short term). Over the longer term, of course, $E(a)$ must also be dependent on the overall scale of the office's operations, e.g., office space, number of employees, and demands for marketing, actuarial, and accounting services are ultimately dependent on the number of policies that the office has to manage. $E(a)$ expenses are usually referred to as the fixed, or overhead, expenses of the office. Here, we will consider them as the expenses that are variable only as the result of management decisions plus, of course, the effect of inflation upon the constituent costs. Hence, $E(a)$ will vary if, e.g., the office decides to expand (or to contract) its scale of operations in order to manage efficiently its expected levels of business, by recruiting more (or fewer) staff, opening new (or closing old) branch offices, and so on.

Provided that commission payments are exactly matched by explicit loadings in the premium basis, as would normally be the case, then we can consider the expense risk as

$$\Pr\left\{ \sum_{i=1}^{n} E_i(p) + E(a) - \sum_{i=1}^{n} (L_i - E_i(c)) > y \right\}$$

or

$$\Pr\left\{ E(a) - \sum_{i=1}^{n} (L_i' - E_i(p)) > y \right\}$$

where y is a certain maximum acceptable expense loss during the year, L_i' is the expense loading for policy i net of commission (i.e., $L_i' = L_i - E_i(c)$); it is assumed that $L_i > 0$.

Assuming $L_i' > E_i(p)$ for all i, then the following events can decrease the risk of expense loss:

1. The expense loadings L_i' in the premium basis could be increased for all new policies. Charges can also be increased under many existing accumulating fund contracts, provided contracts allow charges to be reviewed from time to time. This feature extends to conventional contracts in the U.S., where indeterminate premiums are allowed; i.e., premiums can be varied from time to time in response to significant changes to the company's anticipated future experience. All these features give companies opportunities to control their expense risk. However, it should be noted that any such increase in expense loadings (either for new contracts or during the policy term) may have the effect of producing uncompetitive premium or charging rates, which could depress new sales or encourage discontinuance; this would clearly be counter-productive with regard to the control of expense risk.
2. The number of policies sold during the year is increased. Caution must, however, be exercised regarding the launching of new sales

initiatives, as the additional costs incurred from unsuccessful activities can be highly counter-productive. However, the need to sell business in order to control the expense risk, especially regarding the covering of the fixed expenses $E(o)$, requires the office to have products available that are attractive to the public and which meet their perceived insurance needs. Good product design is thus an important element in the overall control of expense risk.

3. The rate of inflation is lower than anticipated in the expense loadings: both $E_i(p)$ and $E(o)$ will then rise less quickly than L'_i from year to year.
4. The number of terminations of existing policies is reduced, whether by contractual termination (e.g., death) or by voluntary discontinuance.
5. Expenses can be reduced by the management exerting effective cost control measures, particularly regarding the control of overheads. This will improve efficiency and, hence, the productivity of the company.

10.5.1.1   *The Effect of Voluntary Discontinuance on the Expense Risk*

Voluntary discontinuance is a very important aspect of the demography of life insurance funds that has been largely ignored in this book so far. This was not done to belittle its significance, but to enable the fundamental operation of the various contract types to be understood without the added complication of this factor.

The funds described in Chapter 7 and Chapter 8 are, therefore, in reality, all subject to two simultaneous decrements: death and discontinuance. Discontinuance rates can be relatively high, often much higher than mortality rates, particularly at early policy durations. There are essentially three modes of discontinuance:

1. Surrender, in which the contract between insurer and insured is completely severed, and an amount of money (the *surrender value*) may be paid to the policyholder in settlement.
2. Becoming paid-up, in which future premiums under the contract cease but the policyholder remains entitled to a reduced benefit level, commensurate with the reserve held by the office at the date of policy alteration.
3. Policy alteration; a change is made in the terms of the contract, which may take a variety of forms. Note that making a policy paid-up is actually just a form of policy alteration, but is usually considered separately. Increased discontinuance rates will, therefore, nearly always reduce the amount of premium receivable and, hence, will increase the expense risk. Expense risk can, therefore, be improved by minimizing the rate of policy discontinuance.

The rate of discontinuance is affected by the following factors:

1. A change in economic conditions, particularly those relevant to the policyholder. Loss of income through unemployment often leads to policies being made paid-up, with surrender often following due to the need to realize personal assets in the face of financial hardship.
2. The policyholder may have been sold a policy that was inappropriate to his or her needs from the outset, or one which has subsequently become inappropriate. The contract may have the wrong benefit structure or risk profile (e.g., a with-profits policy might have been issued whereas a unit-linked policy would have been more appropriate), or the policy might have been too large for the financial means of the customer.
3. Unscrupulous selling practices by salespersons can lead to policy-holders being persuaded to surrender a contract that they may have with one insurer, and to replace it by a new contract from the salesperson's insurer. This process is called *churning*. Occasionally, such activity may reflect best advice, e.g., where the discontinued policy is highly inferior to the replacement. However, the degree of inferiority would have to be substantial to compensate for the fact that the policyholder would incur twice as much initial expenses and commissions should he or she replace a policy than if the original policy were continued without replacement.
4. Discontinuance may be caused by financial incentives. For example, a policyholder who remains in excellent health and/or has a very low-risk lifestyle while the policy is in force may perceive that his or her life insurance policy is not providing particularly good value for money. This is quite likely to be true, given that the same premium rates are often charged to people presenting varying levels of risk, at least within a limited range. Such people would be more likely to discontinue their policies than others whose health deteriorates, especially as "good health" lives could successfully reapply for a replacement policy in the future on the same terms, while others might be charged extra premiums or declined insurance if they were to make a future reapplication. The net effect of this selective withdrawal is therefore to increase the average mortality of the policyholders who maintain their policies in force, by selectively removing the low-risk lives from the portfolio. (This is another example of adverse selection occurring.) Controlling this aspect of discontinuance is, therefore, also a factor in the control of the insurance risk, which we have described earlier. Other financial incentives may exist where, for example, the surrender value may be generous in relation to the underlying reserve of the policy, in which case the policyholder may be financially better off by surrendering the policy and investing the surrender value and his or her future premiums in another investment medium, which might even be another life assurance policy.

Life offices have very little control over discontinuance where this is due to a worsening in the economic situation of the policyholder, as in (1). On the other hand, offices could have considerable control over their sales practices, particularly where they operate through a direct sales force, and this can have significant influence over (2) and (3) above. Clearly, the more care that is taken to ensure that the policies sold are of the right size and type to meet the means and needs of their clients, then the less likely it should be for the policies to be discontinued. This increased *persistency* would also be reflected by improved trust and confidence that the policyholders will have in the salesman and the insurer, which might improve the prospects of further sales to the same people in the future. This development of "customer loyalty" can be of significant benefit to the long-term new business prospects of any office. Hence, the development of good selling practices can have the double-edged benefit for expense risk in both reducing lapse rates and increasing potential future new business volumes. It can even be triple-edged, in that low lapse rates and good sales records may lead to higher job satisfaction among the sales force, with consequently lower staff turnover and hence less costs required in training new staff. The experience of long-serving successful sales staff is also of considerable value to the insurer.

As mentioned in Chapter 6, selling practices are usually regulated, which should generally help to reduce discontinuance rates. For example, in the U.K., intermediaries are required to research fully the needs of their clients and to provide best advice regarding which products (and providers, where appropriate) would be suitable to meet these needs. While regulations can go some way to improving industry practice, there can remain a wide potential discrepancy between the best and worst practices in the market.

Further control of discontinuance rates, under point (4) above, is by the avoidance of financial incentives for the discontinuance of policies. To meet this need, therefore, the office must attempt to ensure that at all times the majority of policyholders would not be financially better off discontinuing their policies than maintaining them. In the U.S., healthy term assurance policyholders are discouraged from discontinuing their policies by the use of the re-entry feature, described in Section 7.1.1, in which premiums will be reduced each time the policyholder resubmits him or herself to medical screening (provided, of course, that he or she emerges as a select life).

Another strategy adopted in the U.S., with accumulating fund contracts, is to provide policyholders with financial incentives not to discontinue, in the form of *persistency bonuses*. These are additions to the accumulating cash fund which are granted in return for the policy achieving some specific goal, e.g., when it has been in force for a certain number of years, or once a certain level of cash value has been accumulated.

It should also be noted that the existence of any financial incentive to surrender also implies that there is an explicit *discontinuance risk*; this is the subject of the next section.

10.5.2 Discontinuance Risk

The easiest way to consider this risk is to assume that discontinuance is in the form of a complete surrender of the policy, and that a surrender value is payable upon termination. The first thing to note is that, wherever possible, the surrender value should never be greater than the policy's asset share at the same date.

Should a surrender value be paid which exceeds the policy asset share, then there is a physical loss of assets, i.e., a contribution from the estate is required to make good the shortfall between asset share and surrender value. When we first introduced this subject in Section 3.6.2, we stated that most surrender values would reflect changes in asset share value (through changes in the assumed levels of future interest rates used to calculate the surrender values), so that discontinuance risk should normally be minimized. However, surrender values can easily exceed asset shares on the surrender of certain long-term contracts at early policy durations, as, with high initial costs relative to low annual premiums, asset shares can actually be negative during the first policy year, if not beyond. Even if the surrender value at these durations is zero, then any policy lapse will incur a loss. In recent years, much increased public and media attention in the U.K. has been focused on the allegedly poor surrender values paid on early surrender of life policies; as a consequence, many offices have resolved to pay greater than asset shares at early policy durations in order to enhance their public image, thereby further increasing the risk of loss from discontinuance.

However, there is more to determining an equitable surrender value than simply considering the asset share. The asset share would normally only allow for physical transfers of profit made from the policyholders' assets to the shareholders' assets, or to the with-profit policyholders, if the office is mutual. It can be argued that we should assume that, on surrender, the policyholder be entitled to his or her asset share less the contribution that the policy has made to the office's realistic profits. For this to be the case, then the surrender value would need to be equal to a realistic earned-profit type of reserve (as described in Section 8.6.1), i.e., a reserve that avoids crediting the insurer with future profits, thereby leaving the company retaining the accumulated past profit only. A gross premium reserve calculated on the premium basis has a similar effect (for a nonprofit policy); see Section 7.1.3.

Even if surrender values are based on these realistic reserves, there is still a discontinuance risk. As with asset shares, realistic reserves may be negative at early policy durations so that the payment of any surrender value which is greater than or equal to zero will incur a loss (relative to the equitable value, that is). Even at later durations, where the realistic reserve is positive, a relative loss may be incurred. This is because the true value of the realistic reserve at any time is unknown, as it depends upon unknown future events (claim rates, interest rates, and so on). Hence, a surrender value can only be a "best estimate" of the reserve value, and by definition there would then be an approximately 50% probability of its being too high.

There are a number of considerations that would suggest reducing surrender values further:

Who should pay for the losses on early surrender? Unless later surrendering policyholders do so, losses would be borne by the policyholders who maintain their policies and/or by the providers of capital, which does not seem particularly fair.

Using the realistic reserve basis even at later durations implies that, on average, approximately 50% of surrenders will be at the expense of the continuing policyholders, which does not seem equitable. Furthermore, as the date of surrender is at the option of the policyholder, policyholders are more likely to choose to surrender at times that are least favorable to the office (such that the proportion of loss-making surrenders might be substantially higher than 50%).

As the mortality rates of surrendering policyholders are likely to be lower than those who continue, the surrender values should be calculated on the basis of the lower mortality (producing lower values).

The surrender of a policy potentially increases per-policy expense costs in the future for the policies that remain. It might be fairer to attribute at least some of this loss to the surrendering policyholder.

On the other hand, it would seem unfair to penalize unduly a policyholder who has to surrender through no fault of his or her own, e.g., when due to prevailing economic circumstances. Furthermore, any life office that pays apparently penal surrender values risks significant damage to its public image, and such surrender values might well include those based on realistic reserves, particularly at early durations. It should also be borne in mind that there are practical limitations to the frequency with which surrender value bases can be varied by the life office, so that it is important to choose a basis, and a method, which does not go out of date too quickly.

The surrender value basis adopted will take into account the two contributing factors to discontinuance risk: namely the probability of discontinuance occurring and the extent to which the surrender value exceeds realistic reserves. Clearly, if there was no discontinuance at all then there would be no discontinuance risk, whatever the surrender values were. Hence, offices with high persistency would incur relatively little surrender loss even if surrender values were quite generous. On the other hand, offices that suffered poor persistency could incur quite significant losses even with relatively penal surrender values.

Offices can clearly find themselves in virtuous or vicious spirals with regard to discontinuance risk. Ideally, offices should suffer purely economic surrenders. Discontinuance rates should then be low and nonselective, and offices could afford to be generous on surrender (even at short policy durations) which would only serve to enhance its public image and thereby maintain its high persistency rates. On the other hand, an office with high discontinuance rates would need to consider very carefully whether it should increase early surrender values in order "to enhance its public image" if this would result in significant penalties being imposed on the "faithful" policyholders who do maintain their policies.

Determining appropriate surrender value bases is, therefore, a delicate matter, requiring considerable skill and judgement, and further consideration is beyond the scope of this book. A possible practical approach is described by Lumsden (1987).

An alternative approach would recognize the financial option-like feature of surrenders in life insurance contracts, and their similarity to American options in the context of financial derivatives. This approach is beyond the scope of this book, but interested readers are referred to Albizzati and Geman (1994).

10.5.3 New-Business (Valuation Strain) Risk

As described in Section 9.3, the need to meet the statutory solvency requirements means that many contracts incur significant valuation strains in their year of issue that have to be met from the office's free assets. Hence, the issue of relatively large volumes of new business with high initial strains could seriously threaten statutory solvency, particularly of new or rapidly expanding offices.

There are two main solutions to this risk: product design (and product mix) and financial reinsurance.

10.5.3.1 *Product Design and Product Mix*

We have mentioned earlier the need to design products that met the financial needs of policyholders, so that the products would sell in the desired volumes. However, it would be clearly irresponsible to sell policies that did not meet the fundamental financial requirements of the office.

The first of these requirements is that the premium charged must be adequate to meet the benefits, expenses, and risks incurred to the office, while also providing an adequate return on the capital expended in selling and maintaining the policy. We consider how premium rates can be assessed in Chapter 11.

The second requirement is that the product design should incur the office in the minimum amount of new-business strain. The effect on the statutory solvency position (and, hence, the contribution to the risk of statutory insolvency) is best described by the product's statutory profit signature (see Section 9.3). Three examples are shown in Figure 10.1.

Figure 10.1(a), which represents a conventional nonprofits policy, is clearly the greatest contributor to statutory insolvency risk, due to the very large initial strain. Figure 10.1(b) is, on the other hand, a particularly efficient nonprofit design, with relatively little initial strain and with a high degree of *profit acceleration*, i.e., with a large proportion of the product's total profit payable as near to the outset of the policy as possible. This has the dual benefit of providing much relief of valuation strain after the second policy year (i.e., a considerable net statutory surplus), and also in increasing the value of the emerging profit to the office (and, hence, to the providers of capital). Its value is increased for the following reasons:

1. There is less uncertainty (i.e., less parameter risk) about the profit emerging from the policy from the outset, as the experience assumptions over the first 2 years used in the profit test are more likely to be accurate than the experience assumptions later in the policy term.
2. Profit emerging at early durations is less likely to be lost though discontinuance than profit emerging at later durations.
3. Profit earned early in the policy term produces a higher rate of return on capital than the same amount of profit earned later. This is because in the former case the profit itself can be reinvested to earn returns over a longer period of time, or distributed to the providers of capital sooner.

Figure 10.1(c) shows the statutory profit signature of a typical with-profits policy. As in Figure 10.1(b), this policy has produced a net statutory surplus after 2 years, and goes on contributing significantly to the statutory free assets until the year before the policy maturity date. While the ultimate year's strain is very large, this should not produce any overall depletion of statutory free assets provided that total benefit levels have value equal, on average, to the policies' asset shares, as explained in Section 8.4.

Hence, both Figure 10.1 (b) and (c) are examples of product designs that produce desirable statutory profit signatures. However, the sales of contracts with signatures of type (a) may be necessary for other reasons, e.g., in order to provide an important insurance service to the public. However, if the office can balance sales of type (a) by sales of products of types (b) and (c) then the overall risk to statutory insolvency caused by the new-business strain will be neutral. Hence, the mix of business sold with different profit signatures is of vital importance to the maintenance of statutory solvency, and the office should not overlook the need to exercise control in this area.

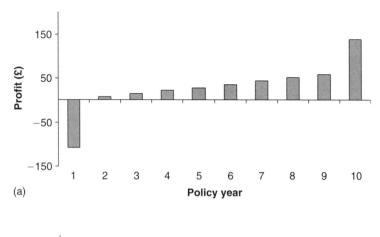

(a)

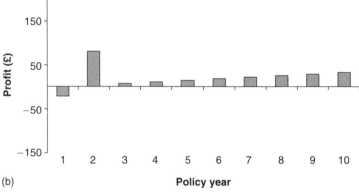

(b)

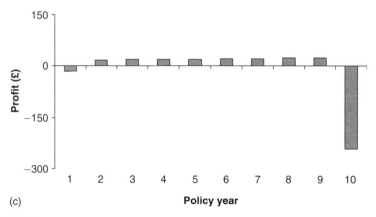

(c)

FIGURE 10.1 (a) Statutory profit signature for a conventional nonprofit policy. (b) Profit signature for a unit-linked policy with nil allocation period. (c) Statutory profit signature for a conventional with-profits policy.

10.5.3.2   *Financial Reinsurance*

Valuation strain can also be reduced by the use of certain reinsurance arrangements. Most appropriate for this purpose is original terms arranged on a quota share basis. Surplus reinsurance can also be used, although it provides less control of the valuation strain than the quota share method, and risk premium reinsurance is also possible but less effective for this purpose. Hence, for example, a new or rapidly expanding office would probably have an $X\%$ quota share treaty with its reinsurer, such that $X\%$ of all its new business, regardless of size, would be reinsured.

The reduction of valuation strain, particularly in the first policy year, is achieved using original-terms reinsurance in two main ways.

Reduction in Statutory Reserves and Solvency Margins

If $X\%$ of the sum assured under each policy is reinsured, then the ceding office only needs to hold $(100 - X)\%$ of the statutory reserve under the policy at any time. The required statutory solvency margin (or the company's statutory risk-based capital) can also be calculated net of reinsurance, although the company should take credit for less than 100% of the potential offset, to allow for the risk that the reinsurer might default (i.e., become insolvent) in the future. Regulations may also exist to enforce limits in this respect.

Payment of Financing Commission

Original-terms reinsurance will normally involve the payment by the reinsurer to the ceding office of a significant commission in the first year. This can be considered as remuneration to the life office for the placement of the business with the reinsurer, and reflects the fact that the ceding office will have incurred nearly all of the underwriting, sales, and acceptance costs associated with the issue of the policy on behalf of the reinsurer. With an automatic treaty, reinsurers have relatively low issue costs and it is therefore appropriate to pay the ceding office a relatively large initial commission for the business. This initial commission can serve as a significant offset against the ceding office's new business strain in the year of issue of any contract.

For an example we will return to our 3-year nonprofit endowment policy last met in Section 9.3 in Table 9.1 and Table 9.2. We will assume that the office has a 40% quota share original terms treaty with a reinsurer, and that the reinsurer pays an initial commission of 60% of its premium in the first year. The fund accumulation is shown in Table 10.2.

It can immediately be seen that the total accumulated profit at the end of the term has fallen from £983,000 to £704,000 as a result of the reinsurance, a reduction of 28%.

Table 10.2 Nonprofit fund accumulation net of 40% original terms quota share reinsurance (£000)

Year	Premium	Commission	Expenses	Interest	Claims	Fund
1	2434	973	(1887)	152	(300)	1372
2	2312	0	(179)	351	(285)	3571
3	2196	0	(204)	556	(5415)	704

Table 10.3 Statutory profit signatures for a nonprofit fund, with and without 40% original terms quota share reinsurance (£000)

Policy year	Before solvency margin		After solvency margin	
	Gross	Net	Gross	Net
1	(256)	87	(364)	22
2	594	277	488	215
3	640	294	886	441

Table 10.4 Cash flow, statutory reserves, and capital provisions at $t = 0+$ with and without 40% original terms quota share reinsurance (£000)

	Cash flow	Reserve before solvency margin	Free assets before solvency margin	Required solvency margin	Free assets after solvency margin
Without reinsurance	2170	2529	(359)	124	(483)
With reinsurance	1520	1441	79	74	5

The effects of the reinsurance on the statutory profit emerging each year, both before and after allowing for the statutory minimum solvency margin, are shown in Table 10.3.

The capital required at the inception of the policy, at time $t = 0+$ (see Section 9.3.1) is shown in Table 10.4.

So, even when we include the required minimum solvency margin, the financial reinsurance has eliminated the new business strain (i.e., a reduction in excess of 100% of the strain has been achieved in return for only a 28% reduction in overall profit).

As noted above, financial reinsurance can also be achieved using risk premium reinsurance. In this case, the reinsurer will again pay high initial commission, which is repaid through appropriate increases being made to the reinsurer's usual risk premium rates.

10.5.4   Other New Business Risks

One of the main new-business risks has already been described in Section 10.5.1: the risk of inadequate sales *volumes* to cover the company's fixed expenses. One-off development and marketing expenses also have to be recovered where these occur, e.g., after the launch of a new product or even when just reviewing an existing one, and again sufficient volumes of sales will be necessary to achieve the necessary coverage.

We also mentioned in Section 10.5.3 that it was important for a company to control its *mix* of business in order to balance out its capital costs. The mix of business creates other risks, in particular in relation to the *size* of policies taken on by the company. So, when a contract is priced (Chapter 11), it is necessary to make an assumption about the average policy size of new business. The company will then be at risk of making an expense loss should the actual average policy size turn out to be less than this.

For example, if the company assumes its average premium will be, say, £1000 per annum, and the company allows per-policy expenses of £30 per annum in its pricing basis, then the resulting premiums will include a loading for per-policy expenses equal to 3% of the premium value. Should the company then issue 5000 policies with an average premium of only £800, then the per-policy expenses will equal $5000 \times 30 = £150,000$ per annum, but the loadings to cover them will only equal $5000 \times 800 \times 0.03 = £120,000$ per annum, a shortfall of 20% of what is required for each premium paid.

The company can mitigate this risk by introducing fixed charges into its pricing structure. This is relatively easy to do for unit-linked contracts, where regular policy fees (e.g., regular per-policy monetary deductions from unit funds) can be charged.

For conventional policies, part of the regular premium could be a flat amount, not varying with policy size. Suppose, in our example above, that the average premium of £1000 bought a benefit (sum assured) of £15,000. This implies a premium rate of $1000/15 = £66.67$ per £1000 of sum assured. Using a flat fee of £30, we would have a premium rate of $(1000 - 30)/15 = £64.67$ per £1000 of sum assured. Table 10.5 shows

Table 10.5 Comparison of annual premiums and premium rates with and without policy fee premium structure.

Sum assured (£)	Annual premium (£)		Overall premium rate (‰)	
	No fee	Policy fee £30	No fee	Policy fee £30
5,000	333	353	66.67	70.60
15,000	1000	1000	66.67	66.67
25,000	1667	1647	66.67	65.88

some specimen premiums that would be charged under the two alternative systems (with or without the flat fee).

The only problem with the policy fee system occurs if the policy fee becomes a very high proportion of the total premium, which could make the premium *rate* being charged seem very high for very small policies (this would be especially acute for "cheap" products, like term assurances). This effect could be prevented by having a minimum policy size and/or by a compromise structure where only part of the per-policy expenses were covered by the policy fee.

10.5.4.1 Matching Cash Flows: Unit-Linked and Accumulating With-Profits Policies

The above shows how risk reduction can be achieved by matching specific outgoes (per-policy expenses) by specific charges (a per-policy fee). This is just one aspect of a general principle: the more precisely that incomes and outgoes can be matched the fewer of many kinds of risks there may be.

The scope for this sort of matching is by far the greatest for unit-linked and AWP product designs. We actually saw another example of this in Section 10.5.3, in Figure 10.1(b); this unit-linked policy had a high initial charge that, by matching the initial expense outgo, greatly reduced the initial capital strain. This design has other beneficial effects; e.g., the unit fund values will be smaller early on in the policy term, making them closer to asset shares and so reducing the risk of making large losses on surrender. (This *can* be achieved with other charging structures too, but only by imposing explicit surrender penalties, and these are more likely to generate policyholder dissatisfaction.)

Another important area is to match annual costs. This includes matching the annual average mortality cost with the appropriate mortality charge; see Section 7.2.1. Other examples include index-linked expense charges and the appropriate mixing of expense charges by nature, e.g., investment management charges proportionate to fund size (the fund management charge), asset transaction charges proportionate to the amount of investment (the bid–offer spread), and administration charges as a regular fee per policy.

The main benefit of matching the regular costs and charges is that it will reduce both the need for and the size of any nonunit reserve, as it will reduce the incidence and extent of future negative cash flows. Note that nonunit reserves may still be needed even where cash flows are well matched, as we explained in Section 9.3.4, but it will still help to reduce their size.

10.5.5 Distribution Risk

If a life office, that has been distributing profits regularly, becomes statutorily insolvent, then with hindsight one can judge that if the office had distributed less of its (estimated) profits in the past then the insolvency may have been avoided. Distribution risk is, therefore, the risk of distributing too much profit to the shareholders and/or policyholders and becoming insolvent as a result.

Probably the single most important way in which a life office avoids paying out too much profit is by appropriate reserving: the more reserves a company holds, the less money it can physically pay out in profit distributions. The statutory reserving requirements help to ensure that overdistribution is prevented.

Holding large reserves, statutory or otherwise, is not the whole story. The company needs to maintain a level of statutory *free* assets in order to provide a financial cushion against future adverse experience, thereby giving it the necessary flexibility to achieve its business aims without risking *future* insolvency, e.g., to enable it to invest more profitably, to perform benefit smoothing on its with-profit policies, and so on. (This is what a statutory dynamic solvency-testing requirement is aiming to achieve; see Section 9.1.3.)

On the other hand, it is possible to hold too much capital. As we saw from our example in Section 6.7.3, while an increase in capital reduces the risk of insolvency, it also lowers the returns paid to the providers of capital, all else being equal. The control of capital, therefore, is one of the most important aspects of the life office's financial management, but also one of the most difficult.

The control of profit distribution can be considered as the main way in which the level of capital in the fund is managed. Profit distribution, or, more correctly, the distribution of a life office's statutory free assets, is made by the payment of bonuses or dividends to the with-profit policyholders and by the payment of dividends to the shareholders, as appropriate. Of particular concern to solvency is the distribution of profits to the policyholders, as policyholders' expectations are that the rates of these distributions will vary only gradually from year to year. Hence, life offices that historically have paid high levels of policy dividends or reversionary bonuses, but which need to reduce their distribution rates in order to reflect less profitable conditions, may find that they consistently overdistribute for a number of years while they reduce their distributions by degrees over each year. This phenomenon has contributed significantly to the risk of statutory insolvency of a number of with-profit life offices in the U.K. over the early 1990s, and again at the beginning of the new millennium.

One of the main responsibilities of the actuary of any life office is to advise its directors regarding the distributions of profit that should be made each year; the importance of this advice to the long-term solvency and profitability of the office cannot be underestimated. We will describe some of the ways by which the actuary can formulate his or her recommendations in Chapter 11.

10.5.6 Risks From Policyholder Options and Guarantees

In Section 3.6.2 we described the difficulties of investing in order to match policyholder options. Many financial options, such as the guaranteed annuity options described in Section 6.7.4, may never come into the money (i.e., they have zero cost for the insurer). The problem is, if they *do* come into the money they will do so for all such policies the company holds on its books, and then they can become extremely expensive.

One approach to their management, suggested in Section 3.6.2, is to buy suitable derivatives (either call options or interest rate swap options will work in this case) well in advance of the guarantees coming into the money. This method is likely to be expensive if the guarantees end up with little or no cost, but a lot cheaper to the company if the guarantees turn out to have value.

For most companies, the secret to surviving financial guarantees when derivatives have not been bought in advance is prudential reserving. Financial guarantees, such as guaranteed annuity options, do not usually come into the money without some warning. Major costs should only occur as a result of major and prolonged shifts in economic conditions, such as occurred over the 1990s when market interest rates fell persistently in many developed economies, and many life insurers in the U.K. were left holding in-the-money guaranteed annuity options. Mortality rates also fell more quickly during this period than was anticipated, thereby further increasing the cost of the guarantees. By building reserves up gradually over the period, many insurers were able to manage these costs, but no doubt with the consequence of significantly reduced returns to policyholders and shareholders in the process. (See Ballotta and Haberman (2003) for a discussion of the pricing and valuation of guaranteed annuity options, with reference to U.K. experience.)

The lesson to be learned is that financial guarantees can be offered by life insurance companies, but those who do so have to be vigilant to their existence and have viable strategies to cover the cost should they come into the money. Reserving is a fundamental priority in managing such risks, whether or not other steps (such as holding matching derivatives) have been taken. On the other hand, any insurer that is not prepared to take on the cost (however unlikely it may be) should not offer such guarantees in the first place.

10.5.7 Data Risks

In Section 10.4, we have defined parameter error: the risk that the parameter used for a variable (such as mortality) in pricing is more optimistic than its true expectation. So one of the main causes of parameter error is due to misestimation of the parameter itself.

Misestimation can arise for a number of possible reasons:

Lack of effort by the company (its actuaries in particular!) in researching the available data and/or allowing for future developments properly (see Section 11.2.1).

Unforeseen developments, causing the rationale of the actuarial assumptions to be incorrect in the event (we could call this a random effect, though this is *not* to be confused with the process error defined in Section 10.4).

Insufficient or inaccurate data.

Insufficient data is a common problem for insurance companies, who would ideally wish to base their future pricing (and other) assumptions on their own experiences. Recourse may be had to other sources (e.g., population, industry, or reinsurers' data) but these are bound to be less relevant to the company. The inevitable result of scarce data is that this reduces the precision with which parameters can be estimated, thereby increasing the parameter risk.

Inaccurate data will also increase parameter risk, so good data management (i.e., taking sufficient care to ensure that data are complete and accurate) is important.

Managing parameter risk first involves making the best use of the available data, but also recognizing the extent of the remaining uncertainty and to reflect this in the margins included in the basis (see Chapter 11 for more details). Reinsurance can also provide protection against parameter risk, simply by reducing the company's overall exposure to the costs of average experience being worse than expected. Quota share reinsurance would normally be used for this purpose.

10.5.8 Taxation Risk

Control of taxation is a rather specialized area and there is, therefore, insufficient space here to deal adequately with this topic. The situation in the U.K. will be described briefly to show an example of the kinds of management possibilities that may exist. In the U.K., taxation is generally levied in proportion to the amount of investment income I, while expenses E can be charged against tax (i.e., E generates relief from the tax payable on the company's I). Hence, tax management is effected largely by:

1. controlling the volumes of business which generate respectively greater or lesser amounts of I or E

2. setting up special reinsurance arrangements that essentially trade *I* from an "*I*-rich" office for *E* from an "expense-rich" office, which can increase the after-tax profitability of both parties

noting that:

1. the I-rich life offices, having $I > E$, pay tax on $I - E$;
2. the expense-rich life offices, having $I < E$ lose (or at least postpone) tax relief on $E - I$.

References

Ainslie, R. R. (2000). Annuity and insurance products for impaired lives. Paper presented to the Staple Inn Actuarial Society, London.

Albizzati, M. and Geman, H. (1994). Interest rate management and valuation of the surrender option in life insurance policies. *Journal of Risk and Insurance* 61, 616–637.

Ballotta, L. and Haberman, S. (2003). Valuation of guaranteed annuity conversion options. *Insurance, Mathematics and Economics* 33, 87–108.

Barnaby, W. (1997). Human genetics: uncertainties and the financial implications ahead. *British Actuarial Journal* 3(Part V), 975–983.

Black, K. and Skipper, H. D. (2000). *Life Insurance*, 13th edition. Prentice-Hall.

Cairns, A. J. G. (1995). Uncertainty in the modelling process. In: *Transactions of the 25th International Congress of Actuaries*, vol. 1, p. 67.

Carter, R. L. (1983). *Reinsurance*. Kluwer, Brentford.

Daykin, C. D., Pentikainen, T., and Pesonen, M. (1994). *Practical Risk Theory for Actuaries*. Chapman & Hall, London.

Diacon, S. R. and Carter, R. L. (1992). *Success in Insurance*. John Murray.

Fisher, H. F. and Young, J. (1965). *Actuarial Practice of Life Insurance*. Institute of Actuaries.

Hooker, N. D., Bulmer, J. R., Cooper, S. M., Green, P. A. G. and Hinton, P. H. (1996). Risk based capital in general insurance. *British Actuarial Journal* 2, 265–323.

Le Grys, D. (1997). Actuarial considerations on genetic testing. *British Actuarial Journal* 3(Part V), 997–1008.

Leigh, T. S. (1990) Underwriting — a dying art? *Journal of the Institute of Actuaries* 117, 443–531.

Luffrum, G. (1989). *Life Underwriting*. Institute of Actuaries.

Lumsden, I. C. (1987). *Surrender and Other Alterations*. Institute of Actuaries.

Smaller, S. L., Drury, P., George, C. M., O'Shea, J. W., Paul, D. R. L., Puntney, C. V., Rathbone, J. C. A., Simmons, P. R., and Webb, J. H. (1996). Ownership of the inherited estate (the orphan estate). *British Actuarial Journal* 2, 1273–1322.

Spedding, A. (1989). *Life Reinsurance*. Institute of Actuaries.

Wilkie, A. D. (1997). Mutuality and solidarity: assessing risks and sharing losses. *British Actuarial Journal* 3(Part V), 985–996.

Chapter 11

The Actuarial Role in Life Office Management

11.1 Introduction

The actuarial role in life office risk management, loosely following the guidance notes published by the Institute and Faculty of Actuaries in the U.K. (2003), can be summarized as follows:

> to monitor all relevant aspects of the office's financial and demographic experience, and all relevant aspects of the company's management policy, and to advise the management of the appropriate courses of action to take with a view to maintaining a sound financial operation for the benefit of the company's policyholders and, where appropriate, its shareholders.

This role encompasses all the elements of risk management that we have described in Chapter 10. The management operations in which the actuary has a legitimate interest, and often a significant advisory role, are, therefore, as follows:

investment policy
product pricing and product design
marketing and new business policy
underwriting policy
reinsurance policy
reserving
profit distribution to policyholders and/or shareholders
sales policy
surrender and transfer terms

The actuary's advice will depend upon his or her estimation of the future experience of the office in terms of its investment returns, mortality, lapse rates, inflation, taxation—indeed any other financial or demographic factor that could affect the office's financial condition.

In order to carry out the actuarial role, the actuary has to construct models of the future, estimate the necessary parameters, and draw conclusions from these models about likely future outcomes, such as with regard to solvency and profitability. For example, the calculation of a reserve requires a model that incorporates assumptions about the future returns on assets, the incidence of future mortality, the incidence and amount of future expenses (including inflation), and so on. In this chapter we deal with what might be called macro life office modeling. Section 11.2, while focusing on product pricing, has also considerable relevance with regard to reserving. In particular, the processes described by the control cycle are equally important for the purpose of setting reserving assumptions as they are for pricing. Other actuarial aspects of life office reserving have already been described in Chapter 7 to Chapter 9, to which you are referred.

11.2 Product Pricing

11.2.1 The Control Cycle

The monitoring and advisory role of the actuary is epitomized by the *control cycle* (Goford, 1985). This cycle represents the continuing process by which a life office seeks to monitor its experience, to set its pricing and reserving assumptions, and make its other major policy decisions. The cycle is summarized in Figure 11.1.

11.2.1.1 Setting Assumptions

The logical starting point for the control cycle would be the initial assumptions. In practice, the cycle should only begin once, i.e., when the company first opens to new business; at all other times the control cycle is always in continuous motion. Hence, when considering the initial assumptions, we assume that we are at the starting point for a new life office. However, the updating of assumptions for existing products involves the same principles of assumption setting, though in different circumstances from when the initial assumptions were set. We can also consider making "initial" assumptions for whenever a new product is being launched by an already existing company.

One of the many vital tasks that the management of a new life office must do is to decide upon the premium rates or charges that it will apply to each of the products that it sells. In order to do this, it must:

1. Make appropriate assumptions about the future demographic and financial experience of the fund.

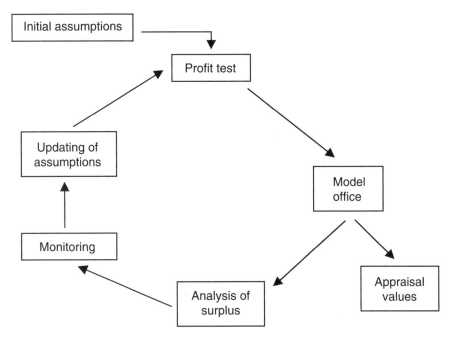

FIGURE 11.1 Representation of the control cycle (after Goford (1985))

2. Anticipate the effect that future management actions will have on the demographic and financial experience.
3. Establish the price (premium and/or charges) to apply to the office's products such that, given (1) and (2) above, the office will anticipate making the level of profit which would meet the reasonable expectations of its providers of capital.

Items (1) and (2) will set the initial assumptions in the control cycle. Item (3) is the process of *premium rating* or, more precisely, *product pricing*. In Chapter 10 we discussed at length how a life office's risks can conveniently be framed with respect to the office's premium basis. Clearly, a prime factor in the control of risk is the setting of the premium basis itself. For example, if the price of a product is high relative to the actual cost, then, provided the product is marketable, the office should make profits most of the time and the risk of "not meeting the assumptions in the premium basis" will be low. Should the product be priced more keenly, in other words where the assumptions upon which the price is based are more optimistic, then the assumed experience may be less likely to materialize with an increased risk that the price will be inadequate. Choosing the assumptions upon which to calculate the prices of the office's products is, therefore, of vital importance to the ongoing solvency and profitability of the fund. This, in essence, is the rationale for the control cycle itself.

The starting point for setting assumptions is an analysis of the relevant past experience data. There are a number of possible sources of mortality data: own company experience, industry data, resinsurers' and population data. These are of variable quality for the purpose, differing in terms of their *relevance* (how similar are the lives involved to those for which the rates are intended) and their *credibility* (the larger the body of data the less random error will be involved in the estimates obtained).

Own-company data of the same class of lives is likely to be the most relevant data, but they usually fall down on credibility. Of course, there will be no own data if we are pricing for a new product, but it may be possible to look at the mortality experience of a different product that has been issued to a similar class of lives in the past, to give some indication. Care must be taken when making inferences between different products, because often the nature of the product itself will influence the type of individual who takes it out, thus affecting the average mortality experience.

Pooled industry data can be available. These are extremely useful, as they relate to insured lives' experiences, usually for the territory required and subdivided by policy type, age, sex, and sometimes even smoker status. These data are much more credible than can be provided by one company alone. The problem can be one of relevance; different companies do target different markets, have different approaches to underwriting, and so on, which will lead to a heterogeneous experience pool whose mortality may not be the same as our company average.

One very common way of allowing for the difference is to investigate own-company data (where this exists) by *reference* to the pooled industry average. We can often use standard tables of mortality, which are based on the pooled industry data, for this purpose. We can then compare the numbers of deaths observed in our own data with the number we would have expected according to the rates in the most relevant standard table(s). This will usually allow us to represent our past experience by some simple adjustment to standard table rates, e.g.

$$\hat{\mu}_x = a\mu_x^S + b$$

where $\hat{\mu}_x$ is the estimate of own-experience mortality, μ_x^S is the mortality rate from the standard table, and a and b are constants estimated from the data.

These analyses would be carried out for different subgroups, such as by policy class, sex, smoker status, and sales channel, so that the company can determine separate premium scales if required for these different risk classes. (Full details of the methods required are beyond the scope of this book, but see Benjamin and Pollard (1993) for example.)

Further adjustments will probably be necessary in order to represent the expected *future* experience for the class of lives we are concerned about for pricing (discussed below).

Reinsurers can be a useful source of data, in that they are usually very willing to provide advice to companies on many kinds of actuarial matters, including product design and pricing assumptions, in return for some reinsurance business. Their broad experience base gives them additional expertise and authority to advise on the kind of market a company works in, or intends to work in. This source of help is particularly useful for a company launching a completely new product (new, that is, to the company), or which is trying to sell to a brand new market or through a new sales channel.

Additionally, there may be published population data, possibly including analysis and even graduated mortality tables. The credibility of the data cannot, of course, be bettered (provided the data are accurate), but they are the least relevant as we are really only interested in the subset of lives who take out insurance policies, and our own company's insurance policies in particular. Adjustments to the standard population mortality tables could be (and sometimes are) used, in the same way as we described using the industry tables. If we were launching a new product in a market that was *new to the industry*, then population data would be essential. Population data are still very important for setting morbidity assumptions in health insurance; see Part V.

The relevance of the various data sources is also affected by how up to date they are. Time delays in data processing usually make industry data less recent than own-company or reinsurers' data, while population data usually take even longer to process.

Finally, most sources of data can shed some light on trends: changes occurring to mortality over time. Care needs to be taken to ensure that any observed trends are not spurious; for example, apparent trends in the industry experience may actually reflect changes in the mix of offices contributing to the (heterogeneous) data over time, rather than to changes in the actual mortality.

Having analyzed the experience, we need to assess what should be the expected *future* experience for the product we are pricing. The task is easier than for general insurance (see Chapter 14) or for health insurance (Part V), because life insurance claim amounts are either known or determined by factors independent of the claim event itself, and claim frequencies are reasonably stable and predictable, at least in aggregate. Nevertheless, the long-term nature of the insurance brings about risks in the pricing process. (These risks are generally worse for long-term health insurance products, though, as described in Part V). In all the main elements of pricing, i.e., interest, expenses, mortality, and persistency, the difficulty is in making appropriate assumptions that are valid for the long term. The source of the problem is essentially one of parameter risk (see Section 10.4), as all of the elements tend to vary over time in a way that is hard to predict accurately. Mortality is perhaps the easiest element to predict, as relatively stable secular changes generally occur. As the general secular trend has involved a decline in mortality rates at nearly all ages, then this is particularly easy for life insurance products to accommodate, the main risk being future lack of competitiveness and loss of business volume as policies based on current (unprojected) mortality rates are likely to have been overcharged. Annuity contracts have to be priced more carefully, however, as lower than anticipated mortality experience leads to a loss (see Section 10.4.1). Hence, it is normal practice to incorporate projections of future mortality improvements in annuity premium bases. In this case it is particularly important to ensure that spurious effects are eliminated from data used for identifying past trends (as described above).

While most secular changes to aggregate mortality have been gradual, relatively sudden changes are possible, such as from an unexpected killer disease (e.g., AIDS), or from the discovery of a dramatic new cure (such as might occur if a reliable cure for cancer were found). Changes due to company management can also affect future mortality experience under the product, e.g., changes to underwriting practice, sales distribution, or product design can lead to a different class of lives taking out an insurance contract than was the case in the past. Such changes should also be allowed for when setting the mortality assumptions.

Turning now to the other elements of the premium basis, the future rate of interest to assume will be dependent upon the company's expected future investment policy, i.e., the nature and types of asset that it expects to purchase in respect of the contracts in question. However, the future rates of return from these assets are extremely hard to predict over the long term due to their dependence on economic and, to an extent, political factors. The expense assumptions are subject to similar uncertainties. We have described the difficulties of producing a useful economic model that embraces both interest and inflation in Chapter 5, which of course also involves potential problems of specification error. Future expenses are further compounded by changes in business volumes and efficiency, as we described in Section 10.5.1, which makes them even harder to predict.

Some of the problems of parameter risk are now alleviated to an extent by modern product design, particularly those in which indeterminate premiums or charging reviewability are incorporated into the contract (see Section 10.5.1). Whether or not this is the case, the control cycle remains of central importance to the pricing process.

11.2.2 Profit Testing

Most product pricing is carried out using cash flow techniques. The basic cash flow model is the *profit test* (see Lee (1985) and Goford (1985)), which generates the expected statutory profit signature of a single tranche of business on the basis of the premium rate assumptions. The process is identical to that shown in Chapter 7 and Chapter 8 using the initial cash flow approach, except with the added details of allowing for discontinuance and, where appropriate, taxation. Using notation defined in Section 7.1.2, the expected average profit in year t for a nonprofit contract can be written

$$pro_t = pm_t + i_t - e_t - c_t + ir_t - \Delta r_t - tx_t$$

where tx_t is the policy's share of tax paid in year t and $c_t = cm_t + cd_t + cs_t$, where cm_t is the claims paid on maturities, cd_t is the claims paid on death, and cs_t is the claims paid on surrender or on other voluntary discontinuance.

Generation of the number of claims by death and surrender would require the use of a double decrement model for the two modes of exit (see Neill (1977) and Bowers et al. (1986)), and appropriate assumptions for the calculation of the surrender values would also be needed in order to estimate cs_t.

The increase in reserves Δr_t would, of course, be the increase in statutory reserves, usually including the policy's contribution to the office's required statutory minimum margin of solvency, along the lines described in Section 9.3.

The expected present value of the statutory profit is then estimated by discounting the projected profits at a *risk discount rate* (RDR) and summing over all future policy years (see also Section 9.3.1: embedded value profits). The expected present value of the future profits (PVFP), often expressed as a proportion of the first year's commission payable under the policy, is then compared against some target profit measure that would be expressed in the same way. The "target profit measure" is usually referred to as the *profit criterion*. It is then straightforward to determine the rate of premium or charges that will exactly generate the profit criterion under the given assumptions.

The RDR represents the return on capital required by the providers of capital. The more risky is the product (e.g., due to parameter uncertainty or risk of marketing failure) the greater would be the return that will be demanded to compensate. The RDR is, therefore, a vitally important element of the pricing basis that, by reflecting the riskiness of the product, has the effect of introducing *margins* into the basis. These margins are then available to protect the shareholders' returns from adverse experience and/or to produce profit. Margins can additionally or alternatively be included in the basis:

as an increase/decrease to individual parameter values, as appropriate
by using a higher profit criterion, e.g., requiring the PVFP to be greater than
    zero

It should be noted that the RDRs used by a company at any one time
could well be different for its different products, and will differ by whether
it is dealing with a repricing of an existing product or with a brand new
product launch. In assessing the RDR, the actuary also needs to consider
changes in the capital markets themselves; for example, if the supply of
market capital becomes scarce, the RDR will need to be higher in order to
reflect the increased cost of that capital.

There are other aspects of a product's profitability that are also usually
considered, and which can form part of the profit criteria. For example:

the number of years it takes for the accumulated statutory profit
    under a tranche of business first to become positive (called the *break-
    even point*)
the overall profit margin for a tranche of business, i.e., the PVFP divided
    by the present value of future premiums, discounted also at the RDR

The profit signature itself is also examined, and this can indicate areas in
which the product design could be improved, as we have described in
Section 10.5.3.

Attempts are then also made to test the sensitivity of the product to
possible changes in future experience, i.e., by carrying out alternative profit
tests under different demographic and financial assumptions. The
assumptions chosen are variously more or less optimistic than the main,
or standard, set of assumptions, so that the range of profit shown by these
investigations will give a fair idea of the range of profit that might
realistically be generated by the business. By varying each element of the
basis in turn, e.g., by varying the assumed rate of return on assets with
all other assumptions equal to the standard basis, a good idea of the
relative sensitivity of the product to changes in each element can be
obtained. As a result of these investigations, further changes to the design
and/or the pricing of the product may be required to produce the desired
profitability.

11.2.3   Whole-Office Pricing

The single tranche profit test is not a sufficient pricing tool. One of its major
shortcomings lies in the treatment of volume and expenses. The profit test

essentially assumes that the share of the office's fixed costs per policy issued remains constant over time. In reality this will vary in the future as a result of future variations in sales of new business and through policy terminations. As the share of fixed expenses will depend upon total business volume, i.e., from all tranches and from all product types, then this is extremely difficult to allow for in a profit test.

Another problem lies in the treatment of the one-off costs of a product launch. The issue of a new product incurs considerable research and development, administrative, and marketing costs. These costs must be justified and, therefore, paid for by the profits generated from all the future new business generated by the sale of the product; again, this is extremely difficult to allow for using a single tranche profit test approach. Chalke (1991) recognizes this problem in his "macro-pricing" approach, and points out that the decision to launch a new product rests upon whether the marginal profit of the whole issue, including all future new business and after allowing for launch costs, exceeds the marginal profit of every alternative decision, including the decision not to develop the product at all. This can also be developed by considering the new issue in the context of a projection of the whole office, i.e., by the use of *model offices*.

Before turning to the use of model offices, we should highlight one further shortcoming of a profit test approach: it does not allow easily for the effect of cross-subsidies between different tranches and different classes of business to be taken into account. One area of cross-subsidy usually involves taxation. For example, as we saw in previous chapters, the tax liability of a U.K. office depends mainly upon the balance between investment income I and expenses E. A policy's share of the office's total tax cost in any future year depends upon the relationship between the expenses and the investment income of the whole office at that time. For example, should E exceed I, then all payments are essentially gross until and unless the position reverses. Without performing a whole-office projection it is impossible to judge the future tax position with any reliability, particularly in relation to timing.

A second and very significant area of cross-subsidy relates to the pooling of insurance risks. Life offices need to hold capital in respect of their whole portfolio of liabilities which, due to the effect of risk pooling, is not the sum of the capital requirements for each policy, or even each tranche of policies considered in isolation. Hence, an increase in business volume in one class of business will reduce the average capital cost for another tranche of business (provided the policyholders involved are independent). A reduction in capital cost will, of course, also increase the expected profitability of the business (see Section 6.7.3). This point not only supports the need for whole-office modeling, but also suggests that a stochastic approach may be necessary to obtain a satisfactory appraisal of profitability.

11.2.4 The Model Office

A model life office is an example of an asset–liability model. It is a tool for projecting the financial condition of a life office at some future date. (See also Ross (1989) and Roff (1992).)

A model office is built up from a number of assumptions:

1. The amount of existing assets subdivided by type of asset class, and possibly also by term of the assets, where relevant.
2. The future investment policy: the assumptions regarding the way that the office will invest its positive cash flows, or disinvest to fund its negative cash flows, which may arise in the future. This model of investment policy may be *dynamic*; for example, it could vary depending, possibly, on the projected level of statutory solvency in the future.
3. The returns generated by existing assets in each asset class, and from future investment in those asset classes.
4. The quantity of existing liabilities subdivided by liability type. All details relevant to the generation of future cash flows should be available:

   a. the type of contract (e.g., with-profits or nonprofit);
   b. policy class (e.g., endowment or annuity);
   c. the level of any existing reversionary bonus;
   d. the expired and unexpired terms of the policy;
   e. the age and sex of the policyholder;
   f. any annual premium payable;
   g. the value of the units held under the policy, if appropriate;
   h. the level of any charges payable and assumptions regarding the future levels of those charges;
   i. any special classification features (e.g., smoker or nonsmoker); and the amount of the existing asset shares.

(It should be noted here that, in practice, the liability structure of the model office can be much simpler than the actual office. It will consist of a number of so-called "model points," where a model point is a representative contract with particular liability characteristics of contract type, amount of benefit, expired and unexpired policy term, sex, age, and so on. Enough different model points will be chosen such that, under appropriate assumptions regarding the number of policies in force at each model point, the overall liability structure of the model office will be broadly the same as that of the actual office. It would also be expected to change, according to the demographic assumptions, in a similar way to the whole office. A simple example of model points would be to assume that existing policyholders only have quinquennial ages.

The number of model points needed will vary depending upon the complexity of the office's actual liability structure and upon the purpose for which the model is required; most points would be needed if the model is being used for short-term prediction (e.g., to predict the financial condition of the office in 1 year's time), while fewer points might be used for a long-term projection for which it would be spurious to suggest greater precision.

On the other hand, if computer-processing time is not a constraint, the actual portfolio may be more easily used to represent the in-force business. This avoids any error that would result from using a simplified model portfolio, and also saves time in not having to determine what the model portfolio should be. However, it is still necessary to determine model points for future new business; see the next point below.)

5. The quantity of future new business in each future year, again sub-divided by liability type, as in (4), and their assumed premium rates.
6. Demographic and other assumptions that will allow future cash flows generated by the liabilities to be calculated, i.e., mortality rates, discontinuance rates, and the methods and assumptions for calculating surrender values.
7. The assumptions for future expenses. Assumptions will be needed regarding:

   a. the current levels of expenses;
   b. future increases in expenses due to inflation;
   c. the way in which future expenses will vary, in real terms, with the volume of business in force (noting that some of the expenses may be considered as "overhead" and, therefore, fixed relative to business volume, while the remainder may be proportionate in some way to volume);
   d. one-off expenses, such as the marketing expenses for a new product launch or for the purchase of a new computer system.

8. The method and bases according to which the office's statutory valuations are expected to be calculated in the future. This will again probably need to be dynamic according to office practice. For example, a link with the current yield on the assets at each valuation date is almost certainly necessary, noting that the maximum statutory valuation rate of interest may be dependent upon this yield.
9. The way in which policy asset shares are calculated for the office's with-profits policies. As noted before, reflecting reality as closely as possible, a regular deduction should be made from the accumulating asset shares to pay for the capital cost of the with-profits portfolio (see Section 8.2.3). Alternatively, the cost of capital can be paid for by the assumed smoothing policy (see (10)).

10. The future distribution policy, i.e., the assumptions regarding the way in which the office will distribute its statutory surpluses to its shareholders and/or policyholders in the future. This must again be a dynamic model, in that the amount of surplus distributed must depend upon the amount of profits earned in each year. Distributions to policyholders will be closely linked to the accumulated value of their asset shares, as we have described in Section 8.2.3, and assumptions will be needed as to how the benefits are to be smoothed in relation to asset shares from year to year. The assumptions may, therefore, include:

   a. target proportions of returns met by regular and terminal distributions;
   b. a formula for the calculation of smoothed dividend or regular bonus rates dependent upon the developing experience;
   c. a formula for the calculation of smoothed terminal distributions similarly obtained;
   d. the way in which shareholders' dividends are calculated (if not directly linked to policyholders' distributions, then a dynamic smoothing formula will also be required for this);
   e. the assumed smoothing method must clearly represent the expected behavior of the office as closely as possible, e.g., assumptions may involve the linking of total payouts in any year to a moving average of asset shares over a number of preceding years; the extent of smoothing assumed may also need to be tempered by the level of statutory solvency in the year in question.

11. The reinsurance policy for new policies issued in the future. This may include assumptions regarding retention levels, which may be dynamically linked to future inflation and also to future business volume because, as we have seen, the risk of relatively large losses reduces if the volume of business increases.
12. Methods of calculating tax, and future taxation rates.

Typically, a model office will be designed to produce the following types of information. For each projection year:

Cash flows (revenue) subdivided by source (premium, investment income, claims, expenses, tax).
All items of the statutory valuation at the end of the year (value of assets, value of liabilities, statutory free assets, statutory minimum solvency margin or risk-based capital).
Statutory profit earned;

Realistic profit earned (may be several different versions, but will include the expected profit on the basis of the current assumptions of the control cycle).

Details relating to profit distribution (asset shares, amount of the estate, the bonus rates declared, amount of transfer to shareholders, and the statutory assets and liabilities after the distribution, all at the end of the year).

A powerful feature of a model office is that we can *project* what the embedded value of the company is expected to be at future dates (usually at annual intervals, though some offices project embedded values on a monthly basis). The embedded value has been defined in Section 9.3.1. In order to project the embedded value of the company at some future time t, the projected in-force liabilities and available assets at time t are required. A submodel must then be run that calculates the embedded value with respect to this projected portfolio assuming no new business is issued *after* time t.

The way in which the embedded value changes over time is, therefore, a key indicator of the profitability of the company from the point of view of the providers of capital. The embedded value profit (as defined in Section 9.3.1) emerging over each time period is, therefore, an important output statistic from the model office projection.

11.2.5   Use of the Model Office in Pricing

The main aim in pricing a new product is to ensure that the office as a whole will generate more profits due to the existence of the new (or altered) product than if the new (or altered) product was not issued. A model office enables the actuary to:

vary the assumed levels of future new business in the new product (this could include the expected effect of competitor activity on new sales at the assumed product price);

vary the assumed levels of future new business (and discontinuance) in other products (e.g., where it is expected that the new product will partly or completely displace other existing products);

allow explicitly for one-off marketing and development costs.

As well as showing the embedded value profits for the whole office, the model will show the expected impact of the assumptions on the office's future statutory solvency position. This is especially important when considering the launch of a product that produces a significant new business strain, as the new business will draw heavily on the office's statutory free assets. There will, therefore, be a limit to the amount of new business growth that is sustainable in such a product, and a model office is the only reasonable way of assessing this. The resulting decision, of course, may be to alter the design of the product. It should be clear from the above discussion that model office testing must be an integral part of any product development process for both pricing and design purposes.

We will return to other uses of model offices in Section 11.5.

11.3   Analysis of Surplus

One of the uses of the model office mentioned above is to project the financial position of the office in, say, one year's time. Such a projection can be used to form a convenient means of comparison against the emerging experience, as part of the control cycle. Hence, we can compare:

1. the actual and projected balance sheets at the end of the year, including the elements of both statutory and realistic surplus;
2. the actual and projected amounts of profit emerging over the year, the latter based upon the current assumptions of the control cycle.

While (1) will show the overall impact that the actual experience has had upon the expected financial position, (2) will identify the contributions to profit over the year from the various different sources of profit and, hence, enable a judgement about the continuing suitability of the office's current realistic assumptions. The method of carrying this out is the analysis of surplus. Traditional actuarial texts (e.g., Fisher and Young, 1965) show the formulae required to calculate the amounts of surplus earned separately from investment returns, expense loadings, and from mortality and miscellaneous sources; the surpluses are calculated with respect to expected values of the relevant parameters, usually equal to the end-of-year valuation assumptions. This kind of analysis (i.e., made by reference to the valuation basis) shows the key sources of statutory profit each year and is important for informing the company's capital management strategy.

In the control cycle, however, the process is greatly facilitated by the existence of the model office. In order to analyze the surplus, it is simply necessary to itemize the revenue items of cash flow and profit for the projected and actual experiences, and to identify the differences between them (Goford, 1985). The differences observed represent divergence from the office's current realistic assumptions (according to which, effectively, the office is currently pricing its products) and this analysis provides a

Table 11.1 Analysis of surplus (£000) (from Goford (1985))

Revenue item	A	B	C	D
Reserves brought forward at start of year	8722	8722	0	0
Premiums received	4255	4255	0	0
Interest and gains received	987	868	119	119
Commissions paid	298	298	0	0
Expenses incurred	351	281	(70)	(70)
Death claims paid	750	500	(250)	(250)
Surrender values paid	720	360	(360)	337
Maturity payments	0	0	0	0
Reserves carried forward at end of year	11562	12318	756[a]	0
Profit for year	283	88	195	195

A: actual experience; B: projected experience from model office; C: actual less expected contribution to profit; D: actual less expected contribution to profit after allocation of released reserves; a = released on: deaths 59, surrenders 697.

very effective way of showing how well the office's assumptions are being borne out in practice.

The example analysis of surplus for a single calendar year shown in Table 11.1 is taken from Goford (1985). In this table, the allocation of the difference in reserves at the end of the year to surrender and death claim surpluses requires explanation. For claims by death or surrender, the expected outgo will be

(Expected claim cost) − (Expected release of reserves through claim)

where the second element is the expected reduction in reserves during the year due to no longer having to reserve at the end of the year for policyholders who claimed during the year. The actual outgo will then be

(Actual claim cost) − (Actual release of reserves though claim)

so that the contribution to profit becomes

$$[(\text{Expected claim cost}) - (\text{Actual claim cost})]$$
$$- [(\text{Expected release}) - (\text{Actual release})]$$

This gives the results listed in Table 11.2.

This analysis of the experience can be extended to produce a full analysis of the embedded value profit for the year. The actual

Table 11.2 Allocation of differences in reserves at end of year

Type of claim	Expected claim less actual claim (1)	Expected release less actual release (2)	Total profit (1) – (2)
Death	(250)	(59)	(191)
Surrender	(360)	(697)	337
Total	(610)	(756)	146

embedded value profit between times $[t-1, t]$ will differ from that expected due to:

1. Different experience during the year from expected (as in Table 11.1).
2. A different value of the *PVFP* at time t compared with anticipated, resulting from:

   different new business, withdrawals, and deaths during the year from that expected, leading to a different liability portfolio at time t;

   the view of the future experience at time t being different from that anticipated, so that some elements of the embedded value basis have been changed (e.g., future interest, mortality, RDR).

11.4 Monitoring and Updating the Assumptions in the Control Cycle

The analysis of surplus will highlight where differences between the expected and actual experience have produced significant increases or decreases from expected profits. It will, hence, demonstrate which are the most sensitive assumptions affecting the profit of the office and, hence, those assumptions that the office will need to monitor most closely.

The significance of each assumption may well vary by product type. Hence, while for endowment assurances the profit will be sensitive to investment returns, for term assurances the profit will be much more dependent upon mortality rates. This would, therefore, suggest an analysis of surplus subdivided by product type, where possible.

In the light of its experience, the office will need to decide whether it needs to change its assumptions for the next year of the control cycle and, if it does, whether or not premium rates or product designs will also need to be changed. In order to be able to update its assumptions, the office will need to carry out, alongside the analysis of surplus, proper experience investigations for the various parameters: for expenses, lapses, mortality, and of course for the returns being obtained on its assets (such as, for example, described for mortality in Section 11.2.1).

Whether or not premium rates will need to be changed will depend on the impact that the revised assumptions are expected to make on the office's profitability and solvency, as seen from the model office projections. In any case, the control cycle imposes good discipline on the office, as it forces it to assess each year the continuing adequacy of the current premium rates in the light of its (updated) experience.

11.5 Further Uses of Model Offices in Actuarial Management

Nearly all aspects of the actuarial role in life office management, as set out at the start of this chapter, can be investigated with the aid of different types of model office. We have already described their integral role in the control cycle and in product pricing; we describe below some of the other major applications.

11.5.1 To Assess Solvency

One of the most important uses of a model office is to enable an actuary to assess the solvency position of the office. In describing the mechanics of this process, one should not overlook the very significant element of judgement necessary for such an assessment, i.e., in the assumptions both of the expected future experience and of the expected future management strategy that the actuary must make in order to use the model for this purpose.

A solvency assessment can be either *deterministic* or *stochastic*. A deterministic approach will generally involve projecting statutory valuations, as at the end of each future projection year, using predetermined (hence "deterministic") assumptions for all of the parameters, such as interest and mortality rates. Assumptions need not be constant between projection years of the model; for example, interest rates may be projected to decline gradually over the projection period, if this accords with some current expectation. The dynamic features of the model are then still necessary in the deterministic framework (for instance, the need for changing asset proportions in response to a deteriorating solvency position, as described earlier).

11.5.1.1 Deterministic Models

A deterministic assessment will indicate whether the statutory solvency of the office is most likely to improve or deteriorate in the future, and will give some indication of the possible rate of change in this position. Key measurements in these observations will be the projected market values of the assets, the statutory value of the liabilities, and the required minimum margin of solvency. Relative solvency is probably most appropriately measured using some ratio, e.g., the ratio of free assets to liabilities, of liabilities to assets, or of free assets to total assets. Changes in these ratios will, therefore, give an indication of changes in relative solvency. However, even the ratios cannot be an entirely adequate guide to solvency; recall our discussion of Section 6.7.3, in which we observed that the relative amount of capital required to secure a given probability of solvency was smaller the greater the number of policyholders. Hence, a constant solvency ratio does not equate to a constant solvency probability under conditions of changing business volumes.

A deterministic solvency assessment should invariably include a *sensitivity analysis*, in order to gain some insight into the range of possible outcomes. This involves making different model office projections on the basis of a range of assumptions for the underlying parameters, some more optimistic and some less optimistic than the central, or "best estimate," basis. The assumed variations would aim to encompass the main range of future outcomes that the actuary would consider as "feasible" or "probable" under current conditions. The sensitivity analysis brings a number of additional benefits:

It focuses attention upon a range of possible outcomes rather than on a single projection, thereby avoiding the undue confidence that would be implied regarding the reliability of any single projection.

It will give some idea of the increasing uncertainty of any projection the further into the future the projection is made (the so-called expanding funnel of doubt).

It enables the actuary to assess the relative importance of each factor to the solvency of the office; for example, for some offices a deterioration in investment return may be most crucial to solvency, while for others it might be an increase in expense levels that is critical, and so on. This is a fundamental benefit of a sensitivity analysis.

This kind of deterministic sensitivity assessment of life office solvency is generally referred to as *dynamic solvency testing*; see Brender (1988). It is also known interchangeably as *dynamic capital adequacy testing, dynamic financial analysis*, and *stress testing*. As we mentioned in Section 9.1.3, this process has been introduced as part of the supervisory regulation in many territories, with Canada being the pioneer in this regard.

11.5.1.2 Stochastic Models

Hardy (1993) has criticized the use of deterministic models for assessing solvency as producing an unrealistically serene set of projections. The alternative approach is to use a stochastic model office. In such a model, the values of the elements of the model (such as the return on assets or the mortality rates) are not predetermined but are allowed to vary, in a pseudo-random way, according to assumed probability distributions for those variables (which, by definition, are therefore allowed to behave as random variables). Hence, assumptions are not only needed for the expected value of each variable, but also for the values of all the parameters which define the probability distribution of each variable. For example, if a particular element of the model were expected to be a normally distributed random variable, then assumptions would be needed for both the mean and the variance of its probability distribution; see Daykin et al. (1994).

The main reason for using a stochastic model is that it can provide an idea of the probability distribution of some complex random variable (such as the future ratio of statutory liabilities to assets), which would be impossible to determine using analytical methods. In Chapter 5 we described how stochastic investment models are constructed, and how they could be used to simulate the distributions of complex variables in an asset–liability model. An example has been given of simulating the distribution of ultimate surplus at the run-off of a closed fund; we return to this example in Section 11.5.4. In Section 5.6, we also stated that stochastic asset–liability models can be used to estimate probabilities of ruin of financial institutions, including life offices. The procedure is straightforward as follows. We might define an indicator variable $W_j = 1$, where in at least one of the future projection years of the jth simulation the statutory free assets are less than the required minimum margin of solvency projected for that year; and $W_j = 0$ otherwise. Then, if n simulations are made, $(1/n) \sum_{j=1}^{n} W_j$ is the simulated frequency of statutory insolvency occurring according to the model assumptions. Assuming the law of large numbers applies, this will be an estimate of the current probability of statutory insolvency.

11.5.1.3 *Parametric Insolvency Measures*

In Chapter 4 we described a variety of approaches to measuring investment risk. Many of these approaches can be applied through stochastic asset–liability modeling in order to assess the solvency risk of life insurers in different ways, including the frequency of ruin just described. The use of downside risk measures, described in Section 4.3, would be particularly appropriate for the purpose of measuring insolvency risk. For example, if we consider a projection of a closed office, there will be some future time (w years from now, say) at which all of the office's policy liabilities will have terminated. If deterministic demographic assumptions are made, then w will be a known constant. It is then possible to project the surplus assets remaining at time w immediately after the final liability has expired. We will express this surplus as a relative amount, such as

$$X_j = \frac{S_j}{M_j}$$

where S_j is the projected surplus at time w in the jth simulation and M_j is a measure of total policy proceeds over the period $[0, w]$ in the jth simulation.

If we assume a stochastic investment model, then the amount X_j is a random variable whose distribution can be estimated from the n simulations performed. A parametric risk measure can then be defined as

$$P(a) = \frac{1}{n} \sum_{j=1}^{n} R_j(a)$$

where

$$R_j(a) = \begin{cases} (-X_j)^a & \text{for } X_j < 0 \\ 0 & \text{otherwise} \end{cases}$$

with $a > 1$. $P(a)$ is therefore an estimate of the expected value of the risk function $R_j(a)$ for any j, which can be shown to be one of a family of risk functions defined by Clarkson (1989).

Essentially, this risk measure is simply the value of the relative loss $-X_j$, taken to a power a, and averaged over the n simulations. Surpluses are ignored. The size of $P(a)$ is, therefore, dependent not only upon the frequency of losses among the n simulations, but also on the relative intensity of the loss. It is also dependent upon the risk aversion parameter a, such that the higher the value of a the greater the relative contribution that large losses will make to the total value of $P(a)$. Chadburn (1996) shows that the relative insolvency risks of different model scenarios as measured by $P(a)$ can appear quite different from those measured only according to the frequency of loss.

This approach to assessing risk is flexible; however, for any particular investigation, the actuary will need to decide which variables are of key interest for his or her purpose and focus on those.

(See also Section 19.4, where a similar downside risk measure is described in relation to pension funds.)

11.5.2 To Investigate Profit Distribution Policy

Model offices are also of key importance in any investigation of distribution policy.

11.5.2.1 Deterministic Models

A deterministic model office, including an appropriate sensitivity analysis, can enable an investigation to be made of the rates of regular and terminal profit distributions that might be supportable over the long term. This can also be performed, for existing business, by carrying out a bonus reserve valuation (BRV) of the liabilities and assets (as introduced towards the end of Section 8.5.3). This can be extended by varying the assumptions of future experience, consistently both in the valuation of the assets and of the liabilities, in order to investigate the extent to which the distribution rates are supportable under different future conditions. This has traditionally formed an important actuarial tool for helping to assess the appropriate amount of profit to distribute with respect to the office's existing business at any point in time. However, a model office does allow greater versatility and, hence, realism than a BRV, as future new business can be included, and variable and dynamic features can be incorporated in the projections as described earlier. Hence, for example, a gradual fall in future regular distribution rates can be assumed (if this accords with expectation); or the model can vary its projected future distribution structure (e.g., the ratio of terminal to reversionary bonus) in response to projected changes to the statutory solvency ratios.

11.5.2.2 Stochastic Models

While a deterministic model will allow an assessment of the future average rates of supportable profit distribution to be made, an analysis of the smoothing element of the distribution policy is not possible. A deterministic model office can project, for example, smoothly progressing values of future asset shares from year to year. However, reality will present a much more erratic progression, as described in Section 8.2, and the policy-holders will receive benefits that equate to smoothed versions of these asset shares. As we described earlier, the difference between benefit levels and asset shares from time to time represents a cost or contribution to the office's capital, and thereby constitutes a solvency risk. The greater is the extent of benefit smoothing, the greater could be the risk to solvency.

The actuary can, therefore, use a stochastic model office, in particular involving a stochastic asset model, in order to investigate the solvency risk (for example) involved with different smoothing policies. The assumed smoothing policy must, of course, be modeled dynamically, as payouts must be allowed to vary in response to the projected asset shares from year to year in each simulation. In this way, the actuary can compare the merits of a variety of smoothing strategies, ranging from high to low degrees of smoothing, and thereby assess the merits of its own current, or proposed, strategy. The merits of any strategy will include not only the relative solvency risk, but also the relative utility of the smoothing strategy to the policyholders and, indeed, to the shareholders where their profit distributions are linked to those of the policyholders. Whether or not the bonus smoothing policy can be modeled realistically and, hence, reliably is, of course, another matter. However, the model should at least give valuable insights into the relative merits of the different strategies, even if the absolute predictions are unreliable. This caution applies equally to any modeling exercise: relative values are more likely to be more useful than absolute values.

There have been a number of published investigations of smoothing strategies using stochastic model offices; e.g., see Needleman and Roff (1996).

11.5.3 To Investigate Investment Policy

One of the key aspects of the actuarial management of life offices is its investment policy, by which we refer to the office's asset allocation or portfolio selection strategy. The task of selecting individual investments falls to the company's investment managers. However, the general direction of asset allocation will require significant input from the actuary, due to the need to select investments that are appropriate to the liabilities. This subject has been discussed fully, including the use of stochastic investment models, in Chapter 3 to Chapter 5.

11.5.4 To Investigate Business Risks

Future new business is one of the most significant factors in the determination of the future financial condition of any life office. It is also one of the least predictable variables from a modeling viewpoint. One of the key features of a model office, which sets it aside from other more traditional valuation methods, is its ability to allow explicitly for new business. It is new business, coupled to a lesser extent with the discontinuance experience, that will determine the nature of the office's future liabilities, and of course its influence will be greater the further into the future the projection is made. New business will affect the future liability profile in the following main ways:

1. by the volume of sales;
2. by the nature of the contracts sold;
3. by the expenses incurred in sales and marketing (including research and development) activities.

A model office is particularly useful for investigating the interactions between new business, expenses, and lapses, as we have already mentioned in Section 11.2.3 and Section 11.2.5. In particular, explicit allowance can be made for the overhead (or fixed) expenses separately from the proportionate expenses in the model. The influence of new business strain on statutory solvency is also an important element of any analysis of business risk, as we have described earlier.

The model can also be used to investigate the effect of declining new business, particularly from the point of view of the coverage of expenses. Declining volumes may soon lead to insupportable levels of fixed expenses. The model will allow the office to assess how rapidly it would need to cut its overhead expenses in the face of such a demise.

A model office allows the impact of so-called "expense overruns" on profit and solvency to be explicitly examined. For example, the cost of unsuccessful marketing and product launches will not be met from income generated by such policies, but will have to be met from other new or existing business. The impact (i.e., the "risk") of such marketing exercises can be estimated through such investigations.

Discontinuance risk can also be easily investigated. For example, where discontinuance at early policy durations produces losses, then the effect on solvency and profit of higher than anticipated rates could be investigated; the effect of lower than anticipated rates would be important where discontinuance produces profits (e.g., at later policy durations).

11.5.4.1 New Business Assumptions

As intimated at the start Section 11.5.4, the inherent difficulty with any model office investigation is the lack of predictability of the new business variables. A possible way forward may be to model new business volumes as a function of simulated market changes, such as to premium rates and/or to other factors linked to economic demand. Daykin and Hey (1990) describe such a model for general insurance. Whether a credible stochastic new business model can be constructed for life insurance applications is an issue that remains to be explored at the present time.

The alternative is to model new business deterministically, but to investigate a wide range and variety of future possibilities. Note that an integral part of the set of assumptions for any particular projection would be the way in which both the volume and type of business sold would vary from year to year. It must always be ensured that expense and discontinuance assumptions remain consistent with the new business assumptions, reflecting the various correlations that may exist between them.

11.5.4.2 Closed Funds

The ultimate deterioration in new business is the example of the closed fund, i.e., where no future new business occurs. This is a particularly vital aspect of any office's solvency assessment, because the actuary's duty is foremost to the office's existing policyholders. Hence, the actuary must ensure that the office's obligations to the existing policyholders will be honored even if the office closes to new business in the future. Apart from where closure is immediately anticipated, it would be usual to model the effect of closure to new business at a variety of dates in the future. An example of such an investigation can be found in MacDonald (1993).

It is particularly pertinent to consider the assumptions that should be made regarding the experience subsequent to closure. Assuming that the life office continues its independent existence, then appropriate assumptions will be required for mortality, discontinuance, expense, and investment experience. Key elements of the assumptions will be with regard to adverse selection of discontinuing lives and expense overruns. Closure to new business will often be accompanied by bad publicity for the office, and many policyholders may choose to discontinue their policies in the period immediately following closure, and to take up new policies from alternative providers. This discontinuance will almost invariably be selective, as only the policyholders who would still satisfy other offices' underwriting standards would leave. The average mortality experience of the remaining policyholders could, therefore, rise significantly.

The expense levels after closure are also highly dependent upon the speed with which the office can dismantle its sales and marketing operations. Contracts of employment with sales staff may mean additional expenses during this process. Ultimate expense levels (those which will be necessary to administer the closed office until its existing business is finally terminated) will need to be assessed, as will the length of time which will elapse before that ultimate level is reached.

In reality, however, closed funds are often taken over by another life insurer who will administer the liabilities until their eventual termination. This usually protects the position of the remaining policyholders, and may reduce the extent of selective discontinuance. The receiving office should have none of the expense-overrun problems that would be faced if the closed fund continued to be administered independently. The assets and liabilities of the closed fund would be transferred at an agreed price, paid by the office receiving the fund, which would be broadly equal to the embedded value of the business. It is unlikely that any goodwill would be paid for the acquisition, bearing in mind that no new business can be anticipated. Hence, when modeling the future effect of closure to new business, assumptions may be needed regarding the price received for the office's disposal of the fund, and the delay between closure and the date of disposal.

A closed-fund approach is also necessary for valuing a mutual office which is considering demutualization.

11.5.5 To Estimate Reserves

A stochastic model office can be used to estimate reserves. By definition, reserves are required for existing business; hence, it will again be necessary to assume a closed fund. The basic procedure is as follows:

1. For a given amount (and mix) of initial assets x_0, perform (say) 10 000 simulations of the model office up to the simulated termination date of the last in-force policy. Record for each simulation $X_j(x_0)$, the amount of assets remaining at the fund termination date in the jth simulation.
2. Calculate

$$\hat{w} = \frac{1}{10\,000} \times \left[\text{Number of simulations in which } X_j(x_0) < 0\right]$$

so that $\hat{w}$ estimates the probability of actual insolvency when initial assets of x_0 are held.
3. Repeat (1) and (2) for different initial assets (both by amount and/or mix) and iterate until initial assets are found that produce $w = \alpha$, where α is the chosen acceptable insolvency probability and w is the realized value of $\hat{w}$. Hence if $\alpha = 0.01$, then we would find by iteration the initial assets x_0 that generates 100 insolvent simulations out of 10 000. This particular value of x_0 is then the reserve (which we might call the *stochastic reserve*) for this level of insolvency risk.

It would be logical to use the same set of 10 000 scenarios (i.e., using the same random numbers) for each iteration. Stochastic reserves can also be based upon a parametric insolvency risk, such as described in Section 11.5.1; see Chadburn (1996).

11.5.6 To Estimate Risk-Based Capital

An office will be particularly interested to assess the amount of statutory free assets it needs to hold at a point in time in order to be able to meet its solvency requirements, with a high level of confidence, in the future. The procedure set out in Section 11.5.5 can be adopted again for this assessment, except that we should use the simulated frequency of statutory insolvency (as defined in Section 11.5.1) for $\hat{w}$, and we are not constrained to assuming a closed fund. From the resulting value of x_0 we can obtain

$$\text{Risk-based capital} = x_0 - {}^sV_0$$

where ${}^sV_0$ is the current statutory value of the liabilities.

11.6   Choice of Stochastic Model

Most stochastic applications of life office modeling will involve stochastic investment models. These models are covered in more detail in Chapter 5, to which readers are referred.

For life offices with large and stable portfolios of policyholders, the impact of random variability in mortality experience upon profitability and solvency will be insignificant compared with the effect of the variability in the asset proceeds, unless the business of the office is predominately in protection contracts with only a small savings (i.e., investment) element. In the case where there is only a small savings element, it may be necessary to perform model office projections with stochastic mortality assumptions. Other instances where it might be necessary to use stochastic mortality models include:

1. where the effect of different volumes of business on the risk-based capital required by the office needs to be investigated;
2. where investigations are being carried out to set the office's reinsurance retention levels (see Section 10.4.3).

11.7   Modeling Risk

In Chapter 10 we discussed the range of risk factors that constitute life office risk, while in this chapter we have described the various models that the actuary can use to assess this risk and, hence, form advice about its control. This process, however, leads to a final and, unfortunately, potentially significant further contribution to life office risk, that the models (and assumptions) used by the actuary in formulating his or her advice may be wrong. This is, by definition, the combined risk from parameter and specification error described in Section 10.4. In fact, we would say that there is almost certainty that, except by rare chance, the actuary's models will indeed be wrong, as that is the nature of any model: it is a simplification of reality and depends upon assumptions made about future uncertain events. The key point is to what extent the model is likely to be wrong and, hence, how significant will the modeling error be in leading to incorrect advice for the company's management. It is imperative, therefore, that the actuary should endeavor to:

1. minimize the extent of modeling error as much as possible conducive with cost constraints;
2. assess the impact of the expected level of modeling error on the reliability of his or her advice;
3. assess the risk to the office of relying on his or her advice given the extent of the modeling error.

Consideration of (2) and (3) will assist the actuary in his or her attempts to minimize modeling error on (1). Modeling error would be expected to be reduced by taking the following steps:

1. Ensure that any model used is consistent with relevant past experience.
2. Ensure that the most relevant data to the projection are used to assess (1).
3. Consider the extent to which future differences in experience may lead to differences from the observed experience in (2).
4. Ensure that the behavior of the model is consistent with an intuitive assessment of the behavior of the variable(s) being modeled.

Indeed, it should be clear from the above that the actual structures of the models themselves, and not just the assumptions (parameters) of those models, should be monitored and updated as an integral part of the control cycle. Hence, in the control cycle diagram of Figure 11.1, "initial assumptions" should be more properly read as "initial model" and "updating of assumptions" should be "updating of the model," thereby encompassing both parameter and specification aspects of the modeling process.

Clearly, the greater the financial significance of any possible modeling error, the greater effort the actuary will expend in attempting to minimize it. Information about the financial significance will be obtained from the analysis of surplus (see Section 11.3). Once the error has been minimized to the actuary's satisfaction, he or she should then proceed to formulate his or her advice, attempting always to quantify the effect of the potential error upon that advice. Good actuarial practice will always involve an assessment of the effect of variation in the actuary's modeling assumptions, and this assessment should also be properly communicated to the recipient of the actuary's advice.

References

Benjamin, B. and Pollard, J. H. (1993). *The Analysis of Mortality and Other Actuarial Statistics*. The Institute and Faculty of Actuaries.

Bowers, N. L., Gerber, H. U., Hickman, J. C., Jones, D. A., and Nesbitt, C. J. (1986). *Actuarial Mathematics*. The Society of Actuaries.

Brender, A. (1988). Testing the solvency of life insurers. In: *Proceedings of the 2nd International Conference on Insurance Solvency*.

Chadburn, R. G. (1996). Use of a parametric risk measure in assessing risk based capital and insolvency constraints for with-profits life insurance. Actuarial Research Paper No. 83, City University, London.

Chalke, S. (1991). Macro-pricing: a comprehensive product development process. *Transactions of the Society of Actuaries* 43, 137.

Clarkson, R. S. (1989). The measurement of investment risk. *Journal of the Institute of Actuaries* 116, 127–178.

Daykin, C. D., Pentikainen, T., and Pesonen, M. (1994). *Practical Risk Theory for Actuaries*, Chapman & Hall, London.

Daykin, C. D. and Hey, G. B. (1990). Managing uncertainty in a general insurance company. *Journal of the Institute of Actuaries* 117, 173–277.

Fisher, H. F. and Young, J. (1965). *Actuarial Practice of Life Insurance*. Institute of Actuaries.

Goford, J. (1985). The control cycle. *Journal of the Institute of Actuaries Students Society* 28, 99–114.

Hardy, M. R. (1993). Stochastic simulation in life office solvency assessment. *Journal of the Institute of Actuaries* 120, 131–151.

Institute and Faculty of Actuaries (2003). Guidance notes GN1 and GN8.

Lee, R. E. (1985). A prophet of profits: an introduction to the theory and applications of profit tests. *Journal of the Institute of Actuaries Students Society* 28, 1–42.

MacDonald, A. S. (1993). What is the value of a valuation? *Proceedings of the 3rd Actuarial Approach to Financial Risks International Colloquium*, vol. 3, pp. 725–743.

Needleman, P. D. and Roff, T. A. (1996). Asset shares and their use in the financial management of a with-profits fund. *British Actuarial Journal* 1, 603–688.

Neill, A. (1977). *Life Contingencies*. W. Heinemann Ltd, London.

Roff, T. (1992). *Asset and Liability Studies on a With-Profits Fund*. Staple Inn Actuarial Society.

Ross, M. D. (1989). Modelling a with-profits life office. *Journal of the Institute of Actuaries* 116, 691–715.

Part III
General Insurance

Chapter 12

Introduction to General Insurance

12.1 Background

General insurance has a wide scope, including any insurance not classified as life insurance. Risk taking is apparent in all human endeavor, and general insurance products offer some compensation should a future event unexpectedly diminish the normal quality of life through injury or damage to property. Risk, in this context, encompasses the timing and the nature of the event, as well as the outcome, although the use of the term is limited to events that might be considered significant. For example, the enjoyment of our possessions or our quality of life could be impaired as the result of a motor vehicle accident, or in some future year due to an existing but currently unknown latent phenomenon, such as environmental pollution. The basis of compensation may be for a *fixed sum* of money, for the *replacement value* of the loss, or for *indemnification*. The latter two arrangements mean that the policyholder should be in the same financial position as before the loss. Estimating the appropriate award is not necessarily straightforward. Although it should not be difficult to assess the replacement cost of a physical good, placing a value on intangibles, such as defamation of character under a libel award or damage to the brand value of a corporation in a commercial legal action, can be a subjective, litigious, and lengthy process.

For insurance to operate effectively, a large group of individual policyholders each contribute a small and usually fixed amount (the premium) into a fund which is used to compensate those few policyholders who suffer a loss. By pooling risk (and the policyholders' premiums), a larger entity is formed which can bear each individual loss more easily and whose relative uncertainty is lower than that of the individual (see Appendix A). Expressed another way, the capital needed collectively to cover, say 95% of the aggregate loss distribution is less than would be the case were each individual policyholder required to protect its potential aggregate loss to the same degree.

Risks exist which are so unpredictable or of such magnitude as to be deemed uninsurable; for example, war risks, terrorism, and nuclear risks are usually excluded from insurance contracts. In certain cases, coverage against such risks is felt to be in the public interest and governments frequently take on the liability for them in some form. In such cases, funding for losses may be by a levy on all insurance premiums as a tax (e.g., on electricity consumption for nuclear risks) or as part of wider fiscal policy.

In general, the amount of the premium and the subsequent management of the fund are determined by an insurance company. Whilst the basic principle of pooled risk applies equally to life insurance and group pension schemes as it does to general insurance, there are some significant differences. The principal contractual difference is in the duration of the cover provided. General insurance products usually have a duration of 1 year, whereas life insurance products can range from 1 year to whole of life. Life insurance cover for durations of less than 1 year is available, but it tends to provide protection against contingencies of a nonlife nature (e.g., temporary cover might be provided for an executive visiting a hazardous region, in which case the peril being insured is violent death rather than death by natural causes). Other important differences between life insurance and general insurance are considered in the following.

12.1.1   Exposure Measurement

Exposure is that quantity to which the *premium rate* is applied.

In life insurance, the unit of exposure is well defined and easily measured, being the life year or sum assured life year.[1] Exposure is well defined because the events that can result in a claim are well defined, and easily measured and verified (usually death or survival). The benefit to be provided, the sum assured, is also usually known in advance (see Chapter 6).

In the case of nonlife insurance, the diversity of perils (i.e., the events which might give rise to a claim) which may be covered by the insurance policy complicate the choice of a suitable exposure measure. For example, in motor insurance the usual measure of exposure is the vehicle year.[2] However, for perils such as moving-traffic accidents a better measure might be distance driven, because it indicates the extent to which the car was actually exposed to accidents. On the other hand, distance driven will

---

[1]The *life year* is the number of lives covered by the policy, multiplied by the period for which the insurer is exposed to risk during the relevant period of measurement. The *sum assured life year* is the life year scaled to allow for the sum assured.

[12]The *vehicle year* operates in the same way as the life year: it is the number of vehicles covered multiplied by the duration in years for which the insurer is exposed to risk.

not properly reflect the exposure to theft claims that occur when the car is not being driven. It might be possible to conceive of a more comprehensive, two (or more) parameter, measure of exposure, including both distance and vehicle years, reflecting more accurately the actual perils to which the insurance policy is exposed. Although distance driven is a reasonable measure of risk, it is rarely used in practice as the cost of obtaining the information accurately outweighs the benefits.

12.1.2 The Evaluation and Identification of Factors Influencing the Degree of Risk

From the point of view of managing the risk, it is important for the insurer to discriminate between policyholders having different risk characteristics. Recognizing the different risk characteristics of its policyholders enables the insurer to attribute premium and other costs fairly between them, and to protect itself from adverse selection by *proposers* (i.e., prospective policyholders).

In life insurance, the factors which influence the degree of risk, or the likelihood of a claim, and the consequent premium charged are well understood, reasonably stable over time, and easily measured. These factors include the policyholder's age, sex, smoking habits, and state of health.

In general insurance, the position is very different. An important distinction is made between risk factors and rating factors. *Risk factors* are those factors that are believed to influence directly the frequency and severity of a claim for a given quantum of exposure. However, risk factors may be difficult to measure reliably. Consequently, insurance companies tend to gather information about those attributes of the proposer and the goods to be insured that are believed to be closely related to the risk factors, but which are easier to measure and to manage. This information is used as a proxy for the risk factors to provide *rating factors*. It is the risk factors that are of real interest, but rating factors are more easily obtained and measured.

Thus, rating factors are not necessarily the same as risk factors. For example, the traffic density in which a car is driven is clearly a significant risk factor, but it is one that cannot be measured directly for the purposes of rating an individual proposer. A crude proxy, which could be used as a rating factor, might be the proposer's address or the purpose for which the vehicle is used, e.g., whether it is used for commuting to work.

12.1.3 The Range of Events Giving Rise to a Liability under the Contract

In life insurance, the circumstances of a claim under the contract are well defined and easily verified. In general insurance, the cover provided by a policy can be very broad and, when a claim is made, events which are deemed to be covered might not have been envisaged at the time that the original contract of insurance was issued. For example, the emergence of latent claims,[3] predominantly from the U.S. has proved very costly. The majority of these claims have arisen due to personal injuries arising from exposure to products containing asbestos and to environmental pollution from a wide variety of industries. The basis of allocating claims to contracts of insurance in these cases, many years after the inception date of the contracts, is a difficult area and usually leads to litigation. Where there is a lack of clarity in the contract language, courts, often after protracted legal debate, tend to interpret the contract wording against the party drafting the contract (in most cases the insurers). Claims may also stem from retrospective changes in the law and the application of strict liability rules. For example, although waste may have been disposed of in accordance with the safety regulations prescribed at the time, the fact that contamination occurs subsequently could render the contributing parties liable. In other words, having used "state of the art" disposal methods might not be a legal defense. Situations may be further complicated by the application of joint and several liability. This means that every participant on, say, a dump site is individually liable for the full cost of cleaning the site. Thus, the damage caused by companies that have since become bankrupt becomes the liability of those companies (and their insurers) who have remained solvent. Issues such as these, arising many years after the issuance of the original policy, are factors that could not have been anticipated when the premium level was established.

12.2   Risks and Their Management

The likelihood of a claim against an insurance policy is influenced by risk factors. Another way of considering the circumstances that make a claim more or less likely is to identify the nature of the hazards to which a policyholder is exposed. Such hazards can be categorized broadly as follows:

Legal hazards, which include changes of laws or the imposition of new conditions under which insurers could become liable. For example, retrospective legislation might introduce broader coverage conditions than were foreseen when the insurer wrote the contract.

---

[3]*Latent claims* is the term given to claims which emerge several years after the date of occurrence.

Physical hazards, which are existing physical or structural conditions that could increase the likelihood of loss. For example, faulty house wiring or outdated safety systems.

Moral hazards, which are deviations from normal behavior made because a policyholder stands to gain financially from taking certain actions within his or her control. In other words, fortuity, a basic principle of insurance, is violated.[4]

Personal hazards, which are closely related to moral hazards but cover the case where a policyholder is particularly accident prone, simply a bad risk. For example, people might be careless or badly qualified for the job they are doing and, therefore, impose an above-average liability on the insurer.

Consideration of exposure and hazards leads to a useful analogy with the physical sciences

exposure measures the *volume* of potential claims;
hazards and rating factors give an indication of the *concentration* of risk within that volume.

Exposure is that quantity to which the premium rate is applied and the premium rate is the expected level of aggregate claims and expense per unit of exposure.

The approaches that individuals or corporate bodies adopt towards the hazards they face vary considerably. The field of risk management combines the design and application of physical controls of risk (e.g., the use of sprinkler systems to reduce the peril due to fire) with insurance packages to provide financial protection. The basic ways that individuals deal with risk can be expressed more formally as follows:

Assumption is also known as retention of risk. A cost–benefit analysis of the alternative risk management options may conclude that the most efficient solution is to take no action concerning the risk; that current measures are already efficient or, in extremis, there is no other choice — the risk is uninsurable.

Elimination involves taking action to remove all the hazards to which one is exposed. For example, pesticides can be used to eliminate the risk of crop failure due to insect infestation. However, physical measures to eliminate the causes of risk often have unpredictable side effects. Entirely new hazards might arise, e.g., pesticides may pollute the environment, causing injury to other, beneficial, animals and to human

---

[4]*Fortuity*, in the context of insurance, means that events giving rise to claims arise due to chance, and are not under the control of the insured.

health. In this context, there is currently speculation as to the implications of genetically modified crops.

Avoidance of risk involves a change in behavior or business practice to avoid the undesirable exposure. For example, one could choose to park one's car in a secure garage or to cease trading in politically volatile regions.

Under *transfer* of risk, the risk is transferred to another individual or organization. This is the basis of insurance and reinsurance contracts. In return for a fixed (i.e., known and certain) premium payable to the insurer, the insured can transfer that part of the financial uncertainty associated with certain events which it would rather not have to sustain.

Insurance contracts usually have elements of all these aspects. In a private motor insurance policy we see the following features:

- The policyholder bears an excess or a retention, e.g., the first £50 of each claim must be met by the policyholder. The insurer, therefore, ensures that the policyholder assumes part of the risk and so is more likely to avoid careless actions.
- The insurance premium might have special discounts if anti-theft devices are fitted; the insurer is encouraging the avoidance of risks by reducing the theft hazard.
- Cover is void if a claim is caused whilst under the influence of drugs or alcohol. The insurer is eliminating an aspect of personal hazard by, again, encouraging the avoidance of risk; cover is also void if the incident occurred whilst in pursuit of illegal activities — an example of society discouraging antisocial behavior
- The proposer is asked to provide an estimate of the value of the car. Compensation should not exceed the true value of the vehicle; thus, the insurer controls the extent of any moral hazard.

The *risk aversion* of individuals varies greatly. Factors determining risk aversion may be objective in the sense that specific criteria are set which have measurable limits. For example, an organization might want to limit its maximum exposure to fire damage claims. Alternatively, risk aversion might be subjective, perhaps based on some vague reluctance to buy insurance cover, or on the grounds that an individual has adequate wealth to absorb unexpected contingencies. In either case, the underlying process of deciding whether or not to transfer risk to an insurer involves evaluating the costs of managing the hazards (including the costs of insurance), and the benefits obtained by exchanging a certain cost (the premium) for an uncertain one (insured claims). Viewed another way the individual operates on a utility curve (see Appendix A) and seeks to maximize the level of utility achieved in the face of the alternative possibilities.

We can now express the choice that insurance offers in a more formal way.

The proposer's perspective. The proposer is faced with the choice of paying a fixed premium or retaining an uncertain liability. In return for the premium the insurer pays claims corresponding to the transferred liability. The decision to pay the premium is driven by the proposer's perception of the cost–benefit trade-off. From an accounting perspective the premium is the cost, or liability, of the buyer and the claim is a contingent asset that only crystallizes when a covered event occurs.

The insurer's perspective. At the individual policy level, the insurer is in the complementary position to the proposer. In this case the premium is the asset and the liability is the financial risk the insurer perceives to be associated with the proposer and the associated policy conditions.

The important distinction between the proposer and the insurer at the individual policy level is the difference in the perception of risk; it is this difference which forms the basis of all mutually acceptable contracts. The foundation of commerce is that the level of utility achieved by both parties is greater after a contract has been agreed than it was before, i.e., increased overall utility is a positive sum game.

Taking a broader view of the insurer's entire portfolio of business, the uncertainty of whether or not an individual will claim becomes irrelevant; the insurer is more interested in the behavior of the claims of the group in aggregate. In theory, predicting the behavior of a group of homogeneous and independent risks is easier than predicting that of individual risks; see Daykin et al. (1994) for a discussion of the properties of aggregate claims distributions. As is discussed in Chapter 15, an aggregation of policies can introduce special hazards that are different in nature to those associated with individual policies, and insurers have alternative mechanisms for dealing with these issues.

12.3 The Contract

The transfer of risk is accomplished via a legally binding contract of insurance whereby the proposer submits details of the goods to be insured together with other details related to the various associated hazards, as demanded by the insurer. With very few exceptions, the principle of utmost good faith places the onus on the proposer to bring to the insurer's attention any factors of which the insurer is unaware and which could affect the insurer's underwriting decision. The policyholder is also normally required to advise the insurer of any relevant changes to the nature of the risk during the course of the insurance.

Having obtained details of the proposer, the cover required, and the goods to be protected, the underwriter will assess the risk and quantify the likely claims, the required contingency margin, profit margin, and administration costs. In large commercial contracts, pricing would normally be carried out on a per proposer basis due to the diversity of risks and lack of adequate statistical data. However, for high-volume "commodity" business, where individual premiums are relatively low, such as private motor insurance, a formulaic approach is adopted, and most insurance companies use a rate-guide book which is periodically reviewed. In sophisticated markets, the "rate book" is a computer algorithm that updates prices on a daily basis.

Having assessed the risk the underwriter can:

- accept the proposer on normal contract conditions
- adjust the normal policy conditions (e.g., to exclude certain unacceptable perils)
- decline the proposer

(as in life insurance; see Chapter 6).

Once the proposer is accepted, a consideration (the premium) is paid by the proposer, the contract is completed, and an insurance policy is formally issued. Details of the risk and the accounting transactions will be recorded on the insurer's information system. In the personal lines[5] classes of general insurance business, temporary insurance contracts ("cover notes") can be issued by agents of the insurance company while the full documentation is being processed.

The insurer is then liable for losses resulting from the perils specified under the contract from the time of inception of the cover until the contract expires (the *policy term*). In general, any event which takes place during the term of the policy, and which is specified in the policy contract, can give rise to an insurance claim, i.e., if the *occurrence date* is within the policy term. In certain classes of business, relating mainly to the services provided by professional groups (professional indemnity insurance, error and omissions, Directors and Officers' cover), cover is usually provider on a *claims made* basis, i.e., only those claims notified during the course of the contract will give rise to a loss.

The flip side of underwriting is in claims management. In making a claim, the policyholder is required to submit formal notification of the loss, usually via a pro-forma claim form. Upon receipt of claim notification, the process of loss adjusting begins. Initially, this involves establishing that the insurer is "on risk," i.e., that the event occurred within the policy period and was a consequence of perils provided for under the contract. The loss adjuster will then evaluate the claim (which may involve engineering or other specialist reports) and establish a reserve estimate as a potential liability within the books of the insurer (the "case reserve"),

---

[5]Personal lines insurance :insurance issued to private individuals not related to business undertakings (e.g., private motor, household).

pending final agreement to pay the claim and, hence, liquidate the liability. Most are paid quickly; however, in complex cases, settlement can suffer considerable delays due to difficulties in estimating the claim size, the possibility of dispute between the insured and the insurer, and due to difficulties in allocating liability. Classes of business principally exposed to property-related claims settle relatively quickly and are commonly referred to as "short-tailed" classes, whereas those classes with a substantial liability exposure can take many years to settle fully, hence the term "long-tailed" classes. The strict procedures for establishing case reserves and of recording the values on the company's systems vary between companies, and obtaining a clear understanding of these procedures is a prerequisite to any actuarial study for pricing or reserving purposes.

12.4   Reinsurance: Insurance for Insurers

Having issued the insurance contracts, an insurance company may feel that the aggregate risk it now bears is in some sense too great. The key benchmarks relate the "free assets" or "surplus" capital to some measure of the risks borne. The main risks which are likely to concern the insurance company are:

- a very large individual claim;
- an event, such as a flood, storm, or other widespread event, that results simultaneously in claims from a large number of separate policies;
- a general adverse development in claims, in effect a miscalculation of the overall premium rate.

The latter two problems are referred to as *aggregation* or *accumulation* risks and may arise from concentration of business within a geographical area. Generally speaking, the variability associated with a group of independent policies is less than the aggregate variability of the individual policies. If policies are not independent then this is no longer the case and insurers can become exposed to potentially catastrophic claims in terms of the impact to the company's balance sheet position, often in highly unpredictable ways. The case of geographical accumulation of exposure (e.g., insuring all the houses in the same flood plain or exposed to the same weather systems) is well understood and can be addressed in a straightforward manner. However, the insurance risks associated with systemic economic factors result from a large number of diverse, though related, phenomena. For example, economic decline, falling house prices, social instability (particularly higher rates of divorce and separation), increasing unemployment levels, high interest rates, and regulatory failure on markets all constitute material and often unfathomable risk factors to underwriters of economically linked insurance (e.g., Directors and Officers, errors and omissions, mortgage guarantee, and other credit-risk contracts). These wider scoped risks are much more difficult to manage because, although they are associated with one another, it is difficult to identify and assess the extent of the connection; however, their combined financial impact can be devastating.

The principle of an insurance company itself seeking insurance is well established in the *reinsurance* market (which is discussed in detail in Chapter 15). The reinsurance market serves to spread individual insurance companies' risks over a much wider capital base. A secondary consequence tends to be that the duration for final settlement of a particular claim is extended, which gives reinsurance companies small additional margins, in that their invested funds are held for longer.

For similar reasons of solvency and stability, reinsurance companies also need insurance, and such risks flow into the *retrocession* market. The higher up the "reinsurance chain" one goes, the more remote the original risk becomes; reinsurance companies rarely obtain information about the individual risk originally exposed. The statistical methodologies used and the philosophical nature of pricing of direct insurers and reinsurers then diverges. There is not a clear distinction between reinsurance companies and retrocessionnaires. Indeed, a reinsurance company will not only underwrite the risks of direct writing companies but might also underwrite the retrocessions of other reinsurance companies. The international flow of claims and premiums between insurer, reinsurer, and the retrocession market, and the mutual interdependence of the finances of a large number of different companies, can inadvertently lead to highly incestuous financial relationships in the market and, hence, dependencies. The nature and implications of these complex relationships often only become apparent when an insolvency in the chain causes financial ripples amongst the other members of the chain.

12.5 Management of General Insurance

Insurance companies face risks other than those arising from the business they underwrite, many of which are subjective and difficult to quantify. The insurance "enterprise risk" can be characterized in many ways for example

$$\text{Enterprise risk} = \text{Core business risk} + \text{Operational risk}$$

The core business risk is made up of insurance risk, investment risk, and reinsurance risk. Operational risk addresses all other aspects of the business risk and management process. Alternatively, enterprise risk has been expressed in terms of the hazard risk, financial risk, operational risk and strategic risk (see CAS report on Enterprise Risk Management,

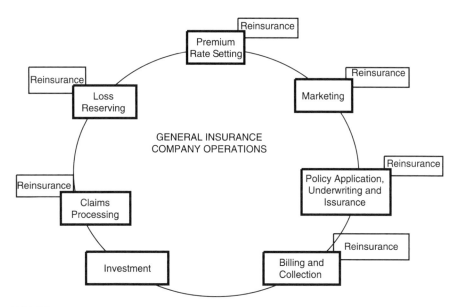

FIGURE 12.1 General insurance company operations

November 2001). Operational risk is a difficult concept to define clearly, and there have been several variations; perhaps the most useful is "the risk of loss, resulting from inadequate or failed internal processes, people and systems or from external events" (see Financial Services Authority (FSA) Integrated Prudential Sourcebook, www.fsa.gov.uk. FSA). A simple working guide might be that core business risks are related to the items shown directly in the balance sheet (principally asset and reserving risks), and operational risks relate to revenue items and all other aspects of the business. Thus, in addition to actuarial controls, it is important that other general management controls are in place to ensure the continuing successful operation of the business, and the existence of clear management control is a specific requirement of most regulatory authorities.

The operation of a general insurance company is summarized in Figure 12.1.

12.6 Providers of Insurance

The main providers of insurance are public limited companies, mutual companies, trade associations, Lloyd's syndicates, captive insurance companies (i.e., insurance offshoots established and wholly owned by industrial companies to provide for their primary insurance needs) and state-owned insurance companies. Reinsurance companies are frequently specialist offshoots of larger insurers, and are often domiciled in territories with favorable tax and regulatory regimes.

12.6.1   The Direct Market

Those companies that provide insurance policies to the general public and to commercial enterprises are known as *direct insurers*. The word "direct" signifies their proximate relationship with the underlying risk and serves to differentiate the business from reinsurance, where the connection to the underlying risk can be many steps removed in the contractual chain.

Most direct insurance is sold through an intermediary. Intermediaries include insurance brokers, who are remunerated through commission; employed agents, who are paid a salary by the insurer; independent financial advisors, who will require a fee; or other businesses, which sell insurance as a sideline (e.g., building societies or savings and loan institutions may sell buildings insurance, and retailers of domestic electrical goods often sell extended warranty insurance). As in all business, the insurer needs to monitor its sales in order to gauge its intermediaries' efficiency. For example, it will analyze the sales made and the commission earned. In particular, it will be concerned with the quality of the business sold, which can be assessed by reviewing the claims history and the renewal rates of the policies sold by individual intermediaries.

For personal lines business, insurers are tending to reduce the use of brokers as a distribution channel, e.g., by using telephone sales, mail shots, and the Internet. In addition, banks and building societies may set up their own insurance subsidiaries rather than operate as agents in order to provide some of the insurance needs of their banking customers. The increasing use of database technology in identifying potentially receptive markets has dramatically altered the way that personal lines business is bought and sold. The increasing use of the Internet as a sales outlet has started to change the relationship between the proposer and the insurer, whereby prospective buyers of insurance become their own brokers, encouraging insurers to bid for their business.

The different classes of general insurance business written on a direct basis include personal lines of business such as:

- private motor insurance
- private medical insurance (but not private health insurance, which is provided by life insurers)
- personal accident insurance
- extended warranty insurance
- travel insurance
- mortgage indemnity

and commercial insurance such as:

- motor fleet or commercial vehicles
- commercial fire insurance
- product liability insurance
- general liability insurance
- professional indemnity
- marine, aviation, and transport insurance
- commercial multi-peril (i.e., package insurance programs for insurance)

12.6.2 London Market

The London Market consists of some 200 companies that operate in the City of London, together with Lloyd's insurance syndicates. It includes specialist reinsurance companies, as well as the reinsurance subsidiaries of direct writers. These companies and syndicates dominate the U.K. insurance market for large international risks, which include marine and aviation business, large property risks, liability, and reinsurance classes. Virtually all of the business is placed by an accredited broker, acting on behalf of the insured, and the market relies heavily on the expertise, experience, and business connections of its brokers.

The London insurance market is unusual, in that the London broker is the hub of information and the vast majority of transactional information is administered through one or two central data bureaux. Underwriters subscribe to these bureaux and construct their in-house management information systems around data interfaces with the bureaux. The broker is normally responsible for initiating a bureau transaction and carries out a considerable amount of the back-office administration that, in other markets, would be performed in-house within the insurance company.

Before a risk can be placed, the broker is responsible for obtaining all of the information which underwriters will need to assess the risk. The broker then presents the information to those underwriters known to be interested in that particular class of business. The information is summarized in a *placing document*, which can vary in size from one page of historical loss experiences to several volumes of claims and exposure information together with a computer database to facilitate the under-writer's evaluation of the data. The terms of the contract are summarized in a document known as a *slip*. This sets out the premium rates and terms of coverage. Frequently, coverage terms are noted as standard clauses, often using acronyms that mean little to those not connected with the market. When the risk has been placed, a policy document is prepared, with the complete text of the terms and conditions.

The broker will negotiate the best terms and conditions available with a leading underwriter in the designated class of business. Once the slip

conditions are agreed the underwriter will decide on his or her participation on the slip, i.e., the proportion of the risk underwritten. The lead
underwriter has the principal responsibility for setting the terms and
conditions of the policy and for the subsequent claims administration.
Because of the large exposures frequently involved in London Market
business, it is rare for an underwriter to accept 100% of slip-based business. A typical lead underwriter's participation would be about 25%,
depending on the size of the exposure involved.

Once a lead has been established the broker will present the slip to other
underwriters known as *followers* or the *following market*. Generally, followers
take smaller proportions of the risk than the lead underwriter. The broker's
objective is to place 100% of the risk. In practice, brokers aim to cover more
than 100% of the risk; they then reduce the exposure of each participant
proportionately to obtain the desired 100%. In some cases risks are not
fully placed due to lack of capacity in the market or to the perceived
unattractiveness of the terms and conditions underlying the risk. In that
case, the proposer must retain a proportion of the risk or get alternative
protection.

The slip mechanism is an example of *co-insurance* (see Chapter 15)
which means that the obligations of the slip participants to the policyholder
are independent; in effect, the policyholder has a separate contract of
insurance with each slip participant.

12.6.3   Other Providers

Insurance can be provided in many other ways than through traditional
insurance. Some large companies form *captive insurers* that are wholly
owned subsidiaries with the sole function of underwriting the various
risks of the holding company and its other subsidiaries. In another example,
companies or individuals with similar interests group together to form
indemnity clubs. These are mutual arrangements whereby certain agreed
risks are pooled together, or shared, between the members of the club.
The Automobile Association in the U.K. is an example of an indemnity
club, and there are also several examples of ship owners pooling risks in
protection & indemnity (P&I) clubs.

12.6.4   The Insurance Broker

The wide range of insurance products available and the complexity of the
market place often mean that those requiring insurance need expert
guidance in order to find the most appropriate policy. This advice is mainly
provided to potential buyers of insurance by insurance brokers. Brokers
act as facilitators, bringing together the two parties to the contract. They can

also give the policyholder assistance when making a claim. They are remunerated by commission payments from the insurer, which is deducted from the premium paid by the insured. In the case of personal lines insurance (e.g., motor or household), the commission can be as much as 30% of the original premium. In commercial lines business, where the premiums are substantially more, the commission rate is less, but will vary according to the amount of work the broker is expected to do. Under most regulatory systems, brokers are required to have a minimum level of competence and to have adequate financial backing, e.g., a financial deposit must be lodged with the market regulator.

The commission system, which effectively "front-end" loads the broker's remuneration, can lead to substantial resource drains on brokers, particularly specialists in long-tailed lines of business, since the timing of income receipts from commission does not match the timing of claims management expenses. This is a particular problem for brokers that have shifted their target market into short-tailed lines of business, or whose transaction volumes have fallen. The mismatch between income and outgo is leading some brokers to accept lower initial commission and to levy a fee on all claims subsequently administered. The nature of the commission system might also be considered unusual, in that the broker's client is the insured but the remuneration is typically paid by the insurer. It could be argued that such a system may be prone to conflicts of interest.

Changes in the personal lines market, with the rise in the market share obtained by direct writing insurers (i.e., companies that do not use brokers), have reduced the influence of brokers in this area. However, in commercial lines business, brokers play a major role in obtaining appropriate cover, even to the extent of designing types of cover and obtaining quotations from underwriters for the suggested coverage.

12.7 The Insurance Environment and the Provision of Capital

Suppliers of insurance operate in a market place that is highly competitive and becoming increasingly sophisticated. The market is subject to an *insurance cycle*. Capital committed to support insurance companies is subject to the usual rules of supply and demand: when profits from insurance are perceived to exceed the profits available elsewhere then insurance will attract more capital. As more capital is dedicated to insurance by a greater number of participants, some of who will be relatively unsophisticated investors, downward pressure is exerted upon premium rates. Owing to the variability associated with the insurance business, unsound premium rates may persist for some time and might even show a profit, so long as the expected claims do not occur. When a large claim does occur, there might be a capital flight into other business areas and some insurers may become insolvent. However, well-established, well-capitalized, and well-managed insurers should be able to withstand insurance cycles, even if they are forced to follow market prices down. Under the right conditions, premium rates will recover to a fundamentally profitable level and might even overcompensate to reflect the need of the remaining insurers to recover losses sustained on past business. The ability of the market to recover losses on past underwriting is a function of the ease of access that investors have to the insurance market and, in particular, the ease with which capital can be withdrawn from the marketplace. If capital can be shifted too easily, then the shorter term insurance investors have a constant and, ultimately, unsustainable option against long-term providers of capital. When rates are low, only the long-term investors are in the market; and when rates increase to profitable levels, the ability of long-term investors to recoup past losses is impaired because of the entry of new capital with no requirement to recoup such losses. The effect of the insurance cycle on premium rates tends to be more pronounced in classes of business associated with rare events, such as earthquake or other catastrophe cover.

The extent of cyclical activity depends on underlying economic activity levels and on the control exercised by the regulatory regime. In the U.S., particularly in the personal lines fields, insurance companies are required to submit rate filing reports in order to justify premium increases. However, in the U.K., for example, premium rates are unrestricted. Instead, regulation attempts to ensure the continued operation of the market by imposing minimum capital and solvency standards on individual companies. The question as to the appropriate amount of capital that an insurer should carry does not have a straightforward answer: regulators, policyholders, and rating agencies prefer higher levels of capital, "free reserves," or "surplus," whilst shareholders seek the best return on equity, which implies providing the least capital needed for the efficient operation of the business. Setting aside the question of the means of establishing the "correct" level of capital for the moment, what is clear is that capital follows risk: the greater the risk, the greater the capital needed to support the business and the greater the reward sought by the investor for taking on that level of risk.

In setting minimum capital standards regulators are tending to adopt a risk-based capital (RBC) approach, which is based around a series of capital charges applied to the components of the enterprise risk, the principal headings are :

Asset-related risks (specific assets or through lack of diversification)
Credit risk (principally failure of reinsurance and the credit rating of
    counterparties)
Reserving risk
Pricing risk
Risks arising from affiliated companies

Since companies have different drivers, and many aspects of risk are in any case highly subjective, it is impossible to establish an RBC requirement that gives regulators and others a completely consistent measure of capital requirement against the risks undertaken; hence, such approaches are attempts at working approximations to the ideal. One of the most difficult issues in applying RBC is the degree to which different aspects of risk are correlated — either between lines of business or between different asset classes and liabilities.

12.8 Conclusions

Insurance exists to provide financial compensation in the case of uncertain events. To be able to do this, insurers need to be able to control their own uncertainty more reliably than an individual policyholder can. In this chapter, we have discussed in broad terms the fundamental operation of general insurance and the complexities of the individual providers of insurance and of the markets within which they operate. The following chapters explain some of the procedures used so that those insurers are able to measure, monitor, and finance their business.

Appendix A

A.1 The Law of Large Numbers

Consider a group of n identical but independent policyholders. Let X_i be a random variable representing the claims made by the ith individual policyholder over the course of a policy year. Then the X_i form a set of independent and identically distributed random variables. Suppose $E(X_i) = \mu$ and $V(X_i) = s^2$, for $i = 1, 2, \ldots, n$. The *coefficient of variance* is defined as $\sqrt{V(X)}/E(X)$. It is a measure of the uncertainty associated with the random variable, relative to the expected size of the risk. For an individual policyholder the coefficient of variance is s/μ.

Let S denote the aggregate claims made by the n policyholders. Then $S = \sum_{i=1}^{n} X_i$. Thus, $E(S) = n\mu$ and, since the policyholders are assumed to be independent, $V(S) = ns^2$. So the coefficient of variance for the insurer is $s/(\mu\sqrt{n})$. Now, the larger n is, the smaller the coefficient of variance will be; in fact, as $n \to \infty$, $s/(\mu\sqrt{n}) \to 0$. This result is a consequence of the *Law of Large Numbers* (Cummins, 1991).

So, we see that, by accepting a large number of similar, independent risks, the relative uncertainty of an insurer's business is reduced. Note that the assumption of independence is important, and the insurer has to take precautions in case this assumption breaks down.

A.2   Utility Theory

"Utility" is a term used by economists to indicate the satisfaction a consumer receives from a product. Economists assume that consumers seek to maximize their total utility. Then, if a consumer has the choice of allocating his or her expenditure to one combination of products or another, we assume that the consumer will select the combination that provides the greater utility. It is possible to assume that certain mathematical functions represent individuals' utilities. The exponential function $U(x) = -Ae^{-Bx}$, where $A, B > 0$ are constants and the variable x represents the wealth of the consumer, is a commonly used function. It satisfies the assumptions that marginal utility declines as total consumption increases and that individuals are risk averse (see Chapter 5.4).

In the case of insurance, the choice is between bearing an uncertain risk which could give rise to an unknown expenditure at some unknown point in the future (but which might give rise to no expenditure) and making a definite, fixed payment at the start of a policy term. Assume a general utility function $U(x)$ and suppose that the premium required for insurance is P and the random variable X represents the individual's perception of the value of the future damage or injury, allowing for its likelihood of occurrence. Then, a consumer with wealth A would purchase insurance if he or she were (at least) indifferent between paying P to the insurer and having the insurer assume the random loss, or assuming the risk him or herself (see Bowers et al. (1997)), i.e.

$$U(A - P) \geq E[U(A - X)]$$

A and P are known, but X is a random variable and $U(x)$ is an unknown function that could be difficult to estimate. If assumptions are made about the form of $U(x)$ and the distribution of X, then the above procedure can be followed and P estimated from the equality, in which case P would be the maximum premium that the policyholder would be prepared to pay.

See Cozzolino (1997) for a discussion of utility from the insurance company's perspective.

A.3   Lloyd's Insurance Market

Historically, the Lloyd's Insurance market was a place where individual members, or *Names*, underwrote insurance on their own account. Each Name accepted unlimited personal liability, and so in effect pledged the whole of their wealth in order to support any insurance written on their behalf. For practical reasons, Names grouped together in *syndicates* that tended to specialize in particular classes of business. However, each Name's legal liability is limited to their own share of the syndicate's business; thus, if a Name in the syndicate is unable to pay the claims that emerge in respect of the business written, the other Names on the syndicate have no direct liability to meet the shortfall that arises, the shortfall being allocated across the whole Lloyd's market (i.e., individual members' shortfalls are mutualized, usually by a market-wide levy).

Whilst the traditional Names continue to provide capital to the market, they have been largely superceded by corporate capital vehicles with limited liability. Such vehicles are generally subsidiaries of insurance and reinsurance companies, a key motive for the establishment of a corporate vehicle at Lloyd's is the Lloyd's brand name (the "franchise value") and immediate access to international operating licenses. The annual premium capacity of Lloyd's is in the region of £12 billion and is primarily in respect of direct and reinsurance business domiciled in the U.S.

The syndicate manager is referred to as the Managing Agent, and the affairs of the remaining individual Names are administered by Members' Agents, who also seek to place the Names' capital with the most profitable syndicates that continue to admit the traditional underwriting Names.

References

Bowers, N. I., Gerber, H. V., Hickman, J. C., Jones, D. A. and Nesbitt, C. J. (1987). *Actuarial Mathematics*. Society of Actuaries, Chicago.

Cozzolino, J. M. (1997). Portfolios of Risky Projects. CAS Forum.

Cummins, J. D. (1991). Statistical and Financial Models of Insurance Pricing and the Insurance Firm. *Journal of Risk and Insurance*, Vol 58.

Daykin, C. D., Pentikainen, T. and Pesonen, M. (1994). *Practical Risk Theory for Actuaries*. Chapman & Hall.

Chapter 13

General Insurance Accounts

This chapter provides an introduction to the items found in the accounts of a general insurance company. A broad understanding of the accounting framework is essential to understand the dynamics of an insurance operation, as each of the major topics of reserving, pricing, reinsurance, and capital management is manifested in the accounts.

In accounting terminology, and in the European Union Insurance Accounts Directive, a distinction is made between "reserves" and "provisions." *Provisions* are defined to be the amount necessary to meet liabilities which have already been incurred; *reserves* are required for unspecified future liabilities that might arise. Actuarial terminology in common usage does not make this distinction and, generally, uses the term "reserves" for both. The nature of the liabilities for which different reserves are established and set aside, and whether they should be termed provisions or reserves, should be clear from the definitions that are given in this and subsequent chapters.

13.1 The Purpose of Accounts

The stakeholders of a company, namely its shareholders, managers, and business partners (e.g., reinsurance companies, brokers, and policyholders), and the regulatory authorities all require information about the company's operations and its financial performance. However, the level of detail and the information the various parties require is different, reflecting their objectives and perspectives

- Insurance regulators[1] are concerned with the solvency of the operation and its ability to meet its claim-paying commitments to policyholders. An insurance company's clients (i.e., those seeking insurance) will

---

[1]The insurance regulator in the U.K. is the Financial Services Authority (FSA); in the U.S. the regulators are the State Insurance Commissioners and the National Association of Insurance Commissioners. In many other countries insurance supervision falls under the remit of the Finance Ministry.

have similar concerns. The more sophisticated of these will vet the company's ability to remain solvent over a long period; in particular, insurance brokers owe a duty of care to their customers in order to ensure that the insurer is suitable for the placement of business. In the U.K., the information required by the regulator is prescribed in the Insurance Companies Acts and their associated regulations.

- The existing shareholders of a company and any prospective investors, whilst concerned about its ongoing solvency, are principally concerned with the rate of return earned on the capital employed by the company, the company's profit and dividend levels, and its short- to medium-term prospects. They will want to be able to assess the opportunity cost of investing in one financial venture as opposed to another. Similar considerations apply to the insurer's competitors. In the U.K., the accounts and information prepared for the shareholders are prescribed by the Companies Acts. Shareholders and particularly prospective shareholders might also be interested in assessing the *appraisal value* of the company (e.g., see Bride and Lomax (1994) or Ryan and Larner (1990)).

- The tax authorities need to assess the amount of tax due for the year. The calculation of profit for tax purposes can be different from the profit calculation required by the regulators and the shareholders' accounts, since some items, which accountants and regulators consider to be liabilities, might not be acceptable as such to the tax authorities. Tax calculations can be complex: for example, reserve items might accrue at different times for tax purposes than for normal reserve purposes; foreign exchange transactions can be subject to international rules and cross-border agreements. The eventual computation of tax is often arrived at after discussion between the authorities and the company management, the result of which might not strictly follow the minutiae of the law.

- Managers are interested in the solvency position of the company, as well as its return on capital. Managers might also have business targets; the measurement of success in achieving these objectives will require detailed investigations of the underwriting and investment experience of the company and evidence of its ability to control expenses and to achieve business growth in a controlled and planned manner.

The analysis of insurance company accounts, benchmarking company performance, assessing creditworthiness, and solvency is a major service industry. The managements of most insurance companies rely heavily on the analysis provided by such organizations. Having established the form that a reinsurance program should take, the choice of reinsurance counterparty normally requires the approval of the company's reinsurance security committee, who would rely in part upon such external assessment agencies. Similarly, each company is concerned with its own security rating, since a higher rating will make it more attractive to potential customers, which gives the insurer more scope to be selective in its underwriting or to underwrite more business.

In most developed insurance markets, there are prescribed formats for presenting accounts and other information to satisfy the interests of insurance regulators and shareholders. There might be differences between the accounts prepared for solvency purposes and those for Companies Act purposes, although they need not imply significantly different financial interpretations. Under U.K. regulations, for example, Companies Act accounts are consolidated, i.e., they include the financial operations of subsidiary companies. The returns required by the regulator, however, are prepared on an unconsolidated basis. The treatment of certain items in the accounts also differs, largely because of the different intentions of the accounts. Most significantly, certain assets cannot be included in the calculation of the Statutory Minimum Solvency Margin[2] (the Statutory Minimum Solvency Margin is defined in Section 13.2.1).

The returns for regulators are usually prepared in greater detail than the Companies Act accounts. In the U.K., for example, the prescribed information required by the FSA is recorded separately for each of 17 accounting classes (accident, health, land, vehicles, railway rolling stock, aircraft, ships, goods in transit, fire and natural forces loss, damage to property, motor vehicle's liability, aircraft liability, liability for ships, general liability, credit, surety ship, and miscellaneous financial and legal expenses). Companies Act accounts are almost invariably not provided in such detail; the main separation is between business accounted for on a 1-year basis and business accounted on a 3-year, or *funded*, basis.[3]

Companies have a wide variety of internal monitoring procedures designed to meet the needs of specific departments, with an information hierarchy leading to a high-level data summary for the chief executive. With good computer systems, all aspects of the business can be monitored and individual targets can be set for those departments within the organization with responsibility for particular aspects of performance, such as underwriting, marketing, claims handling, reinsurance collections (and other credit-control functions), and investment. Management information is likely to be required at least quarterly, and in some cases monthly. For high-volume insurance companies, such as motor insurance business

---

[2]The Valuation of Assets regulations specifies which classes of assets can be taken into account when calculating the Statutory Minimum Solvency Margin, and to what extent they can be included. For example, no more that 5% of the market value of the assets can be made up from the stocks and shares of one company (see Chapter 3.4).

[3]Some long-tailed classes of general insurance business were allowed to develop for 3 years before profits were taken, since it is difficult to assess an appropriate level of reserves after only 1 year's reporting. Three-year accounting has been the traditional method used in the Lloyd's market. However, it is becoming more common for companies to account for profits on a 1-year basis regardless of the class of business.

via telephone sales, management information in various forms is updated continuously.

In order to understand the operation of a company, it is important that the content of the accounts and the relationship between revenue accounts and balance sheets is clearly understood. The basic statements are:

- The revenue account, which summarizes the revenue accrued during the calendar period (usually 12 months), arising out of the insurance business.
- The profit and loss account, which incorporates the insurance result from the revenue account, as well as other revenue accruals, e.g., tax and dividends payments, investment income not attributed to the revenue account, and items not directly related to the underwriting activities.
- The balance sheet, which provides a view of the total assets and liabilities of the company at the end of the calendar period. It can be useful to think of the balance sheet as summarizing the cash flows expected to emerge in the future, in respect of the business in force at the date of the accounts (i.e., a summary of future revenue accounts, ignoring new business and goodwill). This model is not strictly correct, as it ignores future investment income. But it is useful to think in these terms, as it facilitates the construction of business plans and other financial models and highlights certain flaws in the generally accepted accounting framework.

The revenue and profit and loss accounts, therefore, provide the link between the opening (or brought forward) balance sheet and the closing (or carried forward) balance sheet. The accounts prepared in accordance with the Companies Acts are accompanied by notes explaining detailed items of accounting policy and, in particular, the impact of any changes in accounting policy. The notes form an integral part of the accounts and can give significant qualitative insights into the company's operation and status that the quantitative financial figures do not provide. For example, certain long-established companies might have significant liabilities in respect of environmental pollution claims in the U.S. There is no generally accepted basis for calculating the amounts of these claims, and there is a wide range of views regarding the required level of reserves. Two companies with otherwise identical business and reserving policies could be perceived as having significantly different financial performances, simply because of the different views they take on this one issue.

13.2 The Accounting Framework

The following section sets out a basic accounting framework. Figure 13.1 demonstrates the connection between the various parts of the accounts and

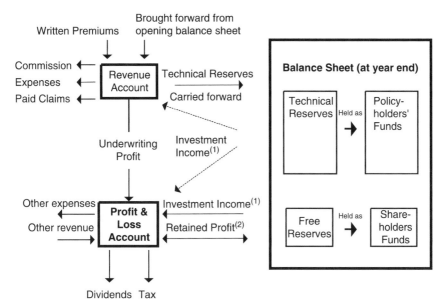

FIGURE 13.1 Connection between the various parts of the accounts and certain accrued revenue items. *Notes*: (1) Investment income in respect of the technical reserves can be credited either to the revenue account or directly to the profit and loss account. If reserves are discounted explicitly, then the investment return expected on the reserves is usually credited to the revenue account and the balance is credited to the profit and loss account. In this case, the balance in the revenue account labeled "underwriting profit" is usually renamed "insurance profit." (2) Retained profits increase the free reserves; losses reduce the free reserves. Source: Taylor (1992)

certain accrued revenue items. Appendix A gives a list of the items found in the accounts of a general insurance company.

13.2.1 The Balance Sheet

The balance sheet gives a view of the financial position of the company according to the accepted accounting practices of the local legislature. A general insurance company has assets, consisting of fixed assets, current assets, and, most significantly, investments. The principal liabilities are represented by the *technical reserves*, which are the company's estimate of the insurance liabilities that have arisen in the past and not yet been settled; these are described in greater detail in Chapter 16. As in any set of accounts, there are incidental amounts of current asset and current liability representing items recognized in the revenue account but for which the cash itself has not moved. The excess of assets over the liabilities is the *free reserve*, or equity, of the company. This is the capital base which provides policyholders with security should claims exceed income, satisfies statutory solvency requirements, and enables the continuing development of the business. (Note that the structure is similar in life insurance; see Chapter 8.1 for a comparison.)

Regulations in the E.U. specify that the minimum level of free reserves (the Statutory Minimum Solvency Margin) must be the greater of:

The premium basis — [18% of the first 10 million units of account[4] and 16%[5] of the remaining written premium income (including reinsurance accepted but excluding reinsurance ceded) for the previous financial year] multiplied by the ratio of net-to-gross claims incurred in the previous financial year, subject to a maximum ratio of 50%.

The claims basis — [26% of the first 7 million units of account and 23% of the average claims incurred for the previous 3 years (7 years for storm, hail, and frost insurance)] multiplied by the ratio of net-to-gross claims incurred in the previous financial year, subject to a maximum ratio of 50%.

Generally, regulators have wide powers to intervene in the operation of an insurance company. Their powers are usually triggered when solvency levels fall below a predefined threshold relative to the prescribed statutory levels.

The E.U. statutory minimum solvency level calculation is a broadly based measure and only indirectly recognizes the effect of different mixes of business and different management practices between companies; it is currently under review. Regulators in the U.S. adopt solvency standards that focus more closely on risk as it relates to an individual company, i.e., the *risk-based capital method* discussed in Section 12.4 and in Section 11.5 in the case of life insurance (see also Hooker et al. (1996)). Alternative means of measuring the degree of solvency is by way of "stress" testing the balance sheet under prescribed standard scenarios, e.g., a given fall in equity values, a standard major loss, shifts in the yield curve, or a reinsurance failure. The production of economic and claims scenario models and extensive corporate stochastic simulation modeling forms the growing area of *dynamic financial analysis*.

13.2.2 The Revenue Account

13.2.2.1 Premiums

The premium due in respect of policies whose coverage began during the accounting period is the *written premium*. Some policies will provide

---

[4]The unit of account is the ECU, subject to a minimum exchange rate.

[5]Certain health insurance contracts have lower percentages of premiums or claims (i.e., one-third of those given in the text).

cover extending beyond the end of the accounting period and, at the year end, the company is contractually obliged to meet claims that arise for the remaining duration of coverage. The reserve set up in respect of the unexpired coverage is called the unearned premium reserve (UPR). The total premium earned in an accounting year, then, will be the

(UPR brought forward from the previous accounting year)

+ (Premium written in the accounting year)

− (UPR carried forward to the next accounting year)

At its simplest the UPR is calculated on a time-earned basis, i.e., the ratio of the policy duration falling beyond the accounting period to the total policy duration is applied to the written premium. This implicitly assumes that risks are "earned" smoothly throughout the policy period.

This formula does not allow for the irregular way in which expenses are incurred. Substantial initial expenses and commissions can be incurred during the acquisition of new business. These expenses are charged to the revenue account in full when the business is written; therefore, strictly, the UPR should be calculated excluding acquisition costs. However, in practice, different approaches are taken. Usually, a separate asset is set up, called the *deferred acquisitions cost*, which is equivalent to the proportion of acquisition costs included in the UPR. More detail is given in Chapter 16.

13.2.2.2 Claims

Under the principle of accruals the insurer must account for the claims *incurred* during the year, not for the claims paid during the accounting year. The claims paid will include amounts in respect of claims incurred in earlier accounting years, which should have been included in the reserve for outstanding claims brought forward from the previous accounting period. Thus, the claims incurred can be calculated as the claims paid together with the increase (or decrease) required to the technical reserves for outstanding claims, i.e., the reserve for incurred claims equals

− (Outstanding claims reserves brought forward)

+ (Claims paid) + (Outstanding claims reserves carried forward)

Outstanding claims include known, notified claims and a provision for incurred but not reported (IBNR; see Chapter 16 for more detail) claims.

13.2.2.3 The Underwriting Result

The items above, together with expenses associated with underwriting activities and commission, are combined in the revenue account to produce the *underwriting result*, i.e.

$$(\text{Earned premium}) - (\text{Incurred claims}) - (\text{Expenses and commission})$$

Provided that the premium basis is fundamentally profitable, the accounting treatment is conservative, since profit recognition on the unearned premium is deferred until it becomes earned. However, when the premium basis is believed to be inadequate, e.g., because of unforeseen inflation of claims costs or because of a fundamental error in the premium rating methodology, it is a requirement to recognize the loss inherent in the premium basis in advance of the UPR being exposed to risk and earned. The additional reserve created is known as the *additional unexpired risk reserve* (AURR). The terminology can be confusing; frequently, the UPR and AURR are collectively known as the *unexpired risk reserve*. The AURR is discussed further in Chapter 16.

If the investment income on the assets backing the technical reserves is identified separately, then the insurance result is calculated as the

$$(\text{Underwriting result})$$
$$+ (\text{Investment income allocated to the technical reserves})$$

This definition can be modified to allow for the taxation charge attributable to the insurance business.

13.2.3 The Profit and Loss Account

The profit and loss account is a continuation of the revenue account, reflecting those revenues and expenses not directly attributable to the underwriting activities. This includes, for example:

- investment income on the shareholders' funds
- investments in subsidiary companies
- extraordinary items
- certain property changes
- senior management (or group) expenses
- tax
- dividends payable to shareholders

The balance of profits, after any charges and after all payments to shareholders, is retained in the business, increasing the free assets (i.e., "equity" or "surplus") of the company. An overall loss is effectively a disinvestment, and the free reserves of the company are reduced by this amount.

In the above outline, no distinction has been made between items before or after reinsurance (*gross* or *net*). For many insurers the difference is dramatic, and statutory regulatory returns normally require all technical items to be separated into the position gross of reinsurance and net of reinsurance, with an accompanying description of the relevant reinsurance policies.

13.2.4   Some Key Analytical Statistics

Accounts provide summarized information and are static statements presented at a point in time. They are particularly deficient in indicating the inherent uncertainty in both the amount and timing of asset receipts and liability outgo. This is fundamental in insurance, and the calculation of risk-based capital is directed at this problem. From a supervisory perspective, there is considerable scope for improving the presentation of information and for requiring supporting information for disclosure purposes. The problem of taking a dynamic stochastic process and expressing it in a static, deterministic, point estimate in the accounts is being investigated in many companies, e.g., using dynamic financial models (see CAS Spring 1996 Call papers).

Notwithstanding the limitations of published accounts, they do provide information prepared using a consistent basis.[6] Thus, the financial characteristics of different insurers can be compared, and the performance of individual companies can be tracked through time, in order to identify features such as premium growth, changes in the asset portfolio, and solvency. Some of the most widely used analytical ratios are shown below. Their usefulness depends on the particular circumstances of individual companies. For stable companies, the ratios are likely to provide reasonable measures for what is intended. However, for companies undergoing major strategic changes, or experiencing major changes in rates of growth, the ratios might need to be adjusted, and in any case are likely to be less meaningful.

- Premium growth statistics, e.g.

$$\frac{\text{Premium written in year } t}{\text{Premium written in year } (t-1)}$$

---

[6]In the U.K., Companies Act accounts have to be prepared in accordance with Generally Accepted Accounting Practices (GAAP), and in the U.S. the equivalent accounts must be prepared according to Financial Accounting Standards (FAS).

- Ratios that provide an indication of the dependence on reinsurance:

$$\frac{\text{Net written premium}}{\text{Gross written premium}} \quad \frac{\text{Net paid claims}}{\text{Gross paid claims}} \quad \frac{\text{Net incurred claims}}{\text{Gross incurred claims}}$$

- A measure of the responsiveness of reinsurers to claims advices:

$$\frac{\text{Reinsurance debtors}}{\text{Reinsurance recoveries due on paid claims}}$$

- The loss ratio, which can be calculated net or gross of reinsurance:

$$\frac{\text{Incurred claims}}{\text{Earned premiums}}$$

- The expense ratio:

$$\frac{\text{Expenses and commission}}{\text{Written premiums}}$$

- The operating ratio:

$$\text{Loss ratio} + \text{Expense ratio}$$

- The investment ratio:

$$\frac{\text{Investment returns}}{\text{Written premiums}}$$

If investment returns have been attributed to the underwriting activity it can also be useful to calculate

$$\text{Operating ratio} - \text{Investment ratio}$$

- Solvency ratio:

$$\frac{\text{Free assets}}{\text{Net written premiums}}$$

- An indication of the proportionate error in the technical reserves that can be absorbed by the free reserves:

$$\frac{\text{Free assets}}{\text{Technical reserves}}$$

- Numerous ratios can be calculated to test the adequacy of the reserves. Most popular are those that provide an indication of duration, e.g.[7]

$$\frac{\text{Closing technical reserves}}{\text{Annual paid claims}}$$

$$\frac{\text{Closing IBNR reserves}}{\text{Annual deterioration in incurred losses}}$$

These ratios are very crude and are most useful when examining a company that has ceased trading and is running off its reserves.

There are several ratios related to investment return, the size of debtors and creditors relative to turnover, and, most importantly, the return on equity, which is

$$\frac{\text{Profit}}{\text{Free assets}}$$

A number of possible definitions of profit arise, for example, depending on whether pre- or post-tax analysis is required or whether the analyst wishes to find the extent to which the company's reinsurers have enhanced the overall return on equity.

Other, more sophisticated and, arguably, more subjective measures can be calculated. For example, by examining the mean duration to settlement of the technical reserves, an appraisal value approach can be taken. Alternatively, a prospective bidder might wish to estimate the implicit margins in a company arising from its policy of, say, not discounting its reserves for outstanding claims or from the perceived conservatism in its basis of setting reserves for outstanding claims.

13.3 Conclusions

A company's financial statements can be used to calculate broadly based measures of its profitability and can provide useful benchmark values for comparisons between other, similar, companies. However, without a detailed understanding of each individual company's circumstances, the conclusions from such comparisons can be misleading. In particular, the

---

[7]Averages can be taken over several years if recent experience is felt to be unusual. Then, strictly, the paid or incurred claims should be adjusted for changes in business volumes.

reserving policy, the impact of reinsurance, and the basis of asset valuation will have significant effects on the results.

The foregoing described the general structure underlying published financial statements. Clearly, for internal management purposes, additional information can be provided in a targeted way that might not normally appear in the published accounts. Each department would expect to receive its own set of accounts focused on measuring the profit drivers within its control and on assisting in the management of the specific department. Important matters, such as performance hurdles used for remuneration purposes, for example, will vary by department and the financial information used in their measurement is not always the same across all departments.

Appendix A

The items that might typically be found in the Companies Act accounts of a general insurance company are listed below.

A.1 Revenue Account

Unearned premium reserve brought forward
+ Deferred acquisition cost brought forward
+ Written premiums
− Unearned premiums carried forward
− Deferred acquisition cost carried forward
$\qquad$ = Earned premium

− Notified outstanding claims brought forward
− Incurred but not reported claims brought forward
+ Claims paid
+ Notified outstanding claims carried forward
+ Incurred but not reported claims carried forward
$\qquad$ = Losses incurred

+ Unexpired risk reserve brought forward
− Unexpired risk reserve carried forward
+ Other contingency reserves brought forward
− Other contingency reserves carried forward
− Commission outgo
− Underwriting expenses
+ Allocated investment income
$\qquad$ = Underwriting profit (loss)

A.2 Profit and Loss Account

+ Balance of profit and loss acount brought forward
+ Investment income
+ Realized capital gains
+ Unrealized capital gains
− Overhead expenses
± Exceptional items
± Underwriting profit (brought down)

        = Overall result

− Tax liability
− Dividend payments

        = Balance of profit and loss account
          carried forward

A.3 Balance Sheet

Assets
Investments
      Equity
      Fixed interest
      Property
Current assets
      Debtors
      Cash

        Total assets

Liabilities
Technical reserves
      Outstanding claims reserves carried forward
      Incurred but not reported reserves carried forward
      Unearned premium reserved carried forward
      Deferred acquisition costs carried forward
      Unexpired risk reserve carried forward
Other contingency reserves carried forward
Creditors

        Current liabilities

Long-term liabilities
      Loans
      Share capital
      Balance of shareholders' funds

        Total liabilities

References

Bride, M. and Lomax, M. W. (1994). Valuation and Corporate Management in a Non-Life Assurance Company. *JIA*, Vol. 121.

Casualty Actuarial Society. (1996). Call Papers on Dynamic Financial Models. CAS.

Hooker, N., Bulmer, J. R., Cooper, S. M., Green, P. A. G., Hinton, P. H. (1996). Risk Based Capital in General Insurance. *BAJ*, Vol. 2.

Ryan, J. P. and Larner, K. P. W. (1990). The Valuation of a General Insurance Company. *JIA*, Vol. 117.

Taylor, J. M. (1992). General Insurance, a Summary of UK Theory and Practice. Institute of Actuaries.

Chapter 14

Premium Rating

14.1 Introduction and Basic Ideas

This chapter concentrates on methods for calculating an "actuarial premium" (Derrig, 1991), i.e., the best estimate of the present value of all the financial components of the insurance contract. We also consider briefly the management decision process by which the premium charged might be chosen to be different from the actuarial premium and the development of models for calculating premiums derived from techniques used in financial economics.

The premium charged depends on:

- the exposure of the policyholder and the insured goods or services to the various insured perils;
- the degree of risk associated with the policyholder;
- the expenses of acquiring and administering the business;
- the profit required by the insurer.

Setting an appropriate rate for the cost of some future contingency of unpredictable timing, frequency, and size is an inexact science requiring estimates of future cash flows, including claims, investment income, expenses (including the cost of reinsurance), profit, and tax. In addition, the pricing structure adopted can profoundly influence the volume and nature of business attracted, and some classes of business are more price sensitive than others: the demand in "commodity" markets, such as personal lines motor insurance, is particularly price elastic. Major participants in the market can influence the pricing decisions taken by their competitors, although in well-informed and fluid markets competitive advantages are often of relatively short duration. So, apart from the underlying economic objective of achieving an adequate level of profitability or return on equity (whether for an individual policy or across a group of policyholders) and ensuring the survival of the company, other corporate objectives might include:

- achieving a minimum market share;
- achieving a target premium growth rate;
- reducing lapse rates and, in particular, retaining the more profitable business;
- writing a given premium volume;
- achieving an overall operating ratio across the whole portfolio;
- obtaining a broader spread of business (e.g., by geographical area, policyholder type, or industry mix).

The premium calculation can, therefore, be regarded as a two stage process:

The costing exercise, i.e., the calculation of a theoretical price for the risk and all associated expenses.

The pricing stage, i.e., the commercial adjustment to the theoretical cost that takes account of broader corporate objectives. These adjustments can be highly subjective. For example, they could be due to the uncertainty associated with the theoretical price, or because the underwriter chooses to accept some business on unfavorable terms in order to obtain other, more profitable, business. The profit margin available may be constrained by market tariffs and is always influenced by the cyclical nature of competitive pressures on premium rates, which is discussed further in Part V.

An essential aspect of premium rating is the design of the insurance product itself. This entails specifying precisely what is and is not intended to be covered (i.e., the extent of the exposure underwritten). It is important that the proposer understands the coverage offered and acknowledges this, since, under many legal systems, unclear contract drafting is interpreted against the drafter and events that were not intended to be insured can become so.[1] There is a fine balance between restricting coverage and keeping the contract language straightforward, and a detailed knowledge of contract wordings and their intention is required. The problem is more significant in commercial insurance classes than in personal lines

---

[1]Claims in respect of environmental pollution in the U.S. provide a dramatic example of this (Bouska, 1994). Many of the dump sites giving rise to claims were polluted gradually, because the containment measures prescribed, although "state of the art" at the time, were in fact not adequate. Underwriters argue that the events insurers intended to cover were of a "sudden and accidental" nature, as specified in many of the contracts at the time. In many situations, however, and after lengthy litigation, very expensive judgements have been held against underwriters due to the lack of clarity in the contract language.

business. It requires input from a number of disciplines: lawyers, underwriters, actuaries, loss adjusters, and claims managers.

If the policy coverage can be controlled, then the insurer's uncertainty over the cash flows emerging from an insurance policy, or a portfolio, will become more manageable. The policy might simply cap the maximum payout in a given calendar period — such provisions are common in finite-risk reinsurance contracts. Careful design can reduce the administrative burden of servicing the claims and improve the profit signature of the business. For example, the frequency of small claims and the associated disproportionate claims handling overhead can be reduced by

- setting an appropriate level of *deductible*, i.e., a retention of part of the claim cost by the policyholder for each claim;
- imposing an experience rating system such that the premium adjusts according to claims experience; in the personal lines business this is often referred to as a bonus/malus system.

Loss prevention measures, such as deductibles, enable insurance companies to modify the coverage offered, and the associated premium level, to suit individual policyholders' needs. Deductibles, in particular, are widespread across many different classes of business. Whilst the expected claim amounts will normally fall as the deductible increases, the premium itself is always subject to a minimum value. The insurer is limited in its ability to reduce overhead costs and the costs of servicing the capital backing the business. As deductibles increase, the frequency of claims will reduce and the volatility in the insurer's account increases; therefore, the insurer has to dedicate proportionately more capital to underwrite a policy and the policy needs to generate an increasing profit margin if the insurer is to meet its return on equity target. The insurer cannot, therefore, accommodate all of the policyholder's needs.

Experience rating, which rewards a good claims history and penalizes a poor history, is common in personal motor insurance policies and is becoming more common in domestic household insurance. In personal lines business, the premium adjustments are designed to discourage small claims, since claim sizes can be less than the additional premium charged as a result of a claim. The bonus/malus method of experience rating, which is used in personal motor insurance, imposes penalties according to the *number* of claims made rather than due to the claim *amount* (see Section 14.5.1). If fewer claims are made, then the reduced cost of administration that results can be used for the benefit of policyholders by reducing premiums. In many countries the bonus/malus system is called the *no claims discount* system. The self-rating character of bonus/malus makes such systems very attractive where the true behavior of the claims is very difficult to establish, with the burden of uncertainty being shifted to the policyholder.

The use of deductibles and experience rating not only controls the claims cost for the insurer but also helps to limit the moral hazard faced by the insurer. Moral hazard can also be discouraged by limiting the maximum sum insured; the aim of such controls is to ensure that the behavior of the policyholder does not change as a consequence of having insurance.

There is some convergence between the products provided by the financial markets and those in the insurance sector. Access to the capital markets' risk management products is usually limited to large commercial clients, or insurers themselves. In such cases the merits of choosing a financial instrument rather than an insurance company for risk management might become more a matter of accounting requirements, rather than a decision based on fundamental technical considerations. One example of such products is bond issues whose repayment terms are linked to the occurrence of certain specified natural catastrophes.

The premium rating process is closely linked with the claims reserving process, and there should be feedback between the two activities within the organization. In reserving for longer tailed classes of business, the claims experience at early development durations is often not a reliable guide to the ultimate level, and the premium pricing or rating is probably the most relevant benchmark, although as far as possible some verification should always be undertaken. This issue relates particularly to reserving methods that depend upon loss ratio assumptions, such as the Bornheutter–Ferguson methods described in Chapter 16. However, the assumptions made for costing purposes might not be the most appropriate basis for establishing reserves due to the commercial influences mentioned earlier. It is important to monitor and distinguish between the technical and the commercial pricing bases if premium bases are to be used for reserving. Regardless of the particular reserving method used, a knowledge of relative pricing levels between accident or underwriting years and the present phase of the rating cycle provides useful cross-checks on the ultimate projected claims.

14.2 Basic Premium Formula

The underlying office premium equation for an individual policy can be expressed in the form

$$(1 - k)\text{OP} = R + lf + mfc + j + \frac{F}{N} \tag{14.1}$$

where OP is the office premium, k is a proportionate factor applied to the gross premium (it might reflect commission payable to intermediaries, and can also include other, proportionate, acquisition costs; the premium structure might be such that it also includes contingency and profit loadings), R is the risk premium (the expected total claim amount, which is the product of expected frequency f and expected severity c). This could be re-expressed as $R(1+d)$, where d is a measure of the variability of the risk. That is, the contingency loading need not be included in k, but might be separately identified or expressed as part of the risk premium. The difference between the two approaches is that in the former the expense loading is indirectly influenced by the contingency loading, whereas in the latter formula it is not. f is the expected number of claims per policy, c is the expected severity of the claim amount (per claim), l is the fixed cost of handling each claim (e.g., the cost of an IT system entry), m is the variable cost of handling claims (dependent on the total amount of claims), j is the per policy set-up cost, F is the total business overhead expense allocated to this class of business, and N is the expected number of policies to be sold over which overhead costs will be allocated.

Whilst it may be desirable to break down the components of the office premium into a fine level of detail, for many commercial classes of business it is not possible to do so, and many of the items explicitly identified above are frequently combined as an overall proportionate addition to the risk premium. For example, claim amounts and those expenses directly associated with settling a claim (the *allocated loss adjustment expenses*) are frequently included in the definition of R. Whilst different companies do not all treat expenses in the same way, there should be a clear and consistent treatment of the cost of indemnity and of claims expenses within each company across lines of business. One could separate R into the indemnity and allocated expense component and, under some legal jurisdictions, this is a statutory requirement.

Each item in the model is expected to generate a cash flow, and so a discounted cash flow calculation can be used to estimate a commercial price. The usual approach is to discount the cash flow using a risk-free rate of discount, but a higher rate might be justifiable. The model can be extended to include the potential impact of reinsurance (whether on the premium, the claims, or the expenses), if it can be estimated in advance. This could have a significant effect if the premium basis used by the reinsurance company is materially different from the cedant's basis (e.g., if the insurance cycles in the direct and reinsurance markets are out of phase). Usually, reinsurance would be expected to add to the cost of the contract. However, in rare circumstances, reinsurance can provide a profit for the ceding company and, hence, a reduction to the theoretical premium.

Rather than including a specific profit margin in the formula for the office premium, an alternative approach is to determine the capital required to support an individual policy and to consider the rate of return required from that capital. For example, Equation (14.1) can be restated as

$$
\begin{aligned}
(\text{Net present value of profit}) = {}& (\text{Net present value of office premium}) \\
& - (\text{Net present value of expenses}) \\
& - (\text{Net present value of claims}) - (\text{tax})
\end{aligned}
$$

Suppose the value of the capital allocated to support a policy is A and the required return on the capital is B. Then, assuming the capital is invested to earn the risk-free rate of return (net of tax), the office premium must satisfy

$$
\frac{\text{Net present value of profit}}{A} + (\text{Risk free rate earned on } A) = B
$$

The more volatile, or risky, the business, the greater the capital required to support it. Hence, to achieve the same return on capital, the net profit required from risky business is greater than that required from lower risk business. The rationale underlying this approach is consistent with general investment theories and the Capital Asset Pricing Model (discussed in Chapter 4.3).

By itemizing the components of the premium, a number of measures can be established for the premium, e.g.,

1. The premium required to achieve the minimum return on equity (normally regarded as the risk-free rate).
2. The premium on a marginal costing basis.
3. Where a premium discount has been offered, the formula is useful as a guide to the theoretical deficiency and the requirement for additional unexpired risk reserves (Chapter 15).
4. Where the premium is known in advance (e.g., because of market tariff agreements), the return on equity being achieved, and hence the management, can decide whether to underwrite the risk.
5. If the contract is negatively correlated with other risks in the portfolio, then a lower capital charge is justifiable and the impact on price can be explored.

An approximate breakdown of the office premium for a personal lines product is given in Table 14.1. Some of the entries in Table 14.1 are discussed briefly below.

14.2.1   Insurance Levies

In many countries, certain charges on premiums are beyond the control of the insurer. In the U.K., the principal levies are those raised by the

Table 14.1 Example of how the components of an office premium are distributed

Risk premium	55%
Claims handling expense	10%
Commission and acquisition expense	20%
Overheads cost	5%
Contingencies, profit, insurance levies, premium, (and other) taxes	10%
Total office premium	100%

Policyholders' Protection Board.[2] In the Lloyd's market, levies can arise as a result of the failure of the members of a syndicate ("Names") to meet their obligations; the consequence is that their remaining liabilities are met by Lloyd's Central Funds, which might occasionally require topping up as a result.[3] Premium taxes can form part of governmental economic and fiscal policy, and other taxes include contributions to local fire brigades, police forces, and other centrally maintained security agencies. Such levies are normally charged as a percentage of premium.

14.2.2   Contingency Loading

Margins are normally included as protection against unforeseen contingencies, such as adverse claims, poor parameter estimation in the calculation of the risk premium, unexpectedly high expenses, or poor investment performance. The loading can be regarded as a contribution to the overall solvency level of the company, over and above the explicit profit loading. It is a moot point whether a policyholder should effectively contribute towards his or her own security fund. One approach to allocating capital to a policy is the "expected policyholder deficit" (EPD) approach. This method uses a conditional probability approach and states that the capital required to back a policyholder must be sufficient to meet, say, 95% of claims given that a claim has occurred; this concept is similar to that described in Part I

---

[2]In the U.K., 90% of the obligations of an authorized insurance company to its private policyholders are guaranteed by law. The Policyholders' Protection Board can require funds to be contributed by the insurance industry to meet the obligations of a failed authorized insurer.

[3]In the U.S. it has been proposed that the cost of environmental pollution clean up, which has hitherto been met from a federal fund called Superfund (i.e., the U.S. taxpayers) or from polluters and in some cases the polluters' insurance companies, will be met by charging a levy across all insurance companies, as well as industrial companies. The levy is calculated roughly according to the written premium income in certain categories of insurance business undertaken during the critical periods when most pollution is deemed to have occurred, together with a loading on any ongoing business. It remains to be seen precisely how such a levy might be apportioned back through time and how it might be allocated between insurance and reinsurance companies.

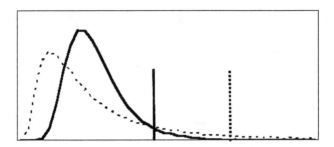

FIGURE 14.1 The 95th percentile points on underlying aggregate claims distributions with different variance

relating to value at risk. Figure 14.1 illustrates the principle. The statistical distribution of claim severity is represented by the curves. The mean is identical for each, but the dotted curve has a larger variance and, hence, the 95% claim value is much larger, as indicated by the vertical line. The further question with this capital allocation approach is the duration for which capital should be assumed to be allocated to a contract; clearly, the longer the duration, the greater the expected profit required from the contract. Formulae explicitly allocating capital under the EPD approach normally provide for an "evaporation" rate of allocated capital based on the delay of claim reporting and settlement.

The size of the contingency loading depends not only of upon the variability of the class of business but also upon the company's risk aversion and the characteristics of the portfolio as a whole. It would be naïve to add a security margin just at the individual policy level. For independent risks, the marginal additional variability of the portfolio reduces as each new policy is added (see Appendix A); for risks that are negatively correlated the combination of two contracts can reduce the total capital required to support the business. The principle of correlation is illustrated in Figure 14.2, where the resultant vector line is the combined capital requirement.

As risks are added to the portfolio, whether as additional policies or other risk sources (asset risk, credit risk, management risk, operational risk, etc.), the principle remains the same: the arrow length designates the quantum and the angle between the current risk set and the additional risk represents the degree of correlation. This approach has been referred to as the "wheel of risk" approach.

Risk aversion can be linked to the *probability of ruin* of an insurer, a measurement of the probability that an insurer will at some time become insolvent (e.g., see Daykin et al. (1994)). Other measures are also used, such as the probability of insolvency during some fixed time period or the average deficit in the event of insolvency. In practice, the precise

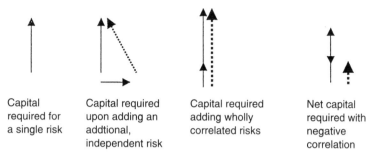

<table>
<tr><td>Capital required for a single risk</td><td>Capital required upon adding an addtional, independent risk</td><td>Capital required adding wholly correlated risks</td><td>Net capital required with negative correlation</td></tr>
</table>

FIGURE 14.2 The principle of correlation

measurement of the probability of ruin is impossible due to the large number of variables involved, which include:

- claims experience of the business
- level of expenses
- adequacy of the reinsurance program
- security of reinsurers
- investment performance
- ability of management

In spite of the uncertainty in the best choice of model and the subsequent parameterization of the model, consideration of the probability of ruin is a powerful theoretical construct, which can focus attention on particular issues, and the assessment of finite time ruin probabilities is straightforward using Monte Carlo simulation (see Daykin et al. (1994)).

Managements tend to adopt a mixture of qualitative and quantitative bases in arriving at a decision regarding the risk aversion of the organization, including:

- disclosed level of solvency
- strength of the reserves held for outstanding and other contingent claims
- quality of current and future business
- possibility of being downgraded by an insurance market analyst
- possibility that profitability of the company will fall in the lower quartile of a peer group of companies
- stability of the profit stream
- availability of reinsurance

As part of the strategic management of a portfolio, it might be possible to achieve a portfolio of negatively correlated risks. This idea extends to holding assets that increase in value should insured events that give rise to severe claims occur (a simple example is holding shares in a local building firm and writing large property risks in the same area, or, in more sophisticated markets, insurance futures as traded on the Chicago Board of Trade). Other strategies may include trading in the debt of counterparties who are also policyholders or reinsurers. Negatively correlated positions can confer substantial competitive advantages and reduce the amount of reinsurance required.

14.2.3   Acquisition Costs

For many classes of personal lines business, the costs of acquiring new business (e.g., advertising, commissions paid to brokers or agents, and, in the case of reinsurance business, ceding commissions payable to the cedant) and establishing the documentation for a new policyholder are high. Business renewal generally involves lower expenses, and the insurer can allow for this when calculating the renewal premium by reducing the appropriate expense items in Equation (14.1); the reduced premium should enable the insurer to retain a greater proportion of business. However, if the premium for renewal business is reduced, then any cross-subsidies between new and renewal business must be fully understood, otherwise there is a danger that the insurer will be priced out of the market for new business.

14.3   Estimating the Office Premium

In the following sections we discuss the procedures followed in analyzing business to establish premium rates. Actuarial approaches are data driven, and insurance data can be of poor quality and may be absent for new products. In such cases, some form of extrapolation from known data or related data coupled with modeling is needed. However, probably more important is the role of underwriting and risk management in ensuring that the exposures accepted in such new areas are adequately contained.

The basic techniques used for estimating the office premium are discussed under the main headings of "risk premium" and of "expenses."

14.3.1   The Risk Premium, and Approaches to Calculating
the Expected Claims

The risk premium is the expected value of claims, usually expressed per unit of exposure. As discussed previously, the extent to which this will include claims handling expenses will vary between companies. There are two basic approaches to calculating the risk premium, namely exposure rating and experience rating.

Exposure rating. This uses the experience of a broad group of policyholders to estimate the expected claims of an individual. Insurers and statistical agencies accumulate data over many years, involving many policyholders and across many classes of business to gain a broad knowledge base. They might also use wider industry or government sources to discern the possible future behavior of a class of business together with the factors correlated to loss experience. This approach results in broadly based rates that can be adjusted to reflect an individual's specific exposure to risk. Such an approach is also known as a *book rate* approach.

Experience rating. Under this approach the premium charged to the policyholder is based on the claims experience of that policyholder. In analyzing the past experience of an individual policyholder, adjustments should be made to reflect changes in the underlying risks, and the environment within which the policyholder has been operating; further adjustments are then made for expected future trends.

In practice, the actual premium charged often incorporates a mixture of the experience and exposure approaches, taking a weighted average or using ad hoc methodologies to give the underwriter some confidence that the major features of the type of business have been incorporated. Exposure rating is discussed in this section; experience rating is discussed in more detail in Section 14.5.

The basic procedure for estimating the risk premium involves the following stages.

1. The policy exposure must be identified and measured. The exposure will depend on the insurance coverage offered and the perils to which the insurance company is exposed. Different perils might require different measures of exposure.

2. The physical attributes of the insured goods or services and the environment in which they operate must be assessed and a view taken on the extent to which they are likely to influence the claims experience. The theoretical pricing mechanism should discriminate between risks and charge different premium levels according to the risk attributes of the individual policyholders.

    For rating purposes, the claim characteristics (or riskiness) are expressed in terms of the expected *frequency* of claims and the *severity*, or average cost, of a claim. The convolution of these two underlying statistical distributions produces the aggregate loss distribution for the policy, and the variance of the distribution provides an indication of the safety margin and the capital commitment needed by the insurer to support the contract. Occasionally, the frequency and severity are not separated, and the focus of the analysis is on deriving the aggregate loss distribution.

3. Once the exposure, scope of coverage, and the riskiness of the insured have been established, a premium reflecting the insurer's view of the risk is usually expressed per unit of exposure (see Section 12.1). For example, the unit of exposure might be

the vehicle year in motor insurance;
the sum insured per year in property insurance;
the number of occupied beds for hospital medical malpractice;
the number of feet drilled for oil wells;
a rate per thousand of the original gross premiums (or *subject premium*) for a reinsurance premium if specific information regarding the underlying risks is unavailable (which is often the case for a reinsurance contract).

Some measures of exposure cannot be known precisely in advance of inception of the cover. In commercial business, the exposure is sometimes measured precisely only once the policy has expired and a retrospective adjustment is then made to the initial deposit premium, e.g., employer's liability insurance is rated on annual payroll or number of exposed employees, which is not known at the outset.

14.3.2   Risk Factors and Rating Factors

Risk factors are key measures of the hazard faced by the insurer. In general insurance, there are often practical problems associated with obtaining reliable measurements of risk factors, and so a distinction is made between risk factors and rating factors. This was referred to briefly in the Chapter 12.

Risk factors are the attributes of the insured goods, the keeper of the goods and the environment in which the goods are used, that affect the likelihood and the severity of a claim. Unfortunately, risk factors can be difficult to measure, and insurance companies might not be able to manage the data, even if an insured could provide it reliably: in extreme cases, it could be illegal to ask for the information. Therefore risk factors cannot always be used directly for pricing insurance and, consequently, it is necessary to find other characteristics of the insured that provide indirect, or proxy, factors that are correlated with the true risk factors. These are the rating factors.

Risk and rating factors vary according to the peril being considered although, in practice, only very large risks would have premiums calculated separately for each peril insured. Also, for many classes it is extremely difficult to define each insured peril precisely. In the example below, on motor insurance, the perils (i.e., the events that give rise to the claims) might include:

- natural perils, e.g., wind, rain or flood;
- manmade perils, e.g., an oil slick on the road or the faulty manufacture of a component.

These perils clearly present a multiplicity of associated risk factors, but measurement at this level of detail is not feasible. The solution in terms of policy design for a diversity of perils is usually to offer cover with specific exclusions, e.g., war, nuclear risks, terrorist activity, earthquake, or hurricane. Separate insurance contracts might then be available for some of the perils excluded from normal cover.

In order to illustrate the problem, consider the case of personal motor insurance policies in more detail. The insured events normally include:

- moving-traffic accidents causing damage to the insured vehicle, to other property, or to passengers or other persons;
- theft of a vehicle or from a vehicle;
- damage caused while the vehicle is stationary.

Broad risk factors connected with these might be:

- distance driven over the duration of the insurance policy (a good measure of the chance of a moving-traffic accident);
- traffic density;
- quality of the driver and the driver's reaction time;
- driver's average speed;
- driver's natural degree of caution (i.e., risk aversion);
- time that the vehicle is parked in areas where car thieves operate;
- value and visibility of goods kept in the car (e.g., a car radio);
- whether the car has an alarm;
- sensitivity of the car's electronic systems to water damage.

Some of these factors cannot be measured objectively (e.g., the quality of the driver or the driver's natural risk aversion); others can, in theory, be measured but it would be impractical to do so (e.g., the driver's reaction time and the traffic density where the vehicle is driven). Note, also, that the relationships between risk factors can be highly complex. For example, a policyholder with a high degree of natural caution would be unlikely to park in high theft risk areas or to drive fast, and the driver's car would probably have an alarm system.

Since many of the risk factors above are difficult to measure directly, it is necessary to find rating factors that are measurable and that correspond in some way to the true underlying risk factor. Rating factors can be broken into three main categories:

Factors associated with the individual proposer.
Factors associated with the risk unit that the proposer seeks to protect.
Factors relating to the coverage being sought.

In the case of private motor insurance, typical factors associated with each of these heading are as follows.
　　The proposer:

The age of the driver. The driver's age gives an indication of the amount of driving experience and general maturity, and there may also be biological reasons for differentiating younger from older drivers. Younger drivers are generally regarded as worse risks than older drivers, although this rule tends to break down at very old ages.
Sex. Males are usually regarded as greater risks than females. Age and sex tend to interact, and the difference in riskiness is greater at younger ages.
Address of policyholder. Address is used to indicate the type of area where the vehicle is most likely to be found. Insurers will use their own experience of motor business and other geographical statistics to give an indication of the claim likelihood as a consequence of this factor.
Car usage. An indication of the annual distance driven can be given by whether the car is limited to social, domestic and pleasure use only, or is also used for business.
Driving record. Driving convictions or penalty points on a driving license, or an advanced driver's license, can be used to indicate the quality of the driver.
No-claims discount (NCD) level achieved. Statistically, there is little doubt that the claims experience and, in particular, the claims frequency that a driver might expect in the future is highly correlated with the past experience. Therefore, the level of NCD achieved is a valid rating factor. There are problems, however, in including the number of claims-free years as a rating factor, as it interacts with other factors and can distort the statistical analysis.
Occupation. This can be used to indicate the policyholders' risk aversion and the distance driven each year.

　　Risk unit (the vehicle):

Vehicle age. A new car is more expensive to repair than an old car, but might be more difficult to steal and, for a short period while very new, will be driven with more care. In general, the frequency of claims tends to decline as the age of the car increases, possibly because older cars tend to be driven less than new cars.

Vehicle make and model. The make and model of the vehicle are usually used to put the vehicle into a designated car group.[4] This can be used to determine the likely cost of repairs, as well as the risk aversion of the policyholder.

Vehicle modifications. If the vehicle has been modified this might influence how it is used.

The scope of cover:

The usual forms of cover for private motor insurance, in order of increasing scope, are:

third party only (which in many countries is compulsory)

third party, fire, and theft

fully comprehensive[5]

The level of deductible (whether compulsory or voluntary) limits the scope of cover and affects the expected level of claims.

The market tends to be governed by price rather than by the efforts of insurers to differentiate their products from those of their competitors by varying service standards or lesser aspects of coverage (e.g., legal cover, windscreen cover, courtesy cars). The policy might cover the insured only, the insured and spouse or another named driver, or it could cover any driver. The data suggest that, in spite of the broader coverage offered under an insured-and-spouse policy, as opposed to an insured-only policy, the claims experience of the former is lighter than for insured only. This might be an indication of the calming effect of marriage. Consequently, marital status could be included as a proposer rating factor.

In order to evaluate the impact of different rating factors on the claim experience, models of the various ways in which these factors influence claims must be developed and tested. This is discussed in Section 14.3.5.

---

[4]Most countries have a vehicle classification or rating group systems depending on, for example, the initial price of the car, its engine size, and performance, and its attractiveness to thieves. In the U.K., until about 1985, six car groups were commonly used. As insurers have become more sophisticated, and in view of the increasing diversity of vehicles, the number of groupings has increased to around 21. Smaller sized cars with powerful engines, which are popular with younger drivers, tend to be in the higher rating groups.

[5]Fully comprehensive cover includes third party, fire, and theft cover, together with own-damage claims.

14.3.3 Exposure Measurement and Analyzing Past Exposure

Exposure has already been described as a measure of the volume of risk. This is linked to the expected claims via the rating factors. There are two principal approaches to exposure measurement: the census method and the exact method.

14.3.3.1 Census Method

The census method takes cross-sections of the exposure base of the portfolio at various dates. The volume of exposure per unit time is calculated as the product of the average exposure between two census points and the time period between the censuses. This assumes that the movements in and out of the portfolio and between classifications within the portfolio occur uniformly over the census period. For large portfolios of many individual policyholders, this approach is generally reasonable, and can be confirmed by a straightforward investigation. The approach can be applied to a portfolio of business and, since the detail required is relatively undemanding, it can also be applied more generally, e.g., to market-wide statistics (although it will become more difficult to check the assumption of uniform movements between census dates). For example, one could research the claims experience of oil rigs in the Gulf of Mexico over past years and obtain approximate estimates of the rig-years exposed. If this is associated with the number of major disasters, then a crude indication of the overall experience, per rig-year, can be obtained. Figure 14.3 sets out a census calculation.

In this example, census snapshots E_i, $i = 0, 1, \ldots, 4$, have been taken at times t_i, $i = 0, 1, \ldots, 4$. Therefore, the exposure volume per unit of time is

$$\frac{[1/2(E_0 + E_1)(t_1 - t_0) + 1/2(E_1 + E_2)(t_2 - t_1)}{+ 1/2(E_2 + E_3)(t_3 - t_2) + 1/2(E_3 + E_4)(t_4 - t_3)]}{t_4 - t_0}$$

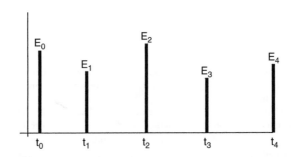

FIGURE 14.3 Illustration of census method for calculating exposure

Normally, census measurements would be taken at regular intervals corresponding with the production of the company's internal management information. In the case of monthly census data, the exposure over a period of 12 months would be

$$E = \frac{1}{12} \sum_{i=0}^{11} \left(\frac{E_i + E_{i+1}}{2} \right)$$

14.3.3.2 Exact Method

Most companies now operate their management information systems in the form of relational databases, and so an exact measurement of exposure is straightforward. Under such a system, the time for which each risk unit has been exposed during the period of the investigation is measured individually, i.e.

$$E = \sum_r E_r t_r$$

where r is summed over all policyholders in the section of the portfolio being studied. In the example below, E_i represents the exposure of an individual policy P_i, $i = 0, 1, 2, 3$, in force during the period of investigation. This calculation is illustrated in Figure 14.4.

In Figure 14.4, policy P_i, $i = 0, 1, 2, 3$, enters the investigation at time t_{2i} and exits the investigation at time t_{2i+1}. If policy P_i has exposure E_i per unit of time, then the exact measure of exposure for the portfolio is

$$E_0(t_1 - t_0) + E_1(t_3 - t_2) + E_2(t_5 - t_4) + E_3(t_7 - t_6)$$

Whilst the principles of exposure measurement are straightforward, in practice the approach presents a number of practical difficulties:

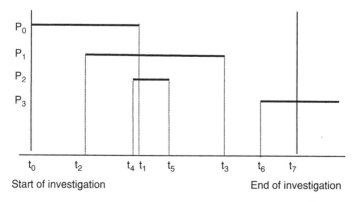

FIGURE 14.4 Illustration of exact method for calculating exposure

- the number of different rating and risk factors within the exposure being measured produces a large number of possible data subdivisions and the data files can become very large, leading to data-processing problems;
- the large number of possible combinations of rating factors can mean that some individual cells have insufficient data to support regression-type analyses;
- data systems change with time and, in any case, the quality of data recorded is rarely consistent;
- for older policies, data on certain rating factors might not have been recorded.

Data integrity and consistency are essential. A particular problem with motor insurance, for example, is the high number of policy endorsements and mid-term changes to exposure, such as the inclusion of additional drivers, changes of address, or changes of vehicle. Ideally, the aim would be to gather enough data in order to test market premium rates and the pricing structure currently used by the company, and also to provide insights into the behavior of the class of business in response to factors that are not currently included in the rating basis. Identifying new rating factors can enable the insurer to achieve further refinement in the portfolio and to focus on profitable areas within the market.

In order to illustrate this problem, consider a greatly simplified example of a private motor premium rating model. The key rating factors in the analysis are policyholder age, sex, area domiciled, vehicle age, and vehicle rating group. Each of these rating factors is subdivided into a number of levels (e.g., policyholder age 17–19 falls into level 1, age 20–24 falls into level 2, and so on) as follows:

Factor	Number of levels
Policyholder age	10
Policyholder sex	2
Area	20
Vehicle age	5
Vehicle rating group	15

The total number of possible combinations of factors is $10 \times 2 \times 20 \times 5 \times 15 = 30\,000$ and the exposure measurement is presented as a table of values (Table 14.2) reflecting the exposure at each level of each rating factor.

If an additional factor were introduced, such as occupation, the $30\,000$ potential combinations would increase by the multiple of the number of different occupations included in the examination, diluting the data further.

Table 14.2 Table of values for exposure measurement

Age	Sex	Area	Car age	Car group	Exposure (vehicle years)
1	M	A	2	1	3.1
1	F	A	2	1	1.5
1	M	A	3	2	0.2
⋮	⋮	⋮	⋮	⋮	⋮
2	M.	A	1	2	4.2
2	F	B	4	2	5.1
⋮	⋮	⋮	⋮	⋮	⋮
3	M	F	3	1	1.5
⋮	⋮	⋮	⋮	⋮	⋮
10	F	H	5	14	0.1
⋮	⋮	⋮	⋮	⋮	⋮

In many situations the data to examine new factors would not have been captured over as long a period as that for which the claims and base exposure data would be available. The lack of data might mean that reliable estimates of the parameters are not available, and so ad hoc adjustments might be required based on market practice, a limited statistical investigation, or the judgement of the underwriter. As with any statistical analysis, a trade-off must be made between obtaining a sufficient volume of data in order to give statistically reliable estimates and using data that are still relevant for the prospective period of interest.

14.3.4 Statistical Modeling

Insurance companies will wish to perform investigations to:

establish a model of the processes underlying claims;
understand the cost structure of the company;
understand the pricing structure of the market in which the company operates;
analyze the phase that the market has reached in the underwriting cycle.

The detail with which a model can be developed depends on the following key factors:

The extent to which the claim process can be identified and analyzed. If it is difficult to correlate the features of claims with the characteristics of the insured goods (the rating factors), then, for all but the largest portfolios, price discrimination between policyholders would be spurious.
The availability of data with which to estimate the parameters.

The market structure and its tolerance to cross-subsidies between individual policyholders (e.g., in medical insurance, it is generally accepted that younger policyholders subsidize older policyholders).

The capacity of the market, i.e., the capital committed to support the insurance business relative to the demand for insurance.

The effect of legislation in controlling insurance terms and premium rates.

Market structure exerts a profound influence on pricing. If the market has a standard pricing mechanism,[6] for example, there is no incentive to build sophisticated pricing models, since insurers do not discriminate risks through price. In such circumstances, insurers are unlikely to maintain detailed statistics beyond those necessary to service the business, since any competitive advantage is obtained through cost savings. Data analysis would, instead, be directed toward a more detailed understanding of the expense structure. Similarly, if profit levels are consistently high, then, even in unregulated markets, companies do not have a strong motivation to refine their rating procedures and pricing structures. Equally, if there is not much competitive pressure, or if the price elasticity of demand of the products on offer is very low, there will be little motivation to develop sophisticated models that exploit variations in risk between policyholders. Expressed another way, provided that policyholders are extremely loyal, then a poor pricing structure should not impair business retention.

Changes in the significance of background influences can have a dramatic impact on the market, especially in the area of adverse selection by policyholders against insurers,[7] and such changes are usually unexpected. Thus, even when a model is not deemed necessary for pricing, there is little justification for not having a detailed knowledge of the true profit structure of the portfolio and the areas of cross-subsidy within the account

---

[6]This could apply if all insurance companies operate under a tariff system, charging the same prices for all relevant risks.

[7]Adverse selection by policyholders stems from two main roots. (1) The quality of data is asymmetric; proposers know more about the nature of the risks they need to insure than the insurer. (2) Insurers can adopt very different pricing strategies. As a result of variations in the pricing structure for similar risks, one insurer could rapidly (and possibly unintentionally) accumulate an overexposure to a given type of insured within a given market. This is a particular problem in the private motor insurance market, where insurance is a standard commodity, price driven, and where information is widely available. In developed markets, policyholders can easily obtain telephone quotations. In the U.K. market, the impact of highly focused price targeting by direct writing motor insurers (i.e., dealing direct with the public and not through insurance brokers), using sophisticated modeling and classification systems, has, over the past 5 years, led to major shifts in the portfolio mix of many of the long-established insurers, as well as to changes in the cost structure of many companies.

that would be the main targets from competitors if the factors listed above were relaxed in some way. Equally, knowledge of any cross-subsidies in the market can give insurance companies an enormous short-term advantage. Good management would always carry out detailed analyses of the available data to be aware of the impact of these influences and to achieve a competitive advantage in other ways, either by improved expense structures or through modified policy coverage.

For most practical purposes, postulating an initial model of the claims process is not too difficult. The problem arises in setting the levels of, and the relationships between, the various parameters. Three simple examples of the basic considerations required are set out below.

Example 1: a simple model for motor insurance business. For a high-volume class of business such as private motor insurance, which has well established rating factors, one of the most common techniques used is generalized linear modeling (e.g., see Renshaw (1989) and McCullagh and Nelder (1989) for a general description of linear modeling, or Brockman and Wright (1992) for more specific application of the techniques to the motor business). In this framework, a hypothetical relationship between the rating factors and the claims is set out in a linear equation and the appropriate parameters estimated from the data using maximum likelihood methods. For example, a simple analysis could investigate the claims frequency in response to three rating factors. In the formula below, a base claim frequency has been obtained and this is being modified according to variations in the three parameters of policyholder age, rating area, and car group:

$$\text{Claim frequency } (ijk) = \text{Base frequency} + \text{Policy holder age}(i)$$

$$+ \text{Rating area } (j) + \text{Car group } (k)$$

where i, j and k represent the levels of the respective parameters. Policyholder age, for example, need not have increments for each year of age; age bands, within which the variations in experience due to age are not statistically significant, may be used. A detailed discussion of the technical aspects of this approach is set out by Coutts (1984), Renshaw (1989), and Brockman and Wright (1992). Modern software packages, such as SAS$^{©}$ and GLIM$^{©}$, provide a variety of interpretative frameworks enabling additive, multiplicative, and other functional forms to be used for the modeled relationship. The key input requirements are:

the units of exposure;
the model that is to be fitted;
the actual responses (i.e., the claims) against which the model will be validated.

Example 2. An insurance product provided to Lloyd's Names (i.e., members of Lloyd's syndicates whose capital provides the security for the policyholders) provides cover so that, in the event of death, the member's liability to the syndicate is taken over by the insurer and the deceased's estate is released from any further obligation to Lloyd's. Without such cover it could take many years to wind up the estates of dead Names. It would be logical to postulate a rating model linking

age and sex are the main parameters associated with the mortality risk
    (which is the contingency covered);
the amount of the Name's participation on each syndicate;
the results expected to be declared in respect of the syndicates.

The first factors determine the expected claim frequency; the latter two factors relate to the exposure to loss and to the potential severity of claims should a death occur.

Example 3. Many insurance companies have exposure to pollution claims in the U.S. that they dispute. The associated lawsuits are lengthy and adversarial, and only a small proportion of cases have been fully resolved. Lloyd's of London has addressed the question of removing such liabilities from individual syndicates, thus releasing the originating Names, and pooling the liabilities in a company with limited liability (the company is called Equitas). In order to assume these liabilities on financially sound terms, a premium basis is required to evaluate the costs attributable to the various syndicates. The basis should allow for such factors as:

the amount of insurance coverage offered to each insured since the
    pollution was deemed to have commenced (an estimate of exposure);
the applicability of pollution and other exclusion clauses in the insurance
    exposed contracts;
the strength of any other defenses available to deny claims;
the strength of the lawsuit against the insureds who are claiming on their
    insurance policies;
the cost of each insured's share of clean up on each particular polluted site;
an allowance for as yet unidentified polluted sites;
the extent of the syndicate's reinsurance coverage.

In such a situation, there are only limited data available upon which to develop a statistical premium rating basis, and the range of possible outcomes is extremely wide. The ultimate cost is mainly a matter of legal precedent, interpretation, and, ultimately, of negotiation between the insured, the insurer, and the reinsurer. The actuary can provide valuable support for this process by designing an information framework and identifying the major factors contributing towards the cost, given certain assumptions about the event that triggered the insurance cover and the size of claim. Once a credible framework is established, then many possibilities arise. For example, claims might be apportioned between the various parties involved in a negotiated settlement.

Example 3 is an extreme case, as the events have already occurred and the exposure has been earned. However, it is important to appreciate the intimate connection between the premium rate charged and the reserve subsequently required. In a perfect world, where reality follows the model, the earned risk premium plus claims-related expenses would equate to the reserve for outstanding claims plus incurred but not reported (IBNR) claims (discussed further in Chapter 16). For Lloyd's, the uncertainties caused by the continuing deterioration on reserves established in respect of business underwritten up to 50 years ago endangered the continuing viability of the market, and so Equitas is a means to separate old years from new. Many insurance companies and individual organizations have similarly developed their own, corresponding, financial structures or obtained appropriate reinsurance protection, and some of those ideas are discussed further in Chapter 15 as "finite" reinsurance.

Finding an appropriate model is a matter of judgement as well as of applying statistical tests to its predictive power. Usually "goodness of fit" tests, based on a comparison of historical observed values with values predicted by the model, are applied. In some circumstances the model might generate unexpected results which are subsequently verified by further research or subsequent claims. Evaluating the insights provided in this case might be more subjective, but, in situations where the underlying exposures are changing, such a model can be better than one that happens to have met past criteria. A model should be parsimonious in the number of parameters used to explain the experience, and the parameters selected should be logically linked to the underlying claims processes.

As discussed in Chapter 23, all models are simplifications of reality and can be improved. Improvements in the parameterization and structure of premium rating models can follow from monitoring company experience as it emerges and from analyzing the reasons for particular features of the experience in relation to the model. Improved pricing models have benefits in terms of reserving and business planning; it is essential to keep a "reality check" on the model and to challenge the logic for preferring one set of assumptions to another.

14.3.5 Claims Data

As far as possible, claims data should be analyzed both as claim counts and as claim amounts, per unit of exposure, since the risk premium can be regarded as the product of claim frequency and claim severity; i.e., as noted in Section 14.3.1, the aggregate claims distribution is the convolution of the frequency and severity distributions.

The claims data represents the response of the claims process to the exposure base. A fundamental prerequisite of the collection of data for building rating models is that there is a precise correspondence between the exposure base and the resultant claims. Claims data, therefore, must be subdivided between rating factors in a way that is consistent with the measurement of exposure. The range of types of claim, the range of perils which cause them, and, more specifically, the options for the specific perils covered or excluded under the contract of insurance means that, in theory, claims data need to be broken down into greater detail than the exposure base. For example, in motor insurance, the different types of claim could include accidental damage to the insured vehicle, windscreen claims, third-party claims, and theft claims; in household insurance, one could subdivide claims into the categories fire, theft, flood, storm, subsidence, and other causes.

Understanding how claim numbers are counted can be difficult if claim definitions change or are not maintained consistently. For example, in the case of motor insurance, a single accident might lead to bodywork damage, personal injury, and windscreen damage. Some companies would count this as three claims, reflecting the different types of payment that might need to be made; others might count this as only one claim, reflecting the single event that gave rise to the loss. Both approaches have their merits, although the latter would make it difficult to cost separately the impact of the different types of payment on the total claim amount. It might be relatively simple to adjust data from a motor insurance account for such features. However, other classes of business will not be as straightforward. In some classes, it might only be possible to analyze the risk premium and not to identify separately the contributions of claim frequency and claim severity to the overall result. Separating these is particularly important if the operating environment is changing, since economic and social trends can affect the frequency and severity very differently.

There are good arguments for including all claim-related expenses in the statistical analysis of the claim amount, and not only external claim costs. Claim-handling expenses vary by type of claim, and a more detailed understanding of their dynamics could result in different premium relativities than if modeling were carried out on claim amounts only. Such information can provide the management with the ability to target niche markets, particularly for those classes of business with relatively high expenses.

An example of a data table corresponding to the measurement of exposure given in Table 14.2 is given in Table 14.3.

Statistical analysis is largely concerned with separating "signal" from "noise." That is, the intention is to separate the normal behavior of the claims from the background of random experience and from certain events that are deemed to be unusual. The larger the source database is, the greater the range of events that would typically form part of the normal claim

Table 14.3 Data table corresponding to the measurement of exposure given in Table 14.2

Age	Sex	Area	Car age	Car group	No. of claims	Total claim amount	Type of claim[a]
1	M	A	2	1	2	25000	AD
1	F	A	2	1	1	1000	THFT
1	M	A	3	2	0	0	
⋮	⋮	⋮	⋮	⋮	⋮	⋮	⋮
2	M	A	1	2	3	6000	AD
2	F	B	4	2	1	75000	PI
⋮	⋮	⋮	⋮	⋮	⋮	⋮	⋮
3	M	F	3	1	0	0	
⋮	⋮	⋮	⋮	⋮	⋮	⋮	⋮

a AD: accidental damage; THFT: theft; PI: personal injury.

process. Small databases that are modeled using linear models fitted with least-squares- and maximum-likelihood-type approaches might need substantial adjustment in order to give results that appear to be meaningful. There is no correct answer as to what constitutes an unusual event, and the adjustments made are a matter of experience and trial and error.

If nontypical claims or events are included within the data that are used to parameterize the model then this could cause an unrealistic distortion of the model, leading to incorrect pricing decisions. So the analyst must exclude, or be aware of, the influence of:

large individual claims;
claims arising from a single event that has influenced the overall experience
    of a particular class of business;
time trends in the data;
shifts in the experience base.

These are discussed individually below.

1. *Large individual claims.* Data cells having large claims will bias the model parameters associated with those characteristics, particularly where the data volumes are small. Some judgement is always needed in adjusting for unusual or infrequent severe events. Certain risk factors will genuinely be associated with large claims, and, if that were the case, it would be wrong to adjust the actual data. However, a large claim is often due to "bad luck" rather than "process." In that case, large claims should be capped and the balance spread across the whole portfolio.

    A separate model can be established for the large losses; this should arrive at a better method of apportioning large claims between the various rating factors. In such a procedure, large losses would be separated out from the main claims table and modeled relative to a less-detailed exposure base in order to ensure a reasonable volume of data in the individual cells across which to spread the large claims. The fine detail might be lost, but a broad basis can emerge suggesting that large claims tend to occur amongst certain subsets of the exposure base, thus enabling limited differentiation between rating factors. If no pattern of large losses were apparent, then the conclusion would be that the claims could equally well have affected any policyholder and their impact should be spread across the entire portfolio with no differentiation between policyholders.

2. *Events that have influenced the overall experience.* Exceptional events, which are not a normal feature of coverage, might have occurred during the period for which data were collected for the premium rating exercise. They might include extreme bad weather or earthquakes, for example, which can lead to a substantial number of claims concentrated in a small subsection of the portfolio. A well-designed data-collection system would enable such losses to be separately identified and excluded from the analysis. As with individual large claims, there is no single correct way of handling these events. One possibility is to evaluate the nature of the peril, the potential return period of the event,[8] and the expected cost per policyholder, and to charge a separate premium for the additional risk from this peril to policyholders living in the affected regions. The additional charge can be made explicitly, and cover for the peril can be made an optional part of the policy or included in the overall premium structure with automatic cover for the peril; the approach, in practice, might depend on the legal jurisdiction and the class of business.

3. *Time trends in the data.* The major trends affecting the data are inflation, the development of open claims to a final value (i.e., *incurred but not enough reported* (IBNER) claims, which are discussed further in Chapter 16) and the emergence of IBNR claims. The latter two are referred to as the *development* of the claims process, as distinct from general inflation. In most cases the effects of these factors would be adjusted out of the data before fitting a model, although it is difficult to separate these effects. However, if the data were sufficiently large and stable then they could, theoretically, be incorporated as additional parameters of the model (Harvey and Fernandes, 1989).

Changes in the level of deductible can introduce particular problems. Increases in deductibles generally lag behind inflation and do not occur smoothly. Changes also arise due to adjustments in management policy. The model could be used to derive parameters for the "ground-up" losses (i.e., the claim paid together with the deductible), which would

---

[8]The *potential return period* is the reciprocal of the claim frequency.

enable the effects of a range of deductible levels to be evaluated. Alternatively, the deductible level could be included as a specific parameter. However, the database is unlikely to include information on losses for amounts less than the deductible, and estimating characteristics of "ground-up" losses from truncated data can be misleading. In addition, the choice of deductible implicitly reflects characteristics of the policyholder, such as risk aversion. Thus, it does not follow that an analysis made on "ground-up" data can be applied generally to different levels of deductible. Strictly, the deductible should be modified in line with the inflation assumption before fitting the model.

4. *Changes in the experience base.* If historical data are unrepresentative of current and expected future business, then inferences based on the model could be misleading. If the differences are small, then the application of the data analysis can be applied in a more limited way, e.g., the model might be applied to specific perils or to claims of a certain size only. Alternatively, it might be possible to adjust the data. Adjustments might arise in respect of:

changes to the risk factors;
unusual or nonrepresentative features of the claims and exposure base period;
the development of open claims to an ultimate value;
changes to the scope of insurance cover;
inflation and other trends.

The idea is illustrated in Figure 14.5.

Data from more recent periods will probably be more relevant in terms of the underlying exposure and policy coverage. However, these claims data are less mature and might require significant adjustments for IBNR

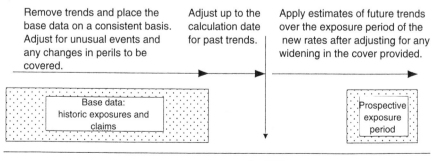

FIGURE 14.5 The idea behind data adjustment

and IBNER claims. If there are sufficient data, specific development factors can be derived at an aggregate level in order to estimate the further development of known case reserves. The likely overall extent of further loss development can be estimated from past claims and will depend on the business class and type of claim. However, in a detailed statistical analysis, it could be spurious to project IBNR claims into specific exposure cells, and this will bias and effectively predetermine the outcome of a statistical model.

Estimates of the amount outstanding for small claims or recent claim notifications will often be generated automatically. The relevant formula might have been developed over many years in order to provide a reliable aggregate estimate of outstanding claims, and its parameters might be different from those for the premium rating exercise. Case estimates (reserves for reported claims, discussed further in Chapter 16) could, therefore, be inconsistent with the proposed model and could distort the final results, especially if the amounts involved are significant. The formulae for producing case estimates are often modified following a change of management or because of a desire to revise the reserving basis (which might be motivated for a number of reasons, prudent and imprudent). Similarly, in cases where reserves are established by individual claims managers, there is a degree of subjectivity involved, and a change of claim manager can have a subtle effect upon the pattern of case estimated reserves. One of the commonest distorting factors is the introduction of new computer systems, which invariably causes a loss of continuity in the data during the period of switching from the old to new systems.

Claim numbers are unaffected by IBNER development and might be more suitable for modeling the rating factors of immature data than claim amounts. Average costs per claim can then be superimposed on the allocation of claim frequency.

It is also important to allow for inflation independently of adjustments to data due to coverage changes and incomplete claims development. Both exposures and claims value should be adjusted to a common currency level. Making such adjustments is not necessarily straightforward: exposure measures might not be of a financial nature. For example, number of employees is used as a measure of exposure for employer's liability insurance, and vehicle-years is used as a measure of exposure in motor insurance, neither of which are financial, but they can still change in an inflationary way that might affect the concentration of risk per unit of exposure.[9] Inflationary pressures will vary according to the types of claim. For example, in medical insurance the cost of drugs will have changed at a different rate to the costs of nursing care. Further complications can also occur if a large proportion of the claims is held as case reserves. Ideally, one would need to know the inflationary assumptions supporting these reserves.

Before any of the above developments can take place, the data used to derive the model of the claims process must be verified. For example, prior to any adjustment, the total claims should be calculated from the database and checked against the statutory accounts or against some other external reference point. Similarly, average costs per claim can be derived and tested for reasonableness and a general examination of the exposures carried out to ensure that the data are correct.

14.3.6 Deriving Expected Claims

Having refined the data as far as reasonably possible, it remains to establish a base premium level and the relationships between the rating factors. Many possible approaches can be adopted but in all situations it is important to formulate a statistical model and to be aware of the dangers of considering a single rating factor at a time. Consider the following example of the experience of a motor insurer.

Suppose the actual database is given in Table 14.4. The one-way analysis by car group is misleading because of the relative numbers of young drivers in car group B and of old drivers in group A. The marginal totals by car group tell more about the experience by policyholder age than about the significance of the car group. In the two-way table (Table 14.4) it can be seen that, in fact, the claim rate is independent of car group. The idea that a one-way analysis can mask information obviously extends into more dimensions: the two-way analysis might mask the impact of a third variable, and so on. It follows that, in performing an analysis, it is important to know how data are distributed within the portfolio (i.e., the heterogeneity of the data) and to understand the possible impact should exposures be concentrated in specific areas. This is a classic example of "Simpson's paradox," also called spurious selection in actuarial textbooks, such as Benjamin and Pollard (1980).

Complete descriptions of the statistical modeling process can be found in many sources (e.g., Renshaw, 1989; Brockman and Wright, 1992). There are a number of key stages in constructing a model, some of which have been discussed in other contexts already:

Specify and fit a model with all reasonable factors and test the contribution that each factor makes to the overall fit, as measured by the residual sums of squares in conventional regression models, or by the deviance in the case of generalized linear models (McCullagh and Nelder, 1989).

---

[9]For example, the number of employees might have remained constant within a particular industry, but the number of hours spent whilst exposed to dangerous processes might have reduced over time (e.g., due to improvements in health and safety legislation).

Table 14.4 One-way and two-way analysis of database

Policyholder age level	Exposure		Claim amount		
	Car group		Car group		
	A	B	A	B	Total
Young	10	100	3	30	33
Old	100	10	10	1	11
Total	110	110	13	31	44

The marginal totals indicate that the claims experience by policyholder age is

Policyholder age	Exposure	Claims	Claim rate
Young	110	33	0.3
Old	110	11	0.1

Whilst the experience by car group is

Car group	Exposure	Claims	Claim rate
A	110	13	0.12
B	110	31	0.28

Test the factors for interaction. For example, one factor might affect
the relative contribution to claims of another factor; in motor insurance,
the effects of age and sex as rating factors are usually correlated.
A model that assumes that age and sex operate independently might
generate relative risk measures that give policyholders the opportunity
to select against the insurer.

Establish a model that relates the rating factors logically to the experience.
Results that appear contradictory can give insights that provide
a commercial advantage but are more likely to be a wrong signal.

The extent to which the different risk characteristics of the different
rating groups are incorporated into a company's pricing plans is dependent
upon the expense structure, the company strategy, the profits in these
segments of the market, and the materiality of this section of the portfolio to
the whole.

The foregoing description gives an empirical derivation of the claim
process: data are used to derive a model with parameters that seem
broadly plausible. Alternative methods have been developed for particular
classes of business, most notably for natural catastrophes and, in particular,
earthquake and hurricane risk in the U.S. and Japanese markets. Such
models use a background model of geological and meteorological data
to describe the exposure to the insured peril in different areas. The
exposures in a particular zone are mapped against the locations of the
natural catastrophes and given specific engineering parameters, e.g., of a

building's ability to withstand earthquake damage. Events can be modeled based on past experience and on geographical areas that are known to be particularly vulnerable to the particular risk, allowing for any special features, such as buildings regulations, of the region (Christofides et al., 1992).

An alternative to taking actual claims histories and expressing the number of claims and the claim amounts relative to an exposure measure is to model the actual process that gave rise to the claims. Explicit modeling of the processes underlying the events that cause claims is not widely used in insurance pricing, although it is more common in risk management. The approach can be particularly useful when designing new products for which there is relatively little past experience. Such models often adopt a fault/event-tree structure, such that all the processes in, say, an industrial plant or in operating a mass transit system are explicitly modeled and potential faults and the likelihood of a fault at each stage evaluated. Associated with each fault is a potential event. For example, a signal failure in a railway network could have different consequences depending upon several factors, such as the number of trains running or the availability of safety backup systems. The severity of the event will depend on other factors, such as its association with other faults or the time of day. Once the expected faults and events have been modeled, estimates can be made of the possible distribution of costs, the need for insurance, and the best design of the insurance product. This is a powerful approach, since the effect of modifications to the process can be evaluated in terms of the changed incidence and cost of events. Many organizations use such approaches to calculate the capital expenditure needed to reduce the number of accidental deaths occurring in the workplace (Pearce and Nash, 1981).

Other forms of regression, e.g., *neural networks*, are being exploited for their ability to identify relationships between a wide variety of rating factors and other collateral data that can be assimilated into an analysis. Techniques that do not depend on a formalized model and identify relationships in a more generalized way are collectively known as *data-mining* methods (see Galfond (1997) and Schrage (1997)).

14.4 Measuring Expenses

A high proportion of a general insurance company's expenditure is for expenses and commission. This is particularly the case in personal lines classes of business, such as household contents insurance, since these classes have a relatively small average premium per policyholder and a relatively high claim frequency and policy endorsement rate. Reinsurance companies and underwriters of large commercial lines of business usually have lower expenses, considered as a proportion of premium income. This is because average premiums are large and there is a smaller number of individual contracts to administer. In addition, the expenses of reinsurance companies are kept down since direct writers incur most of the costs of marketing and acquisition.

The types of expense incurred will vary between insurers. For example, a company that uses brokers to market its policies will pay more commission than one with a direct sales force, which will expect to incur higher salary costs. Broadly, however, expenses (by which we shall include commission) can be divided into two types: fixed costs and variable costs.

Fixed costs are those that are incurred almost regardless of the amount of business handled. This would include, e.g., accommodation costs and salaries. Clearly, these are related to the level of business underwritten, but they can only be adjusted over the long term.

Variable costs are those that vary directly with the amount of business handled. This will include, e.g., commission and claims handling costs. Variable expenses might change in line with:

the number of policies being written (e.g., the administration cost of creating and processing new policy documentation);

the number of claims being handled (each claim will generate administration costs);

the amount of claims (e.g., use of external loss adjusters or legal advice);

the amount of premium received (e.g., commission).

In the same way that a detailed knowledge of the risk profile of a policyholder can confer commercial advantages, a detailed understanding of the expense structure and the factors affecting the expenses gives:

insights into the operational aspects of the business and the opportunities to make the operation more efficient;

improved allocation of expenses between the various classes of business and parts of the organization;

more accurate profit measurement and knowledge of areas being cross-subsidized;

a clearer indication of the marginal cost of putting business on the books and the flexibility that exists within the rating structure to attract new business through lower overhead expense allocation.

We distinguish between costs that can be attributed directly to business activity associated with a line of business and other costs that are necessary, but which are not always easily linked to a contract or a line of business.

14.4.1 Direct Expenses

Some departments will be dedicated to supporting one class of business only, in which case it will be straightforward to allocate their expenses. However, other departments might handle several different classes of business, and their costs must be allocated to the classes or lines of business involved. This can be done using *timesheet* analysis: staff record the time spent on various activities, distinguishing between the class or line of business concerned. However, filling out timesheets is itself time consuming, and substantial inaccuracies and inconsistencies can arise through rounding errors, the treatment of overtime, and flexible working practices and other administrative considerations.

An alternative to using timesheets on a daily basis is *functional costing*. The work done by each department is analyzed in detail using timesheets, or a more formal study of workplace practice, over a relatively short period of time, or over several short intervals of time in order to ensure a representative sample. Each activity is assigned an index value reflecting its relative cost as determined by the study. For example, if Department A processed 100 renewals with internal expenses of £1000 then the index value of renewals for Department A is $I_R = 1000/100 = 10$; if, over the same period, 500 claims were handled with an internal cost of £20 000, then the index value per claim would be $I_C = 20\,000/500 = 40$.

Similar indices can be constructed for the various administrative activities, such as processing new business, cancellations, and endorsements, which will vary by line of business. The indices can be generalized as $^{L}I_K$ where L is the department and K the function or activity. A matrix of indices by function and department is then produced. A company's total expenses can then be apportioned according to the product of the frequency and index level of a particular activity. It is relatively simple to count the number of times any activity is performed and the necessary statistics can be provided by the company's management information systems.

Thus, the relative level of the indices controls the expense allocation. This approach has many advantages over timesheets:

The indices established are objective.

The calculations are simple and not intrusive, once the initial analysis has been performed.

The relative values of the indices are not subject to inflation in the same way as the absolute values, and they can easily be adjusted if factors such as inflation or specific taxes (such as a policy tax) do affect them. Thus, the expense allocation should be more consistent over time. However, a change in the mix of business can result in fundamental changes in the internal procedures of a company, in which case the indices need to be recalibrated.

14.4.2   Indirect Expenses

Indirect expenses can only be allocated approximately between the different classes of business. There are some reasonable rules of thumb that can be used. For example:

1. the costs of a computer department can be allocated in proportion to the amount of processing time spent on each class of business;
2. the costs of the personnel department can be allocated according to the numbers of staff directly related to each class;
3. accommodation costs can be allocated according to the floor area taken up by the different departments;
4. the salaries and other expenses of the directors and senior executives not directly responsible for a class of business can be allocated by premium volume, profits, or other measure reflecting their strategic role within the business.

Different insurers use different approaches to calculate the office premium. Some add a simple percentage loading to the risk premium, whereas others use a more detailed formula, allowing for both direct and indirect costs, and their variable or fixed nature, explicitly as shown in Equation (14.1). Since there can be variation in the expense allocation it can be difficult to compare the fundamental profitability of each class of business, even within an organization. However, as discussed, if marginal costing or a cross-subsidy is deliberately introduced into the rating structure then it is important to have an indication of the implications.

14.5   Experience Rating

We have seen that rating factors are used to divide policyholders into relatively homogeneous risk groups and to quantify the riskiness of each policyholder. However, because of limitations in the data, it is probable that some potentially useful rating factors will not be identified and that there will be errors in estimating the significance of those rating factors that are identified. Thus, considerable heterogeneity will remain. Furthermore, rating factors are, effectively, a static method of premium rating, since, provided their rating factors do not change, policyholders are charged the same premium each year regardless of their actual claims experience. Experience rating enables insurance companies to charge premiums that reflect the actual experience of individual policyholders, thus reducing heterogeneity within the insurers' rating structures, and it is particularly useful if the rating factors cannot be determined.

Usually, experience-rated premiums are calculated by adjusting the "average" premium for the rating group into which the policyholder falls. The difference can be paid in two ways:

1. A *prospective* basis, in which case the renewal premium is adjusted to reflect past clams experience.
2. A *retrospective* basis, in which case an additional premium is due from, or a refund is made to, the policyholder as a result of the experience during the policy period, once such experience can be measured with reasonable accuracy. This approach is similar to profit sharing.

The former method would be used for personal lines business, whereas both could be used for commercial classes of business.

Experience rating can be based either on claim numbers or claim amounts. We shall consider examples of each separately.

14.5.1 Experience Rating by Claim Number

Small individual risks, with a low expected claim frequency, are usually experience rated by claim number. This is for two reasons:

1. it would take a long time to build up the experience necessary for rating according to claim amounts;
2. the variability of the claim cost from year to year would be too great to enable a proper assessment of what was due to the systemic risk of the policyholder and what was due to random events.

In a private motor account operating an NCD system, motorists with no prior claims history purchase insurance at the basic rate of office premium for their particular risk group. For each claim-free year thereafter, the premium paid will be discounted relative to the basic rate of office premium. A scale typical in the U.K. is given in Table 14.5.

If a policyholder makes a claim then he or she falls back down the scale, usually by 1 or 2 years. In the event of more than one claim in a policy year the policyholder will normally lose all entitlement to discount. Discounts can then be re-earned in the usual way, by experiencing claim-free years. Policyholders are not normally penalized for making certain types of incidental claim, such as for windscreen damage, on the grounds that these

Table 14.5 Example of NCDs

No. of claim-free years	0	1	2	3	≥ 4
Discount (%)	0	30	40	50	60

losses are not driven by particular rating factors but are equally likely to affect any policyholder.

There are several reasons why insurance companies use NCD systems:

riskier policyholders gravitate towards lower NCD categories and pay higher premiums;

it reduces the number of small claims, since policyholders might be discouraged from claiming if they risk losing discount, so administration costs are also contained;

it is believed that NCD schemes encourage policyholder loyalty.

Whilst the second purpose is probably achieved by the NCD system, it is possible that NCD systems do not discriminate between risks very effectively. This is partly due to the way the system is operated in the U.K., with a very short scale, and with large steps between each level of discount. A longer and more gradual scale might differentiate between the different risk groups more effectively (Lemaire, 1995).

Because NCD systems fail to discriminate adequately between different risk groups, they give rise to cross-subsidies between policyholders on low and on high rates of discount. They can also create ill will between insurance companies and policyholders if a policyholder is penalized for an accident caused by a third party, perhaps after many years of claim-free driving. Nonetheless, because they do reduce the number of small claims, which give rise to disproportionately large claims-handling expenses, NCD systems are being introduced into other personal lines of business, such as household contents insurance.

14.5.2 Experience Rating by Claim Cost

Risks with a high claim frequency are likely to build up a reliable claims experience more quickly than those risks with a low claim frequency.[10] In order to estimate the true expected claims level of an individual risk, insurance companies need the past claims history of the risk as well as other (collateral) information, such as the book rates (see Section 14.3.1) for the appropriate risk group, the claims experience of other, similar, risks, and more general industry data. The extent of the risk's known claims history will determine whether the premium can be calculated based solely on the experience of the risk, or whether the premium should rely on other claims data, drawn from the appropriate rating group. The assumption is that the actual claims experienced by an individual risk give some indication of the

---

[10]In the context of motor insurance, a motor fleet would be have a relatively high claim frequency compared with a private policyholder, for example.

future experience of the risk. The problem faced by the insurer is to derive a premium that balances the risk's expected claims as indicated by the actual loss experience of the risk with that indicated by a wider body of data. This type of experience rating is known as *credibility rating*. The formula used is called the *credibility formula*

$$P = ZQ + (1 - Z)R$$

where P is the risk premium to be charged, Z is the *credibility factor*, i.e., the weight accorded to the policyholder's own experience and level of exposure, Q is the policyholder's own experience, and R is the expected risk premium based on the collective, or *collateral*, information.

The claims used to calculate Q are often capped, so that the risk's aggregate claims experience is limited. The insurer's premium may include an additional charge (not experience rated) for covering claims above the capped amount.

Whilst relatively simple rules can describe the behavior of the credibility function, the derivation of the function itself can be complex (see Hossack et al. (1990) and Waters (1987)). The theoretical approach is Bayesian and leads to a weighted average of the *prior* (or book) rates and the *posterior* information (the actual experience).

The credibility factor Z is such that $0 \leq Z \leq 1$. Z increases according to the accuracy with which the risk's own experience can be used to estimate the true expected claims relative to the estimate provided by the book rates. If $Z = 1$ then the risk's experience is accorded *full credibility*. Theoretically, it is impossible for an individual risk to achieve full credibility. However, insurance companies might decide that certain risks are "close enough" to full credibility to be assigned $Z = 1$. One possible criterion for assigning full credibility is demonstrated in Appendix B. If the premium is calculated assuming full credibility, then there is no pooling of risks with other policyholders. Under these conditions, the insured achieves a smoothing of results through time and the insurer merely provides an administrative claims-handling and financing function.

Two widely used formulae for Z are

$$Z = \min\left\{\sqrt{\left(\frac{M}{n}\right)}, 1\right\}$$

where M is the observed number of claims and n is the number of claims required for full credibility (this in sometimes referred to as the *limited fluctuation credibility formula*), and

$$Z = \frac{M}{(M + K)}$$

where M is a measure of the volume of data attributed to the individual policyholder and k is a measure of the relative accuracy with which the individual policyholder's true mean is estimated compared with the accuracy with which the "book" true mean is estimated. This is sometimes referred to as the *Bayesian credibility formula*. (Dean, 1997).

M can be generalized to include any reasonable measure of the volume of information, such as:

the number of claims
the amount of incurred losses
the number of policies
the earned premium
the number of insured periods

It is simple to illustrate the retrospective method of charging experience-rated premiums with a credibility premium. The credibility formula can be rearranged as

$$P = R + Z(Q - R)$$

Accordingly, the insurer can charge a deposit R at the renewal date and at the end of the year (or after enough time has passed for Q to be determined with sufficient accuracy) make a *profit sharing* adjustment of $Z(Q - R)$ based on the actual experience of the policyholder during the year. This retrospective pricing arrangement has advantages for both the insurer and the insured:

- By making the end of year adjustment, the insured retains some of the underwriting risk and so might be more cautious about its experience (moral hazard is reduced).
- If the insured is a relatively poor risk, then it is able to earn some benefit from the investment income accruing on the balance it will pay to the insurer at the end of the year. However, this could cause difficulties for the insurer if the insured proves to be a poor credit risk as well.

The credibility rating formula and basic concept is identical to the Bornheutter–Ferguson reserving approach described in Chapter 16.

14.6 Business Planning

The nature of the premium can be viewed in two ways: as a means of charging the "correct" rate for a particular risk, or from a portfolio perspective. The former position is the starting point of most rating techniques and is the one on which we have concentrated in this chapter. However, senior management and shareholders are more concerned with corporate goals, such as the performance of the portfolio as a whole and achieving an overall target loss ratio consistent with the risks undertaken and capital employed. Most strategic corporate decisions are made from this high-level analysis rather than from a consideration of the minutiae of the derivation of premium rates. For example, senior management might require a general premium rate increase and not be unduly concerned with structural changes that discriminate more accurately between risks, or in achieving a balance in the portfolio between different classes of business. An understanding of the market place and the price elasticities of demand can enable management to use the pricing basis in order to control the development and structure of the portfolio.

Prices of insurance products move up and down with the "insurance cycle," whilst insurers strive to differentiate their products by offering additional coverage, effective marketing and advertising, or by developing a good reputation for service. Over the long run, policyholders will buy the policies offering the best value for money and insurance companies will experience portfolio growth in those sections of their account that are priced cheaply relative to similar products offered by other insurers. In the personal lines market, particularly in the motor business, insurance brokers, and to some extent policyholders themselves, rank a company's prices relative to those of its peer group. Relative price is as significant as absolute price in driving the policyholder's propensity to change insurer. Hence, it is important that companies can know how a given price change affects its rank and, using a model of demand based on price rank, then assess the effect of the price change on their expected volume of business. In theory, this information is then available to the insurer's business planning model. However, in practice, the results of this type of analysis are often not sufficiently stable to be incorporated formally into a model. In commercial markets, the products offered are less homogeneous and their prices are less easily available so; whilst brokers and underwriters have a subjective feel for the general market pricing level, the only real test of competitive pressure is when two insurance companies compete on the same risk.

In evaluating the desirability of launching a new product, or the profit implications of maintaining an existing tranche of business at a given price, the basic tool is the profit test model (Cooper, 1996). In this model, volumes of future business are estimated and the revenue items arising as a result of the new and existing business (e.g., expenses, investment returns, and interest charges) are used to generate future cash flows. The emerging cash flows are discounted at a rate consistent with the company's required return on capital, formulated with some consideration given to the risk-free discount rate available in the market place, to give the *net present value* of the business. The net-present-value measure can be used together with other measures of profitability to decide whether the policy will add any value to the company. These techniques are well established amongst life insurance companies (see Chapter 7 and Chapter 8).

The valuation of general insurance companies can then be approached as an extension to profit testing, as for life insurance companies (see Chapter 11). A number of business planning and dynamic financial models are described in Daykin and Hey (1990) and in the Casualty Actuarial Society 1996 Call Papers on Dynamic Financial Models.

14.6.1 Monitoring Experience

As with any enterprise, it is essential to monitor the relationship between prices and costs as the business develops. This requires up-to-date information prepared on as realistic a basis as possible. The information, which should be provided at least quarterly, should include:

portfolio statistics
movement statistics
claims experience statistics
expense statistics

Portfolio statistics are the basis on which exposure can be calculated. They include the numbers of policyholders in each class of business and risk group, together with the premiums paid and the appropriate measure of exposure. These data enable the company to analyze how the size and spread of the portfolio is changing, and to identify any concentrations of risk that might develop.

Movement statistics include the same information as the portfolio statistics, but with regard to policies that are new, or have been renewed, since the previous exercise, or have lapsed or expired. This information should enable the company to identify whether there have been any unexpected gains or losses of business, which can indicate that the premium charged is out of line with the rest of the market and, in particular, whether there is a risk of adverse selection by policyholders.

Monitoring the lapse and new business rates is clearly of value from a marketing point of view in assessing, for example, the effectiveness of an advertising campaign or the impact of bad publicity. It is also critical from an underwriting perspective: reliable loss experience, with which to test the rates in force, will not emerge from new business until a substantial exposure has been put on risk. By the time the data are available, it might be too late to avoid large losses by amending premium rates, revising policy coverage, or by withdrawing from a given area of the market. If the volume of business attracted is in excess of the planned increase, or if the lapse ratio is significantly reduced, then this might indicate that the rates being quoted are significantly cheaper than rates being charged by the rest of the market, where "cheaper" can mean that the coverage is too broad rather than solely a lower premium.

The basic statistics used for measuring transaction volumes are:

- the number of lapses as a proportion of the number of renewals issued;[13]
- the number of lapses as a proportion of the expected number of lapses;
- the actual number of new policies as a proportion of the expected number of new policies;
- the number of new policies accepted as a proportion of the number quoted.

From a marketing perspective, it is also useful to characterize those proposers who declined to take the cover, as there might be some aspect of the rating basis that is discouraging potentially profitable business.

Monitoring actual transaction volumes against the expected values is not straightforward. At an early stage of the development of a class of business, the results are often unsatisfactory: it might take a considerable length of time to develop an adequate feel for the dynamics of the market place. The approach does, however, have numerous applications in evaluating improvements in business arising as a result of strategic management decisions. The principal idea is that there is some background expectation reflecting "normal" management decisions. Revised decisions lead to changes in the transaction volumes, and it is this change, or value added, measured against the "normal" background that is used to assess the effectiveness of management decisions (also referred to as the operational risk), and might be the basis of a remuneration strategy. Bride and Lomax (1992) explore the contribution to shareholder value of such financial and operational decisions.

Because of the importance of this movement information, it is usually provided at least monthly. Delays are inherent in this data, particularly for lapses, and so the total movements for each period are often projected using simple chain-ladder techniques (see Chapter 16).

Claim statistics enable the insurer to monitor the experience of each group and to compare whether the business is developing as expected on the basis used to calculate the underlying risk premium. The type of information needed will be similar to that provided for a reserving exercise, although in less detail:

claim numbers
average cost per claim

---

[13]This gives an indication of the rate at which business is retained.

average cost per unit of exposure
claim cost distributions

This information should allow the insurer to analyze whether there has been any deterioration in experience, beyond what might have been expected, and to investigate the causes of the deterioration. In some cases, the information can be responded to quickly, e.g., by excluding certain perils from the cover provided by the policy or by introducing a policy excess. The effect of these measures can be more immediate than an overhaul of premium rates.

The monitoring of *expense statistics* was considered in Section 14.4.

14.7   Conclusions

Premium rating aims to set an office premium that is both competitive and includes adequate margins for profit and other contingencies. Traditionally, this has involved estimating rating factors, which are closely correlated with the underlying risk factors, allocating claim costs amongst the rating factors, and allocating expenses across the various administration functions of an insurance company. Because of limitations in the available data, the cyclical nature of the insurance market and other management difficulties, there are inefficiencies in the rating structure. The actuary must make the most efficient use of the data possible so that any unintended cross-subsidies within the company are identified as early as possible, and so that the management is made aware of any opportunities to exploit commercial advantages over their competitors.

The techniques applied for modeling claims experience and estimating the appropriate risk parameters are continually developing as data-processing packages become more efficient and are better able to exploit the available information. However, in some classes of business the contracts and policyholders are too diverse for the development of large-scale statistical modeling. In these cases, the expertise of the various professionals employed within the insurance industry can provide invaluable additional insight in order to enable the actuary to complete the premium rating exercise and to link this aspect into reserving and general business management.

Appendix A

Appendix A here extends the simple results of Chapter 12, Appendix A, and demonstrates that the variance, as a proportion of the expected total claim amount, reduces as the number of claims increases.

Suppose that X_i is a random variable representing the total claim amount from policyholder i, and that $E(X_i) = \mu$ and $\text{Var}(X_i) = \sigma^2$. Suppose that all the policyholders are independent, that the X_i are identically distributed, and that each policyholder has the same likelihood of claiming Λ, where Λ is a random variable and $E(\Lambda) = \lambda$ and $\text{Var}(\Lambda) = s^2$. Consider an insurance portfolio with n of these identical policyholders.

Define $S = \Sigma X_i$, where the sum is over all claims. Denote the total number of claims by N. Note that N is itself a random variable, which we assume is independent of the X_i. Because of the assumption about the claim rate of each policyholder, $E(N) = n\lambda$ and $V(N) = n^2 s^2$. Then, using results from conditional probability and noting that for fixed N, $E(S \mid N) = N\mu$:

$$E(S) = E(E(S|N)) = E(N\mu) = \mu E(N)$$

Similarly

$$\text{Var}(S) = E(\text{Var}(S|N)) + \text{Var}(E(S|N))$$
$$= E(N\text{Var}(X)) + \text{Var}(N\mu)$$
$$= E(N)\sigma^2 + \mu^2\text{Var}(N) = n\lambda\sigma^2 + \mu^2 n^2 s^2$$

So the coefficient of variation is

$$\frac{\sqrt{\text{Var}(S)}}{E(S)} = \frac{\sqrt{n\lambda\sigma^2 + n^2 s^2 \mu^2}}{n\lambda\mu} = \sqrt{\frac{\sigma^2}{n\mu^2} + \frac{s^2}{\lambda^2}}$$

Now, as $n \to \infty$, we see that the coefficient of variation $\downarrow \frac{s}{\lambda}$. Thus, by increasing the number of policyholders, the coefficient of variance is reduced.

Note that the first term in the ratio represents the variability caused by random fluctuations in the claim amount between policies. Since we assume policies are independent with regard to the claim amount, this uncertainty decreases to zero as the number of policyholders increases. The second term is the uncertainty caused by not knowing the exact claim frequency for each policyholder. Since we have assumed the claim frequency is the same for each policyholder, so that policyholders are not independent with regard to λ, this uncertainty does not vanish as the number of policyholders increases. In fact, it will only equal zero if we assume $s = 0$, i.e., that we know the claim frequency λ with certainty.

Appendix B

B.1   Criteria for Full Credibility

Suppose the true credibility premium is $\hat{\beta}$ and $\hat{\beta}1$ $(= \bar{x})$ is the estimate of β based on past data only. Then we say the past data are "fully credible (k, π)" if

$$P\left[\left|\beta - \hat{\beta}\right| \leq k\beta\right] \geq \pi$$

where k and π are two numbers such that $0 < \pi < 1$, $k > 0$.

Example

Suppose N_i is a random variable representing the number of claims from an insurance portfolio in year i. Assume that N_i has the Poisson distribution with unknown mean λ. Suppose that n years of experience are available, and that there have been $N_1 + N_2 + \cdots + N_n = M$ claims. Then M is also a Poisson random variable. Using the maximum likelihood estimator for a Poisson random variable we have

$$\overline{X} = \frac{M}{n} = \hat{\lambda}$$

where $\overline{N}$ is the estimated mean frequency in any year, so M has the Poisson distribution with mean λn.

Now, using the data to estimate the number of claims expected to arise in a policy year, the data are fully credible (k, π) if

$$P\left[\left|\hat{\lambda} - \lambda\right| \leq k\lambda\right] \geq \pi$$

Since we have the maximum likelihood estimator $\hat{\lambda} = M/n$, we get

$$P\left[\left|\frac{M}{n} - \lambda\right| \geq k\lambda\right] \geq \pi$$

$$\Rightarrow P\left[\left|\frac{(M - \lambda n)}{n}\right| \geq k\lambda\right] \geq \pi$$

$$\Rightarrow P\left[\left|\sqrt{\left(\frac{n}{\lambda}\right)}\frac{(M - \lambda n)}{n}\right| \leq k\lambda\sqrt{\left(\frac{\lambda}{n}\right)}\right] \geq \pi$$

$$\Rightarrow P\left[\left|\frac{(M - \lambda n)}{\sqrt{(\lambda n)}}\right| \leq k\sqrt{(\lambda n)}\right] \geq \pi$$

Now, since M is the sum of n Poisson random variables, if n is large enough we can use the normal approximation: i.e., M is approximately a normal random variable with mean λn and variance λn. So, e.g., if $k = 0.05$ and $\pi = 0.95$ we would have

$$0.05\sqrt{(\lambda n)}! \geq 1.96$$

$$\Rightarrow \sqrt{(\lambda n)} \geq 39.2$$

$$\Rightarrow \lambda n \geq 1537$$

So we would need the expected number of claims observed in the past experience to be at least 1537; we could say the data are fully credible $(0.05, 0.95)$ if $M \geq 1537$.

Clearly, the data required for "full credibility" depend on the criteria chosen. As a comparison to the above example, if we had chosen $k = 0.05$ and $\pi = 0.9$, then for full credibility we need $\lambda n \geq 1082$. $(0.05, 0.9)$ is often used as the standard.

Note that this example has only considered claim frequency and in practice a compound model of both claim frequency and claim severity is required. A fuller discussion at an introductory theoretical level is given by Waters (1987).

References

Benjamin, B. and Pollard, J. H. (1980). *The Analysis of Mortality and Other Actuarial Statistics*. Heinemann.

Bouska, A. and McIntyre, T. (1994). *Measurement of US Pollution Liabilities*. CAS Forum.

Bride, M. and Lomax, M. W. (1994). *Valuation and Corporate Management in a Non-Life Assurance Company*. JIA, Vol. 121.

Brockman, M. J. and Wright, T.S. (1992). *Statistical Motor Rating: Making Effective Use of Your Data*. JIA, Vol. 119.

Christofides, S., Barlow, C., Michaelides, N. and Miranthus, C. (1992). *Storm Rating in the Nineties*, GISG Convention. Institute of Actuaries.

Cooper, D. R. (1996). *Measuring Profit in General Insurance*. The Actuary, October.

Coutts, S. M. (1984). *Motor Insurance, An Actuarial Approach*. JIA, Vol. 111.

Daykin, C. D. and Hey, G. B. (1990). *Managing Uncertainty in a General Insurance Company*, Vol. 117.

Daykin, C. D., Pentikainen, T. and Pesonen, M. (1994). *Practical Risk Theory for Actuaries*, Chapman & Hall.

Dean, C. G. (1997). *An Introduction to Credibility*. Casualty Actuarial Society Forum, Winter.

Derrig, R. A. (1991). *The Development of Property-Liability Insurance Pricing Models in the United States 1969–1989*. Casualty Actuarial Society Forum, Fall edition.

Galfond, G. (1997). *Data Mining Creates Advantages*. National Underwriter, Life and Health – Financial Services Edition, June.

Harvey, A. C. and Fernandes, C. (1989). *Time Series Models for Insurance Claims*. JIA, Vol. 116.

Hossack, I. B., Pollard, J. H. and Zehnwirth, B. (1987). *Introductory Statistics with applications in General Insurance*. Cambridge University Press.

Lemaire, J. (1995). *Bonus-Malus Systems in Automobile Insurance*. Kluwer Academic Publishers.

McCullagh, P. and Nelder, J. A. (1989). *Generalized Linear Models* (Second Edition), Chapman and Hall.

Pearce, D. W. and Nash, C. A. (1981). *The Social Appraisal of Projects*. Macmillan.

Renshaw, A. E. (1989). *Chain Ladder and Interactive Modelling (Claims Reserving and GLIM)* JIA, Vol. 116.

Renshaw, A. E. and Verrall, R. J. (1992). *Claims Reserving: Statistical Treatment of the Chain Ladder Technique*. Dept. of Actuarial Science and Stats, City University, London.

Schrage, M. (1997). *Data Mining in a Vicious Circle*, Computerworld, June.

Waters, H. (1987). *Special Note: An Introduction to Credibility Theory*. Institute and Faculty of Actuaries.

Chapter 15

Reinsurance

15.1 Introduction

Reinsurance is the principal mechanism that insurance companies and reinsurance companies use to transfer part or all of the risk assumed through their own underwriting activities. The insurer is said to *cede* risk to the reinsurer, and is described as the *cedant*. When a reinsurer itself reinsures a risk, this is called *retrocession*.

15.1.1 Why Buy Reinsurance?

The premium a ceding insurer pays to its reinsurer reflects the risk that the reinsurance company perceives to be intrinsic in the business being transferred, together with loadings for profit (or return on allocated capital), expenses, commission, and contingencies. The total premium will also reflect, to some degree, the market level of premium rates, the extent to which past losses are being recouped from expected profits on current and future business, and the less tangible value of maintaining a business relationship with the client or the broker. The question must be asked: If the risk was profitable enough to insure in the first place, then why pass it on to reinsurance companies?

The reasons for buying reinsurance can be both operational and financial, and these are set out below. Note that insurers are also motivated by the same considerations that encourage individuals to purchase insurance, as discussed in Chapter 12.

15.1.1.1 Balance Sheet Protection

Reinsurance is purchased principally to protect the solvency of the insurance company. This is sometimes called *balance sheet protection*. Theoretically, the reinsurance program might be chosen with the aim of minimizing the probability of ruin. Another formal consideration, which is related to the probability of ruin, will be the way in which the reinsurance program contributes to shareholder value (Bride and Lomax, 1992). Such solvency protection is provided:

In the form of catastrophe reinsurance cover, which provides high limits of cover against the risk of an accumulation of claims from many policies all affected by the same event;

By controlling the risk that a very large single claim might present to the cedant. This form of reinsurance is *excess of loss*.

By providing protection against losses (no matter how incurred), should they exceed some predetermined aggregate level; this is usually expressed as a multiple of premium income. This is called *stop loss* reinsurance.

15.1.1.2 *Stability of the Net Profit Stream*

Publicly quoted insurance companies, which would generally place a high value on a stable emerging profit stream, might be prepared to accept lower long-term underwriting profits in exchange for keeping the variability of the expected profit stream within a target range. In this context, it should be borne in mind that reinsurance is only one aspect of the financial dynamics of the insurance operation. The investment performance achieved on assets (including both income and capital growth) can itself lead to substantial variability in the overall financial results, and capital market instruments may be used to hedge out some investment volatility.

Consideration must be given to the type of reinsurance to buy and the way it matches the exposures and resulting claims. For example, buying reinsurance to protect sections of the account with relatively little variability would not be productive. An insurance company, ceding the more variable elements of the account in order to stabilize the retained account, will have to pay a larger premium to its reinsurers in order to reflect the greater uncertainty being transferred. There is a point at which the ceding company must strike a balance between the marginal additional cost of reinsurance and the marginal reduction in variability in the retained account. This will vary between companies according to numerous considerations, including, for example, the size of the company, its corporate structure, its financial strength, its trading history, and the types of business written. It might also be a function of the local regulatory environment, for example, depending on whether the insurance company is permitted to set up claims equalization reserves (see Chapter 16) from pre-tax profits.

15.1.1.3 *Increased Risk Capacity*

The exposure that an insurance company can reasonably accept is restricted by the size of its capital base. Reinsurance provides access to additional capital, or "capacity," and gives companies the ability to take on larger risks than their free assets would normally support. The ability to accept large risks makes the company more attractive to insurance brokers and their clients and enables it to contemplate expansion or more risky strategies than would otherwise be possible. Increased participation in risks, or the ability to accept larger risks, gives the insurance company more prestige and a greater degree of control over the underwriting terms and conditions and over the subsequent handling of claims.

15.1.1.4 Gaining Experience in New Markets and Product Areas

The facility to cede a large proportion of risks written in a class of business or in a geographical region that is unfamiliar enables a company to gain expertise in the underwriting and administration of that business with reduced risk. The choice of reinsurance company is likely to be influenced by the expertise that the reinsurer has in a particular class of business, since insurance companies often receive technical assistance from their reinsurers, particularly in handling claims.

15.1.1.5 Advance Commission and Cash Flow Assistance

Reinsurance companies sometimes provide cash in advance of an insurance company writing the business covered by the reinsurance contract. This is often called *ceding commission*. Such cash flow is of particular assistance for smaller and newly formed companies. Profit-sharing arrangements are sometimes negotiated such that, if the claims experience is favorable, a "profit commission" is paid back to the ceding insurer.

15.1.1.6 Management of the Statutory Solvency Margin

Under most regulatory regimes, the statutory solvency margin requirement is proportional to the volume of business written: the more business written, the larger the amount of free assets required by law. Regulators do not tend to recognize negative correlation between the various items. Under the U.K. and the E.U. regulatory system there is scope for reducing the volume of business used as a basis for statutory solvency purposes by using reinsurance. Reinsurance can be a useful mechanism, particularly in a small, rapidly growing company, for reducing the growth rate of the net account and, thus, the additional strains of financing the growth of the statutory solvency margin. However, the statutory credits permitted for reinsurance would normally have an upper limit. The upper limit is imposed because, as an insurer's dependence on reinsurance increases, so does its exposure to the risk that one of its reinsurers will fail. Furthermore, a heavily reinsured could be regarded as simply an agent for the reinsurer.

15.1.1.7   Arbitrage

Pricing anomalies can occur in any market. When the price of reinsuring a risk is less than the premium received by the cedant for insuring the risk then the cedant can make a "turn" by reinsuring. The insurance company faces the risk that the reinsurance company might not pay claims and, before entering into such an arrangement, it is crucial to understand the circumstances which have created the pricing anomaly, as well as the credit rating of the reinsurer. Other forms of arbitrage arise as a result of regulatory rules concerning, principally, the discounting of reserves (see Section 16.2). Where one regime permits discounting and another does not, it becomes possible for an insurance company operating under the nondiscounting regime to cede business to a reinsurance company operating under the discounting regime: the premium payable will reflect the discounted reserve value, but the benefit assumed in the accounts will be at the undiscounted level. Large transactions of this character, which have little true transfer of risk, are broadly classified as "finite risk." Many insurance regulators no longer allow such reinsurance contracts to be valued at the full undiscounted value of the recoveries, and the financing and true risk components must be valued separately (see Financial Accounting Standard 113).

15.1.1.8   Balanced Portfolio

If the business has expanded in one section of the portfolio more rapidly than planned (e.g., a concentration in a particular line of business, or a geographical concentration), the management might feel that the account is "unbalanced" and might seek to adjust the composition of the account. One mechanism to achieve this is by swapping tranches of business with other insurance companies in an arrangement known as *reciprocity.*

15.1.2   Some Background

The reinsurance market is complex, and many of the participants in the market are financially interdependent. Since, at the time of purchase, most reinsurance is a contingent asset, the extent of the interdependency is only truly known in the event of a claim and will depend both on the size and type of event that triggers a cycle of reinsurance claims activity. Some interrelationships and potential dependencies are illustrated in Figure 15.1.

The insurance market has three levels: the primary (or *direct*) market, the reinsurance market, and the retrocession market. In practice, the obligations of the parties involved are defined in the contract of reinsurance and can be many stages removed from the original insurance contract which led to the cycle of reinsurance collections. In Figure 15.1, arrows have been used to

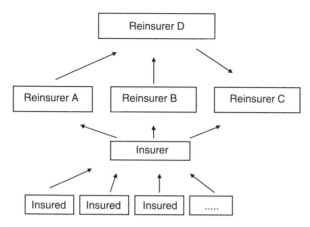

FIGURE 15.1 Reinsurance interrelationships and potential dependencies

denote the direction in which risk has been transferred. We assume that, in any year, the insurer has a large number of policyholders (the "insured") and a relatively small number of reinsurers, companies A, B, and C. Reinsurance companies A and B themselves purchase reinsurance with reinsurance company D. The three reinsurance companies A, B, and C appear to have a clear relationship with the insurer. Similarly, D has an apparently clear relationship with A and B. However, D purchases its own reinsurance protection from company C. The dependencies now become more involved, since C has a much greater exposure to the insurer than its initial involvement as a reinsurer. It might be unaware of the additional exposure, since D would not necessarily know that the exposure ceded by A and B originates with an insurer to which C is also exposed, or D might not pass the information on to C.

A and B believe that, by reinsuring with D, they have diversified their exposure to the insurer. However, since they depend upon D, which in turn relies on C, which is itself exposed to the claims of the original insurer, the extent of the diversification is diluted. Consider the consequence of a large claim that triggers a round of reinsurance collections. A loss, which is large enough to erode seriously the capital bases of A, B, and C on a net of reinsurance basis, could lead to the insolvencies of A and B. This could happen since A and B anticipate making a recovery from D, which, in turn, expects to make a recovery from C. Since C has a weakened capital base, due to the recovery made by the insurer, it could be unable to meet its obligations to D. Thus, D's capital would be required to pay A and B. If D has insufficient capital then the result could be that A and B are also rendered insolvent.[1] It is important to note that the reason for the failure

---

[1]This chain of events is sometimes referred to as a domino effect.

need not be a catastrophic single event; there might have been persistent underreserving for past losses by any of the parties involved which is only exposed when a modestly large loss requires settlement. This example also illustrates the consequences of the sometimes incestuous financial ties between reinsurance companies called the *spiral*. Risks are passed between a virtually closed group of interconnected reinsurance companies and the resultant claims will rest largely with the first reinsurer to exhaust its reinsurance program. The spiral effect has several adverse consequences:

Gross losses from one event can inflate dramatically to many times the size of the original claim. It is axiomatic that the reinsurance asset must similarly increase, since the total retentions in the system cannot exceed the value of the original loss. Thus, whilst the net position can appear stable, the gross values can be highly variable and demanding on capital — which is frequently aimed at the gross liability and not the net.

Since the process is a means of allocating the claim, the transactional costs quickly become substantial as each "turn" of the spiral reduces the individual transaction size and further expands the numbers of transactions in the system. As an allocation method, it is highly inefficient.

The above greatly simplified example examines a somewhat artificial situation, but it serves to illustrate two key aspects of reinsurance:

the necessity to monitor the financial strength of reinsurance partners;
the need for individual reinsurance companies to be aware of potential accumulation of exposure from a single source.

In practice, the dependencies between insurance companies can be far more involved. They are built up over long periods of time and often through a complex international web of mutually interacting reinsurance contracts. The intricate nature of reinsurance arrangements can add considerably to the frictional costs of administering the business.

Insurance companies and reinsurance companies normally have a department with responsibility for monitoring the financial strength of existing and potential reinsurers which provides assessments of the creditworthiness of reinsurers. Since insurance companies rarely fail overnight, in the example above most companies would have anticipated part of the nonrecovery from reinsurance company D by reducing its credit rating and setting aside appropriate provisions for nonperformance of this reinsurance asset. The financial impact would, then, not have been as sudden as suggested. Also, if the financial strength of a reinsurance company were perceived to be deficient, it would be unlikely to be used as security for more recent business. Thus, in the example above, company C's dependence on D would be reduced and so the impact across the business as a whole should be minimized.

The precise circumstances intended to be covered under a reinsurance contract are frequently only loosely specified and the information available to reinsurance companies is often not specific to the original risk. For example, reinsurance contracts sometimes provide "whole account" protection, or the contract might state that the reinsurance company "follows the fortunes" of the ceding company. Reinsurance agreements are legal contracts, but in some ways they are viewed more like "gentlemen's agreements." Reflecting the somewhat loose nature of many arrangements, the reinsurance contract will usually stipulate that reinsurance disputes should be resolved by arbitration rather than recourse to the courts. Arbitration has a number of advantages over formal legal proceedings: it keeps the facts of cases confidential to the parties involved; it leads to quicker and less expensive resolutions; and it means that the outcome in one case does not set a precedent for any other cases. However, arbitration is not legally binding, and its success depends on the ability of the arbitrator and the willingness of both parties to reach a compromise.

The operation of reinsurance can be viewed from two basic perspectives:

Normally, exposures and claims are shared amongst many reinsurance companies and, hence, across a broad spread of capital bases,[2] so that reinsurance serves to transfer and diversify risk across the insurance market.

When claims occur that satisfy the terms of the reinsurance contract, capital is passed from one reinsurance company to another. Reinsurance can, therefore, be regarded as a line of credit (as can insurance), or as a financial instrument with its value linked to claims experience (i.e., a derivative). The premium can be regarded as the cost of having the credit facility available or of purchasing the financial instrument.[3]

The idea of reinsurance foremost as a financial transaction, rather than as a risk transfer mechanism, has led to a number of innovations in the reinsurance market. For example:

The duration of contracts is frequently extended beyond the normal 12 months to as long as 5 years.

Policy conditions have been introduced limiting the cash flow payable in any given time period.

In the U.S., a futures market has been established (most notably those regulated by the Chicago Board of Trade; CBoT) with contracts linked to the experience of catastrophe claims.

---

[2]In exceptional cases, the circumstances illustrated in Figure 15.1 can result in losses being carried by a surprisingly narrow capital base. This occurred in the Lloyd's insurance market in the late 1980s (see the summations by Mr Justice Phillips (1994) on the Gooda Walker risks).

[3]We will see later the similarity between a call option and excess of loss reinsurance.

In the same way that the value of a traditional insurance contract to a policyholder is correlated with a claim, one can devise capital market instruments with characteristics that are, in various ways, similarly correlated with events of concern to insurers and reinsurers. CBoT futures are examples of instruments of interest across an entire market place, but alternatives of more specific interest to individual reinsurers include capital raising through the bond markets, with bond redemption and interest profiles dependent upon local phenomena or linked to underwriting performance. Clearly, the economic question in such cases is the relative cost of raising capital and servicing it through a bond versus the cost/benefits of a more traditional reinsurance arrangement.

Some of these innovations are discussed below, but first we consider reinsurance from the more traditional perspective.

15.2 Contractual Relationships: A Comparison of Co-Insurance and Reinsurance

In general insurance the terms co-insurance and reinsurance refer to significantly different ways of arranging insurance cover. Large risks, such as an oil-drilling platform or a jumbo jet, are individually of such magnitude that few single insurance companies could consider insuring the whole risk without some method of sharing the exposure. A reinsurance arrangement is one option. Whilst the insured has a contract with a single insurer, the insurance company cedes part of the risk to its reinsurers. An alternative possibility is for the party seeking insurance to arrange cover with several insurance companies, each independently taking a share of the risk. Such an arrangement is called *co-insurance*. The significant difference between co-insurance and reinsurance is in the contract that the insured has with its insurer(s). This is illustrated in Figure 15.2.

Under a co-insurance agreement, the separate contracts of insurance with different insurance companies mean that if one of the co-insurers is unable to meet its financial obligations, then the insured effectively bears the portion of the claims that would otherwise have been recoverable from the insolvent insurer. Thus, in Figure 15.2, if the insured had co-insured its risk, and placed 40% of the risk with insurance company X, which had subsequently become insolvent, then the insured would only be able to recover 60% of the claim from insurance companies Y and Z. In the reinsurance example, the insured has only one contract of insurance and a failure of reinsurance companies B or C is first and foremost the problem of insurance company A; it could ultimately become a problem for the insured if A is also financially impaired as a result. Occasionally, protection against the failure of A is afforded by a "cut through" to B and C

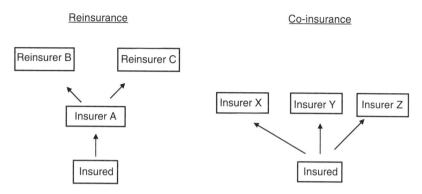

FIGURE 15.2 The difference between co-insurance and reinsurance

either contractually or as imposed by statute, which may vary with the class of business and the legal jurisdiction.

Traditionally, policies issued through Lloyd's of London are co-insurance arrangements. The Names that make up the syndicates operate financially on the basis of "each to his (or her) own." In principle, if one Name were to become insolvent (possibly for reasons unconnected with Lloyd's underwriting activities), then the other Names in the syndicate would not be liable for the failed Name's share. This principle has rarely, if ever, been tested, since such losses are, in practice, spread across the entire Lloyd's community by means of a levy. The shift in capital base in favor of limited liability corporate vehicles means that the issue of co-insurance is now less significant. However, the principle of mutualization of the impact of an insolvent corporate Lloyd's member is applied in the same way as for an insolvent individual Name.

Reinsurance contracts can also be placed on a co-insurance basis. If a cedant wished to purchase reinsurance of a given specification, the broker might attempt to place the risk with various reinsurance companies who could respond by participating in the reinsurance program to varying degrees. It is possible that the broker might fail to place the entire risk, e.g., 20% of the cover required might be unplaced. Then the cedant retains that part of the risk; effectively, the cedant acts as a co-insurer. Alternatively, the cedant itself might wish to retain a certain proportion of the layer, or the reinsurance company might insist that the cedant retains a proportion. Then, the reinsurance contract will have a specific co-insurance clause stating that the cedant is obliged to retain a prescribed percentage of all claims that are recoverable from the reinsurer. Whilst the net result appears similar, the response of higher levels of reinsurance to a contractual versus an accidental co-insurance can be materially different and would need to be clarified. Furthermore, co-insurance through nonplacement is less desirable from the cedant's point of view because of its unplanned nature and the consequent additional exposure on a net of reinsurance basis.

15.3   Reinsurance Arrangements

Risks can be ceded either on an individual, or *facultative*, basis or under a blanket arrangement applicable across an entire class of business, known as a *treaty* arrangement.

Under a facultative arrangement, each risk is underwritten by the reinsurer on its own merits with a separate reinsurance contract and policy terms. Occasionally, the broker will offer business with facultative reinsurance already in place. Facultative reinsurance is usually the first line of reinsurance defense and serves to standardize the risk profile of the insurer prior to applying the treaty reinsurance contracts. Difficulties in finding appropriate facultative reinsurance can lead to delays in the placement of risks with insurers and to increased costs. In some circumstances, it might not be possible to reinsure a risk facultatively and the insurer might be forced to retain more of the risk than it had intended; it could even have to decline to underwrite the proposed risk. The use of facultative reinsurance involves additional administration and can compli-cate the process of making reinsurance collections. However, administration systems would carry a "facultative flag" on the policy database indicating that claims could be subject to facultative reinsurance. Provided the structure of reinsurance is not complex, then the insurer's computer systems would automatically perform the relevant calculations and prepare appropriate advice to brokers and reinsurers.

Under the treaty system, a reinsurance contract includes certain stipulations regarding the business to be reinsured and the allocation of claims to a particular treaty. Provided these are met, the reinsurer is obliged to accept the business ceded to it under the treaty and, with limited exceptions, the ceding company is obliged to cede the business. Where the cedant can choose whether or not to pass on part of a risk, the treaty is referred to as "facultative obligatory" or, more commonly, *fac/oblig*. Where the cedant must cede the business covered by the contract the treaty is called *oblig/oblig*.

Typically, the treaty document would include the following terms and conditions:

the inception and expiry dates of the treaty;
the basis of operation (e.g., whether it protects policies incepting or losses
    occurring during the period of the treaty);
the territories covered by the treaty;
the classes of business covered;
any exclusions to the cover;
the retention of the ceding company;
the limits of cover;
when cover will be granted automatically by the reinsurer;
arrangements for paying premiums;

the ceding commission payable;

arrangements for profit commission;

terms for the payment of claims and the rendering and settlement of accounts;

a currency clause;

provisions to ensure the reinsurer has adequate access to the records of the cedant;

conditions for the termination of agreement and notice of cancellation;

an arbitration clause and the governing law.

In addition, the treaty document sometimes gives the reinsurer the power to manage the settlement of the original claim and occasionally prohibits an assignment of rights to the collections under the contract to other parties.[4]

Reinsurance is applied in two basic functional forms: proportional and nonproportional. Examples of each type are described in Section 15.4.

In general, proportional reinsurance acts to reduce the scale of the exposures and subsequent losses experienced. Thus, the shape of the statistical distribution of the claims at the level of application of the cover is the same net as it is gross, but with appropriately scaled parameters depending on the proportion ceded. The cedant would expect the mean and standard deviation of its net experience to be less than that of the gross experience in absolute terms, but would expect the coefficient of variation to remain the same (see Appendix A for proofs of these results). However, since the capital of the company is largely fixed (assuming the cost of reinsurance is not disproportionately high), the result of the reinsurance is that the variation of the claims experience relative to the capital base will be reduced.

Under nonproportional reinsurance the result for the cedant is usually equivalent to truncating the gross losses at some value: the deductible or excess point D. The truncation point could be a fixed value or a variable value that reflects some pre-agreed index, such as the Consumer Prices Index, or an experience-related index. The amount of claims transferred is usually restricted by a limit on the cover and, again, this can be fixed or variable; in some cases, there will be no upper limit restricting the amount of the reinsurance recovery. The protection can be applied on a per-risk basis, per-occurrence basis, or on some aggregate basis such as whole-account stop loss (see below). Under most conditions the effect of nonproportional reinsurance is to reduce the residual variance in the net account. However, the success of nonproportional reinsurance in achieving the basic objectives of reinsurance depends on suitable matching of the

---

[4]Under most financial contracts the owner of an asset can sell the benefits arising out of the contract to another party. Such trading is common in the capital markets and in commodity trading, and is becoming a more common transaction with reinsurance receivables as a way of managing credit risk exposure to particular counterparties.

gross exposures to the reinsurance protections purchased. This problem does not arise with proportional reinsurance since, in most cases, matching is automatic.

The statistical analysis of the effect of a nonproportional reinsurance is not as straightforward as that of a proportional cover (e.g., see Hogg and Klugman (1984)). A simple example of the effect of nonproportional reinsurance is given in Appendix A.

15.4   Descriptions of Each Type of Reinsurance

The principal types of proportional cover are quota share and surplus, and of nonproportional cover these are risk excess of loss, catastrophe excess of loss, and stop loss.

15.4.1   Proportional Reinsurance

It is normal for quota share and surplus reinsurance to be placed as treaties rather than facultatively.

15.4.1.1   *Quota Share*

Under *quota share* reinsurance all claims emanating from the category of business protected by the treaty are shared between the insurer and reinsurer in the same predefined proportions. Premiums are also shared in the same proportion, subject to an initial adjustment to reflect the initial commissions that the cedant will have paid to its agents or brokers and to reflect the additional expenses the cedant will incur through underwriting, processing policy applications, and handling claims. There might be further adjustments, such as a return commission, paid by the reinsurer to the cedant, or a profit sharing arrangement, which might only be instituted after an agreed period. In essence, however, the reinsurer is "following the fortunes" of the reinsured portfolio exactly and trusting that the underwriting and reserving of the insured is adequate.

Quota share reinsurance enables the ceding company to obtain exposure to a wider spread of risks than would otherwise be the case. For example, suppose an insurance company considers retaining a proportion α of a portfolio. If individual gross claims are denoted by the random variable X then the net claims X_N, will have a distribution defined by $X_N = \alpha X$. Assuming all policies are independent, the insurer effectively has the choice of retaining 100% of n (say) policies, or $\alpha \times 100\%$ of n/α policies (considering its exposure to claims, for example, the choice is between expected aggregate claims of $nE(X)$ or $(n/\alpha)E(\alpha X) = nE(X)$). The latter choice gives exposure to more policies, and so quota share reinsurance can enable greater diversification.

The exposure of a book of business to specific areas, such as a class of business or a geographical area where the management feels an accumulation of risk is becoming a threat, can be rapidly reduced using quota share reinsurance. It can also be useful if there is a need to reduce the statutory minimum solvency margin.

The administrative process of making reinsurance collections for quota share reinsurance is relatively straightforward. Collections would normally be made by advising the reinsurer of the total gross claims net of any reinsurance inuring[5] to the benefit of the quota share reinsurer (such as underlying facultative reinsurance). However, from the point of view of the reinsurer, the quality of data is frequently poor. Often, details of individual claim amounts and dates of occurrence are not provided. However, large claims, or claims arising from one large event, would normally be separately identified to help the reinsurer aggregate the claims arising from specific events in order to generate its own reinsurance recoveries.

Proportional treaties sometimes have a "cut-off" provision specifying that, at some predetermined time, the reinsurer must pay an amount to the ceding company in order to liquidate its liabilities under the treaty. Such arrangements are referred to as *clean-cut* treaties. The amount paid is determined by the cedant and will include a provision for incurred but not reported (IBNR) claims (see Section 16.2). This idea of a cut off is similar in principle to that of commutation, i.e., the early termination for commercial reasons, although when included as a contractual provision it is simpler to negotiate and to account for.

Clean-cut treaties tend to occur when there is a long-term, continuing relationship between the ceding and reinsurance companies. The payment made by the reinsurer to the cedant then forms part of the premium paid by the cedant for the succeeding year's reinsurance. The rest of the premium is the usual percentage of the ceding company's premium income for the year. Thus, for a continuing relationship, the result is that the liability outstanding from the previous year's reinsurance is transferred into the succeeding year's. This arrangement has advantages where the reinsurance is provided by a pool of reinsurance companies (rather than a single company) which changes from time to time, as it simplifies the administration for the ceding company.

Accounting for clean-cut treaties and interpreting their claims development can be difficult, as a single year of origin is actually an amalgamation of the claims arising from a number of past years of origin, each of which will be at a different stage of loss development. Added complications may also arise under certain accounting conventions, as the cut-off premium is paid to protect events that have already occurred, and so, for example, there is no unearned premium reserve element.

---

[5]"Inuring" reinsurance is a term referring to reinsurance that applies prior to the application of the reinsurance in question.

15.4.1.2 *Surplus Reinsurance*

Whereas quota share reinsurance scales down each risk in the same proportion, a surplus treaty is more flexible: the proportion of each individual risk reinsured might vary and, within the parameters specified in the treaty, is under the control of the ceding office. Under most surplus treaties, the reinsurer is obliged to accept the allocation made by the insurer. This is an example of fac/oblig cover, since the insurer can reinsure losses on an individual basis, at its discretion, whereas the reinsurer is obliged to accept the risks ceded (provided that the general provisions of the treaty are satisfied).

The parameters that define the amount that might be ceded are:

R, the maximum amount that the insurer may retain.

L, the maximum number of lines that may be ceded under the treaty. One *line* is equal to the amount retained; the maximum amount that can be ceded is LR.

It follows that the maximum exposure that the insurer can underwrite to be covered under the terms of the treaty is the retention R plus the amount ceded LR, i.e., $(1 + L)R$.

Having expressed the retention and the amount ceded as values, the proportionate allocations are calculated. It is these proportions that will be applied to the claims. For example, if the maximum amount LR is reinsured, this represents a proportion $LR/(1 + L)R = L/(1 + L)$ of the claims and premiums.

For a risk with an exposure E, say, where E is less than $(1 + L)R$, the insurer has some flexibility in the amount it retains. If the insurer wishes to cede the minimum amount, then, as the maximum retention is R, it follows that the amount ceded should be $E - R$. The proportionate allocation between the insurer and the reinsurer is R/E retained and $(E - R)/E$ ceded; the number of lines used is $(E - R)/R$. Conversely, in order to maximize the amount ceded to the reinsurer, the proportion $L/(1 + L)$, as defined in the treaty (using L lines), can be applied to the exposure, with $1/L$ the retained proportion.

Where the initial exposure exceeds $(1 + L)R$, an insurer wishing to use the surplus treaty must first reduce the exposure so that the residual amount is no more than $(1 + L)R$. The exposure can be reduced either through specific facultative cover purchased per risk or as part of another surplus treaty.

The flexibility available to the ceding company presents the reinsurer with moral hazard. The natural tendency of the ceding office would be to cede large proportions of the bad risks and retain the maximum proportion of good risks. Whilst the reinsurer might expect this to a limited extent, there is a high degree of trust implicit in the agreement. This assumes that the facility will not be abused and that the underwriting standards and claims management ability of the ceding office are adequate. Moral hazard can be controlled up to a point by imposing minimum retentions on the insurer in addition to maximum retentions. If the reinsurer feels that the ceding office is in some way acting outside the spirit of the treaty (e.g., by deliberately accepting and reinsuring some underpriced business from brokers in order to gain some highly profitable business which it retains), then the reinsurer can decline to renew the scheme. If abuse can be proved, then the reinsurer might seek some form of redress as provided under the terms of the treaty.

Measuring the exposure to be ceded to a surplus treaty depends on the class of business concerned. Surplus reinsurance cover is most commonly found in conjunction with commercial property fire accounts where there can be a wide spread of individual sums insured and types of business being covered, and a simple mechanism is needed to standardize the exposure profile. The most common measures are the sum insured or the *expected maximum loss* (EML). EML can be defined as an estimate of the most serious loss that can reasonably be expected from a single event arising from any insured peril. In a commercial property account, for example, this will vary according to the type of construction of the property, the neighboring buildings, the activities carried out on the premises, the standards of any fire prevention measures (including proximity to a fire station), and other physical risk controls. For example, the sum insured of a 15-story block of flats might be £20 million, but the EML (allowing for fire prevention measures) might be £5 million, reflecting the judgement made by a risk manager that it is highly unlikely that more than three floors would ever be seriously affected by a single occurrence.

No matter what the actual claims, the ceding company and the reinsurance company remain liable for their proportions. If a claim occurred which was substantially in excess of the EML, then the surplus reinsurer has the right to inspect the underwriting and evaluation standards of the ceding office and can deny coverage if there has been misrepresentation.

15.4.2 Nonproportional Reinsurance

The types of nonproportional reinsurance available vary widely and are constantly being refined by the major interested parties, i.e., brokers, insurance companies, and reinsurance companies. Changes are motivated by reinsurance companies' needs to improve their ability to set a correct price for the risk and to persuade the cedant that the structure and scope of coverage give an optimal balance between cost and benefit.

Nonproportional reinsurance is provided with the following contingencies in mind:

Coverage per risk. Protection in the event that a claim that is larger than could reasonably have been expected arises from an individual policy — the "wrong claim".

Coverage per event. Protection against an event such as a hurricane giving
   rise to claims on many individual contracts, i.e., protection against
   an accumulation risk;
Aggregate cover for the entire portfolio. An all-embracing protection for the
   portfolio, which might include cover in the case of inadequate premium
   rating, reserving risk, as well as the adverse claims scenarios described
   above.

The traditional forms of nonproportional reinsurance cover are known as
excess of loss and *stop loss*. Both provide cover once claims exceed a certain
level and they usually have a limited insured amount.

15.4.2.1   Risk Excess of Loss Reinsurance

Risk excess of loss is the common name for "per-risk" cover and is written as
a treaty. "Per risk" means that cover applies to individual claims arising
from any policy covered by the treaty.

   Excess of loss reinsurance provides cover for claims which exceed a
certain *excess point* or *retention* (sometimes referred to as the *attachment
point*), up to an agreed limit. The region from the excess point to the upper
limit of cover is known as a *layer of cover*. The cedant might arrange several
layers of risk excess cover in succession, each stacking on top of the other;
and the top layer might be unlimited.[6] Lower layers of cover will be
exposed to a higher frequency of loss, and in cases where the reinsurer is
regularly involved on a high proportion of claims the layer is referred to as
a *working layer*. Very high layers, relative to the exposure underwritten
by the ceding company, within which neither party expects a claim, are
sometimes referred to as *sleep easies*. Excess of loss reinsurance is often
arranged in several layers — this may be because it is possible to achieve
better prices at different attachment points and limits, or the layers might
be imposed upon the ceding company by the market if no individual
reinsurer were prepared to expose its own account to claims from the
cedant beyond a given value.

   A particular excess of loss reinsurance contract might be described as "£1
million in excess of £500 000." This means that the company retains £500 000
of each loss and can recover claims that fall in the region £500 000 to
£1 500 000, the maximum recovery in respect of any one loss being
£1 000 000, the limit. In general, if an excess of loss reinsurance contract is
bought with an excess of D and a limit U, say, then, if the gross claim is X,

---

[6]Certain classes of compulsory insurance require insurance companies to provide unlimited
cover. In the U.K. this is the case for third-party motor insurance.

claims net of per risk reinsurance are calculated as follows:

Gross	Net
$X \leq D$	X
$D < X \leq D + U$	D
$D + U < X$	$X - U$

The two main parameters which define the extent of the cover are, therefore, the excess point and the limit of coverage. Other characteristics which would be covered by the treaty include:

the number of reinstatements[7] and the premium payable on reinstatement;
the aggregate deductible (i.e., the number of losses, of value U, to the layer before cover is triggered);
the definition of the event that triggers the cover;
whether the losses covered are in respect of losses occurring during a given period, policies commencing during a given period, or claims made in the period;
the period for which the contract is exposed;
the amount of any explicit co-insurance or co-reinsurance;
whether there are conditions which the claims must satisfy before coverage is triggered (*loss warranties*);
in some circumstances, the limits could be indexed to some predefined formula, most commonly an inflation-linked index.

An explanation of some of the terms is given below.

Reinstatement Clause and Aggregate Limit — A layer of excess of loss reinsurance might have a maximum total collection, i.e., an *aggregate limit*; if the claims to the layer exceed the aggregate limit they revert to the cedant.

A *reinstatement clause* operates similarly. A number of reinstatements is specified in the treaty, restricting the total amount recoverable under the reinsurance contract. It is usually expressed as a multiple of the limit U and is a key factor in defining the scope of coverage. For example, suppose a contract provides protection for losses occurring during the period 1 January 2002 to 31 December 2003 and has a deductible of £1 00 000 with a limit of £3 00 000 and four reinstatements, i.e., the layer provides

---

[7]Frequently, excess of loss reinsurance contracts limit the total value of claims that can be made against the policy; this is expressed in terms of the number of reinstatements available. A reinstatement clause means that the policy will continue to be applied to subsequent claims under the conditions specified in the treaty, until the number of reinstatements permitted has been exhausted. The total financial limit is therefore $(1 + \text{Number of reinstatements})U$.

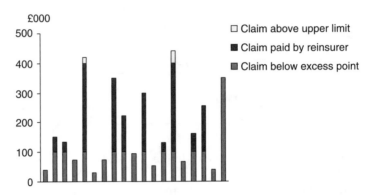

FIGURE 15.3 Claims and their ultimate values relative to the reinsurance layer

protection for five total losses to the layer. Then the maximum amount recoverable from the reinsurer is

$$£3\,00\,000 \times (1 + \text{Number of reinstatements}) = £3\,00\,000 \times 5 = £15\,00\,000$$

Figure 15.3 shows 19 claims, and their ultimate values relative to the reinsurance layer. A number of claims have not reached the attachment point, two used an entire reinstatement to themselves and the others are at an intermediate stage. The first 17 claims have exhausted the maximum recoverable; therefore, the last claim must be retained by the ceding company unless additional reinsurance is purchased. Insurers will monitor the performance of the reinsurance program throughout the course of the year and, if losses are perceived to be eroding the layer at a much faster rate than expected then additional sideways (i.e., at the same layer, rather than at higher layers), reinsurance protection (generally referred to as "backup layers") might be purchased part of the way through the year.

Usually, working layers of excess of loss reinsurance would have a large, or possibly an unlimited, number of reinstatements. The number of reinstatements purchased tends to reduce as the attachment point increases. For example, in a property account the number of reinstatements at higher layers would typically be no more than one or two, reflecting the low expected frequency of severe losses.

There is considerable variation in the premium conditions for contracts with reinstatements. For example, the contract may specify that reinstatements are "free," in which case no additional premiums are payable; but, more usually, the reinstatement of a policy requires an additional premium that is expressed as a proportion of the original premium. In the example above, a condition such as "two free, two at 50%" means that no additional premium is required for using the first two reinstatements, but a premium of 50% of the original premium is required for claims that fall into each of the third and fourth reinstatements. Under most treaties, claims are deemed to use the reinstatements first and the original cover purchased last, so that if reinstatement premiums are due, then the first claims to be collected under the treaty trigger the premium payment. Premium payment conditions also vary according to the degree to which a claim exceeds the retention, i.e., whether the full premium for the reinstated coverage is payable on the first claim under a reinstatement or whether reinstatement premiums are payable pro rata to the claims made in that reinstated layer. Also significant is the period the reinsurance treaty has left to run. A reinstatement triggered the day before the treaty expires may not represent good value, for example, and so, in some cases, the unexpired duration of the treaty is used as a further adjustment factor.

There are many other industry conventions, e.g., regarding currency conditions for claims arising in currencies different than the denomination of the layer in the treaty and the further complications if a claim arises in more than one currency. These details are beyond the scope of this text (e.g. see Kiln (1991)).

Aggregate Deductibles — This is a risk retention mechanism that defers the operation of the risk excess of loss reinsurance protection purchased. The concept is the complement of the aggregate limit, meaning that claims falling into the risk excess layer are retained by the cedant until their aggregate value exceeds a fixed sum — the *aggregate deductible*. The amount of the aggregate deductible can also be expressed in terms of the number of total losses to the layer retained by the ceding company before the excess of loss protection is triggered.

Contracts are available in which the deductible applies in a variable way, e.g., as a backup protection to other layers, being applied to the first layer suffering reinsurance exhaustion (known as "top and drop" protections or "cascade" covers).

Basis of Coverage — There are various bases under which a reinsurance contract can provide coverage:

Claims made basis. The claims made basis covers losses that are reported to the cedant during the contract period and which are promptly advised to the reinsurer. Usually a claims made basis has a retroactive date of inception which excludes claims that occurred prior to this date in order to avoid exposure to claims from business written in earlier time periods. Under certain conditions claims would still be allowed for a limited time after the expiry date, e.g., if cover is not renewed.

Losses occurring basis. Under a "losses occurring during" (LOD) basis, the reinsurer will pay claims if the date of loss falls within the calendar period covered by the reinsurance policy. Thus, a reinsurance company's experience of a contract written on a losses occurring basis should reflect the earned exposure of the cedant.

Policies incepted during basis. Under a "policies incepting during" (PID) basis the reinsurance treaty protects against losses on business written during the treaty period. The duration of exposure of a company reinsuring on a PID basis can be several years, and loss development is likely to be considerably longer than the other bases. The length of the exposure will depend on the terms of the policies written by the cedant.

As the claims that can be made against a layer of reinsurance on these different bases occur over different time periods, the loss development characteristics of the layer will be different under each basis. Under the claims made basis the events giving rise to a claim must be reported to the insurer during the term of the reinsurance policy, although the occurrence date could be many years prior (depending on the retroactive date). Thus, all claims are known about when the policy expires and settlement can be expected to occur relatively quickly. The LOD basis covers events that occur during the term of the policy, so there is an IBNR element that is not present under the claims made basis. Consequently, losses can be expected to take longer to settle than under the claims made basis. The PID basis is longer tailed again, since some of the claims that arise under the reinsurance policy will not occur until after the term of the policy has expired, i.e., there is an unearned exposure. The time until all losses emerge under a PID contract can be greatly extended if some of the underlying contracts written by the cedant have durations longer than 12 months.

It is important to ensure that a reinsurance program matches the gross exposures that the cedant expects to write. This is especially so if the cedant's reinsurance companies change the basis under which claims are allowed. For example, changing from a LOD basis to a PID basis could leave the cedant without reinsurance cover for losses arising in respect of policies written but unearned at the time of changeover; conversely, changing from a PID basis to a LOD basis could result in an overlap of coverage in respect of risks which are unexpired at the date of changeover.

The most dramatic financial consequences tend to occur when an insurer ceases trading, having purchased LOD reinsurance. In this case the reinsurer has unexpired risk reserves, effectively unreinsured; since the natural tendency of run-off managers of (solvent) companies is to protect the balance sheet, they are likely to seek some form of reinsurance. In such distressed circumstances, purchasing reinsurance for the open exposure is likely to be problematic and expensive.

Co-Insurance Clause — For layers with higher attachment points, there is sometimes a condition whereby the ceding company is required to retain a proportion of claims that would otherwise be recoverable. The motivation for such conditions is twofold:

It reduces moral hazard, since the cedant continues to retain exposure to claims in excess of the retention and will have a financial interest in ensuring that it maintains careful claims control.

As claims increase in value a greater proportion tends to fall to the reinsurance market. The London reinsurance market suffered badly in the late 1980s because claims arising from major catastrophes that occurred at that time spiraled from one reinsurer to another with little leakage out of the market until one of the participants exhausted its excess of loss reinsurance protections with disastrous financial consequences. Having a co-insurance clause ensures that large losses "leak" out of the spiral at each circuit, thus reducing the scope for such a "claims spiral."

Loss Warranties — A reinsurance policy might make payment of a claim under a particular layer of excess of loss reinsurance dependent upon the original loss satisfying certain conditions, in particular that the gross claim exceeds some prescribed value. For example, for aviation losses in the London Market, contracts with original loss warranties (OLWs) might require that the total loss exceeds $50 million, or that the hull value alone exceeds $50 million. Such conditions are designed to restrict claims to those arising from the sorts of event that the reinsurer intended to cover, in the latter case a large aircraft crash.

In order to minimize disputes on the size of the gross loss, a recognized external reference source is usually used.

Stability Clause — Claims inflation can have a highly geared effect on claims recovered under excess of loss reinsurance. For example, if a claim has an expected present value of £100 and inflation is expected to be 10% per annum, then the expected value of a claim arising at the year end will be £110. The resultant claim under an excess of loss reinsurance program with a deductible of £50 has increased from £50 to £60, i.e., by 20%. In this example, all the liability arising due to inflation has fallen on the reinsurer. A stability clause ameliorates this problem by increasing the deductible in line with inflation. In this simple example, if the deductible increases in line with expected inflation it will have increased from £50 to £55 should the claim arise at the year end, so that the effect of inflation is shared equally by the ceding company and reinsurer.

15.4.2.2 Aggregate Excess of Loss Reinsurance

Aggregate excess of loss reinsurance recognizes that the fundamental cause of a claim is not the policyholder, as in the case of the "per-risk" contracts, but a specific event. The per-risk excess of loss basis of cover probably satisfies most of the reasons for buying reinsurance when policyholders are independent. However, as a body of exposure grows it can become exposed to circumstances that cause claims from groups of policyholders to be highly correlated; in other words, insurers can build up exposure accumulations, sometimes unwittingly, across a wide range of classes of business. The significance of risk accumulation can be appreciated from Figure 15.3. Suppose, for example, that each loss arose due to a single event and, as a result, each "per-risk" claim can be aggregated together. Then the ceding company would have only a single retention of £1 00 000, rather than the total which was the result under the per-risk basis, of £10 00 000. In addition, because claims are aggregated, claims that previously fell below the retention would become recoverable from the reinsurer. Having suitable reinsurance, with sufficient vertical range to match these risk accumulations, restores the basic risk-spreading concept of insurance.

Events that give rise to accumulations of risk include:

Man made disasters:
  airplane crashes, which can give rise to claims under a variety of classes of policy, including airline property and liability policies, ground-based property policies, aviation product liability, and personal accident policies;
  an oil rig explosion, which could give rise to claims under marine hull, liability, and personal accident policies;
  a motorway crash or an industrial fire affecting neighboring property are examples of relatively minor situations where many policyholders might be affected; neither of these examples is likely to be of the scale of the first two.
Natural disasters:
  hurricanes, which can lead to claims under property insurance policies, marine policies, and aviation policies;
  earthquakes, which can lead to claims across a similar range of classes of business;
  storms and floods;
  tidal waves.

Events such as an earthquake in California or a hurricane making landfall in New York can potentially cause damage of hundreds of billions of US dollars.

The above list concentrates on the potential for claims to arise across a range of different classes of business for a direct writing insurer. For a reinsurer, these exposures are difficult to monitor, since many claims arise through writing retrocessional business where access to the original exposures may be impossible. In theory, aggregate exposure monitoring should be possible for those reinsurance policies issued to direct insurance companies, if access to the ceding companies' databases is allowed.

The event triggering the reinsurance must be clearly defined. In the case of property-related classes, this does not usually present a problem. For example, for aviation disasters, the date and time of a crash are unambiguous. However, for natural catastrophes there can be some ambiguity, as hurricanes, earthquakes, and winter freezes can last many days or weeks and the relevant reinsurance contracts usually restrict losses deemed to have arisen from a single event to claims occurring during a specified time period (see later).

In liability classes, it is far more difficult to find a precise definition that satisfies the basic principle of fortuity and which limits the insured event to be sudden, unexpected, and unintended. Insurance companies have developed contract language over many years to define words such as "occurrence," "event," "happening," and "common cause." However, law courts have interpreted the definitions inconsistently, generally broadening the intended scope of coverage.[8] In product liability claims resulting in mass tort actions, it is understandably difficult to identify the negligent "event" that caused injury. For example, it could be considered to be:

the management's decision to manufacture a product (which would aggregate all claims to a relatively early time period);

the manifestation of injury (which could lead to a spreading of claims through time as affected individuals show injury);

one of many other possibilities, each with dramatically different implications for insurers.

In many situations, the "event" definition in a reinsurance contract simply states that it follows the definition of the underlying insurance contracts, thus providing cedants with a matched reinsurance cover. However, in some cases the "event" definitions are subtly different, resulting in mismatches of reinsurance cover with the original exposure. The problems and arguments resulting from particular policies are often unique, so that existing legal precedents are not necessarily relevant. Also, many of the arguments are constantly changing; it is certain that many millions of dollars a year will continue to be invested in obtaining legal opinions that appear counterintuitive but which satisfy a particular purpose.

From the above discussion it can be seen that a typical excess of loss contract can be represented by the four parameters:

per event or each loss deductible;
aggregate deductible;

---

[8]It has been said that, in the case of claims arising due to certain major latent diseases, the scope was broadened cynically to ensure that the maximum amount of money was available to provide compensation.

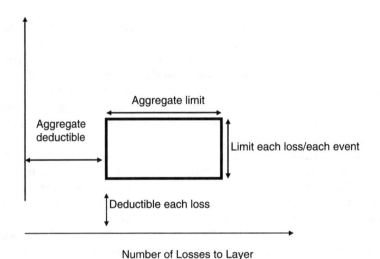

FIGURE 15.4 Graphical representation of a typical excess of loss contract

per event or each loss limit;
aggregate limit.

It can be represented schematically as set out in Figure 15.4.

15.4.2.3 Catastrophe Excess of Loss Cover

Catastrophe reinsurance, as its name suggests, gives the cedant some cover against claims arising from a catastrophe. By their nature catastrophes are not expected to happen very often, and so the contract is unlikely to have free reinstatements, although it might have automatic reinstatements, subject to a premium calculated under a formula specified in the original contract. The policy, which is available under treaty only, will provide for cover should an aggregate of claims exceed the (often very high) excess point, usually with a maximum limit. The aggregation will be of all claims arising from a specified event, which must be carefully defined so that there is little room for disagreement between the cedant and the reinsurer. The basic conditions of the treaty are:

the class or classes of business covered, e.g., domestic household insurance, whole account;
the perils covered, e.g., hurricane, flood, or earthquake;
the geographical scope, e.g., the east coast of the U.S.;
the time period over which damage occurs to constitute one event, e.g., damage caused in any 72-hour period (this time period may vary).

The other conditions are essentially those specified above for per-risk excess of loss reinsurance.

The period of cover starts at a time designated by the ceding company. Should the "event," such as a winter freeze, last longer than the 72 hours, and should the reinstatement provisions permit it, the reinsured may choose to trigger a second "event."

Catastrophe cover involves losses to a very large number of individual policyholders and, as handling the claims is beyond the capacity of catastrophe reinsurers, this falls to the ceding company. The reinsurer's function in such circumstances is to boost the capital base of the cedant. Since the ceding company has little financial interest in the cost of claims once the aggregate amount exceeds the deductible, there is a danger that they might settle claims generously in order to maintain good relations with their policyholders. Therefore, it is important to consider the moral hazard in catastrophe loss claims handling and for independent measures to be available to verify the quality of claims management. In an ideal world, the ceding company and the reinsurance company would have a long-term relationship and reinsurance premiums would average out any such misdemeanors by the ceding company.

15.4.2.4 Stop Loss

Stop loss reinsurance is usually the final part of a reinsurance program to operate; all other reinsurance policies inure to its benefit. It may be applied to the aggregate claims arising from a specific class of business or on a whole-account basis. The upper and lower limits are expressed as proportions of the total net earned premium during the period, i.e., in terms of the *loss ratio*:

$$\frac{\text{Net incurred claims}}{\text{Net earned premiums}}$$

For example, stop loss reinsurance might be triggered at a loss ratio, net of all other reinsurances, of 110% and have a limit of 20%; in other words, being exhausted when the loss ratio exceeds 130%.

The moral hazards faced by excess of loss reinsurers are amplified in stop loss reinsurance, as it is effectively a "catch all" protection which encompasses risks to which reinsurers at lower levels of the program are not exposed, such as:

reserving risk;
failure of underlying reinsurers;
poor management of expenses and brokerage;
inadequate pricing bases;
other management control and operational risks.

As a result of the extra perils to which the stop loss reinsurer is exposed, stop loss is rare other than as part of arrangements between a parent company and its subsidiary, where the parent may be supporting a new subsidiary, or to provide a mechanism for protecting the solvency of a subsidiary following some local financial problems.

15.4.3 Order of Application of Reinsurances

The process of reducing a gross claim to its net level can be represented by a system of mathematical operators. For example, suppose:

$f(X)$ represents the facultative reinsurance operator
$q(X)$ represents the quota share reinsurance operator
$e(X)$ represents the per risk excess of loss reinsurance operator
$c(X)$ represents the aggregate excess of loss reinsurance operator

Each operator may itself be a function of several different operators, particularly in the case of excess of loss reinsurance contracts. Now, ignoring stop loss reinsurance, if X is the gross claim or exposure and X_N the net, it follows that

$$X_N = c\big(e\{q[f(X)]\}\big)$$

The order in which reinsurance treaties operate can have a significant effect on the net result of the cedant, as well as on the recoveries made from the various reinsurers. To a certain extent, the process of reinsurance notification and collection can be automated, provided there is no ambiguity over the recovery route and functional operation of the program. However, frequently, claims can be presented to reinsurers in different ways and judgement is needed on the optimal choice. For example, the reinsured company might choose not to make a collection, since this might be better for longer term profitability. The retrocession program of a reinsurance company is, in many cases, its biggest asset, and whether reinsurance is processed purely automatically, or largely manually, the program in place must be well documented, controlled, and understood.

Reinsurance programs are complex, hierarchical structures, and succeeding layers of reinsurance benefit from the protection afforded by preceding layers. If a reinsurance company becomes insolvent and cannot pay a claim, then the impact on the remaining layers of reinsurance is the financial responsibility of the ceding company, with the possible exception of the whole-account stop loss. In effect, the ceding company not only underwrites risks at the gross level, but also has to guarantee the financial soundness of the reinsurance companies whose policies inure to the benefit of its other reinsurers.

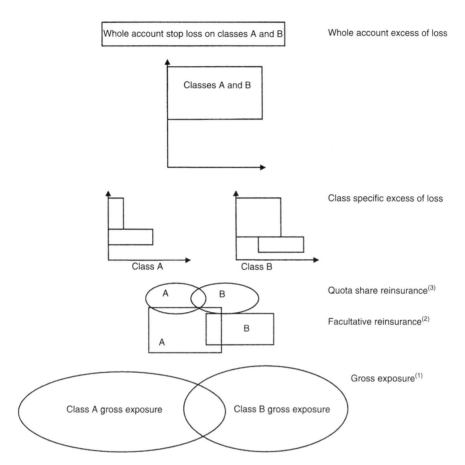

FIGURE 15.5 Schematic representation of a reinsurance program. *Notes:* (1) The figure assumes the insurer writes two classes of business, A and B, and that some policies in classes A and B are exposed to the same events. (2) Facultative reinsurance could be either proportional or nonproportional, or a mixture of both. It could respond to the overlapping exposure to certain events assumed to be experienced at the gross level. (3) The effect of quota share reinsurance is depicted as having the same "shape" as the gross exposure, but this would depend on the effect of the facultative cover.

A reinsurance program may be represented schematically as in Figure 15.5.

Considered from the perspective of an individual policy, within a class of business, the sequence of reinsurances transforms gross exposures or claims into net claims in a series of ordered steps, as shown in Figure 15.6, where each bar represents the transformation of an individual claim by each stage of the reinsurance process:

Facultative reinsurance is the first to apply. In this example, nonproportional facultative protection has reduced the gross claim. This tends to

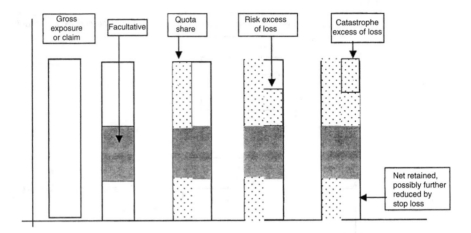

FIGURE 15.6 Transformation of gross exposures or claims into net claims in the reinsurance process

occur when a gross exposed to risk is greater than has been allowed for in the treaty reinsurance program; the residual risk is then covered by the treaty.

The loss or exposure has then been reduced further by a 50% quota share treaty.

The next to operate are the excess of loss treaties. The structure of excess of loss protections can be complex. There might be many layers of excess of loss protection on a per-risk basis, following the general structure of the portfolio of business. This example includes one layer above the facultative cover.

A sequence of "event" excess of loss protections or "whole-account" reinsurances might be placed above the per-risk excess of loss program. It is not possible to tell in advance how an aggregate excess of loss reinsurance will apply to an individual claim since, by definition, aggregate cover is only triggered when total claims in respect of an "event" exceed the deductible level. It is a moot point whether the subsequent recoveries should be attributed proportionately to all claims arising from the event or only to those claims recorded after the deductible level is achieved.

After allowing for these reinsurance recoveries, the net retained claim would be amalgamated with the balance of the portfolio. At an agreed point, often coinciding with the accounting year-end, the financial result of the whole operation will be assessed and, if appropriate, any stop loss reinsurance recoveries can be made.

15.5 Practical Considerations of Reinsurance Program Design

A portfolio of risks, both net and gross of reinsurance, can easily be modeled, given estimates of the statistical distributions of claim frequency, claim severity, and the correlations between the various lines of business coupled with the functional form of the reinsurance program. For planning and reinsurance design purposes, estimates of future experience, volumes of business, the likely exposures in the various classes of business, and the perils commonly insured are required. Some of these factors may themselves be treated as parameters in an all-embracing model. Detailed modeling can provide insights and useful feedback to the business control cycle, but the detailed frequency and severity analysis and the impact of alternate reinsurance structures does not normally feed directly into the wider business planning models, which tend to be based on much higher level business scenarios (e.g., see Hodes et al. (1996), and other papers in the same journal). The factors that must be taken into account in the design of reinsurance programs include:

past practice and the structure of the existing program;
underwriting guidelines, e.g., on maximum line sizes and exposures, and changes in these guidelines;
changes in the basis on which cover has been underwritten (e.g., changing from a claims made to a losses occurring basis);
expected business volumes and new business areas;
the availability of reinsurance cover, the cost of reinsurance, and the market cycle;
the desire to develop relationships with certain reinsurers or cedants and opportunities for reciprocation;
operational and financial considerations, including the probability of ruin, the need for profit stream stability, external legislative changes, tax changes, and the perceived return on equity;
relationships with the reinsurance broker;
changes in management structure, particularly for an organization that is part of a larger group.
wider commercial motives of the reinsured (rapid growth, capital raising, sale of the business, etc.).

Most reinsurance treaties are specified and purchased prior to the cedant's accepting risks on its own account. Therefore, it is important for both parties that the expected exposure and resulting gross premium income which is to be subject to reinsurance cover is estimated accurately, or that the arrangements for payment of the reinsurance premium have adequate safeguards to protect against poor estimation. The premium payable to the reinsurer is usually expressed as a percentage of the gross premium in the class of business to be reinsured (the *subject premium*) and, in addition, a minimum and deposit premium are also specified. The *deposit premium* is the initial payment made by the ceding company to the reinsurer. The final payment is based on the total volume of subject premium, which will be adjusted at agreed accounting dates during the term of cover, such as every 90 days. The total reinsurance premium payable is subject to a *minimum premium*, which is specified in the initial contract. The minimum and deposit terms imposed by reinsurance companies are usually flexible, the two running in an inverse relationship: the lower the deposit the higher the minimum, and vice versa.

We now consider the factors listed above in more detail.

15.5.1   Past Practice and the Structure of the Existing Program

Continuity and stability are important considerations for the management of a well-established and mature account. However, maintaining stability for its own sake would be a mistake, and reinsurance arrangements should be reviewed periodically in order to ensure that they continue to achieve the basic objectives. Reinsurance programs evolve over time to match the core attributes of the cedant's business; the language and intent of reinsurance programs will have developed so that there should be no misunderstanding between the ceding company and reinsurer.

The development of internal procedures to handle reinsurance recoveries and the understandings reached with brokers and reinsurers over the treatment of particular types of claim or individual contentious cases are also important practical concerns.

15.5.2   Underwriting Guidelines, Maximum Line Sizes and Exposures and Changes in the Guidelines

Individual underwriters operate within a control framework determined by the management of the company. For example, the maximum exposure which they are authorized to write on any individual risk is limited. Such limitations reflect many influences:

the expertise of the individual underwriter;
the position of the company in the market for that type of risk (e.g., whether the company is perceived to be a leader in that field);
the profitability of that class of business a whole;
the reinsurance protection in place to protect the class.

Accumulations of exposure to common perils can be monitored at class level or measured centrally and, when they reach pre-set levels, underwriting might have additional restrictions, e.g., the profit requirement could increase once risk accumulation thresholds are reached.

If underwriting limits are closely specified (and adhered to), then reinsurance requirements can similarly be closely specified.

15.5.3 Changes in the Basis on Which Cover Has Been Written

Reinsurance can be regarded as a contingent asset matching the contingent liabilities being written, so it is essential that, should the terms and conditions of the inwards book of business change, this is recognized in the outwards protections. Changes to the reinsurance program cannot always immediately follow the inwards business, but sufficient controls should be in place to avoid writing business that has material exposure and is completely mismatched with the reinsurance available. For example, facultative reinsurance could be purchased to cover perils not included under the existing treaty arrangements. As part of the planning process, trends in the demands of the insured public should be anticipated and reflected in the design of the outwards reinsurance arrangements.

15.5.4 Expected Business Volumes and New Business Areas

Since reinsurance treaties are usually purchased before the business covered has been written, some flexibility is required by both parties. For example, the reinsurer must be prepared to accept more business than expected and maximum limits would be agreed in advance. If the subject premium exceeds the agreed maximum, additional reinsurance premiums would be payable (or else claims on the underlying risks would be excluded from the reinsurance cover). Increased volumes of business will have a disproportionate effect on nonproportional reinsurers in a manner similar to the impact of inflation when there is no stability clause.

Existing reinsurance arrangements might be restricted to established classes and areas of business and might not extend to new areas of business. For example, business in certain geographical areas might be excluded or the reinsurer might have political reasons for not covering certain risks, particularly if the reinsurer is state owned.

15.5.5 The Availability of Reinsurance Cover, the Cost of Reinsurance, and the Market Cycle

The business cycle and pricing levels in reinsurance markets can be different from those experienced by the direct market. Consequently, there will be points in the cycle at which ceding companies do not feel that they are getting value for money from certain reinsurance covers and they might decide to retain more business, reducing the amount of reinsurance purchased for certain sectors of the account. Similarly, reinsurance may appear cheap and the cedant might choose to increase the volume of inwards business to gear up, or leverage, its profitability at the expense of reinsurers; in these circumstances, reinsurance companies' security is paramount.

15.5.6 Desire to Develop Relationships with Certain Reinsurers or Cedants and Reciprocation Opportunities

A successful relationship with a reinsurance company can have several advantages. For example:

it can offer strategic business opportunities, such as alliances and partnerships in certain markets;
it gives the cedant the resources to develop technical solutions to common
   problems;
the cedant and reinsurer can develop new products of benefit to both.

15.5.7 Operational and Financial Considerations

The probability of ruin, profit stability, and return on equity all offer benchmarks against which the other factors mentioned can be measured.

Most insurance and reinsurance business is transacted in highly regulated markets and the impact of legislative changes can be wide rang-ing with often unpredictable consequences and impact upon the best form of reinsurance:

Legislation of the insurance market is primarily aimed at ensuring the
   solvency and good management of a company. The credit permitted
   for reinsurance in calculating the required solvency levels is usually
   restricted both in terms of the overall reliance and in the amount of
   overdue debt that can be held as a current asset. Changes in the rules
   may materially impact companies operating at their limit.
Legislators can impose premium taxes that alter the financial efficiency of
   alternative reinsurance strategies.
Revisions to the rules under which compensation awards are made can
   affect both the frequency and severity of claims, with consequent
   implications to the level of reinsurance required.

A requirement for equalization reserves could lead insurers to purchase less reinsurance [9] and have further implications upon the tax strategy of the company.

15.5.8 Relationships with the Reinsurance Broker

The relationships that insurers have with their brokers range from one where the program, the price parameters, and any security requirements are entirely specified by the ceding company, so that the broker's role is purely to carry out instructions, to one where the company relies heavily on the broker who effectively takes the role of a financial consultant, developing reinsurance strategies alongside the company's management. The broker might have technical skills and access to information beyond the means of individual insurers and can provide considerable added value to the reinsurance process.

The broker is fundamentally a sales operation driven by commissions earned from selling insurance and reinsurance; the broker's natural position, therefore, is to sell more, rather than less, insurance. This is a position that is not always aligned with the client's best interests.

15.5.9 Changes in Management Structure

The insurance industry is becoming increasingly consolidated as companies restructure and merge with or take over competitors; companies are constantly searching for more efficient financial strategies, e.g., in order to make group capital available for supporting localized profit opportunities. These trends lead to reinsurance being purchased on a group-wide basis so that improved economies of scale can be achieved whilst replicating the reinsurance requirements of local business units.

Internally, companies frequently "re-engineer" the way they operate, or change their business focus. For example, a company organized around product lines (i.e., based around its classes of business such as property, marine, and aviation), might deem that it is more appropriate to restructure along the business classifications of its customer base (e.g., industrial or leisure). It might then be desirable for the reinsurance program to match the new structure in order to assist internal profit measurement. In practice, this requires reinsurers either to restructure along similar lines or to package products with the appearance of a matching structure.

---

[9]The third EU Non-Life Directive has made the provision of equalization reserves from pre-tax profits mandatory for some classes of business.

15.5.10 Alternatives to Reinsurance

Reinsurance can be viewed in the same way as direct insurance, i.e., as providing security against the uncertain outcome of specified events. However, in some ways the reinsurance market is more sophisticated than direct insurance, and can be viewed differently. A reinsurance contract can be regarded as a contingent asset that matures when an event covered under the terms of the contract arises. Under this definition, a reinsurance program is an asset that perfectly matches the claim liabilities.

Reinsurance companies and reinsurance brokers (to some extent) consider the reinsurance policy as providing cedants with the equivalent of a bank account, which remains in credit unless claims exceed premiums. Similarly, reinsurance can be viewed as a line of extended credit whereby premiums are paid into the bank and withdrawn when claims occur.[10] The reinsurer, knowing that claims will not be payable immediately, might discount the premium charged to reflect the real value of the claims: effectively, an interest-bearing bank account.

The difference between the discounted and undiscounted value of liabilities coupled with a degree of control over the cash flows has been exploited with the introduction of various forms of "financial" reinsurances (Craighead, 1993). In particular, the *time and distance policy*, which in its simplest form is an annuity certain, provides a fixed, guaranteed income for n years in return for a "reinsurance" premium payable immediately. The premium is based upon a compound interest calculation based on a matched portfolio of fixed-interest securities (usually government stocks) and has the effect of capitalizing future investment returns on the premium in the year in which the contract was effected.

The cedant is faced with a cash flow issue, in that the premium for the reinsurance is paid out immediately. The capitalization of future investment income will boost the underwriting result and may lead to the triggering of a dividend or other additional cash outgo. In addition, the cedant has not matched the claims payment profile (although the design of the payment profile would normally be based on the cash flow expectations of the protected portfolio). This form of reinsurance initially evolved by exploiting differences between insurance regulations in different jurisdictions governing the permissibility of discounting general insurance liabilities for solvency purposes. This combination of factors caused regulators rapidly to discourage the use of such contracts for these purposes, and the design of such "finite" risk or "alternative" risk transfer arrangement has evolved considerably to overcome regulators' concerns whilst offering the cedant certain benefits. In general, finite-risk contracts must exhibit "true risk transfer" in order to qualify for accounting as a traditional reinsurance

---

[10] These ideas are just as true of direct cover as they are of reinsurance, but they have not been developed for the direct market.

contract. Contracts that do not meet certain criteria of risk transfer must normally be accounted as deposits and, hence, the cedant would lose the direct benefit of explicit discounting built into the finite contract. However, it should be noted that this does not necessarily mean that the finite reinsurance is not worthwhile — if the taxation position or the market perception of the company is enhanced, then the contract may have achieved its objectives.

Finite reinsurance has evolved into a mixture of traditional format reinsurance coupled with wider scale balance sheet support that may involve capital market instruments. There is, therefore, no clear definition of a finite contract, but the following are characteristics:

Future investment income is explicitly recognized.

Policy durations are usually for multi-year deals.

There may be many lines of business covered.

The terms may cover timing risk, underwriting risk, credit risk, and the asset risk aspects of the cedant's exposures, i.e., nonunderwriting risks can be integrated into the structure.

The deals are usually large financial transactions and will have been evaluated in great detail with complex modeling involving extensive simulation assessment.

The underwriting disciplines will include expertise in taxation, actuarial science, accounting, and corporate finance.

Careful consideration will have been given to the regulatory treatment of the contract, and in some cases the regulators will have been consulted beforehand.

Risk transfer for regulatory purposes will also have been carefully considered. In general, to avoid being accounted under deposit accounting rules (i.e. the benefit being included in the accounts at net present value), the probability of a significant financial loss to the reinsurer (i.e. 10 to 15% of the total premium) must be at least 10%.

There is frequently a bonus/malus adjustment built in to the terms of the deal.

Figure 15.7 summarizes the spectrum of reinsurances and where they can be characterized as "traditional" or "finite" reinsurance (Davis, 2001).

Considering reinsurance as a contingent asset has led to the development of a Catastrophe Futures and Options market on the Chicago Board of Trade. Derivatives contracts have been a recognized tool of risk management for some time, but it is only recently that financial futures have become commonplace. The Catastrophe Futures are designed to reflect the catastrophe losses experienced in areas of the U.S., based on estimates provided by a subsidiary of the American Insurance Association. By simultaneously buying and selling call options on a futures contract, at different strike prices, insurance companies can effectively hedge,

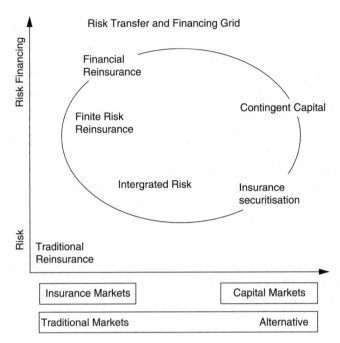

FIGURE 15.7 The spectrum of reinsurance contracts

or underwrite, their insurance risk. For example, by buying at a strike price consistent with a loss ratio of 60% and selling at a strike price consistent with a loss ratio of 70%, the insurance company is able to hedge the exposure within a tranche of risk in much the same way as effecting stop loss reinsurance.

Financial derivatives give insurance companies the opportunity to reduce their exposure to underwriting risk, just as reinsurance does. However, reinsurance companies provide important additional services, such as expert underwriting advice and reserving assistance. In addition, reinsurance can be matched perfectly to the cedant's experience, whereas the Catastrophe Futures depend on an estimate of average industry experience. The mismatch between an individual insurer's experience and the losses underlying Catastrophe Futures is an example of "basis risk"; this introduces uncertainty as to whether a hedge will be effective. The amount of basis risk (or the degree of mismatching) is fundamental to the success of insurance derivative contracts (Black, 1986).

In order to benefit from an insurance derivative contract, the insurance company must have a loss ratio no greater than the industry average. On the other hand, passing risk to a reinsurance company can reduce the incentive to the cedant to select policyholders with low expected claims. Reinsurers deal with this moral hazard by devising appropriate contracts and business practices, such as using underwriting audits and risk sharing. Futures contracts remove the moral hazard.

Other insurance derivative markets are emerging in the U.S. The State insurance department of New York has given approval to an exchange to facilitate insurance swaps, so that insurance companies will be allowed to swap blocks of policies in different parts of the country. For example, a company with too much exposure to hurricane losses in Florida could swap with another insurance company heavily exposed to earthquake losses in California. This type of risk management is similar to reciprocation in the reinsurance market.

A further innovation is in the over-the-counter market.[11] JP Morgan has written an option for a U.S. insurance company under which the company can draw on a $400 million surplus note facility. The money can be used to meet claims following a large loss, or to replenish the capital base of the insurer. The facility is similar to a bank credit line, but the insurance company views it as part of its reinsurance program. In a similar development, reinsurance companies are investigating policies with a "double trigger." Under these, for a claim to be paid, two events must occur simultaneously, e.g., a catastrophe loss occurring when the insurance company has a depressed solvency margin.

15.5.11 Practical Issues in Managing the Reinsurance Asset

The process of collecting cash under contracts of reinsurance demands close management of the relationships with brokers and reinsurers, good administration systems, and clear strategies by the cedant in the event of slow or recalcitrant payment, security downgrading, or other factor that may impair the value of the reinsurance asset.

Many regulatory regimes prohibit credit being taken for aged debt beyond 90 days overdue. Late cash recovery clearly has adverse cash flow implications and consequent investment income implications. It is axiomatic that slower debt recovery rates are linked with increased credit risk.

Debt recovery is a delicate matter, with many underwriters reluctant to take legal action, in the form of arbitration proceedings against counterparties, if there are continuing commercial relationships elsewhere within the portfolio and often also for personal reputational motives.

---

[11]The over-the-counter market is the phrase used to describe individually tailored derivative contracts.

Close monitoring of the balances due from the various counterparties, including the performance of brokers, is the most important feature of ensuring that overdue debt does not get out of control. The most common approach to terminating arrangements with "problem" reinsurers is by way of commutation (i.e., an early termination of the contract for an agreed value that incorporates unpaid debt, case reserves, and IBNR). Under some circumstances reinsurance contracts have prescribed commutation provisions, and, in extremes, the basis of the commutation is specified in advance — for finite reinsurance contracts such an arrangement is relatively common. Otherwise, such cases tend to be for classes of business having predictable characteristics, such as health care and other annuity-style benefits. Debt management has more possibilities where the reinsurer counterparty is also a cedant to the company, in which case amounts due to and from the counterparty can be "set off." This is particularly useful if the reinsurer becomes insolvent, as the insolvency practitioner is normally obliged to allow a 100% set off against counterparties who are both debtors and creditors of the insolvent estate (i.e., where the debts are mutual), whereas pure creditors will get a proportional payment depending upon the eventual solvency of the estate. The capital markets can provide further hedges against default that are of interest to the creditors of near-insolvent (and occasionally insolvent) estates of publicly listed companies. The possibilities of coupling the capital markets with the counterparty relationships amongst a group of companies is a powerful means of mitigating the impact of financial failure of a reinsurer, and there is a growing market in debt factoring and in sophisticated reinsurance debt assignment for these purposes.

Commutations, both in respect of inwards contracts and of reinsurance contracts, carry their own particular problems if additional parties are involved. For example, an inwards commutation may be in respect of contracts which would normally be protected under a given outwards reinsurance program. The problem faced by the cedant in this case is persuading these other reinsurers to contribute towards the cost of the inwards deal, as they would have eventually done had the commutation not been transacted. The fundamental issue is that, in a compromise deal, particularly one involving the early settlement of IBNR, the reinsurers involved are unlikely to be legally compelled to share the cost of IBNR claims and they may also query the basis of settlement of case reserves. In these cases, it is crucial to effect the commutation within a legal framework that preserves, as far as possible, the cedant's ability to collect the associated reinsurance asset or to have agreed a settlement basis with the reinsurers prior to finalizing the inwards deal.

An unfortunate reality of insurance operations is that the skill and imagination applied to the management of the reinsurance asset is invariably far lower that that applied to an investment portfolio of equivalent value.

15.6 Conclusions

Reinsurance policies are at least one step removed from the original risk, so it is almost inevitable that information about the extent of risk will be less than that available to direct insurers. As one moves through layers of reinsurance it becomes increasingly necessary to make broad general-izations of the exposure of underlying accounts. This makes it difficult to establish statistically reliable models, and so reinsurance companies rely also on the following, general, considerations:

the past experience of the portfolio being protected;
the rates available in the market and, in particular, the ease with which they
    can cede their own risks;
the diversity of the inwards book of business.

Many long-term relationships between cedants and their reinsurers are built on the understanding that, if claims are higher than expected, future premiums will need to be increased; and if claims are lower than expected, the reinsurer will compensate the cedant in some way.

However, in some areas, such as the design of contracts, the reinsurance market is quite sophisticated. To some extent this has arisen because of the increasingly complicated world in which insurance companies operate, but it is also in response to competition from emerging capital markets. Reinsurance companies have to demonstrate that they can continue to provide a useful and valuable service to insurance companies. As more efficient data-handling packages are developed, more information will be available and the market is likely to become more responsive to its consumers, as well as to its own experience.

Appendix A

A.1 Proportional Reinsurance

Let X be the random variable denoting the size of the gross claims. Suppose that X has the distribution function $F_X(x)$ and the density funct-ion $f_X(x)$ and mean and variance μ and s^2 respectively. Assume the insurer cedes the proportion $(1 - a)$, so that the amount retained is the random variable $Y = aX$. Y will have the distribution function $F_Y(y)$ and the density function $f_Y(y)$.

A.1.1 The Relationship between the Mean and the Variance

$$\text{Mean} : E(Y) + E(aX) = a\mu$$

$$\text{Variance} : V(Y) = V(aX) = a^2 s^2$$

Thus, the standard deviation of Y is $sd(Y) = as$ and the coefficient of variance of Y is $sd(Y)/E(Y) = s/\mu$, the same as for X.

A.1.2 The Distribution of Y

We have that $Y = aX = g(x)$, say. Then $X = Y/a = h(y)$, say. Now

$$\Pr(Y \le y) = \Pr(\alpha X \le y) = \Pr(X \le y/\alpha)$$

or, more formally

$$F_Y(y) = F_Y(g(x)) = F_X(h_Y))$$

Thus

$$f_Y(y) = \frac{\mathrm{d}}{\mathrm{d}y} F_Y(y) = \frac{\mathrm{d}}{\mathrm{d}y} h(y) \frac{\mathrm{d}}{\mathrm{d}x} F_X(h(y)) = \frac{1}{\alpha} f_X(h(y))$$

And, consequently, $f_Y(y) = f_X(y/a)/a = f_X(x)/a$.

That is, Y has a similar density and distribution to X, but scaled by the parameter a.

A.2 Nonproportional Reinsurance

Let X be the random variable denoting the size of the gross claims. Suppose that X has the distribution function $F_X(x)$ and the density function $f_X(x)$ and mean and variance μ and s^2 respectively. Assume the insurer cedes claims in excess of the deductible D and with a limit U. Then the amount retained is the random variable Y where

$$
\begin{aligned}
Y = X \quad &\text{if } X < D \\
D \quad &\text{if } D \le X < D + U \\
X - U \quad &\text{if } X \ge D + U
\end{aligned}
$$

From this it is possible to calculate the moments of Y. For example

$$E(Y) = \int_0^D x f(x)\,\mathrm{d}x + D \int_D^{D+U} f(x)\,\mathrm{d}x + \int_{D+U}^{\infty} (x - U) f(x)\,\mathrm{d}x$$

In some cases it is also possible to derive the distribution (see Hogg and Klugman (1984)). "Closed"-form solutions, i.e., pure algebraic expressions, are increasingly unlikely to be useful as more sophisticated forms of reinsurance are designed (e.g., traditional reinsurance contracts coupled with capital market instruments), and greater reliance is placed on the results of simulation models with formulaic approaches used as backup checks on the results obtained.

References

Black, D. (1986). *Success and Failure of Futures Contracts: Theory and Empirical Evidence.* NY Salomon Brothers Center for the Study of Financial Institutions.

Bride, M. and Lomax, M. W. (1994). *Valuation and Corporate Management in a Non-Life Assurance Company.* JIA, Vol. 121.

Craighead, D. (1993). *Financial Reinsurance.* JIA, Vol. 120.

Davis, J. W. (2001). *A stochastic approach to recognizing profits of finite products.* Casualty Actuarial Society Reinsurance seminar.

Hodes, D. M., Neghaiwi, T., Cummins, J. D., Phillips, R., Feldblum, S. (1996). *The Financial Modeling of Property/Casualty Insurance Companies.* Casualty Actuarial Society Forum, Spring.

Hogg, R. V. and Klugman, S. A. (1984). *Loss Distributions.* John Wiley & Sons.

Kiln, R. J. *Reinsurance in Practice.* Willoughby.

Chapter 16

Reserving

16.1 Introduction

The *technical reserves*[1] represent the principal liabilities of an insurance company. They are established to enable the company to meet and administer its contractual obligations to policyholders. Specific reserves are required to meet indemnity or other compensatory payments to policyholders plus the associated administration costs. In addition, reserves of a contingent nature might be carried (e.g., *claims equalization* or *catastrophe* reserves) in order to provide a further buffer against adverse development of claims and to smooth the emergence of profit. The reserves of general insurance companies are not usually discounted, and this is generally regarded as a further safety margin.

The apparent profitability and solvency of a business is highly dependent upon the reserve level and the reserving philosophy. Most of the key financial performance statistics used by insurance company analysts depend in some way upon the reserve level. Reserving is therefore a fundamental aspect of business management. The insights that the reserving process provides into past claims performance and policy exposures can influence the terms and conditions offered on future business and are usually the basis of decisions to cease underwriting certain classes or to withdraw from insurance entirely and support alternative enterprises which are expected to offer better rates of return on capital.

Estimates of reserves are required for a variety of reasons, each of which might require a different approach and degree of conservatism. The main

---

[1] We use "reserves" to mean the amount required by an insurance company in order to be able to meet its expected insurance liabilities. Thus, reserves are required in respect of business written, both earned and unearned. This is different from the definition used by accountants, which has been discussed in Chapter 13.

purposes of reserving exercises are:

for the published accounts and disclosure of profit;
to provide management information and internal accounts;
for tax purposes;
to assist in sale and purchase negotiations;
to advise on portfolio reinsurance;
the negotiation of contract commutation;
for rate making.

In any actuarial work it is essential to identify the client and the terms of reference of the assignment. In most reserving work the actuary works with a range of possible outcomes and, depending on the actuary's commercial role, it is legitimate to argue a case within the range that best suits the purpose at hand, e.g.:

An actuary who is employed by a prospective bidder for a company will seek to obtain the best terms for his or her client. This would usually mean adopting a negotiating stance requiring higher, or more conservative, reserves than the actuary who is acting on behalf of the vendor might propose. Similar situations exist for contract commutation and portfolio reinsurance.

An actuary involved in the preparation of year-end reserves might be required to explain the reserve level to a number of different parties, each with a different financial perspective. For example, tax authorities might desire lower reserves and, consequently, higher profits in order to achieve larger tax income; this might conflict with the objectives of solvency regulators, who would prefer insurers to retain more conservative reserves and to delay profit releases.

The degree of conservatism in estimating reserves also depends on the nature of the client. For example, an insurer with an overriding obligation to ensure that policyholders' valid claims will be paid, operating in a highly sensitive market (e.g., some form of guarantee fund), would tend to have larger margins than might otherwise be the case.

16.2 Reserves Carried

The reserves held in respect of insurance-related liabilities fall into the following categories:

Reserves in respect of unexpired or unearned exposure:
unearned premium reserve (UPR)
deferred acquisition costs (DAC)
additional unexpired risk reserve (AURR)

Contingent reserves
  catastrophe reserves
  claims equalization reserves (CERs)
Reserves in respect of earned exposure
  notified outstanding claims, which can be subdivided into
    notified (open) claims
    reopened claims
    incurred but not reported (IBNR) claims
    incurred but not enough reported (IBNER) on existing notified claims
Provision for claims handling costs

These categories of reserves are discussed below.

16.2.1 Reserves in Respect of Unexpired or Unearned Exposure

A liability for unexpired cover arises because, in general, the accounting period for reporting purposes does not coincide with the end of the policy period for the business accepted. The insurer, therefore, has contractual obligations to provide coverage beyond the accounting date. UPRs relate almost exclusively to business accounted for on a 1-year basis and will be substantial for lines of business covering, for example, extended warranty guarantee or large construction projects that have a policy duration of greater than 1 year. Funded accounts do not have a UPR.

Figure 16.1 illustrates this concept. The beginning and end of the accounting period are denoted by $t = 0$ and $t = 1$. Three policies have been identified, two of which have expiry dates falling beyond the accounting date, $t = 1$, and the potential claims that might subsequently arise are provided for by the establishment of a reserve for the unexpired exposure.

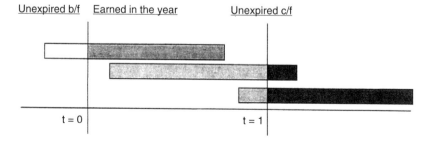

FIGURE 16.1 Unearned premium reserves

Circumstances can arise when no unexpired cover exists at the accounting date, for example:

a company providing holiday insurance cover, where the contract duration is generally short, and having a financial year end outside the holiday period;

certain types of commercial insurance and reinsurance contracts have cover beginning at particular dates that can be chosen to coincide with the insurer or reinsurer's year end;

it can suit a captive insurer to provide cover to its parent company such that no UPR is accountable at year end. The motivation is likely to be one of accounting simplicity.

The nature of the reserve for unexpired cover is twofold. First, it represents the provision for claims and associated expenses once cover becomes earned. Second, under a break-up or policy cancellation approach, it can be thought of as representing the refund due to the policyholder if the company ceased trading at the accounting date. These two concepts lead to the sometimes confusing use of the terminology "unearned premium" and "unexpired risk" reserve. The essential difference is that the reserve for unexpired risk could exceed the UPR if the premium basis upon which the original business was written proves to have been inadequate. In this situation the UPR is increased to a level consistent with the true risk profile of the business through an AURR. In other words:

$$URR = UPR + AURR$$

If the UPR is likely to exceed the insurer's liabilities on the unexpired cover, then URR = UPR and the terminology is interchangeable. Note, however, that any insurance profits implicit in the UPR are not usually recognized until the premium has been earned; this follows the accounting principle of prudence.

The basic approach to establishing the UPR is to use a proportionate approach:

$$UPR = \frac{\text{Period of unexpired cover}}{\text{Duration of original policy}} \times \text{Original premium}$$

This is clearly simplistic and certain additional features need to be considered.

The perils to which the unexpired cover is exposed might not occur uniformly throughout the calendar year, nor indeed throughout the policy period. For example:

Crop insurance is subject to its worst claims experience during bad weather seasons, particularly in climates where these coincide with harvest time.

Holiday insurance and extended warranty insurance tend to have claims correlated with the age of the policy. Extended warranty policies for electrical goods tend to have peaks of claim activity shortly before the policy expires, when there is an element of moral hazard and a desire on the part of the policyholder to ensure that any potential problems are solved whilst the expense can be passed to the insurer.

The nonuniform nature in which exposure is earned (by calendar date or policy duration) has to be analyzed in relation to the historic frequency and timing of the perils and the severity of the resulting claims, and a functional relationship established. For most practical purposes, simple models are used.

The cost of acquiring business, in the form of commissions paid together with policy setting-up costs, can be a substantial proportion of the written premium (possibly as much as 25% of the original premium, depending on the class of business), and the UPR is normally calculated excluding these initial expenses to reflect their nonrecurring nature.

The UPR in respect of an individual policy can be then expressed as

$$URP = P(n)F(n)(1 - k)$$

where n is the date of policy inception, $P(n)$ is the premium received in respect of a policy written at time n, $F(n)$ represents the proportion of the cover not exposed for this policy at the accounting date, and k is the proportion of the gross premium lost through acquisition expenses that has been treated as earned (i.e., a revenue account expense) when the policy was written.

It would be unusual to calculate a UPR on an individual policy basis; generally, the premiums received are aggregated within similar classes of business by month or quarter and the UPR calculated on the assumption that the premiums have been received, on average, halfway through the period. For example, for business monitored on a monthly basis with a 12 month policy duration and with intensity of risk spread uniformly throughout the policy period, this approach gives rise to the "24ths" method, i.e.

$$F(n) = \frac{2n - 1}{24}$$

where n is the month in the accounting year of the inception of cover ($n = 1, 2, \ldots, 12$). Similarly, quarterly accounting will give rise to the "eighths" method.

It is common accounting practice to make explicit the impact of acquisition costs upon the UPR. This is achieved by expressing the UPR as if based upon the premium gross of acquisition costs and to show a separate asset representing the unearned acquisition cost. Using the break-up concept, this can be thought of as the amount of commission recoverable from intermediaries if the company had to refund premiums to all its policyholders. Under this scheme

$$\text{UPR} = P(n)F(n)$$

$$\text{DAC} = P(n)F(n)k$$

$$\text{UPR (net of acquisition costs)} = P(n)F(n)(1 - k)$$

If the company feels that it will be unable to recover its DAC upon break up, it might reduce the allowance made for the assumed DAC in order to reflect its security of recovery. The extreme case would be to assume a fully expensed UPR with nil DAC. Then the underwriting result would, effectively, suffer a deduction of the acquisition costs twice on the UPR. The expense element of the UPR thus calculated would be released in the next year as the UPR is earned.

The above sets out a straightforward calculation based on individual or grouped policies gross of reinsurance. For practical purposes, the UPR is set up net of reinsurance. It is not always possible to identify the particular reinsurance premium or unexpired reinsurance cover that corresponds with either an individual policy or group of policies. For example, reinsurance that protects losses occurring during a given period relates to exposure earned, whereas UPR relates to unearned exposure. In such a case, the rein-surance in respect of unearned exposure might not have been purchased at the valuation date. If reinsurance is purchased to protect policies beginning during a period, unless it is placed on a facultative basis or as a proportional treaty, then the rate at which the reinsurance premium is earned, and can therefore be used to net down the UPR gross of reinsurance, must be decided. Otherwise, the UPR will not reflect the true unearned (net) liability at the start of the following accounting year. Where the reinsurance purchased does not match the exposure period of the original business, the unexpired risk reserve can be thought of as the unexpired risk reserve of the gross written business plus the cost of buying the corresponding reinsurance, less the expected benefits of the reinsurance. In normal circumstances, this results in a value less than the UPR on the gross business and the preceding formula remains valid. However, the cost of reinsurance, particularly high-layer excess of loss protection, can be volatile, reflecting the recent experience of large losses. If market conditions are changing, or if there have been significant events after the balance sheet date but before the accounts have been finalized (such as major catastrophes which could influence the cost of reinsurance), then this information should be used to adjust the calculations. Further difficulties can arise in determining the net UPR when reinsurance is mismatched relative to the gross business (e.g., protecting losses occurring during the period) and particularly if the company ceased trading.

The foregoing uses the premium basis, with adjustment for acquisition costs, as a measure of exposure to risk in calculating the unexpired risk reserve. In multi-year finite-risk deals, the issue of profit recognition is more complex and not dissimilar to the approach in life office valuation. No consideration has been given to the other elements of the premium (profit and contingency loading) and these are assumed to be earned at the same rate as the exposure. In a profitable portfolio, this approach is conservative, in that it does not anticipate profit on unexpired cover. However, business is not always profitable, and it is necessary to recognize situations when the premium basis is inadequate and additional amounts are required in order to cover the deficit in the UPR. Inadequacies in the UPR might be due to:

underestimating the expenses of administering and acquiring the business;
poor premium rating ;
underestimating the coverage really being provided or the true exposure.

The additional amount for unexpired risks is particularly difficult to estimate, since the claims experience most relevant is that of the earned portion of the same cohort of business, which is immature. It is unlikely that the true characteristics of the claims distributions can be measured with any accuracy in anything other than short-tailed business. However, reasons for reserve inadequacies can sometimes be identified, e.g., if certain parameter assumptions in the premium basis are known to be different to the expected values. This might be the case if investment experience has been markedly different to that assumed, or if acquisition expenses were underestimated. It might then be possible to make reasonably detailed calculations of the extent to which the UPR is too low. If the problem relates to excess claims, then a detailed investigation of claims experience might give some insights into an appropriate level of reserve in respect of the unexpired cover; the appropriateness of this type of investigation will depend upon the class of business. A distinction should be made between adverse claims due to some specific event in the calendar period (e.g., as a result of a change in legislation) and a generally higher than expected level of claims due to a pricing or product design error. For liability classes, it is extremely unlikely that a deficiency in the UPR in respect of adverse claims experience would be identified at an early stage of the development of the cohort of business. The additional amount required will invariably be a crude approximation, and working solutions based on observations of claims trends in the portfolio relative to trends in premium rates, knowledge of the wider market, and the experience of other similar companies would then be applied.

The additional amount for unexpired risks can be expressed as

$$P(n)F(n)(1 - k)(L - 1)$$

where L (>1) is the true claim plus claim expense ratio, excluding acquisition costs.

In this formula, the premium is used as an exposure measure; but, in the underlying calculations, more relevant exposure measures, such as vehicle-years or payroll, would be used, particularly if the premium rate per unit exposure has been changing during the accounting period. It is normal to calculate L on a discounted basis when calculating the AURR so that it is consistent with the premium basis. Note that expected surplus (i.e., profit) on the UPR in one class of business can normally be used to offset deficits on other classes of business, i.e., under most regulatory regimes the AURR is considered at the corporate level rather than at the line of business level.

16.2.2 Contingent Reserves

Insurers occasionally set up reserves of a precautionary, nonspecific nature in order to provide additional funds should the emerging experience differ adversely from the assumptions underlying the main reserves. Such reserves are generally known as catastrophe reserves, CERs, or adverse deviation reserves. In the published accounts, contingent reserves are usually regarded as free capital, although the reasons for earmarking funds tend to be more specific than general "free assets." They could also be regarded as a form of internal reinsurance.

There are several reasons why contingent reserves might be desirable. For example, an insurance company that sets reserves on a best estimate basis might feel that some areas of the account are exposed to greater risk of underreserving than others due to the inherent uncertainties of estimation. Similarly, the insurance portfolio might not have achieved the desired corporate level of diversification, in which case dependencies between different sections of the account could lead to "portfolio"-level claims. In either case, establishing contingent reserves in profitable years means that such risks can be met, should they arise, and a smooth dividend stream can be maintained.

CERs (or fluctuation reserves) are usually established to enable insurance companies to smooth results over the duration of their business cycle, which stretches over several years. In addition, they can be required if there is an insufficient spread of risk across policyholders in one calendar period, reflecting the extended return periods of certain events and the fact that it takes longer than the usual 1-year account for a true measure of the profitability of the business to emerge. A potential solution to this problem is to achieve the spreading of risk through time by writing business with policy periods of longer duration, reflecting the expected return period of the insured perils (as in many finite-risk reinsurance arrangements). However, this approach is not always practical and has not yet been widely adopted, although it is becoming more common. The perils of concern are, for example, property catastrophe events such as hurricanes, floods, and earthquakes.

Consider an insurer writing earthquake insurance and assume that an earthquake has a 5-year return period that is reflected in the premium basis. An example of a simple revenue account is given in Table 16.1. Expenses, profit, etc. have been ignored.

The normal 1-year accounting basis requires the insurer to recognize a profit of 100 in each year when no earthquake occurs. In the fifth year a large loss occurs; unless the insurer has been prudent and retained part of the apparent profits generated in the previous years, then it might be unable to pay the claim.

If the insurer has accumulated a CER from "profits," which could be used to fund the eventual loss, using a basis consistent with the underlying claims pattern, then the overall profit stream would suffer less violent swings. This is shown in Table 16.2.

This table demonstrates how a fund can be accumulated at a rate consistent with the underlying expected claims when apparent profits are available and a recovery made from the fund in the year of a substantial claim. The example is an extreme one, in that a nil profit is declared in each

Table 16.1 A simple revenue account

Year of account	1995	1996	1997	1998	1999
Premium	100	100	100	100	100
Claim	0	0	0	0	500
Balance	100	100	100	100	−400

Table 16.2 Accumulation at a rate consistent with the underlying expected claims

Year of account	1995	1996	1997	1998	1999
Premium	100	100	100	100	100
Claim	0	0	0	0	−500
To/from the equalization reserve	−100	−100	−100	−100	400
Balance	0	0	0	0	0

year; hence, the results have been perfectly smoothed. In practice, residual profits would remain subject to variation, the extent of which will depend upon the objectives of the equalization procedure.

Many developed economies actively encourage or even prescribe the use of CERs, offering tax incentives to assist companies in their establishment. The rules and regulations regarding transfers to the reserves and the level permitted of such reserves follows from the philosophical position that was adopted when the legislation was introduced. For example, it might be felt that catastrophe losses should be protected by reinsurance and that equalization reserves are targeted at more general premium or management deficiencies or fluctuations related to the business cycle or, alternatively, equalization reserves might be intended to remove the impact of extreme individual events only. The propensity of insurers to establish separate contingency reserves depends on two main factors:

whether the law demands them;
the tax treatment of CER contributions.

Specific rules governing CERs generally address:

the maximum tax allowable contribution in a year;
the treatment of investment income attributable to the CER fund;
the maximum level of the reserve itself;
the conditions under which transfers are made out of the fund.

Insurers are faced by conflicting pressures as a result of CERs. Clearly, there can be conflicts between the solvency regulators and the tax authorities, the former possibly legislating for compulsory equalization reserves on certain classes of business and the latter denying any tax relief on the contributions. Also, if the CER is regarded as a technical provision, then the solvency level disclosed in the balance sheet might appear to fall relative to other technical provisions. Then, in order to maintain solvency levels, the free assets, excluding the CER, must be increased by drawing money from profits. However, if the CER is regarded as part of the free assets then a company might be prepared to reduce the residual free assets to reflect this situation, notionally maintaining the same overall free-asset level. The differing interpretations are of significance to insurers, since insurance company analysts might perceive these reserves differently; in general, companies with higher free-asset ratios obtain better security ratings from analysts and, consequently, attract more business and are able to charge higher premiums because of the greater security provided.[2]

---

[2]The development of the European system of CER is described more fully in Morgan et al. (1992).

As well as explicit contingent reserves, many companies hold implicit margins. In other words, there are certain factors that have been ignored or stated on very conservative bases so that, if made explicit or restated to a more realistic level, they would reduce the reserve requirement or improve the apparent solvency margin. The most significant of these are:

The tendency of general insurers not to allow for investment income and the time value of money in setting reserves.

The practice of many companies to show assets at the lower of purchase price and market value.[3] Many continental European insurers have substantial asset margins in their long-standing property portfolios.

Circumstances might also arise where there could be a conflict of opinion on the reserves needed for a particular claim. For example, an insurer might have obtained legal opinions on the potential outcome of a contentious court case with widely different financial impact (e.g., concerning the date of a loss or the number of occurrences relating to an event), and, when confronted with an either/or situation, insurance companies might consistently adopt a "worst outcome" reserving basis. It would be unlikely that every contentious case would go against a company and, consequently, some of the case reserves when settled will release a surplus.

There might also be situations where a company adopts a position of denying that coverage exists for the particular event concerned and carries no case reserve. This could be due to the legislative regime in a particular territory, which biases outcomes in favor of plaintiffs if an insurer has established case reserves. In other words, the mere existence of a case reserve can be deemed by the courts to be a tacit acceptance of liability. This situation could lead to substantial negative implicit margins, although, provided that case reserving has been consistent through time, IBNR provisioning should reflect the degree to which case reserves have been over- or under-stated. Insurance companies can circumvent this problem by calculating "block" reserves, not attributable to specific claims and, for internal management information, notionally allocating part of the free assets to the reserve. The residual free assets not allocated would then be increased to a level considered appropriate.

The issue of implicit margins cannot be considered in isolation. It can seem that insurance companies' treatment of such items is often illogical, but one needs to be aware of all facets of the liabilities and assets in order to form an opinion. There is, however, little doubt that organizations do effectively smooth results by managing the level of implicit margins from one year to the next. In practice, fundamental changes in the treatment of

---

[3]In the U.K. the returns required by the regulators (the Department of Trade and Industry) must state the market value of the assets (see also Chapter 3).

implicit margins are very difficult to reverse. For example, switching from an undiscounted basis to a discounted basis and back to undiscounted when funds become available could be difficult to justify to the taxation authorities.

16.2.3 Reserves in Respect of Earned Exposure

The process underlying the appearance of a claim in the insurance company's books involves:

the occurrence of an insured event causing a loss to the policyholder;
the policyholder being aware of the loss and subsequently advising the insurer via a claim form;
the insurer processing the claim form and establishing a case reserve which might lead to a payment.

In most classes of personal lines business (e.g., motor and household insurance), these stages are straightforward. However, in longer tailed classes, such as employers liability insurance, each of these stages can be subject to considerable delay and uncertainty. For example, in a claim for deafness induced through exposure to noisy working conditions, the occurrence causing deafness could have been one very intense burst of sound or protracted exposure to moderately elevated noise levels, possibly over many years; the event is not necessarily clear or easy to prove. The injury, impaired hearing or deafness, might not manifest itself until the injured party gets older, perhaps many years after the original event: the attribution of hearing impairment due to the normal aging process and the impairment due to the industrial injury can be subjective and medical opinion can vary considerably. Claim advice delays can be decades and, whilst it is imperative to ensure that genuine claims are paid, a proprietary company cannot be overgenerous with claim settlement, since it might need to recover money from its reinsurers, who could reject claims that are not valid. Also, an overgenerous claims settlement will impair profitability and ultimately the solvency level and the security of the company's other policyholders. It follows that the company will itself need to examine the validity of such claims, establish the basis upon which insurance will respond (typically, in such a case, whether a "manifestation" or "exposure" trigger is adopted[4]), and possibly seek legal redress to test the validity of the claim, which could introduce additional delays.

---

[4] A manifestation trigger means that the claim would be allocated to the insurance policy in force at the time the injury was known or diagnosed; an exposure trigger means an allocation of claims to those policies in force over the time of the exposure (in this case, to injurious noise levels).

Note that the establishment of a case reserve is not necessarily the same as accepting liability. The insurer could set up a case reserve and continue to dispute the claim aggressively; in many situations claims are agreed and payments made with no acceptance of liability. Insurance companies adopt different philosophies in recognizing disputed claims for reserving purposes. The principal concern is that, in a court action, the legal process producing evidence ("discovery") may weaken an insurer's defense if an explicit case reserve is found to have been established for the disputed claim, as noted earlier.

Delays in the claim recognition process are reflected in the reserves for outstanding claims. The basic categories of reserve consist of:

reserves in respect of advised losses;
reserves in respect of losses that have been incurred but not yet reported (IBNR).

Amounts in respect of advised losses can be categorized as:

specific case reserves;
reserves for reopened claims;
additional reserves in respect of case reserves believed to be inadequate (reserves for claims "IBNER" or IBNER).

The distinction between notified claims and IBNR claims is often not clear. Many commercial insureds advise their insurers of events that might never materialize as claims, i.e., such advice is made on a precautionary basis. The practice of insurance companies varies widely in the treatment of such claims. One company presented with a precautionary advice by its clients might set aside a specific case reserve, whereas another might set case reserves only where liability is reasonably certain. In the latter case the insurer might establish a nominal case reserve (of $1, say[5]) in order to ensure that the date of loss and the date first advised is recorded and to have an entry on its database should a real transaction arise. The IBNR element is established separately, which might give a total reserve broadly similar to the former situation, but the ratio of IBNR to case reserves could appear different. Movements in reserves from one period to another might be because of changes in existing case reserves (IBNER), the reopening of previously closed cases, and the delayed reporting of claims (IBNR). The precise terminology to describe the movement in reserves requires careful definition and, in practice, interpretations vary. For example, in many commercial classes and in reinsurance business the term "IBNR" is

---

[5]Claims handlers often record precautionary advice claims for unusual low amounts, for example, $1, $1.99 or $2.99 which carry particular meanings relating to the claim that are not recorded elsewhere.

frequently used to describe all reserve deterioration and includes new advices as well as IBNER and reopened claims (Chamberlain, 1989). It is important, therefore, that the claims-handling procedures and the philosophies behind the establishment of case reserves are understood when the reserves are being reviewed.

Specific reserves are established by claims adjusters examining the individual circumstances of a claim, or by statistical methods. Reserves for closed claims that might subsequently be reopened, for IBNER require a statistical treatment based on the expected trends. They are, by definition, beyond the scope of the case reserve.

The case reserve normally reflects the expected ultimate settlement value remaining of an individual claim against the insurance company. Usually, this would include expenses paid to external parties (e.g., fees for lawyers or external loss adjusters) involved in the settlement, although these amounts would occupy separate entries in the claims file. The basic processes of arriving at a case reserve are:

claim notification
file establishment
reserve estimation
loss adjustment
claim settlement
salvage and subrogation

These stages are discussed below.

16.2.3.1 Claim Notification

Under property insurance, claims notifications are generally advised by the policyholder using a standard claim form, often provided as part of the policy documentation. Liability claims are frequently advised by third parties or their representatives, who allege injury through the negligence of the insurance company's policyholder. Such claim advices might be extremely varied and their handling requires considerable skill.

16.2.3.2 File Establishment

The receipt of a claim form, or a more detailed claim report, leads to the establishment of a claim record on the computer system. The claim reference number and the associated policy number are key fields for subsequent statistical analysis and administration. Usually, the paperwork would also lead to the creation of a paper correspondence file.

16.2.3.3 Reserve Establishment for Individual Claims

The reserve estimation process varies greatly according to the class of business concerned, the severity of the claim, and the insured peril. In classes of business which are subject to high volumes of broadly similar claim types, such as motor, physical damage or household contents, companies might adopt a "formula" reserving approach whereby a case reserve is calculated automatically by the computer system depending upon the particular circumstances of the loss, e.g., type of car, nature of damage, or geographical area. A formula method automatically applied can include a consistent set of global variables, such as expected delay to payment, inflation, and interest rate assumptions. Such global variables can be changed centrally as expectations change, or to explore the sensitivity of the reserves to changes in the base parameters. The final settlement of a claim would be reviewed by a claims handler in order to ensure that excessive payments are not made and as a control on the formula estimation procedure. The formula approach has advantages in homogeneous and high-volume cases because the expenses of operating the claims department are kept to a minimum.

For bodily injury claims and liability claims, the complexity and potential amount of damages (both indemnity and punitive) awarded against the company mean that these claims are allocated more personal attention. Estimates might be sought from medical practitioners and other relevant experts in order to assist with the quantification of damages should the company concede the case. Insurance companies usually have "authority levels" for handling claims: as the amount of a claim increases, more senior claims managers become involved in evaluating and managing the claim. The settlement of highly contentious cases frequently involves some negotiation, and occasionally compromises are reached to avoid large legal expenses. For large commercial cases in which the insurance policy was jointly written by many insurers, such as in the London Market (including Lloyd's), claims negotiation and primary handling of the claimant is usually carried out by the lead underwriter on the original policy. The other co-insurers on the slip (the "following market") will, in general, accept the decision reached by the leader. Large claims might require insurance companies to liaise with other insurers participating on the insurance contracts of the policyholder, e.g., in order to agree the allocation of external expenses.

One of the problems of case reserving is that it is prone to subjectivity and, potentially, an inconsistent approach towards implicit inflation and other variables of a broader nature. Case reserves provide a useful basis for negotiating settlements for individual live claims, but by definition they cannot provide information directly on IBNR or IBNER claims. Individual case reserves are required to advise reinsurers of potential recoveries, although, for natural catastrophes such as flood or windstorm, initial indications of potential claims against the reinsurers would normally be based on formula estimates or more crude aggregate methods based on the severity of event and the sums insured or exposures involved. Such an approach is necessary due to the large numbers of individual claimants and the potential processing delays caused by these high volumes; it is important that the reinsurer is "put on notice" with an initial estimate as soon as possible. Late notice to reinsurance companies of outstanding claims potentially recoverable from them can prejudice the collection.

16.2.3.4 Loss Adjustment

The process of loss adjustment is often an intermediate stage between file establishment and case reserve estimation. For more straightforward cases, this might amount to no more than ensuring:

that the policy was in force at the time of the loss;
that the peril that caused the loss is insured under the terms of the contract;
that any estimates submitted for repairs or replacement costs are reasonable.

More complex cases might require an inspection of the damage or the commissioning of an investigation by external consultants.

16.2.3.5 Claim Settlement

The information available regarding a claim can at first be sparse and, consequently, the estimated claim amount might be subject to considerable uncertainty. As the claim matures, more data become available and a point is reached at which the claim can be settled equitably on the part of the insured and insurer. In general, the circumstances surrounding first-party property claims are far more clear cut than those of third-party liability claims. Contentious cases can arise where an insurer must use its discretion in deciding whether to agree a settlement prior to a court judgement or to defend a suit and take the risk of receiving a more costly adverse decision. The calculation of an appropriate amount involves many factors that depend on the class of business, e.g., in cases of personal injury the amount would be broken down between economic loss, such as medical expenses, loss of income, and legal fees, and noneconomic, such as "pain and suffering," "loss of company," and the "value" of the injury. For certain injuries, standard tables have been compiled to assist courts with the quantification of damages. Under certain legal jurisdictions, a system of punitive damages or awards against insurers for not acting in "good faith," which can be many multiples of the amount for compensatory damages, are a major incentive for insurers to pursue early settlement. Insurance companies sometimes include policy exclusions against punitive awards as a measure of protection against (occasionally arbitrary) adjudications, although, in practice, they are not always successful.

Claims in respect of property damage are usually settled by payment of a single lump sum. However, payments for personal injury or payments for large amounts are frequently made in stages, e.g., as a building is reconstructed or as an income benefit payable to an injured party. Such staged payments are referred to as *partial payments*, and it is important to ensure that the claims database identifies these payments as such. Normally, the case reserve would be recalculated every time a payment is made and, unless there is information to indicate otherwise, the revised case reserve will equal the old case reserve less payments made. A particular problem with staged payments is the basis on which the case reserves are established once the payment formula has been agreed. In many cases, the formula amounts to an annuity and discounted cash flow techniques are used, with the case reserve effectively equal to the net present value of the annuity. This need not be a problem in analyzing data, but, when the overall company reserving exercise is carried out, it is important to be aware of the extent of discounted reserves, particularly if the company sets its overall reserves on an undiscounted basis.

The pattern of claims becoming reactivated after being deemed "closed" is a particular problem with liability claims. For example, personal injuries can worsen over time, or the full extent of the injury might not have been appreciated when first "settled."

16.2.3.6 Salvage and Subrogation

In settling a claim, the insurer reserves the right to pursue other parties who were liable in causing the original loss. For example, in the case of a motor insurance vehicle total loss, the insurer indemnifies its own policyholder and might pursue a third party implicated as causing the accident for compensation; such a procedure is known as subrogation. In addition, by indemnifying the policyholder, the insurer acquires the vehicle and can sell it for scrap metal; this principle is known as salvage. The rules of salvage vary considerably by class of business, and for marine business, in particular, some of the governing principles date back many hundreds of years.

16.2.4 Reserving Exercise: The Basic Steps

Fundamental to the reserving process is an understanding of the uncertain nature of the technical reserves and the establishment of a framework to monitor progress as claims become paid and as new reserves are established.

The reserving process has the following principal stages:

Establish the purpose of the exercise, the terms of reference given, the client, and other likely recipients of the results.

Obtain background information and data sets for performing projections. This will include numerical data and descriptive information regarding the development and objectives of the business.

Analyze and check the data to identify any unusual features and reconcile the data to the published accounts or other reference points.

Clarify points of detail on the data and, if necessary, obtain more extensive data. A discussion will probably be required with the claims manager, underwriter, loss adjusters, IT staff, and, possibly, with the principals of the organization.

Perform projections, possibly using several reserving techniques.

Analyze and interpret the projection results and obtain feedback from the client. The interpretation of results should include, wherever possible, a comparison against an external reference or benchmark.

Finalize the projections and document the calculations and rationale for making specific assumptions, paying particular attention to the more subjective areas.

The wide range of contingencies that can give rise to claims, and the influence that factors beyond the insurer's control (such as taxes, social inflation, legislative changes) have upon claims, means that the ultimate level of claims can never be known with certainty until the last claim has been finally settled. The inherent error of reserve estimates can be thought of as comprising three principal elements:

Process error. Claims might, on average, be distributed by amount and through time according to some well-defined pattern or structure. However, since they are inherently variable, it might only be possible to express their ultimate level in terms of an estimate of the underlying distribution.

Parameter error. Since policy terms and conditions cannot, in practice, restrict the behavior of claims to predefined distributions, the estimation of the parameters of the distribution are not known precisely and, hence, are themselves subject to statistical estimation error.

Specification (or model) error. In describing the above error components it is implicitly assumed that, through some process of prior knowledge, the general structure of claims development is known (e.g., some models assume that incremental payments in respect of a particular underwriting cohort follow a lognormal pattern). However, there is no guarantee that a selected structure is fundamentally the correct one. Some other structure might be equally likely (e.g., see Daykin et al. (1994)).

A close fit between actual and expected claims year by year indicates that the model used to establish the reserves is a good one. Significant departures from the model require explanation and might lead to a revision of the model or the method used to make estimates. However, note that close correspondence to historic data does not necessarily imply that the model is adequate for forecasting future claims.

Whilst statistical methods can be applied to the majority of circumstances, if data volumes are low and if there are unstable trends, then the errors in statistical estimation can be significant. The actuary might be required to make subjective judgements and to provide an opinion that does not necessarily fall within the strict scope of what can be analyzed in a purely numerical sense. He or she might have to form a view or seek the opinions of other experts (e.g., lawyers) on the possible outcomes of issues that have not yet manifested themselves, nor are ever likely to, in a form that can be directly measured. Therefore, it is important to establish a logical framework that allows scope for both objective numerical work and subjective, judgmental, input. Detailed numerical calculations can provide a useful indicator of the inherent variability of a well-established and well-behaved class of claims, whilst a "what-if" investigation of the more subjective considerations gives added insights on the factors or processes that might be material in the ultimate development of claims. Uncertainty reduces as a class of business matures and subjective opinions become hard facts: the "funnel of doubt" surrounding the ultimate level of claims shrinks in absolute terms. For example, for North American environmental pollution claims, as legal disputes are settled and case law defines the specific circumstances of liability in the various U.S. legal jurisdictions, the claim process can be identified more closely and the uncertainty of the parameter values reduced so that the confidence with which estimates can be made should improve.

There are usually three types of problem in a major reserving exercise:

reserving a general class of business of reasonably homogeneous characteristic;

reserving for a specific event or an individual (usually large) claim;

reserving for a "phenomenon" (e.g., mass tort claims, asbestos, and pollution claims).

Unless there has been a consistent mix of experience in the past (which is expected to continue into the future), then the amalgamation of claims arising under these general headings could materially distort the projections for the account as a whole. The claims data required for each purpose should be identified separately and the most appropriate reserving approach taken.

16.3 Form of the Data

The analytical approach taken depends on the data that are available. If the data are inaccurate, or have been recorded in an inconsistent manner, then false signals can be generated that can cause any reserving method (most of which extrapolate past trends) to give a materially misleading view of the true behavior of the underlying claims.

Apart from problems of correctly identifying risks or claims as belonging to one class of business rather than another, one of the commonest problems of analysis is linking reinsurance recoveries correctly to gross claims. Sometimes this does not matter, e.g., when analyzing a motor or other large personal lines category where any catastrophe experience can be excluded from the data and examined separately. In other situations, there is no natural connection between an individual claim and a reinsurance recovery, e.g., for stop loss reinsurance or for other forms of reinsurance that are triggered once total claims accumulate to agreed levels. In these cases, the issue of reinsurance should be dealt with independently of the basic projections.

Data formats vary considerably, depending upon the particular organization, its systems, and the issues that it perceives to be most relevant. The information available to the actuary will depend upon the relationship between the actuary's client and the organization in question. An actuary who is employed by an insurance company might expect to have complete access to all the company data. At the other extreme, the actuary working for the bidder in a hostile takeover is only likely to have access to data that are publicly available from the target company together with statutory returns for other companies and other industry data that can be used to establish benchmarks.

Other constraints imposed might include the time available to deliver the analysis. Special care must be taken to document the limitations imposed on the analysis through this type of constraint.

Data can be loosely classified as either hard or soft.

Hard data are principally of a numerical type that is not a matter of opinion. Data might be specific to the portfolio of a particular company or might be information of a wider nature that could influence the development of the company's loss experience (e.g., inflation figures, general market conditions, exchange rates).

Soft data can be invaluable for putting the experience of a particular company in context, but it is qualitative rather than quantitative. It might include conjecture on the implications of political change, prospective legislative changes, or legal outcomes on individual claims. Assertions, made by the managers, about company procedures, which it might not be possible to prove from the data, also fall into this category, e.g., that the rate of claims settlement is changing.

Sometimes there is little to distinguish between hard and soft data. The extent to which information can be directly incorporated into the determination of the numerical parameters used to project future losses is a matter for the actuary to decide. Soft data are often useful to demonstrate to external scrutineers, such as auditors, that there is some justification for adopting a given reserving stance.

The hard data available within a company should include claim numbers and amounts, separated according to whether they are paid, closed, reopened, settled at nil cost, or outstanding. The payments in the period might similarly be subdivided into payments on previously existing claims or payments on newly advised claims. These values would normally be available by reference to a period of origin and a calendar period for the transaction. The claim amounts might be subdivided by type of payment (e.g., indemnity or expenses) and by claim type (e.g., physical damage or liability).

Ideally, the data will provide sufficient detail to enable a breakdown into homogeneous categories of business, i.e., classifications that might be expected to have broadly similar loss development characteristics. The data should also be provided in sufficient detail to enable any unusual items, such as items that are not representative of the class and items that lead to a material distortion of the projection result, to be excluded and addressed separately. For reserving purposes the data would normally be grouped by class of business, but they might also be gathered with specific reference to individual large policyholders or by reference to particular events (such as a hurricane) or by type of claim (such as asbestos-related claims or industrial deafness claims). As far as possible, the data groupings for reserving purposes should be designed to be compatible with the other operational requirements of the company (e.g., business planning groups, profit centers, or regulatory reporting groups).

Careful consideration must be given to the quantity of data required to justify a separate classification. Too little data could mean the projection method will not necessarily provide a representative indication of the true behavior of the category. But too large a grouping might include sub-classifications of business that behave very differently, distorting the parameterization of the model and leading to incorrect estimates. Reasonable estimates could be achieved for data grouped over classes of business where the business mix, the relative volumes of business from year to year, and the external factors operating on the account have been constant. However, a stable operating environment is the exception rather than the rule. Factors that might distort the data include:

changes in the scope of coverage which could lead to new insured perils; the new perils covered might lead to claims subject to different inflation
    rates which will disturb the development of the business;
changes in the deductible levels or other policy conditions;
external market changes, e.g., because of new legislation.

Changes in the classes of business written will have a similar, though more profound, effect on loss development as changes in coverage, but they are generally more straightforward to eliminate since the new class of business might be isolated and treated separately.

In addition to claims information, exposure-related information, such as premium income, numbers of policies, sums insured, or other measure as appropriate to the class of business, might also be available.

The corresponding reinsurance recoveries, by category of reinsurance (e.g., facultative, quota share, or treaty excess of loss), would enable historical data to be assembled on a net and gross basis. Further information detailing the reinsurance protection relevant to a particular category of business giving an indication of the program structure, deductibles, limits, reinstatement conditions, and premium adjustment terms and the balance of coverage remaining provides further insights into the likely behavior of the net account.

The data upon which projections of future claims are based are usually presented as a loss development triangle in the form of Table 16.3.

The observation period of origin is usually a year, but may depend on the purpose of the analysis and on the available data. The origin period is the opening reference point for a claims cohort and is commonly an accident or underwriting period (see Section 16.4). Exposure in respect of the accident or underwriting year i would typically be the premiums earned or written, but might include more sophisticated measures such as vehicle-years or turnover, depending upon the class of business. Further refinement of the data can be carried out at a later stage to obtain data that are more relevant to the particular reserving situation under consideration.

Notwithstanding the fact that this data format might not be suitable for all reserving situations, the results of all projections can be expressed in this form. The objective is to "complete the rectangle," i.e., to estimate values for the "Future" section of Table 16.4.

Projections reflect those items included in the historical loss development triangle. For example, if the data combine both claim indemnity and the

Table 16.3 Loss development triangle

| Period of origin (i) | Development period (j) | | | | | |
	1	2	3	...	j	...
1	$C_{1,1}$	$C_{1,2}$	$C_{1,3}$	.	$C_{1,j}$	
2	$C_{2,1}$	$C_{2,2}$	.	.	.	
3	$C_{3,1}$	.	.	.	.	
.	.	.	.			
L	$C_{i,1}$	.			$C_{i,j}$	
.						.
.						
.						

Note: $C_{i,j}$ represents the claims (numbers or amounts) for period of origin i and development period j.

Table 16.4

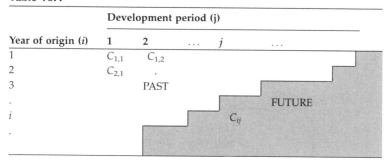

Year of origin (*i*)	Development period (j)				
	1	2	...	j	...
1	$C_{1,1}$	$C_{1,2}$			
2	$C_{2,1}$	.			
3		PAST			
.					
i				C_{ij}	
.					

associated loss adjustment expenses, then so will the estimates of the future experience. Conversely, if the base data represent only indemnity amounts, then the projected values need to be increased by a factor to reflect the expected cost of loss adjustment. Similar comments apply with regard to projections based on data net of reinsurance or net of late premium receipts (e.g., premiums processed after, say, 3 years being deducted from the claim payments after 3 years, as might typically be the case in many Lloyd's syndicates).

The validity of estimates based on development triangles of composite items is a matter of judgment. It depends on the stability of each component of the data, both by the period of origin and by the development period, and presupposes that past patterns will be continued in the future. For example, it would not be unreasonable to project claims net of quota share reinsurance, as quota share does not fundamentally change the distribution of claims through time. However, caution should be used when projecting triangles of claims net of excess of loss reinsurance. The net loss development pattern after nonproportional reinsurance can be profoundly different from the gross pattern, and small changes to the reinsurance program can lead to large differences in the development patterns by period of origin. In particular, such development triangles can produce an impression of stability that could be completely misleading as a basis for projecting future net claims if the reinsurance program is close to exhaustion.

A more general format for the loss development triangle is a rhombus (e.g., Table 16.5).

The data triangle could be truncated for several reasons:

data might not have been collected or might not be available by reference to the selected year of origin and development period before some calendar date, e.g., the date when a computer system was introduced;

Table 16.5 An example of the general format of a loss development triangle

Year of origin (*i*)	Development period (*j*)				
	1	2	3	4	5
1				$C_{1,4}$	$C_{1,5}$
2			$C_{2,3}$	$C_{2,4}$	$C_{2,5}$
3		$C_{3,2}$	$C_{3,3}$	$C_{3,4}$	$C_{3,5}$
4	$C_{4,1}$	$C_{4,2}$	$C_{4,3}$	$C_{4,4}$	$C_{4,5}$
5	$C_{5,1}$	$C_{5,2}$	$C_{5,3}$	$C_{5,4}$	
6	$C_{6,1}$	$C_{6,2}$	$C_{6,3}$		
7	$C_{7,1}$	$C_{7,2}$			

Table 16.6 Fundamental influences in the loss development triangle

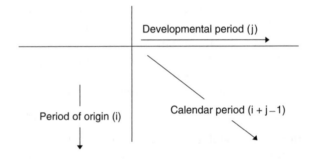

data might not be monitored beyond a certain development period on
   the grounds that the class of business under consideration is fully
   developed, or early years might be incorporated into successive years
   after a given development duration;
the insurer might have ceased trading for the selected class.

There are three fundamental influences represented in the loss development triangle and, ideally, these should be considered separately when
making projections. These are shown in relation to the loss development
triangle in Table 16.6.

Trends by period of origin predominantly reflect changes in underwriting
conditions, either specific to the insurer and class concerned or to the
more general underwriting cycle and the market within which the
insurer operates. The impact of the underwriting cycle is a matter of
scale, i.e., a unit of premium in the down cycle ("soft" market) is less
profitable (i.e., produces more claims) than a unit premium in the up
cycle ("hard" market). Changes in coverage, even seemingly small changes,
can materially alter the characteristics of the claims from one period of

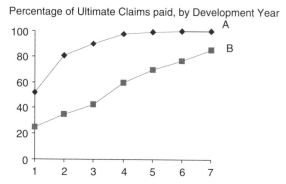

FIGURE 16.2 Percentage of ultimate claims paid, by development year

origin to the next (e.g., introducing a policy excess or altering the penalty on windscreen damage claims in a motor account). This will also be the case with changes in reinsurance, for claims being examined net of reinsurance. In many cases changes in coverage are a consequence of the cycle; in the down cycle an insurer would typically broaden coverage in order to maintain premium income volume and then narrow the coverage in the up cycle. Such practices could exacerbate the peaks and troughs of the cycle.[6]

Patterns by development period are broadly characteristic of the class of business; shorter tailed business reaches its ultimate value quickly, whereas longer tailed business develops much more slowly. Development and origin periods operate at right angles to one another in the data triangle and, therefore, in most models, they are assumed to be independent. Calendar period trends, however, operate on both development periods and periods of origin.

A starting point of projection methods is to derive from the data, perhaps allowing for external knowledge, a characteristic pattern of the development of claims for a typical year of origin. A profile of the shape given in Figure 16.2 might be expected to represent the development of cumulative claims. The percentages are calculated from claims paid, adjusted to remove distortions to the development caused, in this case, by inflation. Inflationary trends can arise due to several causes, including general price and wage increases, and must be related both to the period of origin of the claim and the development year in which payment takes place. However, data, particularly for case reserves, can be affected by administrative changes which might exhibit a "shock" rather than a trend: sudden changes to the development pattern could be caused by a change of claims manager,

---

[6]For personal lines business, coverage tend to remain reasonably constant and deductibles and sums insured would tend to move with prevailing inflation rates. This is not the case for commercial lines of business or reinsurance and retrocessional business.

changes of internal procedures for calculating case reserves, or backlog processing (i.e., a catching up which might be seasonal, such as in the first quarter following the company's financial year end, or might be in respect of accumulated backlogs of claim case files that might not have been the subject of review for many years), and purges of dormant case files by the claims department revealing historical errors which are then corrected all at once. Thus, it is important to develop a working knowledge of the internal procedures of the company and the environment in which it operates.

In Figure 16.2, the claims for class of business A are shorter tailed than for B, i.e., it reaches its ultimate level more quickly. Generally, property and first party claims are shorter tailed than liability or third party claims.

A general model of a claims "triangle" can be represented as

$$C_{ij} = F(N_{ij}, R_j, S_i, X_{i+j-1}) + \varepsilon_{ij}$$

where N_{ij} is a parameter for period of origin i and development period j, R_j is a parameter reflecting the development period (e.g., it could represent the proportion of ultimate claims observed at the end of development period j), S_i is a parameter associated with the ith period of origin (a scale parameter), X_{i+j-1} is a parameter reflecting the calendar point in time, and ε_{ij} is an error term. This formula is put into context in Section 16.5, where different reserving methods are discussed.

16.4   The Choice of Year of Origin or Claim Cohort

The basic groupings of claims into cohorts that define the period of origin are:

reporting period
accident period
underwriting period

Projection methods extrapolate the loss development within a period of origin to an ultimate value. The results of the projections and, in particular, the meaning of the projected future development (or, expressed another way, the projected ultimate less the cumulative position to date) depends on the choice of period of origin.

16.4.1   Reporting Period

Claims are grouped according to the period (usually a year) in which they are reported to the insurer. By definition, therefore, once the period is over, no new claims can be added to the cohort. The projection achieved by "completing the rectangle" represents the ultimate level of claims reported in the period of origin. The movement between the current level of claims and the projected ultimate level, therefore, represents the extent to which current case reserves have been over- or under-estimated (IBNER), together with the cost of reopened claims, with reopened claims attributed to the period of origin in which they were originally reported. Projections based on these data will make no allowance for IBNR claims or for unexpired risks, so that separate estimates would be needed for each of these in order to obtain a complete picture of the technical liabilities of the company. Special caution is also needed with the data, since any cohort will be a mixture of claims from a variety of underwriting and calendar periods, so that the cover provided and the specific perils included are unlikely to have remained the same over the exposure periods giving rise to the reported losses. Similarly, the environment (whether legal, social, or economic) in which policies gave rise to claims might also have changed.

Such a grouping of claims does not have a corresponding or straight-forward premium or exposure measure. Statistics relating the claims to the original policy's premium or exposure details might be available, but if a single policy gives rise to multiple claims with different reporting dates then it is not clear how the policy exposure should be allocated to the different origin periods (e.g., pro rata to the claims, or by allocating the full exposure to each claim, thus counting the exposure more than once), and care is needed in the interpretation of the resulting frequency and loss ratio statistics.

Grouping in relation to reporting year has a number of drawbacks. It is only appropriate for short-tailed classes and business written on a claims-made basis, or to provide information on the quality of case reserves and the potential IBNER component.

16.4.2 Accident Period

If the period of origin is the *accident period*, then claims are grouped by the period in which the accident occurred. This grouping is consistent with the usual 1-year accounting basis (earned premium) and reflects the experience of all policies that were exposed (or earned) over the same period. Claims development within an accident period reflects IBNER development on case reserves, reopened claims, and delayed advice (IBNR) claims. It follows that the projected ultimate level for each accident year includes estimates for all these items.

The claims recorded within the accident period all stem from the same exposure period, so that the broader economic and environmental influences on the propensity to make insurance claims is the same.

However, the claims themselves could arise from policies issued over a period of several years, depending on the policies' duration. (For policies with a duration of 1 year, the period over which policies included in the exposure could have been issued is two calendar years.) The coverage offered might have changed over this period, so that the actual claims themselves might not be completely consistent.

16.4.3 Underwriting Period or Policy Period

If the period of origin is the *underwriting period*, then claims are grouped in relation to the period in which the policy giving rise to the claim was underwritten.

Traditionally, triangulations based on an underwriting year have been used by Lloyd's syndicates, companies operating in the London Market, and reinsurers who have used the 3-year funded accounting approach. The underwriting year was allowed to mature for 3 years before requiring a definitive estimate of reserve requirement. Technically, Lloyd's syndicates are 1-year ventures whose composition changes each year and, therefore, the underwriting year is the basis for obtaining the profits for the business written by the members of the syndicate. Claims reflect a consistent underlying policy structure; however, in contrast to the accident period approach, claims arise from policies which are exposed over a period of up to two calendar years, or longer, depending on the duration of the contracts. Thus, the broader environment might not be as consistent as for the accident period definition. Furthermore, since the premium exposed (or earned) in the first development period can be low (relative to the premium written), the claims reported by the end of the first development year might also be low and, possibly, not representative of the ultimate level of claims.

The claim development within the underwriting period includes all the liabilities arising from the business written, i.e., IBNER, reopened claims, IBNR, and unexpired risks. This approach is therefore useful in evaluating the ultimate result for a policy rating series. Since the ultimate level includes an element in respect of unexpired risks, it is essential to test the implied cost of the unexpired risk. From an accounting perspective, it would not be prudent to declare a profit on the UPR, and this is normally prohibited by regulation.

Table 16.7 illustrates the points made above, and the development triangles in Table 16.8, Table 16.9, and Table 16.10 show how the different definitions of year of origin affect the result.

Note the delay of claims under each presentation: the development triangles represent the same information, but appear significantly different. The interpretation of reserve estimates that are calculated using the methods discussed in the next section depends fundamentally on the construction of the original data.

Table 16.7 Claims delays

	Calendar year		
	2000	2001	2002
Policy 1	▨	*	+
Policy 2		*	+
Policy 3			* +

Note: The shaded areas represent the duration of each policy; *indicates the date of accident; + indicates the date reported. The three claims were £100, £150 and £200 for policy 1, policy 2 and policy 3 respectively.

Table 16.8 Reported year basis

	Development year		
Reporting year	1	2	3
2000	0	0	0
2001	0	0	
2002	450		

Table 16.9 Accident year basis

	Development year		
Accident year	1	2	3
2000	0	0	0
2001	0	450	
2002	0		

Table 16.10 Underwriting year basis

	Development year		
Underwriting year	1	2	3
2000	0	0	250
2001	0	200	
2002	0		

16.5 Reserving Methods

This section considers different approaches to estimating reserves required in respect of outstanding claims.

The methods described are:

basic chain ladder
inflation-adjusted chain ladder
separation technique
average cost per case methods
loss ratio and Bornheutter–Ferguson method
operational time model
bootstrap method
a distribution-free method (Mack)
approaches to reserving using formal statistical models

Reserving methods can be expressed formulaically and there is frequently a judgmental overlay in the final selection of results. Methods should not be applied mechanically, since the portfolios being considered can vary for many reasons, not least being the basis of data capture, and the actuary should obtain a good understanding of this and of the underlying claims processes.

16.5.1 The Basic Chain Ladder Method

The most widely used reserving method is the chain ladder or link ratio method. There are many variations, but they have the same objective: to extract from the loss development triangle a pattern for the claims run-off that can, possibly with some additional manual adjustment, be used to extrapolate the less mature years of account. The method is very simple and is based on the assumption that proportionate relationships experienced between values in successive development periods in the past will repeat in the future.

The model underlying the basic chain ladder method can be expressed as

$$C_{ij} = S_i R_j + \varepsilon_{ij}$$

where S_i represents the ultimate level of claims for year of origin i and R_j is the proportion of the ultimate that has emerged by the end of the jth development period. C_{ij} here represents the cumulative claims to development period j.

In the basic chain ladder the R_j are derived using weighted averages of the ratios of cumulative values of claims in successive development periods. The method assumes that the cumulative position at development period j is the most appropriate basis to estimate the accumulated position

Table 16.11 Incremental paid claims

Period of origin	Development period					
	1	2	3	4	5	6
1	52 546	28 729	9 186	7 816	4 885	3 102
2	62 285	36 210	11 601	8 250	5 336	
3	72 173	41 126	11 041	8 543		
4	86 135	41 224	11 050			
5	97 068	53 408				
6	1 28 982					

Table 16.12 Cumulative paid claims

Period of origin	Development period					
	1	2	3	4	5	6
1	52 546	81 275	90 461	98 277	1 03 162	1 06 264
2	62 285	98 495	1 10 096	118 346	1 23 682	
3	72 173	113 299	1 24 340	132 883		
4	86 135	127 359	1 38 409			
5	97 068	150 476				
6	128 982					

at development period $j+1$ and, by deduction, the incremental movement from period j to period $j+1$. The method of calculating ratios of values in successive periods could equally be applied to incremental values; however, since the values become progressively smaller as an account matures, the ratios derived can also become volatile compared with the ratios based on cumulative data. Many practitioners argue that the use of ratios based on cumulative values makes future estimates overreliant on the early experience of any given period of origin applies. This issue is examined further in Section 16.5.8.

Consider the example shown in Table 16.11 and Table 16.12. The data are taken from Ackman et al. and are in respect of paid claims. The periods of origin have been indexed from 1 to 6.

It is useful to begin by plotting the data in order to obtain a visual indication of the development pattern, to see the consistency of the development for each year of origin and to identify any unusual features. The development of the cumulative claims is shown in Figure 16.3.

It is also useful to calculate the ratios between successive cumulative payments, within the period of origin (Table 16.13). This shows the proportionate relationship between periods of origin at different development points and is a useful diagnostic check on the factors of the chain ladder model.

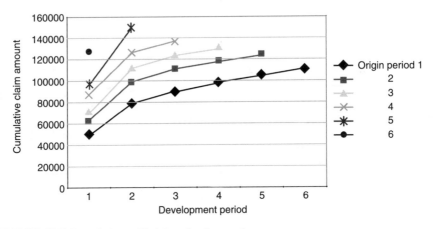

FIGURE 16.3 Cumulative paid claims development

Table 16.13 Ratio of cumulative payments in successive development periods

Period of origin	Development period				
	1–2	2–3	3–4	4–5	5–6
1	1.547	1.113	1.086	1.050	1.030
2	1.581	1.118	1.075	1.045	
3	1.570	1.097	1.069		
4	1.479	1.087			
5	1.550				
6					

Table 16.14 Calculation of chain ladder ratios

	Development period					
	1	2	3	4	5	6
Column sum	4 99 189	5 70 904	4 63 306	3 49 506	2 26 844	1 06 264
Column sum (Excluding latest value)	3 70 207	4 20 428	3 24 897	2 16 623	1 03 162	0

The volume-weighted development factors are calculated as

$$\text{Development factor } (j \to j+1), \ f_j = f_j = \frac{\sum_{i=1}^{n-j} C_{i,j+1}}{\sum_{i=1}^{n-j} C_{i,j}}$$

This calculation is shown in Table 16.14 and Table 16.15. The calculated average development factors are then used to complete the development triangle of period-to-period development ratios as in Table 16.16 and this

Table 16.15 The weighted average succession ratios (f_j)

	Development period				
	1–2	**2–3**	**3–4**	**4–5**	**5–6**
Ratio	$\dfrac{5\,70\,904}{3\,70\,207}$	$\dfrac{4\,63\,306}{4\,20\,428}$	$\dfrac{3\,49\,506}{3\,24\,897}$	$\dfrac{2\,26\,844}{2\,16\,623}$	$\dfrac{1\,06\,264}{1\,03\,162}$
	$=$	$=$	$=$	$=$	$=$
	1.542	*1.102*	*1.076*	*1.047*	*1.030*

Table 16.16 Ratio of cumulative payments in successive development periods (Table 16.3 completed)

	Development period				
Period of origin	**1–2**	**2–3**	**3–4**	**4–5**	**5–6**
1	1.547	1.113	1.086	1.050	1.030
2	1.581	1.118	1.075	1.045	*1.030*
3	1.570	1.097	1.069	*1.047*	*1.030*
4	1.479	1.087	*1.076*	*1.047*	*1.030*
5	1.550	*1.102*	*1.076*	*1.047*	*1.030*
6	*1.542*	*1.102*	*1.076*	*1.047*	*1.030*

leads to the completed cumulative period claims triangle[7]: in Table 16.17. The estimated reserve, Table 16.18, is found by taking the difference between column 6, the estimated ultimate claim value, for periods of origin 2 to 6, and the current cumulative paid claims. Note that period of origin 1 is assumed to be fully developed. The period-to-period development factors can be multiplied together to give factors that take claims from a given level of maturity to ultimate. The inverse of these factors represents the proportion of ultimate that is expected at each development period (Table 16.19).

From the base model of the chain ladder we have the estimates

$$C_{ij} = S_i R_j \quad \text{and} \quad S_i = \frac{C_{ij}}{R_j}$$

---

[7]Note that, if readers try to reproduce the calculations, there are likely to be rounding errors. Although calculated factors have only been presented to three decimal places, the exact factors have been used in the calculations.

Table 16.17 Cumulative paid claims

Period of origin	Development period					
	1	2	3	4	5	6
1	52 546	81 275	90 461	98 277	1 03 162	1 06 264
2	62 285	98 495	1 10 096	1 18 346	1 23 682	1 27 401
3	72 173	1 13 299	1 24 340	1 32 883	1 39 153	1 43 337
4	86 135	1 27 359	1 38 409	1 48 893	1 55 918	1 60 606
5	97 068	1 50 476	1 65 823	1 78 383	1 86 799	1 92 416
6	1 28 982	1 98 906	2 19 192	2 35 794	2 46 920	2 54 344

Table 16.18 Estimated reserve

Period of origin	Reserve
2	3 719
3	10 454
4	22 197
5	41 940
6	1 25 362
Total	2 03 673

Table 16.19 Calculation of development period factor R_j

Development period	1–2	2–3	3–4	4–5	5–6
Development factors (f_j)	1.542	1.102	1.076	1.047	1.030
	1–ultimate	2–ultimate	3–ultimate	4–ultimate	5–ultimate
Development period Cumulative factors	1.972	1.279	1.160	1.079	1.030
Inverse (R_j)	0.507	0.782	0.862	0.927	0.971

S_i is usually estimated using the latest calendar period information, i.e., in this case

$$S_1 = 1 06 264/1 = 1 06 264$$
$$S_2 = 1 23 682/0.971 = 1 27 376$$
$$\vdots$$
$$S_6 = 1 28 982/0.507 = 2 54 402$$

(Note that Table 16.20 gives $S_2 = 1 27 401, \ldots, S_6 = 2 54 344$. The differences are due to rounding the factor R_j to three decimal places in the above calculations.)

Table 16.20 Estimated ultimate claims at each development period

Period of origin	Development period					
	1	2	3	4	5	6
1	1 03 617	1 03 928	1 04 969	1 06 009	1 06 264	1 06 264
2	1 22 822	1 25 947	1 27 753	1 27 656	1 27 401	
3	1 42 320	1 44 877	1 44 281	1 43 337		
4	1 69 853	1 62 856	1 60 606			
5	1 91 412	1 92 416				
6	2 54 344					

Table 16.21 Incremental paid claims

Period of origin	Development period					
	1	2	3	4	5	6
1	52 546	28 729	9 186	7 816	4 885	3 102
2	62 285	36 210	11 601	8 250	5 336	3 719
3	72 173	41 126	11 041	8 543	6 270	4 184
4	86 135	41 224	11 050	10 484	7 025	4 688
5	97 068	53 408	15 347	12 560	8 417	5 617
6	1 28 982	69 924	20 286	16 602	1 111 26	7 425

If the trends within the data are stable, then the estimates of S_i should be constant regardless of the development period at which the estimate is made. The triangle in Table 16.20 gives estimates of ultimate $(S_{ij} = C_{ij}/R_j)$ using the above R_j at each point in the triangle.

This simple analysis provides initial insights into trends within the data. It also provides a crude indication of the stability with which the ultimate claims would have been estimated in the past and hence, crudely, the reserving error that could be inherent in using this method.

The model also provides a simple means of estimating the incremental future payments, Table 16.21, which are the differences of the successive values of Table 16.17.

The incremental cash flow triangle can be rotated through 90° to give projected future payments by calendar period (Table 16.22).

Having estimated future payments, estimates of reserves on a discounted basis can also be calculated. When discounting claims, a view is taken of the returns achievable on the funds that match the technical liabilities; clearly, the duration of the settlement pattern has a direct bearing on the effect of discounting. Generally, insurance regulators permit discounting provided that all future costs have been recognized, that the selected discount rate is conservative, and that there has been full disclosure in the company's

Table 16.22 Estimated future payments by calendar year

Period of origin	Calendar year of payment				
	1997	1998	1999	2000	2001
1					
2	3 719				
3	6 270	4 184			
4	10 484	7 025	4 688		
5	15 347	12 560	8 417	5 617	
6	69 924	20 286	16 602	11 126	7 425
Total	1 05 744	44 055	29 707	16 743	7 425

Table 16.23 Back-fitted incremental claims

Period of origin	Development period					
	1	2	3	4	5	6
1	53 888	29 214	8 475	6 936	4 648	3 102
2	64 607	35 025	10 161	8 316	5 573	
3	72 689	39 406	11 432	9 356		
4	81 446	44 154	12 809			
5	97 577	52 899				
6	1 28 982					

accounts (there may be specific statutory requirements, e.g., that the mean duration of the discounted business must exceed a predetermined level). In practice, most companies choose not to discount outstanding payment liabilities, taking the view that this provides a useful, implicit, margin against claims or expenses deteriorating significantly. Notwithstanding the fact that a company might choose not to discount, estimates of future cash flows are a fundamental aspect of business management resource planning and investment strategy.

As well as projecting future incremental payments, the results of the chain ladder model can be used to calculate the past incremental claim values that satisfy the model. The difference between the actual incremental values and the back-fitted incremental values calculated using the model gives the *residual errors*, a basic measure which can be used to test how appropriate models are, given the underlying data. In particular, residual errors can highlight specific regions of the triangle that are a particularly poor fit. This information can then be used to improve the estimation of the most appropriate loss development profile. The feedback approach can be driven by formal statistical methods or subjectively. The back-fitted incremental values and the residual errors are shown in Table 16.23 and Table 16.24 respectively.

Table 16.24 Residuals (actual minus expected)

Period of origin	Development period					
	1	2	3	4	5	6
1	−1 342	−485	711	880	237	0
2	−2 322	1 185	1 440	−66	−237	
3	−516	1 720	−391	−813		
4	4 689	−2 930	−1 759			
5	−509	−509				
6	0					

It was assumed in the calculations above that $R_j = 100\%$ at development period 6. That is, the claims process has fully matured at this point and so there will be no further movement. The "tail" assumption is a key assumption in the basic chain ladder, since all years of origin are subject to it. The assumption is particularly weak, since the development factor for the last development period might be determined by only two data points (as in this example). The sensitivity of the result and the lack of data supporting the assumption mean that caution is required in selecting the tail factor. The best approach is to arrive at a value using as many reasonable alternative measures as possible, for example:

Validate the assumption against the experience of other much larger companies with more historical experience.
Use the average factors in the preceding development periods as a basis for extrapolating beyond the pole positions of the earliest year. In many situations, the development period factors decay exponentially, so a satisfactory result can be obtained by using linear fits of the logarithms of the claim values. The model is often extended by fitting a smooth curve so that the chain ladder factors R_j follow a continuous function. The families of curves commonly used include:
  the power curve, i.e., $\log R_j = b^j \log a$;
  the Weibull curve, i.e., $R_j = 1/[1 - \exp(-aj^b)]$;
  exponential decay, i.e., $R_j = 1 + a\,e^{bj}$;
  the inverse power curve, i.e., $R_j = 1 + a(c + j)^b$.

Advantages of such curve fitting include the ease with which R_j can be determined at any duration and the ability to retain a family of development curves within, say, a reserving group with the possibility of reflecting small differences by flexing the parameters. A set of parameterized curves enables additional modeling for planning and other purposes to be simplified.

The data used in the above example are in respect of paid claims. A projection based on incurred claims (i.e., paid claims together with case reserves) contains more information and might be expected to reach an ultimate level at an earlier development period than the paid claims. So, projections should also be made using incurred claims. A comparison of the results and an analysis of the differences between the two provide insights into the factors that are influencing the patterns of loss development. Significant differences between the results are often more useful than agreement, since this can form the basis of joint investigative work with the claims and underwriting departments that enhances the detailed knowledge of the account.

The basic chain ladder method described is essentially a deterministic device and is not a rigorous statistical approach in the usual sense. However, it is a useful and simple framework, giving a practitioner some feel for the historic development and indication of the future (Verrall, 1991). Renshaw and Verrall (1994) have analyzed further the stochastic model that can be regarded as underpinning the chain ladder method.

A number of simple variants to the claim-weighted approach are possible. For example, it could be that certain features of the data in Table 16.13 can be explained in terms of factors that are not representative of the "normal" development pattern and, as such, might require manual adjustment (e.g., an acceleration of claim settlement in one calendar period). In such circumstances, the data may be refined to remove the feature with such items subsequently added back with separately determined estimates. Alternately, the average development factors can be recalculated using nonweighted averages or with problematic periods of origin excluded from the average factor estimation. This is straightforward when the calculations are set up on a spreadsheet, and it is a useful method of generating a range of results based on various "what-if" scenarios. Other variations include:

using the latest factors only to test the impact of developments in the latest period on the ultimate claims;

using all factors excluding the latest to examine the ultimate claims calculated using the previous year's projection basis;

if there are trends in the development factors, then the weights applied can be changed in order to give the more recent periods greater significance.

16.5.2 Inflation-Adjusted Chain Ladder Method

The basic chain ladder method does not explicitly allow for any calendar-year effect. The development factors calculated are based on payments from many different calendar periods; so, whilst inflation is implicit in the factors derived, it is not clear what assumption for future inflation is actually made, other than that it is a weighted average of past values. The loss development profile calculated, therefore, mixes together the true features of a class of business and external factors, which are independent of the pure claims process. If expectations for future inflation are very different from past trends, then the implicit allowance for inflation might be inappropriate for projections of future claims.

Making adjustments for calendar-period influences, such as inflation, is relatively straightforward if the inflationary rates that the claims have experienced are available. However, in stating the model it is necessary to consider incremental, rather than cumulative data, so that the calendar-year influence is applied to the correct datum.

The basic chain ladder model can be expanded into the form

$$c_{ij} = S r_{ij} \lambda_{i+j-1} + \varepsilon_{ij}$$

where S_i again represents the ultimate level for origin year i. However, in this model the currency value of S_i is set at a value excluding the impact of inflation; c_{ij} is the incremental payment made in development period j, due to year of origin i; r_j represents the proportion of S_i (in real terms) observed in development period j and λ_{i+j-1} is an inflation index for the calendar year.

In the example, assume that past inflation rates for this class of business were as follows:

1–2	12.4%
2–3	22.0%
3–4	21.9%
4–5	15.9%
5–6	13.2%

In order to complete the projection there are two more stages. First, paid claims must have the impact of inflation removed by placing the incremental payments on a standard money basis. Then, assumptions are required for future inflation rates, which are applied to the projected future incremental values. The results of these calculations are set out below. The inflation-adjusted incremental payments are set out in Table 16.25 (based on the incremental figures in Table 16.11). Their cumulative values are in Table 16.26, along with the development ratios; the method of their calculation is the same as for the basic chain ladder method.

Payments were assumed to have been made at the end of the calendar year in question, e.g.

$$c_{2,2} = 36\,210 \times 1.219 \times 1.159 \times 1.132 = 57\,911$$

These development factors calculated here are lower than those calculated using the basic chain ladder method. This is because the basic

Table 16.25 Incremental paid claims, having been adjusted for inflation

Period of origin	Development period					
	1	2	3	4	5	6
1	115 239	56 055	14 691	10 255	5 530	3 102
2	121 528	57 911	15 221	9 339	5 336	
3	115 427	53 957	12 498	8 543		
4	113 008	46 666	11 050			
5	109 881	53 408				
6	128 982					

Table 16.26 Cumulative paid claims, having been adjusted for inflation

Period of origin	Development period					
	1	2	3	4	5	6
1	115 239	171 294	185 985	196 240	201 770	204 872
2	121 528	179 439	194 660	203 999	209 335	
3	115 427	169 384	181 882	190 425		
4	113 008	159 674	170 724			
5	109 881	163 289				
6	128 982					
Development period	1–2	2–3	3–4	4–5	5–6	
Ratio (f_j)	1.466	1.079	1.050	1.027	1.015	

chain ladder factors make an implicit allowance for future inflation, based on past trends, whereas the factors calculated in Table 16.26 have no built-in element in respect of future inflation.

These factors are then applied to the current money cumulative values given in Table 16.26 to give the future incremental payments and the reserve requirement, in current money terms (Table 16.27). The reserve requirement in real terms may be thought of as a discounted reserve, where the discount rate equals the future rate of inflation (Table 16.28).

In order to calculate the nominal reserve, the incremental payments must be inflated at an appropriate rate. If we assume that, following an analysis of past inflation trends and a review of future expectations for the particular class of business, future claims inflation is expected to be 12% per period,[8] then the future payments are as given in Table 16.29.

The net present value of future payments can also be calculated (Table 16.30 and Table 16.31).

---

[8]The rates of interest used for these calculations were chosen to be consistent with the calculations in Ackman et al. There is no suggestion that they are appropriate in current economic conditions.

Table 16.27 Future incremental payments (current money)

Period of origin	Development period					
	1	2	3	4	5	6
1						
2						3 218
3					5 170	3 007
4				8 539	4 867	2 831
5			12 841	8 810	5 021	2 920
6		60 107	14 870	10 202	5 814	3 382

Table 16.28 Estimated reserve in current money

Year of origin	Reserve
2	3 218
3	8 177
4	16 237
5	29 592
6	94 375
Total	1 51 599

Table 16.29 Incremental payments, with future inflation of 12% per annum

Period of origin	Development period					
	1	2	3	4	5	6
1						
2						3 605
3					5 790	3 772
4				9 564	6 105	3 977
5			14 382	11 051	7 054	4 595
6		67 320	18 653	14 333	9 149	5 960

Table 16.30 Incremental payments, with future inflation of 12% per annum, discounted at 6% per annum

Period of origin	Development period					
	1	2	3	4	5	6
1						
2						3 400
3					5 462	3 357
4				9 023	5 433	3 339
5			13 568	9 835	5 923	3 640
6		63 510	16 601	12 034	7 247	4 454

Table 16.31 Comparison of reserves allowing for 12% future inflation and a rate of discount of 6%

Period of origin	Reserve		
	Current money	Nominal	Net present value
1			
2	3 218	3 605	3 400
3	8 177	9 562	8 820
4	16 237	19 646	17 795
5	29 592	37 082	32 966
6	94 375	115 415	1 03 845
Total	1 51 599	1 85 310	1 66 826

The inflation-adjusted chain ladder model adjusts the basic chain ladder to include an explicit allowance for inflation. When setting reserves on a discounted basis, it should be remembered that investment returns are generally correlated with inflation. As for most actuarial valuation bases, the choice of discount rate should be consistent with the assumed annual rate of future inflation. In general, a conservative basis would assume a low real rate of return.

Whilst the method of adjusting for calendar-year influences has been straightforward, their effects can be complex.[9] It is important to remember that the various causes of inflation are not the only factors that have a calendar-year effect on the data: changes in claims processing or in claim settlement philosophies, for example, can also appear as calendar-year influences. Testing alternate inflation/interest scenarios is straightforward in a spreadsheet.

In the calculations for the inflation-adjusted chain ladder method, it has been assumed that payments in the period reflect the inflationary factors in the same period. In longer tailed classes of business, and even for particular types of claim within a generally short-tailed class of business, this is not necessarily so. Payments can be extended over many years, e.g., the amount of a claim in respect of a motor bodily injury accident might be set by a court in one time period but, following appeals and other delays, a different amount might be paid in a different time period. The court might award interest on any payments delayed due to the appeal process, but this will not necessarily match inflation.

Fixed deductibles in the underlying insurance contracts can further distort the impact of inflation. For example, assume an insured loss occurs of real value 100. Assume that, as a result of economic and other more

---

[9]Inflation-adjusted projections are even more complex if they are based on incurred claims, since consideration must be given to the way inflation has been reflected in case estimates throughout the claims triangle.

Table 16.32 Effect of inflation on the claim paid by insurer

Deductible	Claim payable		Inflation rate experienced by insurer (%)
	Before inflation	After 20% inflation	
0	100	120	20
10	90	110	22
50	50	70	40
70	30	50	67
95	5	25	400
100	0	20	Infinite

specific inflationary forces, the claim becomes 120 when it is eventually settled. From the point of view of the insured, the loss has increased from 100 to 120, i.e., by 20%. The effect that inflation has on the claim paid by the insurer, considering policies with different deductible levels, is shown in Table 16.32.

So, where a policy has fixed deductibles, we see that the transference of inflation, or of any other factor that operates in proportion to the actual loss incurred by a policyholder, is amplified. Further, the inflation rate experienced by the insurer will vary (inversely) with the size of the actual loss. So, unless the insurer subdivides data in an account by size of loss and deductible, many different effective rates of inflation will be operating within each year of origin and each development period.

Chapter 15 described how some excess of loss reinsurance treaties have variable limits that share the impact of inflation between the cedant and the reinsurer, but these are not widespread.

In less well-defined circumstances, it might be possible to link the "ground up" (see Section 14.3.5) loss values with a similar price index and to estimate the impact of nonproportional policy conditions upon the portfolio of the insurer based on a study of the deductibles in the portfolio and the original ground up claims. In a reinsurance account, the complexity can be much greater than in a direct account because the intervening insurers will have their own deductibles and particular policy conditions and the ground-up loss might not be known.

16.5.3 Separation Technique

The separation technique is, in essence, very similar to the chain ladder method, but it focuses on the derivation of calendar-year factors from the data.

The model is identical to the inflation-adjusted claim ladder, namely:

$$c_{ij} = S_i r_j \lambda_{i+j-1} + \varepsilon_{ij}$$

Table 16.33 Numbers of claims handled in each development period

Period of origin	Development period					
	1	2	3	4	5	6
1	963	405	52	14	4	2
2	1 010	445	55	13	5	
3	991	430	46	12		
4	932	370	41			
5	915	394				
6	1 029					

The year of origin effect is removed by scaling the data using an appropriate exposure measure L_i. So:

$$\frac{c_{ij}}{L_i} = \frac{S_i}{L_i} r_j \lambda_{i+j-1} + \frac{\varepsilon_{ij}}{L_i}$$

The ultimate aggregate claim amount per unit of exposure is assumed to be constant, so S_i/L_i is constant for all i and can be incorporated into the price index factor λ_{i+j-1}.

The separation technique is not widely used, but it can be useful in a high-volume, stable account. The choice of L_i must be consistent from year to year and will depend on the type of business and the basis on which it is written. Premium income might be an appropriate measure for certain classes, for example, but, where premium rates fluctuate due to the influence of the insurance pricing cycle, they will need to be adjusted to remove this extraneous factor. For an employers' liability account, which has been written on a consistent basis in the past, the salary roll or number of employees might be suitable measures of exposure; for stable, high-volume, classes of business, such as private motor, the number of claims expected in a year could be used.

In the sample data set from the basic chain ladder model, we have the numbers of claims corresponding to the claim amounts (Table 16.33).

If we assume for our purposes that the number of claims handled by the end of the first development period are an accurate proxy measure of the scale of total claims, then the data can be standardized by dividing the data of Table 16.11 by the claims in column 1 of Table 16.33.

The standardized payments $r_j \lambda_{i+j-1}$ are given in Table 16.34, together with some subsidiary calculations. So, for example, $r_1 \lambda_1 = 54.56$. If we further assume that the class of business in question is fully developed after 6 years, then it follows that

$$\sum_{j=1}^{6} r_j = 1$$

Table 16.34 Standardized claim payments

Period of origin	Development period					
	1	2	3	4	5	6
1	54.56	29.83	9.54	8.12	5.07	3.22
2	61.67	35.85	11.49	8.17	5.28	
3	72.83	41.50	11.14	8.62		
4	92.42	44.23	11.86			
5	106.09	58.37				
6	125.35					
Total	512.92	209.78	44.03	24.91	10.35	3.22
	Calendar period (sum of diagonals)					
	1	2	3	4	5	6
Total	54.56	91.50	118.22	153.53	174.90	212.70

Table 16.35 Parameters of the separation method

Calendar period	λ_{i+j-1}	Development period	r_j
1	93.76	1	0.5819
2	107.87	2	0.2664
3	129.48	3	0.0648
4	160.20	4	0.0453
5	177.39	5	0.0265
6	212.70	6	0.0151

The calculations of the parameter of the model are

$$(r_1 + r_2 + \ldots + r_6)\lambda_6 = 125.35 + 58.37 + 11.86 + 8.72 + 5.28 + 3.22 = 212.70$$

so $\lambda_6 = 212.70$. Then, since $3.22 = r_6\lambda_6$, we can calculate $r_6 = 3.22/212.7 = 0.0151$.

Next, notice that $(p_1 + p_2 + \ldots + p_5)\lambda_5 = 106.09 + 44.23 + 11.14 + 8.17 + 5.07 = 174.70$. But

$$(r_1 + r_2 + \ldots + r_5) = 1 - r_6$$

so $\lambda_5 = 174.70/(1 - 0.0151) = 177.39$.

Continuing in this manner we obtain the values in the Table 16.35 for r_j and λ_{i+j-1}. The factors λ_{i+j-1} are defined to be an index, such that the ratio $\lambda_{i+j+1}/\lambda_{i+j}$ represents the calendar period effect in the period $i+j$. The calendar period effects calculated from the data in Table 16.35 are as follows:

1	15%
2	20%
3	24%
4	11%
5	20%

Table 16.36 Estimated future standardized payments

Period of origin	Development period					
	1	2	3	4	5	6
1						
2						3.61
3					6.32	4.04
4				10.78	7.08	4.53
5			15.43	12.08	7.93	5.07
6		63.45	17.28	13.52	8.89	5.68

Table 16.37 Estimated future incremental claims

Period of origin	Development period					
	1	2	3	4	5	6
1						
2						3644
3					6267	4004
4				10048	6602	4218
5			14116	11049	7259	4638
6		65290	17780	13917	9143	5841

The calendar period factors calculated using the separation technique should be compared with actual economic trends and with other external indices appropriate to the class of business, to investigate whether they are consistent with the actual experience of the class.

Assuming a future rate of inflation of i per period, $\lambda_{6+k} = 212.70(1+i)^k$, $k = 0, 1, 2, \ldots$. The estimate of future payments requires two further stages. Table 16.36 shows the standardized future payments, calculated using the formula $r_j \lambda_{i+j-1}$, and Table 16.37 shows the full value of expected payments by scaling the standardized payments by L_i. The calculations have been carried out assuming future inflation of 12% per period (Table 16.38).

16.5.4 Average Cost per Claim

The separation method described used claim numbers as a consistent indicator of the overall scale of claims between different periods of origin. The combination of numbers and amounts of claims leads naturally to the consideration of average claim amounts and projections based on average costs per claim. The method requires three steps:

the average cost per claim, in real terms, is estimated in each development period;

Table 16.38 Comparison of reserves allowing for 12% future inflation

	Reserve	
Period of origin	Nominal	Current money
1		
2	3 644	3 253
3	10 272	8 788
4	20 868	17 237
5	37 062	29 526
6	1 11 971	91 500
Total	1 83 817	1 50 304

Table 16.39 Estimate of future claim numbers

Period of origin	Development period					
	1	2	3	4	5	6
1	963	405	52	14	4	2
2	1 010	445	55	13	5	2.5
3	991	430	46	12	4	2
4	932	370	41	10.45	3.48	1.74
5	915	394	46.32	11.81	3.94	1.97
6	1 029	437.18	51.40	13.10	4.37	2.18
Ratios		0.42486	0.11758	0.25490	0.33333	0.5

the number of claims settled in future years is estimated using the basic chain ladder technique;

the average claim amount is then adjusted to make an appropriate allowance for inflation and the amounts multiplied by the corresponding projected future numbers of claims.

The stability of the calculation of claim numbers between development periods and between periods of origin is critical. Similarly, a consistent correspondence between the period in which a payment is made and the period in which the claim is recorded as paid in the claim counts is important.

These calculations have been made using the same data as the earlier examples. The results are shown in the following tables. In Table 16.39 the numbers of claims settled in future development years are estimated using the weighted average of past experience, using the same basic chain ladder method that has been used for estimating claim amounts in

Table 16.40 Average cost per claim (current money)

Period of origin	Development period					
	1	2	3	4	5	6
1	119.67	138.41	282.52	732.46	1382.46	1551
2	120.32	130.14	276.73	718.38	1067.20	
3	116.48	125.48	271.70	711.92		
4	121.25	126.12	269.51			
5	120.09	135.55				
6	125.35					
Average	120.53	131.14	275.12	720.92	1224.83	1551

Table 16.41 Estimate of future claim amounts

Period of origin	Development period					
	1	2	3	4	5	6
1						
2						3 877.50
3					4 899.31	3 102.00
4				7 534.34	4 266.88	2 701.58
5			12 744.8	8 512.86	4 821.04	3 052.44
6		57 332	14 141.6	9 445.83	5 399.41	3 386.98

Section 16.5.1. For example, $0.33333 = (4 + 5)/(14 + 13)$ and $3.48 = 10.45 \times 0.33333$.

The next step is to calculate the average cost per claim. The entries in Table 16.40 are calculated by taking the incremental claims used to calculate Table 16.26 (i.e., the past claims adjusted to allow for inflation) and dividing by the number of claims settled in the corresponding year of origin and development year.

The method assumes that the average claim settled in each development period is stable. So the estimated future claims are calculated by multiplying the average claim calculated in Table 16.40 by the estimate of future claim numbers in Table 16.39. This is presented in Table 16.41.

The estimated reserve required, in nominal terms, can be obtained by multiplying the projected incremental claims by a factor for inflation as in the previous examples (Table 16.42).

A number of operational factors need to be borne in mind when considering the appropriateness of this method. In particular:

If nil claims are recorded in the numbers then the average cost will fall. Ideally, nil claims should be excluded from the claim counts and/or

Table 16.42 Comparison of reserves allowing for 12% future inflation

	Reserve	
Period of origin	Undiscounted	Discounted
1		
2	4 343	3 878
3	9 378	8 001
4	17 586	14 503
5	36 529	29 131
6	1 09 608	89 656
Total	1 77 444	1 45 169

analyzed separately in order to assess their stability (e.g., as a proportion of total claims; and their pattern in relation to the development period). Similar consideration must be given to the administrative procedures for recording precautionary advices.

Partial payments and, in particular, changes in the patterns of partial payments within periods of origin or between periods of origin may invalidate the assumption that past patterns of average claims will be maintained into the future. It might be preferable to accumulate partial payments and record them as a single payment to correspond with the claim count.

Changes in the mix of claim by claim type (e.g., in a motor account a changing mix of bodily injury and physical damage claims) would distort averages. When more than one type of claim arises from a single event, the systems of the company might record a claim number for a loss under each section of the policy rather than record a single claim count for the original accident. In circumstances where there are distinct claim types, a separate projection by claim type might be preferable. This can reduce the distortions caused by a changing claim type mix and enable more appropriate allowances to be made for the effects of inflation, which might affect the various claims types differently.

In the later years of development (the "tail" of the account), the numbers of claims settled might be very low and the average cost per claim can be highly variable from one year of origin to another. The assumed average cost in the tail is therefore largely a matter of judgment based on the types of claim that appear in the tail and an extrapolation of average costs in the preceding development periods.

The average cost per claim approach is intuitively appealing and can be applied with a high degree of flexibility. In addition, the information that it can provide on the severity of the individual claim types provides useful feedback to the premium rating process.

16.5.5   The Loss Ratio and Bornheutter–Ferguson Method

The loss ratio of a cohort of business is defined as the ratio of ultimate losses to the relevant premiums.[10] In the case of an accident-year cohort, premiums will be those earned; if the underwriting year is used, the appropriate premium will be that written over the period. A naïve calculation of reserves could be made by multiplying premiums by the expected loss ratio, to give an estimate of the ultimate claim amount. The loss ratio can be estimated after consideration of past experience and market statistics together with discussion with the underwriting and claims management teams that should focus on the underwriting cycle and the way premium rates have changed between periods of origin.

New companies might have no choice other than to set reserves for outstanding claims based on loss ratios, since they might have too little actual claims data to support any other analysis. The selected loss ratios might be those used in the company's business planning exercises, provided that the conditions expected when the business plan was drawn up have not changed significantly.

The true loss ratio clearly is not known until the last claim has been paid and some way of adjusting the initial estimate, as claims emerge, is required. The Bornheutter–Ferguson method gives a way of combining the prior expectation of losses provided by simple loss ratio estimates with the actual rate of emergence of claims (Bornheutter and Ferguson, 1972). By doing so it moderates the influence that the most recent years of account have on the reserve estimate, which, as has already been noted (see Section 16.5.1), can have an undue influence on chain ladder estimates. Because of this, the Bornheutter–Ferguson method is particularly useful when experience is volatile.

We will now investigate how the method works with the data set prepared for the inflation-adjusted chain ladder method. The key items of information required are

the initial expected ultimate loss ratio;
the premium or exposure base to which the loss ratio is applied;
the expected development pattern of claims.

In the example we shall assume a loss ratio of 80% of premium for each period of origin. We can calculate a "naïve" estimate of ultimate claims by multiplying the written premium by the initial expected loss ratio (IELR).

---

[10]Loss ratio methods are generally applied to premiums because this is a familiar accounting concept; however, a more fundamental approach would be to relate expected losses to a more accurate measure of exposure. For example, for motor vehicles this might be vehicle-years or annual distance driven.

Table 16.43 Written premium and "naïve" estimate of ultimate claim

| | Period of origin | | | | | |
	1	2	3	4	5	6
Written premium	2 92 674	3 03 647	2 83 718	2 67 086	2 75 545	3 19 082
"Naïve" ultimate claim estimate	2 34 139	2 42 918	2 26 974	2 13 669	2 20 436	2 55 266

Table 16.44 Calculated development patterns

Period of origin	Naïve estimated ultimate claims	Expected proportion observed R_j	Expected proportion outstanding, $1 - R_j$	Estimated reserve
2	2 42 918	0.985	0.0151	3 678
3	2 26 974	0.959	0.0412	9 345
4	2 13 669	0.913	0.0868	18 556
5	2 20 436	0.847	0.1534	33 820
6	2 55 266	0.577	0.4225	1 07 858
Total				1 73 257

Note: R_j is calculated from the period-to-period development factors in Table 16.26. For example, $R_1 = 1/(1.466 \times 1.079 \times 1.050 \times 1.027 \times 1.015) = 0.577$ (with minor allowance for rounding errors) and $3678 = 2\,42\,918 \times 0.0151$.

This is presented in Table 16.43. Note that the loss ratio is the ratio of the expected undiscounted claims to premiums.

The next step is to calculate development patterns using some method such as the chain ladder or, if in-house data are not available, suitably adjusted industry statistics. We shall use the inflation-adjusted figures in Table 16.26. The proportion of ultimate claims observed to the end of development period j is given by R_j. By multiplying the estimate of ultimate claims in Table 16.43 by $1 - R_j$ we can estimate the claims (i.e., based on LR) still outstanding at the end of development period j. The results of these calculations are shown in Table 16.44.

The estimated ultimate claims for year of origin i and development period j is

$$P_i(\text{IELR})_i(1 - R_j) + C_{ij}$$

where P_i is the premium income received and $(\text{IELR})_i$ is the IELR for year of origin i. The calculation of the estimated ultimate claims in Table 16.44 is a mix of a prior expected value and the actual ("posterior") claims, similar to the credibility rating discussed in Chapter 14.5.2. As the year of origin matures, the prior weight $(1 - R_j)$ naturally falls and the estimate rests, ultimately, at the true ultimate claim level. In practice, the prior expectation is adjusted as claims experience emerges, and once $(1 - R_j)$ falls below 50% the prior component is often removed completely and an alternative projection method is used for future claims.

The structure of the methods set out so far is based on averages and, to a greater or lesser extent, the application of informed judgement in dealing with highly variable or unusual elements of the data or methods. No use has been made of regression models for providing a statistically justified indicator of the goodness of fit of the models, nor of the statistical reliability of the projection bases and the subsequent estimates of outstanding liabilities that have been derived.

We shall consider in Section 16.5.7 to Section 16.5.9 three methods of estimating reserves that also include an estimate of their variability. Each of them uses the chain ladder method with only minor modification. For the detailed theory underlying the methods the reader is referred to Mack (1993), Wright (1990), Lowe (1994), and England and Verrall (2002).

16.5.6   Operational Time Model

The chain ladder methods assume that the claim amounts, the numbers of claims incurred, and the rate of settlement remain stable from year to year. The method can be adapted to allow for the variations in the former, but not the latter. *Operational time* models attempt to deal with variations in the speed of settlement.

Operational time refers to the speed of settlement of a cohort of claims. Business in the cohort starts to be written at operational time 0, and the final claim is paid at operational time 1. A fundamental assumption of operational time reserving models is that claims paid at a particular moment of operational time have similar characteristics. For example, claims settled within a short time of business being written are likely to be small, whereas those settled toward the end of the development period are likely to be large. By using operational time, the difficulty with different categories of business having different development periods is avoided.

The methods require data that should be readily available: claim amounts, numbers of claims settled, and numbers of claims incurred. Although the latter might not be known, it would be reasonable to estimate the number using some proxy, such as claims reported.

An example of a reserve calculation using operational time follows. The data used are the same as have been used in the previous triangles, i.e., Table 16.26 of the inflation-adjusted chain ladder calculation and Table 16.33 of the separation method. If the data are not all in constant money terms, then they must first be adjusted to remove the effect of inflation. The first step is to calculate the average operational times for each development period. To do this we need to estimate the numbers of claims incurred in

Table 16.45 Estimated numbers of claims incurred

Period of origin	Estimated number of claims incurred
1	1440
2	1530.1
3	1485.6
4	1360.9
5	1372.9
6	1537.7

Table 16.46 Average operational time

| Period of origin | Development period | | | | | |
	1	2	3	4	5	6
1	0.3344	0.8094	0.9681	0.9910	0.9972	0.9993
2	0.3300	0.8055	0.9689	0.9911	0.9970	
3	0.3335	0.8118	0.9720	0.9915		
4	0.3424	0.8208	0.9718			
5	0.3332	0.8100				
6	0.3346					

each year of origin. Using the link ratios of the basic chain ladder method, applied to the cumulative values of the incremental figures given in Table 16.33, the estimated numbers of incurred claims, are given in Table 16.45.

The average operational time depends on the proportion of total claims settled by the end of each development period. Thus, for example, referring to Table 16.33, in period of origin 2, by the end of development period 2, 1455 claims have been settled whereas only 1010 had been settled at the end of development period 1. Thus, the average operational time for development period 2 in period of origin 2 is

$$1/2(1010 + 1455)/1530.1 = 0.8055$$

Next, we need the average claim paid in each development period. This can be taken from Table 16.40 of the average cost per claim method. Table 16.46 shows the results of all the calculations.

Our assumption is that the claims paid at a given operational time are drawn from the same distribution. In order to estimate the relationship between the average claim paid and operational time, a model is hypothecated and a regression analysis performed. In this example we use the formula

$$y = bm^x$$

Table 16.47 Calculations of claims statistics

Period of origin	Number of claims		Average future		Reserve
	Settled	Outstanding	Operational time	Claim amount	
1	1440	0	1	456.76	0
2	1528	2.13	0.99931	456.03	969
3	1479	6.56	0.99779	454.44	2 983
4	1343	17.93	0.99341	449.89	8 065
5	1309	63.87	0.97694	432.93	27 652
6	1029	508.72	0.83459	312.01	1 58 728
Total					1 98 397

where y is the average claim amount, x the operational time, and b and m are constants determined by the data.

This formula was chosen since the variation of average claim size by development duration suggested an exponential relationship. For this example we transformed the equation to give $\log(y) = \log(b) + x \log(m)$ and performed a linear regression (where $y =$ Table 16.40 and $x =$ Table 16.46) and the resulting coefficients were exponentiated to give $b = 45.61$ and $m = 10.01$. The result is clearly sensitive to the assumptions of the functional form of the relationship of average cost to operational time; in practice, alternatives should also be tested.

Now we calculate the average future operational time. The formula for this is the average number of claims settled to date and the estimated total number of claims incurred, as a proportion of the estimated total number of claims incurred. That is, we are assuming that, on average, claims are settled halfway through the remaining operational time. For example, for period of origin 3, the average future operational time is

$$1/2(1479 + 1485.56)/1485.56 = 0.997\,79$$

We can then substitute this into the regression line for average claim paid, to calculate the value of the average claim outstanding. Finally, the average outstanding claim amount is multiplied by the number of claims remaining to be settled, to give the required reserve. These calculations are shown in Table 16.47.

16.5.7 The Bootstrap Method

Bootstrapping is a simple and powerful method which enables the calculation of a number of different estimates of a random variable, using empirical data as an approximation for the true distribution of a (related) random variable (Efron, 1982). It provides a simple approach for calculating estimates of the error associated with the claims reserves.

As we have demonstrated in Table 16.20, the chain ladder development factors can be used to estimate the ultimate claims at each development period and for each year of origin. Table 16.23 shows the estimated past incremental claims, and Table 16.20 shows the residual error. We shall assume that the residual errors are random. This assumption should be tested in order to ensure that there is no systematic component, but it is consistent with the chain ladder model. The significance of this assumption for the bootstrap method is that each error could equally well have arisen as the residual error from any other development period and year of origin.

In order to use the bootstrap method we draw a random sample (allowing for replacement) from the set of residual errors. The values from the random sample are added to the actual incremental data to produce an alternative, but equally likely, outcome of the actual claims, called *pseudo data*. The incremental "pseudo claims" are then cumulated, the chain ladder method applied as normal, and a second reserve estimate obtained.

A new set of residual errors is calculated using the second set of estimates and used to produce a new set of pseudo data and a new reserve estimate. After a sufficient number of estimates have been produced, the method assumes that, taken together, they provide a random sample of the distribution of the true reserve value, and can be used to calculate estimates of the moments of the distribution.

If the incremental claims have been affected by inflation, then the residual errors calculated on unadjusted data will not be random. Ideally, therefore, data should be adjusted for inflation before applying the bootstrap method, and the inflation-adjusted chain ladder model used to calculated the reserve estimates. However, if the data are not inflation adjusted, it could be argued that the effect of inflation is amalgamated in the final result as an additional random component.

Before sampling, the residual errors should be adjusted for the scale of the period of origin with which they are associated. For example, the residual errors could be divided through by L_i (see Section 16.5.3) and, once sampled, adjusted back up to the correct period of origin scale. A formal approach to standardizing residuals for these purposes is provided in England and Verrall (2002).

A simple example, using the same data triangle as the inflation-adjusted chain ladder method (Section 16.5.2) and with no scaling adjustment made to the sampled residual errors, is given in order to demonstrate the method. The residual errors calculated using the inflation-adjusted chain ladder method are set out in Table 16.48; Table 16.49 gives a set of residuals randomly sampled from those in Table 16.48. The resulting pseudo data are set out in Table 16.50.

The chain ladder reserve estimate for each triangle of pseudo data can be calculated. Each reserve estimate is a random sample from the population

Table 16.48 Residual errors calculated using the inflation-adjusted chain ladder data

Period of origin	Development period					
	1	2	3	4	5	6
1	3068.0	−922.3	−1051.7	−897.2	−196.9	0
2	1214.7	−711.4	−1069.3	369.2	196.9	
3	−740.5	−511.3	723.8	528.0		
4	−5044.2	3647.0	1397.2			
5	1502.0	−1502.0				
6	0					

Table 16.49 Sampled residuals

Period of origin	Development period					
	1	2	3	4	5	6
1	3647.0	−196.9	−511.3	−5044.2	1214.7	3647.0
2	−1069.3	1502.0	3647.0	−897.2	−5044.2	
3	−511.3	−922.3	−740.5	196.9		
4	−1069.3	−5044.2	1502.0			
5	1397.2	−5044.2				
6	196.9					

Table 16.50 Pseudo data resulting from sampled residuals

Period of origin	Development period					
	1	2	3	4	5	6
1	118 886	55 858	14 180	5 210	6 744	6 749
2	1 20 459	59 413	18 867	8 442	282	
3	114 916	53 034	11 758	8 740		
4	111 939	41 621	12 552			
5	111 278	48 364				
6	1 29 179					

of the distribution of the "true" reserve, and so the samples can be used to calculate estimates of the moments of this distribution. After 100 simulations, for this particular example, the estimate of the population mean (the "true" reserve) was 153 077 with a sample standard deviation of 9922.

For simplicity the calculations have been carried out assuming that the error structure of the model is the same throughout the triangle. Thus, the pseudo data could be calculated by randomly sampling from the complete set of residuals for the whole triangle. This is probably an oversimplification, since the incremental payments and their associated errors are likely to depend on the development period. Incremental settlements for short-tailed classes of business are likely to be larger in the first two development years and smaller later on; for longer tailed classes of business there might be a peak of settlement activity around development years four or five, for example. Unless the residual errors are scaled in some way, the sampling for particular development years might have to be confined to the residual errors arising in those development years. This would lead to difficulties sampling for errors in the tail of the triangle, as the amount of data are limited, so that some judgment might be required to extend the sample or to standardize the sampling residuals to remove these distortions.

16.5.8 Distribution-Free Approach

This approach was set out in Mack (1993), and extends the basic chain ladder method so that more formalized estimates of the reserving error can be calculated.

The distribution-free method for estimating reserves is based on the following three assumptions which can be shown to underlie the chain ladder method:

$$E\left(C_{i,j+1}|C_{i1},\ldots,C_{ij}\right) = C_{ij}f_j, \qquad 1 \leq i \leq n, \quad 1 \leq j \leq n-1$$
$$\left\{C_{i1},\ldots,C_{in}\right\} \text{ and } \left\{C_{j1},\ldots,C_{jn}\right\} \text{ are independent, for } i \neq j$$
$$\mathrm{Var}\left(C_{i,j+1}|C_{i1},\ldots,C_{ij}\right) = C_{ij}\sigma_j^2, \qquad 1 \leq i \leq n, \quad 1 \leq j \leq n-1$$

where C_{ij} denotes the total claim amount paid by the end of development year j arising from year of origin i, as before, f_j is the development factor for development year j to $j+1$, and σ_j is a parameter.

A corollary of assumption (3) is that the development factors are not correlated. So, for example, if the development factor is high in one period, then it does not follow that it should be high (or low) in the next period. This will not always be reasonable; e.g., if a company decides to change its claims-handling procedure so that a large number of outstanding claims are settled in one development period, then it is likely that the following period will have a low development factor.

Since the reserve required (which, for period of origin i, we shall denote by Q_i) is just a function of the claim amounts

$$Q_i = C_{in} - C_{i,n+1-i}, \qquad i = 2,\ldots,n$$

it is possible to use the three assumptions above to derive formulae for the mean squared error (MSE) of the reserve estimate for each year of origin, as well as for the overall reserve Q.

Mack (1993) shows that the estimate of the parameter f_j calculated using the chain ladder method is unbiased. An unbiased estimator of σ_j^2 is also suggested:

$$\hat{\sigma}_j^2 = \frac{1}{n-j-1} \sum_{i=1}^{n-j} C_{ij} \left(\frac{C_{i,j+1}}{C_{ij}} - \hat{f}_j \right)^2 \qquad 1 \le j \le n-2$$

where the hat denotes estimated values. For $j = n-1$ Mack uses loglinear regression to estimate

$$\hat{\sigma}_{n-1}^2 = \min(\hat{\sigma}_{n-2}^4 / \hat{\sigma}_{n-3}^2, \min(\hat{\sigma}_{n-3}^2, \hat{\sigma}_{n-2}^2))$$

If $f_{n-1} = 1$, then we could have put $\sigma_{n-1}^2 = 0$.

The following formulae are given for the mean squared error of Q_i and of Q:

$$\text{MSE}\left(\hat{Q}_i \right) = \hat{C}_{in} \sum_{j=n+1-i}^{n-1} \frac{\hat{\sigma}_j^2}{\hat{f}_j^2} \left(\frac{1}{\hat{C}_{ij}} + \frac{1}{\sum_{k=1}^{n-j} C_{kj}} \right)$$

and

$$\text{MSE}\left(\hat{Q} \right) = \sum_{i=2}^{n} \left[\left(\text{s.e.}(\hat{Q}_i) \right)^2 + \hat{C}_{in} \left(\sum_{k=i+1}^{n} \hat{C}_{kn} \right) \sum_{j=n+1-i}^{n-1} \frac{2\hat{\sigma}_j^2}{\hat{f}_j^2} \left(\frac{1}{\sum_{k=1}^{n-j} C_{kj}} \right) \right]$$

The formulae use only the data from the chain ladder triangle, and the development factors and reserves calculated using the method. Thus, we can apply the method to the data used so far.

Using the inflation-adjusted data, we have already estimated the development factors and can also estimate σ_j. These are given in Table 16.51 and the estimated reserves and standard errors are given in Table 16.52.

Implicit in the Mack model is the regression formula

$$C_{i,j+1} = C_{ij}f_j + \varepsilon$$

Table 16.51 Parameters

Development period	1–2	2–3	3–4	4–5	5–6
f_j	1.466	1.079	1.050	1.027	1.015
$\sigma_j^2/1000$	106.0	11.25	3.69	0.409	0.045

Table 16.52 Estimated reserves and standard errors

Period of origin	Reserve Q_i	Standard error	s.e.$(R_i)/R_i$
1			
2	3 218	139.1	0.043
3	8 177	368.7	0.045
4	16 237	1 008.9	0.062
5	29 592	1 947.2	0.066
6	94 375	5 281.8	0.056
Overall	1 51 600	7 004.5	0.046

where $\text{Var}(\varepsilon) = C_{ij}\sigma_j^2$. This suggests that the incremental claims in period j are a fixed proportion of the cumulative claims up to period j (as in the case of the basic chain ladder model), an assumption which might prove weak if the amount of observed claims is small. Instead, the model can be modified so that

$$C_{i,j+1} = \alpha + C_{ij}f_j + \varepsilon$$

(Murphy, 1993). Under this model, the fundamental assumption of the chain ladder method, that incremental payments in any period depend entirely on the cumulative payments to date, is challenged. If f_j is small, then the implication is that incremental payments are (nearly) independent of payments to date, so that the best estimate for future payments is obtained by averaging payments at the same development periods for prior periods of origin. This could require a scaling adjustment to reflect the relative exposures in the different origin periods.

16.5.9 Approaches to Reserving Using Formal Statistical Models

Increasing access to computing power over recent years has enabled the application of techniques that had, in the past, only been of theoretical importance. In particular, statistical modeling has become more flexible and widespread. This section gives brief introductions to some models used for claims reserving (e.g., see Kremer (1982), De Jong and Zehnwirth (1983), Taylor (1988), Christofides (1989), Renshaw and Verrall (1992), or England and Verrall (2002)).

The basic chain ladder model combines a "development year effect" with a "year of origin effect," i.e.

$$C_{ij} = S_i R_j + \varepsilon_{ij}$$

Kremer (1982) noted that the model can also be expressed multiplicatively. Suppose c_{ij} represents the incremental payment (i.e., $c_{ij} = C_{1j}$ and $c_{ij} = C_{ij} - C_{i,j-1}$, $i > 1$). The multiplicative model is

$$c_{ij} = S_i r_j E_{ij} \qquad (16.1)$$

where S_i represents the "year of origin effect" and r_j is the "development-year effect." That is, r_j is the proportion of the ultimate claims payable in development year j and ($r_j = 1$, where the sum is over all development years and S_i is the total ultimate claims for period of origin i.

Now, by virtue of Equation (16.1), taking logarithms of the data we can make the model additive rather than multiplicative. Then the theory of conventional or generalized linear modeling can be applied to the data.

Taking logarithms we get

$$\log c_{ij} = m_{ij} = \log S_i + \log r_j + \varepsilon_{ij}$$
$$= a_i + b_j + \varepsilon_{ij}$$

Note that the condition $\Sigma r_j = 1$ does not convert into log space (since $\log \Sigma r_j \neq \Sigma \log r_j$). At the simple level it is reasonable to obtain estimates and seek the exponential of the parameters derived so that they sum to unity, in which case the condition will be satisfied. However, it is more convenient to drop the condition, although this introduces an indeterminate matrix into the model solution.

Using this model and presentation of the data, the reserving problem reduces to a two-way analysis of variance (ANOVA). In a conventional ANOVA experiment a square of data would be available: the row and column effects are found by calculating the sums of squares, drawing up an analysis of variance table and looking at the resulting F-statistics.

A detailed explanation and simple examples of Kremer's approach are given by Christofides (1992). For a 4×4 data triangle the layout of the table for calculation purposes is as in Table 16.53.

To avoid the matrix singularity b_1 is set equal to zero, which changes the interpretation of the parameters but does not affect the eventual result.

Table 16.53 Table layout for a 4 × 4 data triangle

I	j	c_{ij}	Y	Design matrix X						
				a_1	a_2	a_3	a_4	b_2	b_3	b_4
1	1	c_{11}	m_{11}	1	0	0	0	0	0	0
1	2	c_{12}	m_{12}	1	0	0	0	1	0	0
1	3	:	:	1	0	0	0	0	1	0
1	4	.	.	1	0	0	0	0	0	1
2	1	c_{21}	m_{21}	0	1	0	0	0	0	0
2	2	:	:	0	1	0	0	1	0	0
2	3	.	.	0	1	0	0	0	1	0
3	1			0	0	1	0	0	0	0
3	2			0	0	1	0	1	0	0
4	1			0	0	0	1	0	0	0

The model parameters can be fitted using a spreadsheet regression package. For a least-squares fit, assuming the errors have the standard normal distribution, the parameter estimates are given by the matrix

$$\left(X^T X\right)^{-1} X^T Y$$

where X is the design matrix and Y is the data vector in log space.

The advantage of using statistical modeling is that, in estimating the parameters, errors of the estimated values can be calculated, although since the original calculations are in log space a transformation is required to obtain the error estimates for the untransformed values (examples are described in papers by Renshaw and Verrall). The variance of the predicted future payments is obtained from the variance–covariance matrix

$$\sigma^2 X_F (X^T X)^{-1} X_F^T$$

where σ^2 is the model variance (a scalar), X_F is the design matrix of the future payments and $(X^T X)^{-1}$ is the model information matrix.[11]

In Table 16.51 we see that a 4 × 4 data triangle gives 10 data points; with this data the chain ladder model and this model estimates seven parameters. It can be argued that these models have too many parameters. If a model is *overparameterized*, the parameters themselves are increasingly determined partly by the random error component in the data and the model is too general. Such models have a tendency to produce large estimation errors. An *underparameterized* model runs the risk of some

[11]If this model is applied to the data used for the examples throughout Section 5.5, the result is an estimate of outstanding claims of 202 146, with a standard error of 10 114 or, using the inflation-adjusted data, outstanding claims of 149 555 with a standard error of 6109.

systematic bias in the estimates it produces. Thus *parsimonious*[12] models are generally preferred. The framework described permits modeling using fewer parameters, e.g.

$$m_{ij} = a_i + bj + \varepsilon_{ij}$$

Here, log payments follow a linear scale depending on the development period. Each year of origin, therefore, has a unique level and incremental payments decay exponentially in real space. However, it might be felt that some years of origin depend on the volume of business written and that there is a fixed relationship between years. The simplest amendment would then be to add a further linear component, thus

$$m_{ij} = f(i)a + bj + \varepsilon_{ij}$$

where $f(i)$ is a linear function of i. In this model the data are required to fit only two parameters. The number of fitted parameters is reduced significantly from the basic chain ladder model. The linear model for this model could be set out as

$$\begin{bmatrix} m_{11} \\ m_{12} \\ \vdots \\ m_{ij} \\ \vdots \end{bmatrix} = \begin{bmatrix} f(1) & 1 \\ f(2) & 2 \\ \vdots & \vdots \\ f(i) & j \\ \vdots & \vdots \end{bmatrix} \begin{bmatrix} a \\ b \end{bmatrix} + \varepsilon_{ij}$$

A model described by De Jong and Zehnwirth (1983) introduces calendar period influences, where

$$m_i(d) = \alpha_i + \beta_i \log(1 + d) + \gamma_i d$$

with $d = j - 1$ treated as a continuous regressor variable. The parameters are estimated recursively using the Kalman filter.[13] The untransformed model is

$$\exp(m_i(d)) = (1 + d)^{\beta_i} \exp(\alpha_i + \gamma_i d)$$

and the condition $\gamma < 0$ ensures claim amounts ultimately decay. An attraction of such models is that prediction beyond the development period observed in the data is straightforward.

---

[12]A parsimonious model is one that has the least number of parameters possible, when still satisfying the other requirements of the model, such as good fit and explaining the trends in the data.

[13]The Kalman filter is a set of algorithms that generate successive and updated best estimates of parameters (see De Jong and Zehnwirth (1983)).

Regression techniques open rich possibilities in model building and forecasting. However, it is still necessary to test the appropriateness of a given model. A useful approach is to plot residual errors and see whether it is possible to identify whether the pattern is random. Standardized residual plots by reference to calendar period, development and origin periods can give insight into each of the three "directions" which are assumed to influence the results.

The practicalities of using logged data and linear models are not entirely straightforward, and even when the model is based on the basic chain ladder the estimate for the ultimate claims is likely to be different. If the assumptions underlying the approach are valid, then, theoretically, the maximum likelihood estimates derived using the linear model are the best (e.g., in the sense of having least bias). In addition, the usual assumption that the errors have a normal distribution implies that the data have a lognormal distribution. Thus, there are problems involved in the reverse transformation, which have to be carefully considered. Note however, that generalized linear models do not necessarily require the assumption of a normal distribution for the error term.

The log framework described requires incremental payments to be positive. Zero or negative amounts can occur in practice, e.g., in incremental data triangles based on net of reinsurance data or in classes of business with large subrogation or salvage recoveries. Thus, pragmatic approaches are needed to overcome this problem, e.g., by adding a level parameter to all incremental values and subtracting at the end (Zehnwirth, 1985), or by simply excluding negatives (Verrall, 1993).

Models such as those described in this section are useful in identifying trends in the data, in particular for large homogeneous business groupings such as motor accounts. The variability of the data in a reinsurance account makes it more difficult to interpret the outcome of such models, and the results might be less reliable.

16.5.10 Summary of Reserve Calculations

	Reserve		
Reserving method	Undiscounted	Discounted	Standard deviation
Basic chain ladder	203 673		
Inflation-adjusted chain ladder	182 310	151 599	
Separation technique	183 817	150 304	
Average cost per claim	177 444	145 169	
Bornheutter–Ferguson	168 047		
Operational time	198 397		
Bootstrap	153 077		9 922
Distribution-free	151 600		7 004
Linear model	202 146		10 114

16.6 Conclusions

The models described in Section 16.5 make no attempt to analyze the underlying claims process. For a limited set of data, such as a claims triangle, this would be a very complex operation and would give rise to highly spurious results. Instead, the models described provide a crude explanation at an aggregate level of the way in which the underlying claims process has manifested itself. Ideally, a number of different approaches would be used and tested to identify whether the implications of the various models were consistent; frequently, it is the differences that are most revealing.

The results of reserving exercises in general insurance are often subject to substantial estimation error. An intrinsic part of a reserve review is the establishment of a reasonable range of estimates; in other words, a quantification of the estimation error. The ability to make formal statistical estimates of error requires adequate data and an understanding of the underlying model. In practice, the data may be prone to numerous distortions and the underlying claims processes, which might give some insight as to the true underlying model, are often opaque to analysis. The reserving procedure, therefore, requires extensive professional judgment. As in the other fields of actuarial work, it is important to review any assumptions made, to compare actual developments against the expected, and to perform a reconciliation of the current valuation results to those from the prior period analysis, i.e., a valuation of surplus.

References

Ackman, R. C., Green, P. A. G. and Young, A. G. *Estimating Claims Outstanding*, General Insurance Monograph, Institute of Actuaries.

Bornheutter, R. L. and Ferguson, R. E. (1972). *The Actuary and IBNR*, Proceeding of the Casualty Actuarial Society, 59.

Chamberlain, G. F. (1989). *Claims Reserving Manual*, Vol. 1. Institute of Actuaries.

Christofides, S. (1989). *Regression Models based on Log-incremental Properties*, Claims Reserving Manual, Vol. 2, Institute of Actuaries.

Christofides, S., Barlow, C., Michaelides, N. and Miranthus, C. (1992). *Storm Rating in the Nineties*. GISG Convention. Institute of Actuaries.

Daykin, C. D., Pentikainen, T. and Pesonen, M. (1994). *Practical Risk Theory for Actuaries*. Chapman & Hall.

De Jong, P. and Zehnwirth, B. (1983). *Claims Reserving, State Space Models and the Kalman Filter*, JIA, Vol. 110.

Efron, B. (1982). *The Jacknife, the Bootstrap and other resampling plans*. Society for Industrial and Applied Mathematics, Regional Conference Series in Applied Mathematics.

England, P. D. and Verrall, R. J. (2002). *Stochastic Claims Reserving in General Insurance*, BAJ, Vol. 8 (iii).

Kremer, E. (1982). *IBNR Claims and the Two Way Model of ANOVA*, Scandinavian Actuarial Journal, 1.

Lowe, J. (1994). *A practical guide to measuring reserve variability using: bootstrapping, operational time and a distribution free approach*. GISG, Institute of Actuaries.

Mack, T. (1993). *Distribution free calculation of the standard error of chain ladder reserve estimates*. ASTIN Bulletin.

Morgan, K. et al. (1992). *Equalisation Reserves on a European Basis*. GISG, Institute of Actuaries.

Murphy, D. M. (1993). *Unbiased Loss Development Factors*. Proceedings of CAS.

Renshaw, A. E. and Verrall, R. J. (1994). *The Stochastic Model Underlying the Chain Ladder Technique*. Actuarial Research Paper No. 63, Dept. of Actuarial Science and Stats, City University, London.

Renshaw, A. E. and Verrall, R. J. (1992). *Claims Reserving: Statistical Treatment of the Chain Ladder Technique*. Dept. of Actuarial Science and Stats, City University, London.

Taylor, G. C. (1988). *Regression Models in Claims Analysis (II)*. W. M. Mercer, Campbell, Cook & Knight, Sydney, Australia.

Verrall, R. J. and Li, Z. (1993). *Negative Incremental Claims: Chain Ladder and Linear Models*, JIA, Vol. 120.

Verrall, R. J. (1991). *Chain Ladder and Maximum Likelihood*. JIA, Vol. 118.

Wright, T. S. (1990). *A Stochastic Model for Claims Reserving in General Insurance*, JIA, Vol. 117.

Zehnwirth, B. (1985). *Interactive Claims Reserving Forecasting System (ICRFS)*. Benhar Nominees Pty. Ltd., Tarramurra, N. S. W., Australia.

Part IV

Pensions

Chapter 17

Types of Pension Plan

17.1 Classification of Plans

17.1.1 Introduction

A pension plan is a financial contract between a pension provider and the member(s) of the plan, established for the purpose of providing an income in retirement for the member(s).

Many pension plans also provide other types of benefit, sometimes called *ancillary benefits*. The most common type of ancillary benefit is the provision of life assurance, often in the form of an income payable to the member's spouse (or family) on the member's death.

Pension plans can be categorized in several different ways.

17.1.1.1 Coverage

First, there is the question of who can belong to the plan. At one extreme we have *single-member plans*, which are often insurance contracts taken out by an individual for the purpose of saving for retirement.

Then we have *group pension plans*, which cover a number of individuals who share a common interest; usually they work for the same employer, who sets up the plan on their behalf. Group plans can have anywhere from two members to several hundred thousand members, depending on the size of the employer.

Lastly, we have *state pension plans*, also referred to as *social security plans*, which usually aim to cover the citizens of an entire country. An important feature of such plans is that membership is usually compulsory for people who work in the country.

17.1.1.2 Benefits and Contributions

Every pension plan must have rules for the calculation of benefits (the money paid *from* the plan to its members) and contributions (the money paid *into* the plan by its members and/or their employer).

There are basically two types of plan: a defined-benefit plan and a defined-contribution plan.

A *defined-benefit plan* is one where there is a fixed rule for calculating the benefits. The simplest type of defined-benefit plan would be the payment of a flat-rate pension of £x per month to all members when they reach a specified retirement age. More usually, the pension is calculated according to a formula based on the member's salary and *service* (period of pensionable employment) before retirement. The contributions paid into the plan are whatever is required to meet the cost of providing the defined benefits, and have to be estimated periodically by the plan actuary. They usually vary over time.

A *defined-contribution plan* is one where there is a fixed rule for calculating the contributions paid into the plan. The simplest approach would be to require a contribution of £x per month for each member until retirement. More usually, the contribution is calculated as a defined percentage of the member's salary. In most defined-contribution plans, each member has an individual account in which the contributions paid for the member accumulate until retirement. The fund at retirement can then be used to purchase a whole-life annuity.

17.1.1.3 Funding

The funding of a pension plan refers to the timing of the payment of contributions to the plan.

A *funded* plan is one in which the cost of providing a member's pension is met by accumulating a fund over the same member's period of service before retirement. A funded plan, therefore, accumulates a pool of assets on behalf of the working members (or *active members*) in order to provide for their future benefits. An important issue is how these assets should be invested.

An *unfunded* plan is one in which the cost of pensions is met directly by the contributions paid *at the same time*. Such a plan, therefore, involves a transfer of money between different generations of members: the contributions of the active members pay for the pensions of those who have retired. There is no pool of assets in an unfunded plan, because all of the contributions are immediately used to pay benefits. Unfunded plans are also referred to as *pay-as-you-go* plans.

It is possible for a pension plan to be *partially funded*, which means that the contributions to the plan are timed so as to accumulate a fund which is always significantly less than the present value of the benefits accrued from past service.

17.1.2 Single-Member Plans

These are perhaps the easiest types of pension plan to understand. An employee saves for retirement by paying regular contributions into his/her own personal fund. At retirement, the accumulated value of the fund is used to purchase a whole-life annuity from an insurance company. If desired, the member can also purchase life assurance benefits, to provide an income for the spouse on death before and/or after retirement. However, providing such ancillary benefits will reduce the fund available for the member's own pension.

Single-member plans are therefore funded plans. A fund is accumulated and must be invested in suitable assets. Often, the plan consists of a contract with an insurance company, which invests the fund on behalf of the member.

In theory, a single-member plan could be defined contribution or defined benefit. In practice, it would be difficult for a member to guarantee a defined benefit without being prepared to increase contributions greatly should investment returns on the fund be less than expected. So, most single-member plans are defined contribution, and the pension at retirement will depend on:

the accumulated investment returns up to retirement;
the management expenses deducted from the fund up to retirement;
the terms on which a whole-life annuity can be bought at retirement.

17.1.3   Group Pension Plans

Group pension plans can be either defined contribution or defined benefit. Defined-benefit plans are more common for large employers. The type of plan favored varies from country to country: the U.S. and U.K. have traditionally had both types of plan, although recently there has been a move towards defined-contribution plans in both countries.

Group pension plans are normally set up by an employer on behalf of its employees, as part of their remuneration packages. It is customary for the employer to pay a substantial part of the contributions to the plan. The employees may also have to pay contributions, but in *noncontributory* plans the employer meets 100% of the cost.

Defined-contribution group plans are nearly always managed like a collection of single-member plans (described above). The plans are funded, and the contributions paid are notionally segregated into separate accounts for each member. The members share the investment returns and expenses of the group fund and, at retirement, can use their personal account to purchase a whole-life annuity. Defined-contribution plans run in this way are called *money purchase plans*, because each member uses the assets accumulated in his/her personal account to purchase the desired benefits.

Defined-benefit group plans are more complex. In theory, they can either be funded or unfunded, although funded plans are more common in the private sector. In a funded defined-benefit plan, the assets are not

segregated into separate accounts for each member. The important feature is
the formula used to calculate the pension at retirement. This varies from
plan to plan, but certain types of benefit formula have been adopted by
most plans. One of the earliest defined-benefit pension plans established in
the U.K. was for employees of the British government. The pension was
calculated according to the following formula:

$$\text{Pension} = (\text{Service}/60) \times \text{Final salary}$$

"Service" refers to the number of years that the employee has worked for
employer (or has been an active member of the plan). "Final salary" refers
to the employee's salary over some period close to retirement. Thus,
a member with a career service of 40 years would get a pension equal to
two-thirds of final salary, and this became the target pension for many
defined-benefit pension plans subsequently set up in the U.K.

Although the cost of a defined-benefit plan is unpredictable, the
employee contributions are usually set at a fixed percentage of salary. The
employer then shoulders the burden of uncertainty by paying the balance
of the cost of the plan. In an unfunded plan, the employer contribution
is simply the current amount of benefit payments less the employee
contributions. In a funded plan, the employer contribution is periodically
estimated by the actuary, and depends on many assumptions, such as:

the investment return on the fund;
the growth in employees' salaries;
the mortality of pensioners.

Advice given on funding defined benefit plans is probably the most
important area of actuarial work in relation to pension plans.

17.1.4 State Pension Plans

The importance of state pension plans varies from country to country. In
some industrialized countries the state plan has been the main source of
retirement income for pensioners, whereas less developed countries may
have no state pension plan at all. Both the U.S. and U.K. have state pension
plans, although the trend in both these countries is towards greater
provision outside the state system.

In state pension plans, membership is usually compulsory for persons
working in the country. Both workers and employers normally pay
contributions in the form of social security taxes levied on the earnings of
each employee.

Traditionally, state pension plans have been unfunded defined-benefit plans. The pension is often linked to career salary rather than final salary, or may simply be a flat-rate pension for all employees. Many unfunded state pension plans have become a financial burden because of unfavorable demographic trends in developed countries (see Section 18.4). Certain countries are phasing out their unfunded defined-benefit plans in favor of compulsory defined-contribution plans.[1] The future development of state pension systems is likely to move in this direction.

17.2 Defined-Benefit Plans

17.2.1 Benefit Formulae

Detailed descriptions of typical benefit structures are provided in Lee (1986) for U.K. plans and McGill et al. (1996) for U.S. plans. A brief outline is given below.

The simplest type of defined-benefit plan gives all retired members the same flat-rate pension. Part of the U.K. state pension is delivered in this form, although the pension is reduced for those without an adequate record of social security contributions.

In employer-sponsored defined-benefit plans, there is a strong presumption that the retirement pension should be proportional to the length of service with the employer. The pension can then be seen as a form of deferred remuneration, with the employee earning a defined amount of pension from each year of service.

The simplest benefit formula consistent with this aim would be a pension equal to a fixed monetary income for each year of service. This type of benefit formula is now unusual, but used to be more common in the U.K. and U.S. before the advent of the high inflation rates of the 1970s. It is now generally accepted that defined-benefit plans should provide a pension linked to the member's salary.

17.2.2 Salary-Linked Benefits

Reasons for linking benefits to the member's salary are:

to protect benefit levels from being eroded by inflation;
to enable the member to maintain a standard of living in retirement commensurate with that experienced during employment.

Pensionable salary is the definition of earnings used for the calculation of benefits and contributions. It often differs from the total remuneration of the

---

[1]Most notably Chile, which closed its defined-benefit plan in 1981 and replaced it with a system of mandatory individual accounts. The Chilean experience has been adopted as a model for reform by other countries, particularly in Latin America.

employee, possibly because its definition excludes certain categories of pay (e.g., bonuses and overtime pay). In addition, the rules of a plan may apply a fixed deduction in the calculation of pensionable salary, so that:

$$\text{Pensionable salary} = \text{Basic salary} - \text{Deduction}$$

The aim of applying such a deduction would be to reduce the pension at retirement by an amount roughly equal to any flat-rate pension provided by the state plan. Another adjustment to basic salary might be required if the government imposes an upper limit on pensionable salary for tax-approved plans.[2]

17.2.2.1 Final Salary Plans

A *final salary plan* provides a pension proportional to *final pensionable salary*, which is the pensionable salary over a defined period close to retirement, typically the year preceding retirement. In some occupations, however, employees' earnings fluctuate considerably from year to year, especially when bonuses or overtime pay are a significant part of earnings. In these plans, final pensionable salary might be defined as the average pensionable salary over a period of several years prior to retirement. Averaging periods of 3 or 5 years are common in U.K. plans. Some plans allow each salary figure to be revalued by an inflation index up to retirement.

The pension formula of a final salary plan is of the form

$$\text{Accrual rate} \times \text{Pensionable service} \times \text{Final pensionable salary}$$

The accrual rate varies from plan to plan. In the U.K., most private-sector final salary plans have an accrual rate of $1/60$, whereas public-sector plans usually provide a pension based on an accrual rate of $1/80$ plus an additional lump sum benefit. *Pensionable service* is essentially the number of years spent working for the sponsoring employer (normally including fractions of a year). It may differ from actual service if the employee has spent some time working for the employer while not a member of the pension plan.

17.2.2.2 Career Salary Plans

A *career salary plan* provides a pension proportional to some definition of average salary over the member's period of service. The simplest type of career salary pension formula for an employee retiring with N years of

---

[2] In the U.K., there is an "earnings cap" for tax-approved plans, which was £99 000 per annum in the 2003–2004 tax year.

pensionable service is

$$\text{Accrual rate} \times (S_1 + S_2 + S_3 + \cdots + S_N)$$

where S_k is the pensionable salary received in the kth year of service. A problem with the formula given above is that the earlier salary figures may be worth very little in real terms at retirement. The solution is to revalue each salary figure by an index of wage or price inflation, so that the pension formula becomes

$$\text{Accrual rate} \times (S_1/Q_1 + S_2/Q_1 + S_3/Q_3 + \cdots + S_N/Q_N\} \times Q_N$$

where Q_k is the value of the inflation index at the end of the kth year of service.

Career salary benefit formulae are typically found in state pension plans. They are less common in employer-sponsored defined-benefit plans, where final salary benefits tend to predominate.

17.2.3 Retirement Benefits

The *normal retirement age* is the age at which members can retire with a pension calculated according to the formula given in the rules. But the plan rules are normally flexible enough to allow members to retire at younger ages, in which case the method of calculating the pension payable on voluntary early retirement must be specified. There is normally a minimum age for such early retirements, which may correspond to the legal minimum age for tax-approved plans.[3]

Defined-benefit plans sometimes provide early retirement pensions based on the formula used for normal retirement. Hence, the same benefit formula would be used at different retirement ages (although members retiring early would generally have smaller pensions because of their lower service and salary at retirement). It is not difficult to show that an early retirement pension calculated in this way is worth more, per year of service, than the pension expected at normal retirement. For this reason, it is usual for a reduction factor to be applied to the normal retirement benefit formula.

The calculation of early retirement reduction factors is one of the duties of the plan actuary. These reduction factors are tabulated by age of retirement and depend on actuarial assumptions, such as the discount rate and salary growth. Two approaches to the calculation of these factors are described in Section 18.3.2. Similar methods can be used to calculate enhancement

---

[3]In the U.K., tax-approved plans may not allow members in normal health to retire before age 50.

factors for members who choose to retire late (i.e., above the normal retirement age specified in the plan rules).

17.2.3.1 Ill-Health Retirement

If an employee must stop working because of ill health, the rules of the plan may provide for an immediate pension based on the member's salary at the date of stopping work. Most plans provide a more generous pension than would be payable on voluntary early retirement, although the trustees of the plan usually have discretion in deciding whether the member's illness is severe enough to qualify for the benefit. There is usually no minimum age for ill-health retirement.

A relatively modest enhancement to the benefit payable on voluntary early retirement would involve waiving the early retirement reduction factor. Many U.K. plans, however, go further than this. For example, the ill-health pension formula may include both past service and potential service up to normal retirement age. Another approach would be to provide an ill-health pension equal to a fixed fraction of final pensionable salary, irrespective of actual or potential service. Other variations are common, and there may also be eligibility conditions related to past service.

The provision of enhanced ill-health retirement benefits creates additional costs for the plan, which must be allowed for by the actuary when recommending a funding strategy (see Section 18.2.3).

17.2.4 Death Benefits

Although death benefits are generally less costly to provide than pensions, they are an important part of the benefit package and are highly valued by members with dependants. They can be subdivided into those payable on death before retirement and those payable on death after retirement.

17.2.4.1 Death before Retirement

On the death of an active member, the rules of the plan may provide for a lump-sum benefit and/or a pension payable to the deceased member's spouse (or other dependant). The lump-sum benefit might be a fixed multiple of pensionable salary at death, whereas the spouse's pension is often based on the same formula used for ill-health retirement multiplied by a "spouse's fraction."

In U.K. plans, the lump-sum benefit is typically two to three times the pensionable salary at death, whereas the spouse's fraction is usually one-half.[4] The spouse's pension may be payable for life or may terminate on remarriage.

---

[4]For tax-approved U.K. plans, the maximum permitted multiple of pensionable salary is four and the maximum permitted spouse's fraction is two-thirds.

17.2.4.2 Death after Retirement

Defined-benefit plans usually provide for a spouse's pension on death after retirement equal to the deceased member's pension at death multiplied by the spouse's fraction (see above). This is generally a more costly benefit to provide than the pension on death before retirement, and may be reduced if the spouse is younger than the member by more than a specified number of years (typically around 10 years). The benefit is usually only payable to a spouse who was married to the member at the date of retirement.

Most U.K. plans also provide a "5-year guarantee." If the member dies within 5 years of retirement, then this benefit provides for the continuation of the member's pension until 5 years from the date of retirement. The benefit is sometimes provided in lump-sum form in addition to the spouse's pension. The cost of providing this benefit is not very significant.

17.2.5 Withdrawal Benefits

The provision of benefits to members who voluntarily leave service before retirement is not normally a high priority for the sponsoring employer. Until legislation was enacted in 1973, the only withdrawal benefit provided by U.K. plans was a refund of the member's own contributions (if any). As employee contributions typically cover less than half the cost of funding benefits, such refunds were poor value compared with the actuarial reserve held for the member prior to leaving the plan.

In both the U.K. and U.S., defined benefit plans must now provide "early leavers" who have completed a minimum period of service with a deferred pension based on salary and service at exit, using the normal retirement pension formula. Unlike the early retirement pension, this benefit does not come into payment until the member reaches normal retirement age. As a result, the deferred pension is worth less, per year of service, than the expected pension had the member stayed in service until normal retirement.

In the U.K., early leavers' rights have been further strengthened by legislation that requires:

the indexation of the deferred pension between leaving service and retirement;

an option for the member to take a cash equivalent of the deferred benefits on condition that it is invested in another tax-approved pension plan.

17.2.6 Pension Increases

There are two types of increase to pensions:

increases to pensions which are currently in *payment*;
increases to pensions which are currently in *deferment*.

The first type of increase is applied to the pensions of retired members and spouses in the course of payment. The second type of increase is applied to deferred pensions over the period of deferment.

Both are granted for the same reason, i.e., to protect the real value of pensions from eroding because of inflation. U.K. legislation specifies a minimum rate for each type of increase, which is essentially the lower of price inflation and 5% per annum compound in each case.

17.2.7 Cash Benefits on Retirement

Defined-benefit plans sometimes provide retirement benefits in lump-sum form rather than as income. Depending on the country's legislation, the member may then have to use all or part of the lump sum to buy a whole-life annuity from an insurance company. In the U.S., an increasing number of employers are setting up *cash balance plans* (see Section 17.4), which provide a lump-sum retirement benefit based on revalued career salary. In the U.K., defined-benefit plans are allowed to offer a commutation option, where part of the retirement pension (approximately one-quarter) may be taken as a tax-free cash sum.

17.2.8 Member Contributions

A *contributory plan* is one in which the cost of benefit provision is shared between the active members and the sponsoring employer. In a *noncontributory plan* the cost of benefit provision is met entirely by the employer.

The member contribution rate is usually defined as a fixed percentage of pensionable salary, leaving the employer to meet the balance of the cost of funding the plan. A typical member contribution rate for a U.K. plan would be 5% of pensionable salary. In addition to compulsory member contributions, U.K. plans must also allow active members to pay additional voluntary contributions in order to obtain extra benefits above the basic entitlement set out in the plan rules.

Defined-benefit plans in the U.S. tend to be noncontributory, because member contributions are not deductible against income tax (as they are in the U.K.).

17.2.9 State Pension Benefits

The benefits provided by state pension plans often have features that are relatively uncommon in private-sector defined-benefit plans. For example:

Some social security plans pay a flat-rate pension that is the same for all pensioners. If contributions are linked to earnings, then this implies a redistribution of income towards poorer pensioners.

If the pension is earnings-linked, then it is likely to be based on career-average earnings rather than final earnings. Otherwise, employers could take advantage of the plan by inflating their workers' salaries just before retirement.

Pensionable earnings are likely to exclude earnings above some upper limit, so that benefits are limited to the coverage of basic needs.

Pension benefits are normally fully protected against inflation, both in payment and deferment.

Workers who have had career breaks to look after dependants may not be penalized for their missing social security contributions.

17.3 Defined Contribution Plans

17.3.1 Types of Contribution Formula

The rules of a defined contribution plan must indicate the method of calculation and the frequency of payment of the contributions allocated to the account of each active member. In a single-member plan, the contributions are entirely from the individual member, who can usually vary them at will. In a group defined-contribution plan, it is customary for both the active members and their employer to contribute to the plan, and the rules normally specify a minimum member contribution rate and a more rigidly defined employer contribution rate. The main types of contribution rate formula are outlined below.

17.3.1.1 *Fixed Monetary Contribution*

The simplest contribution rate formula provides for a fixed monetary payment into each member's account. The disadvantage of this approach is that the real value of the contributions will diminish over time if the rate of inflation is positive, making it unlikely that a satisfactory fund will be accumulated at retirement. This method is used mainly for individual pension policies, where the member has discretion to increase the rate of contribution as the need arises.[5]

17.3.1.2 *Fixed Percentage of Salary*

The most common approach in group defined-contribution plans is for both the active member and the employer to pay fixed percentages of pensionable salary. These contribution rates are not necessarily equal. In a survey of 236 U.K. organizations by Watson Wyatt in 2000, the average employer contribution rate was twice the average member contribution rate.

---

[5]Some insured pension policies allow for the automatic indexation of contributions in line with price inflation.

The member contribution rate specified in the plan rules is typically a minimum that the member can choose to increase. If so, it is possible that the rules may require the employer to match any additional member contributions up to some maximum level.

17.3.1.3 *Variable Percentage of Salary*

Another approach used in group defined-contribution plans is for the percentage of pensionable salary paid by the member and the employer to increase with the age or service of the member. As contributions paid close to retirement have less time to accumulate interest, an age-dependent contribution rate could be calculated so as to produce a uniform expected rate of benefit accrual as a fraction of final salary, as in a final salary plan.

17.3.2 Money Purchase Principle

Defined contribution plans normally operate on the "money purchase" principle. This means that the money accumulated in the member's account is used to purchase the same member's benefits. The money purchase principle ensures that there are no cross-subsidies between different groups of members (as can occur in a defined-benefit plan) and allows the plan to give its members a degree of choice in the investment of their fund and in the types of benefit they purchase. The disadvantage for the members is that they are subject to investment and annuity rate risk, making the real value of their future benefits rather unpredictable.

17.3.2.1 *Retirement Benefits*

On attaining normal retirement age, the member has access to the fund that has accumulated in his or her account. Depending on the country, legislation may require that all or part of the fund is used to buy a whole-life annuity. If so, the member may be allowed to buy the annuity from the insurer of choice (to obtain the most competitive rates available in the market). The member may also be permitted to choose the type of annuity product purchased (see Section 17.3.3). The same principles would apply on voluntary early retirement.

17.3.2.2 *Death Benefits*

A member who requires a spouse's pension on death after retirement must use part of the fund accumulated at retirement to buy a reversionary annuity, which reduces the money available for the member's own pension. This is an important difference compared with a defined-benefit plan, as the latter provides the spouse's pension as an additional benefit.

On death before retirement, the money purchase principle implies that the fund accumulated at death should be used to provide spouse's benefits. However, this would result in benefits far inferior to those typically offered by defined-benefit plans and clearly inadequate for members with little past service. For this reason, many defined-contribution plans provide additional benefits on death before retirement, sometimes based on pensionable salary at death (as in defined-benefit plans). This is a departure from the money purchase principle and involves extra costs for the sponsoring employer. However, the cost of funding such benefits through a group insurance policy may be relatively stable.

In a single-member plan, the member could purchase life assurance benefits from the insurer providing the pension policy.

17.3.2.3 *Ill-Health Benefits*

As for death before retirement, providing ill-health pensions purely on a money purchase basis would result in unsatisfactory benefits for most members. Some defined-contribution plans, therefore, provide additional benefits based on pensionable salary at the date of stopping work. Although most defined-contribution plans in the U.K. do not provide additional ill-health benefits, the sponsoring employer often provides income protection insurance benefits outside the plan by means of a group insurance policy.

17.3.2.4 *Withdrawal Benefits*

Members who leave service before retirement receive benefits based on their fund at the date of withdrawal. In some countries, a direct refund may be permitted. Alternatively, legislation may require the plan to provide a deferred benefit based on the fund at exit accumulated up to the normal retirement age.

The important feature of the withdrawal benefit in a defined contribution plan is that it is equal in value to the fund accumulated for the early leaver. Moreover, if the member's fund is left in the plan, it receives the same rate of accumulation that would have applied if the member had remained in service. In this respect, a defined-contribution plan provides superior withdrawal benefits to a final salary plan, where accrued benefits are revalued at a lower rate (if at all) for members who leave service.

Although the provision of "portable" benefits is viewed as one of the main advantages of a defined-contribution plan, it should be noted that this feature is also shared by revalued career-salary plans. Thus, the "early leaver problem" is a feature of final salary plans rather than defined-benefit plans in general. For further discussion of these issues refer to Cooper (1997).

17.3.3 Member's Pension Choices

As each active member has an individual account, it is feasible to allow members some degree of investment choice. The usual approach is to allow the member to choose between different investment funds managed by the insurance company contracted to administer the plan.

The other important area of choice is in the type of annuity product purchased at retirement. It is important here to allow the member to choose between different insurance companies, as well as between different types of annuity contract.

17.3.3.1 Pre-Retirement Investment Choices

As the investment risk in a defined-contribution plan is borne by the members, it is desirable for the plan to allow them some control over the investment of their fund. In theory, the members could then choose assets suited to their personal tolerance for risk and their personal need for retirement income.

The problem is that many members may lack the expertise to make informed choices. Evidence from the U.S. and Australia, where most private-sector plans are defined contribution, suggests that members have tended to adopt overcautious (or "myopic") investment strategies (see Gordon et al. (1997, pp. 45–66)).

This indicates a need for investment advice as well as investment choice. The approach normally recommended is for members to maximize expected returns when far from retirement and progressively switch into more defensive assets on nearing retirement. This is known as a "lifestyle" investment strategy, often used by defined-contribution plans in the U.K. as the default investment strategy for members who are unable to make their own choices.

An example of how utility-maximization models can be used to determine optimal lifestyle investment strategies is given in Section 20.4.

17.3.3.2 Annuity Choices

Defined-contribution plans provide a lump-sum retirement benefit. How this lump sum is utilized depends on the preferences of the retiring member, subject to restrictions imposed by the legislation of the relevant country.

The main argument for the purchase of a whole-life annuity is longevity insurance, as the member is assured an income until death. Legislation may require members of tax-approved plans to purchase whole-life annuities in order to ensure that they cannot squander their retirement fund or exhaust their wealth by living too long. In the U.K., at least three-quarters of the retirement fund in a tax-approved plan must be used to purchase a whole-life annuity, although this purchase may be deferred until age 75 by members who take the "income drawdown" option (see below).

The main aspects of annuity choice for the member relate to the spouse's reversionary annuity and the level of increases in payment, although these may also be subject to legal restrictions. Whether to buy a reversionary annuity will obviously depend on the marital status of the individual and also on whether the spouse has an independent income (e.g., from another pension plan). The question of whether to buy a level or increasing annuity is more difficult. Although an annuity providing increases linked to price inflation is the least risky option, the initial level of income is normally far below that from a level annuity. Methods of comparing these alternatives are described in Section 20.3.

17.3.3.3   Income Drawdown

An alternative to the purchase of an annuity is for the member to draw income from the retirement fund while retaining the freedom to invest the fund in the assets of choice. The main advantage of this approach is that the retirement fund can be invested in assets offering higher expected returns than those held within annuity funds (which invest mainly in government bonds). The member's next of kin can also inherit the capital remaining in the fund on the member's death.

However, the income drawdown approach provides no longevity insurance, and the return of capital on death means that higher investment returns are required simply to match the income available from a whole-life annuity. This required extra return, sometimes called "mortality drag," increases with the age of the member. Attempting to compensate for mortality drag by investing in riskier assets exposes the member to investment risk as well as longevity risk.

In the U.K., legal restrictions apply to the level of income that may be drawn from the fund, and the member must not delay the purchase of a whole-life annuity beyond age 75. Methods for comparing the income drawdown option with a whole-life annuity are described in Section 20.3.

17.4   Hybrid Plans

Defined-benefit and defined-contribution plans represent opposite extremes in pension plan design. Investment risk is borne wholly by the employer in a defined-benefit plan and wholly by the members in a defined-contribution plan.

From the members' point of view, both plans have positive and negative aspects. The flexibility, transparency, and choice in a defined-contribution plan must be weighed against its uncertain benefits. The provision of secure and predictable benefits in a defined-benefit plan must be weighed against the adverse effects of cross-subsidies on particular groups of members (particularly early leavers).

As a group pension plan is a way of remunerating the workforce, the sponsoring employer has an interest in ensuring that its employees place a high subjective value on the benefits provided by the plan. This aim might be better achieved by a hybrid plan that attempts to combine some of the features of defined-benefit and defined-contribution plans. The main varieties of hybrid plan currently in existence are described below.

17.4.1 Benefit Underpins

Plans with benefit underpins provide the greater of two benefits: one calculated on a defined-benefit basis and the other calculated on a defined-contribution basis. These plans are normally designed so that the higher benefit will usually be of one type (i.e., either defined benefit or defined contribution), with the other type of benefit providing a guarantee or "underpin." In the U.S., these types of plan are referred to as "floor plans" (see McGill et al. (1996, pp. 304–305)).

A final salary plan would provide a money purchase underpin to ensure that early leavers and retirees who have experienced low salary growth receive reasonable benefits. The least generous form of underpin would be based on the member's own contributions accumulated with interest. A more generous underpin might use a multiple of the member's accumulated contributions to make an implicit allowance for the employer contributions as well. On withdrawal or retirement, the present value of the defined benefits normally provided by the plan would be compared with the money purchase underpin. If the latter is greater, then the benefits would be enhanced to make them equal in value to the underpin.

A defined-contribution plan would provide a defined-benefit underpin to protect retiring members from the extreme consequences of investment risk. The defined-benefit underpin could, in principle, be a career salary benefit, a final salary benefit, or a flat-rate benefit revalued over time by an inflation index. In a survey of 236 U.K. organizations by Watson Wyatt in 2000, a final salary underpin was provided in 22% of the defined-contribution plans covered by the survey.

17.4.2 Cash-Balance Plans

A cash-balance plan can be thought of as a revalued career salary plan providing a lump-sum retirement benefit. In the U.S., where these plans are most common, the rate of revaluation is usually linked to the market interest rate on cash instruments rather than an inflation index. This results in a plan that looks like a defined-contribution plan to the members: over each year they accrue a benefit equal to a fixed proportion of salary, which then earns interest up to retirement. The effect is the same as allocating this proportion of salary to an individual account and investing it in cash deposits.

However, the individual accounts are entirely notional and the member has no investment choice. The assets of the plan are invested as a single fund with centralized decision making, as in a defined-benefit plan. Furthermore, there is no requirement to invest the fund in the cash instruments used to determine the revaluation rate. By investing in assets offering higher expected returns, it is possible to provide a rate of benefit accrual (as a proportion of salary) that exceeds the expected contribution rate required to fund the plan. In these circumstances, the sponsoring employer bears the investment risk and must pay extra contributions to amortize any deficits that occur. It is customary for cash-balance plans to be noncontributory (i.e., no member contributions are required).

As cash-balance plans provide revalued career salary benefits, the cross-subsidies that adversely affect early leavers (and members with low salary growth) in final salary plans are avoided. The members are exposed to some investment risk, because market interest rates are unpredictable, but past experience indicates that real returns on cash deposits have been significantly more stable than those from stock-market securities. As the retirement benefit is a lump sum, retiring members are subject to annuity rate risk, as in a defined-contribution plan, but can also be allowed the same annuity choices.

Although the sponsoring employer bears the risk of adverse experience, as in a defined-benefit plan, these risks are more manageable. The investment risk depends on the extent to which the actual investments of the fund differ from the cash instruments used to determine the revaluation rate. The sponsoring employer can, therefore, decide how much investment risk to assume in order to reduce the expected contribution rate. The provision of lump-sum retirement benefits means that longevity risk is transferred to the members and the career salary benefit structure means that promotional salary increases have no effect on the required contribution rate. For further information about cash-balance plans, refer to Gordon et al. (1997) and Bolton Offutt Donovan Inc. (2000).

17.4.3 With-Profits Fund

A pension plan can operate in a similar way to a participating or "with-profits" life assurance fund. From the members' point of view, this is a defined-contribution plan in which the rate of accumulation is smoothed over time, with a guaranteed minimum rate of accumulation. As in a cash-balance plan, however, there is no individual investment choice and the assets are invested in a single fund with centralized decision making.

If the plan closely follows the with-profits design, then the return on each member's notional account would be a discretionary bonus rate declared by the plan manager. This bonus rate would be amended from time to time, under the advice of the plan actuary, in order to reflect the investment performance of the fund. However, the plan would aim to provide a bonus rate more stable than the actual return on the fund, with a guaranteed minimum rate (e.g., no negative bonuses).

From the sponsoring employer's point of view, the liabilities of the plan would be the sum of the members' notional accounts, and would generally differ from the market value of the fund. The surplus of the assets over the liabilities would be held as a reserve in order to enable the smoothing of returns and the maintenance of guarantees in the event of adverse experience. If this surplus became a deficit, however, the sponsoring employer would be responsible for providing extra resources to maintain the solvency of the plan. Thus, investment risk is shared between the members and the sponsoring employer.

Some plans have adopted more transparent methods of investment smoothing than the provision of discretionary bonuses. One such method revalues each account by the average of actual past returns on the fund over a suitable period (e.g., 3 years). Smoothing methods of this kind are quite common in Australian group plans (see Humphreys and Newman (1993)). Another method revalues each account by the annual return on the fund subject to upper and lower bounds — returns above the upper bound are used to build up reserves that help to ensure that the lower bound is maintained. This approach has been adopted by the mandatory defined-contribution plan in Chile, where the upper and lower bounds are defined relative to the average real return on all the funds participating in the plan. Members may prefer explicit smoothing methods of these kinds, and stochastic simulation can be used to investigate the properties of different methods. A comparison of different types of investment guarantee is provided in Turner (2001).

References

Bolton Offutt Donovan Inc. (2000). Actuarial aspects of cash balance plans. Submitted to Society of Actuaries on 7 July 2000. www.soa.org/research.

Cooper, D. R. (1997). Providing pensions for U.K. employees with varied working lives. *Journal of Actuarial Practice* 5(1), 5–47.

Gordon, M. S., Mitchell, O. S., and Twinney, M. M. (1997). *Positioning Pensions for the Twenty-First Century.* University of Pennsylvania Press.

Government Actuary's Department (2003). Occupational pension schemes 2000: eleventh survey by the Government Actuary. www.gad.gov.uk.

Hannah, L. (1986). *Inventing Retirement.* Cambridge University Press.

Humphreys, J. and Newman, C. (1993). Crediting rates for superannuation plans. *Transactions of the Institute of Actuaries of Australia* 229–345.

Lee, E. M. (1986). *An Introduction to Pension Schemes*. Institute of Actuaries and Faculty of Actuaries.

Mason, J. J. (1993). Design of company pension arrangements. Presented to Staple Inn Actuarial Society, 30 March. www.sias.org.uk

McGill, D. M., Brown, K. N., Haley, J. J., and Schieber, S. J. (1996). *Fundamentals of Private Pensions*. University of Pennsylvania Press.

National Association of Pension Funds (2003). Annual survey. NAPF 2003.

Turner, J. (2001). The design of rate of return guarantees for defined contribution plans. *Journal of Pensions Management* 7(1), pp 55–63.

Turner, J. and Watanabe, N. (1995). *Private Pension Policies in Industrialized Countries: A Comparative Analysis*. W E Upjohn Institute for Employment Research.

Chapter 18

Actuarial Modeling of
Defined-Benefit Plans

18.1 Cash-Flow Projections

The financial obligation undertaken by the sponsor of a defined-benefit plan is to ensure that the promised benefits are paid to current and future generations of members (and their dependants). This obligation can be better understood by estimating the benefit payments due in any future time period, so as to obtain a cash-flow profile for the expected benefit outgo over time.

Having projected the expected benefit outgo, alternative strategies for funding the plan can be considered. All feasible funding strategies must satisfy the following general funding equation:

Current value of fund + Expected present value of future contributions

= Expected present value of future benefit outgo

This equation is consistent with a wide variety of funding methods, ranging from "pay-as-you-go" when the value of the fund is always zero, to "initial funding" when no further contributions are required. The funding methods adopted for employer-sponsored plans normally lie somewhere between these extremes, often aiming for a stable contribution rate throughout the lifecycle of the plan. It is likely that further constraints on the funding strategy would be imposed by the regulatory requirements of the relevant country.

18.1.1 A Simple Model Plan

We shall derive formulae to project the benefit outgo of a final salary plan that pays only pension benefits to members who retire at normal retirement age. It is assumed that the active members of the plan form a stationary population, as given by a survival model in the form of an actuarial life table. The benefit structure can be summarized as follows:

Normal retirement age $= 65$
Pension $= k \times$ service $\times$ final salary (paid continuously)
Annual rate of increase to pensions in payment is c

It is assumed that the plan is established for an existing group of employees, continuously distributed between ages 25 and 65, in accordance with our life table. A stationary active member population is maintained by replacing retiring members with new entrants who join at age 25.

18.1.2 Projection of Benefit Outgo

18.1.2.1 *Projection of Pensionable Salaries*

In projecting the benefit outgo of the plan, we must allow for two types of salary increase:

general salary increases which reflect the growth in wages over time;
promotional salary increases based on the seniority of the employee.

If the plan is set up at time $t=0$, then the average pensionable salary of active members aged x at time t can be modeled by a function $S(x, t)$ given by

$$S(x, t) = \left(1 + j\right)^t s_x$$

where j is the rate of general salary escalation and s_x is the age-related promotional salary scale. The pensionable payroll $TS(t)$ is obtained by the summing the active member salaries over all ages; hence

$$TS(t) = \left(1 + j\right)^t \int_{25}^{65} s_x l_x \, dx$$

where $l_x \, dx$ is the number active members between ages x and $x + dx$.

18.1.2.2 *Change in Benefit Outgo over Time*

How does the benefit outgo of this plan change over time? To begin with it must be zero, because there are no pensioners. The first members who retire will have very little pensionable service, so the benefit outgo will grow slowly at first. The rate of growth will increase as more members retire with larger pensions. Eventually, the plan will reach a position where:

there is a stationary population of pensioners up to the maximum age in the life table;
all pensioners have retired with the maximum possible pensionable service of 40 years.

At this stage, the plan would be described as a *mature* plan.

The benefit outgo can be modeled mathematically as follows. The past service at retirement of a pensioner aged $65 + z$ at time t will be the lesser of $t - z$ and 40 years. Thus, the pension paid to a pensioner aged $65 + z$ at time t is given by the function $b(z, t)$, where

$$b(z, t) = k(t - z)(1 + j)^{t-z} s_{65}(1 + c)^z \quad \text{for } 0 \le t - z \le 40$$
$$b(z, t) = 40k(1 + j)^{t-z} s_{65}(1 + c)^z \quad \text{for } t - z > 40$$

The total benefit outgo of the plan $B(t)$ is given by integrating $b(z, t)$ over the values of z representing all the retired members of the plan. If we write $e^{\beta} = (1 + j)/(1 + c)$, we obtain the following equations for the total benefit outgo at time t:

$$B(t) = ks_{65}(1 + j)^t \int_0^t l_{65+z}(t - z) e^{-\beta z} \, dz \qquad \text{for } t \le 40$$

$$B(t) = 40ks_{65}(1 + j)^t \int_0^{t-40} l_{65+z} e^{-\beta z} \, dz$$
$$+ ks_{65}(1 + j)^t \int_{t-40}^t l_{65+z}(t - z) e^{-\beta z} \, dz \qquad \text{for } t \ge 40$$

If the highest age in the life table is w, then the pensioner population will become stationary when $t \ge w - 65$. However, all pensioners will not have retired with 40 years service until a later time, when $t \ge w - 25$. At this later time, we will have a mature plan in which the benefit outgo is given by

$$B(t) = 40ks_{65}(1 + j)^t \int_0^{w-65} l_{65+z} e^{-\beta z} \, dz$$

Thus, for a mature plan, the benefit outgo increases at the same rate as the general rate of salary escalation. It follows that the benefit outgo as a fraction of the pensionable payroll tends to an upper limit as the plan matures. Figure 18.1 shows how the benefit outgo for a typical defined-benefit plan might develop over time.

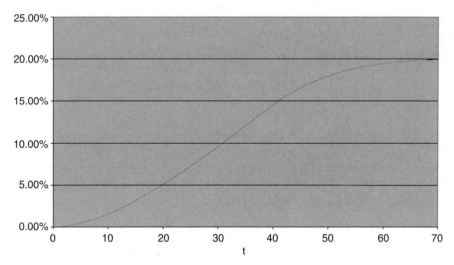

FIGURE 18.1 Benefit outgo as a percentage of pensionable salary

18.1.3 Cash-Flow Equation

For a funded pension plan, the change in the value of the fund over any time period is equal to the investment return on the fund plus the contributions received less the benefits paid. We now define the following variables in continuous time:

$F(t)$ is the value of pension fund assets at time t
$C(t)$ is the contribution income at time t
$\delta_i(t)$ is the force of return on pension fund assets at time t

The cash-flow equation can be represented as the following ordinary differential equation:

$$F'(t) = F(t)\delta_i(t) + C(t) - B(t)$$

If we replace the time-varying force of return $\delta_i(t)$ with a constant force of interest δ, this differential equation has a simple solution:

$$F'(t) - F(t)\delta = C(t) - B(t)$$

$$\Rightarrow \frac{d}{dt}\left[F(t)\,e^{-\delta t}\right] = C(t)\,e^{-\delta t} - B(t)\,e^{-\delta t}$$

$$\Rightarrow \left[F(t)\,e^{-\delta t}\right]_{t_0}^{\infty} = \int_{t_0}^{\infty} C(t)\,e^{-\delta t}\,dt - \int_{t_0}^{\infty} B(t)\,e^{-\delta t}\,dt$$

The upper limit of the left-hand bracket gives the limiting value, as $t \to \infty$, of the projected fund discounted by the assumed force of return. In this limit, we would expect to have a mature plan in which both the benefit outgo and the fund are increasing at the same rate as the pensionable payroll. If the expected force of return exceeds the expected increase in the pensionable payroll (something which pension actuaries normally assume), then the upper limit of the left-hand bracket is zero.

Thus we can write

$$- F(t_0) \, e^{-\delta t_0} = \int_{t_0}^{\infty} C(t) \, e^{-\delta t} \, dt - \int_{t_0}^{\infty} B(t) \, e^{-\delta t} \, dt$$

$$\Rightarrow F(t_0) + \int_{t_0}^{\infty} C(t) \, e^{-\delta(t-t_0)} \, dt = \int_{t_0}^{\infty} B(t) \, e^{-\delta(t-t_0)} \, dt$$

The above equation is a mathematical restatement of the general funding equation.

18.1.4 Stationary Fund Equation

For a mature plan with a stationary population, where the benefit outgo is a constant fraction of the pensionable payroll, we can derive a simple formula connecting the contribution rate and the fund. We again assume that the force of return is a constant given by $\delta_i(t) = \delta_i$ and express the benefit outgo and contribution income as

$$B(t) = B_M TS(t) = B_M TS(0)(1+j)^t$$
$$C(t) = C_M TS(t) = C_M TS(0)(1+j)^t$$

where B_M and C_M are constants for the mature plan.

It then follows that the cash-flow equation can only be satisfied if the fund is also growing in line with general salary escalation; hence

$$F(t) = F_M TS(0)(1+j)^t$$
$$\Rightarrow F'(t) = F_M TS(0)(1+j)^t \delta_j$$

where $\delta_j = \ln(1+j)$.

On substituting all of the above relationships into the cash-flow differential equation we obtain

$$B_M = F_M(\delta_i - \delta_j) + C_M$$

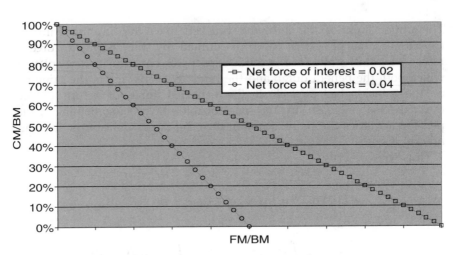

FIGURE 18.2 Stationary fund equation

B_M, C_M and F_M are respectively the mature plan benefit outgo, contribution rate, and fund, each expressed as a fraction of the pensionable payroll, and $\delta_i - \delta_j$ is equal to the force of interest net of salary escalation. This is the stationary fund equation, which is illustrated in Figure 18.2.

When $C_M/B_M = 1$ we have an unfunded or "pay-as-you-go" plan in which the fund is zero. As the fund increases, the required contribution decreases linearly. The slope of each line is the force of interest earned on the fund net of general salary escalation. When $C_M/B_M = 0$ we have a plan in which no contributions are required because the investment income alone is sufficient to pay the benefits. In practice, this would require a very large fund to have been accumulated. Mature defined-benefit plans are usually in an intermediate position, where a mixture of contributions and investment income is used to meet the benefit outgo.

The stationary fund equation refers to an idealized plan in which the benefit outgo and contribution income are fixed fractions of the pensionable payroll and the investment return net of general salary escalation is a constant. As shown above, this must be a mature plan with a stationary population of members and a stationary distribution of pensionable salary by age. Although such idealized conditions do not apply to real defined-benefit plans, the stationary fund equation is a useful model for projecting the cash flows of a mature plan. In particular, it can be used to determine the new equilibrium values that the fund and contribution rate will approach in the event of a change in the plan's investment experience or funding strategy; see Thornton and Wilson (1992).

18.1.5 Example of Cash-Flow Calculations

We now show how the cash-flow method could be used to control the funding strategy of a hypothetical defined-benefit plan. It is assumed that the plan starts off with a stationary population of active members who have no past service credits. The actuary projects the future benefit outgo at the following durations from the establishment of the plan.

t (years)	$B(t)/TS(t)$
20	0.05
40	0.15
≥ 60	0.20

The above projection indicates that the plan is expected to attain maturity after 60 years with a benefit outgo of 20% of the payroll. As one would expect, the rate of increase in the benefit outgo is greatest during the middle years between $t=20$ and $t=40$. We will make the simplifying assumption that the benefit outgo increases linearly over each of the 20-year periods before maturity.

Having projected the benefit outgo, the only further assumptions required to derive a funding strategy for the plan are the investment return on the assets and the rate of general salary escalation. As in the stationary fund equation, the relevant parameter is the investment return net of salary escalation, for which we shall assume

$$\frac{1+i}{1+j} = 1.02$$

The following variables will be used in the cash-flow calculations:

C_p is the required level contribution rate as a fraction of payroll
$PVC_p(t)$ is the present value of contributions paid between time zero and time t as a multiple of payroll at time zero
$B_p(t)$ is the benefit outgo at time t as a faction of payroll at time t
$PVB_p(t)$ is the present value of benefits paid between time zero and time t as a multiple of payroll at time zero
$SF_p(t)$ is the target fund at time t as a fraction of payroll at time t
$F_p(t)$ is the actual fund at time t as a fraction of payroll at time t

18.1.5.1 Required Contribution Rate

Our first task is to determine the level contribution rate required to fund the plan over an indefinite period. If there is no initial fund, then the general funding equation requires that

$$PVC_p(\infty) = PVB_p(\infty)$$

Under our assumption that the benefit outgo is linearly increasing over each of the 20-year periods before the plan matures, the present value of the benefit outgo is given by

$$0.05\left(\frac{(\bar{I}\bar{a})_{\overline{20|}}}{20|}\right) + 0.05v^{20}\bar{a}_{\overline{\infty|}} + 0.10v^{20}\left(\frac{(\bar{I}\bar{a})_{\overline{20|}}}{20}\right) + 0.10v^{40}\bar{a}_{\overline{\infty|}}$$

$$+ 0.05v^{40}\left(\frac{(\bar{I}\bar{a})_{\overline{20|}}}{20}\right) + 0.05v^{60}\bar{a}_{\overline{\infty|}}$$

All the above compound interest functions must be calculated at 2% per annum, i.e., the assumed investment return net of salary escalation. The above expression simplifies to

$$PVB_p(\infty) = \left((\bar{I}\bar{a})_{\overline{20|}}\big/20 + v^{20}\bar{a}_{\overline{\infty|}}\right)\left(0.05 + 0.10v^{20} + 0.05v^{40}\right)$$

The present value of the future contributions to the plan is given by

$$PVC_p(\infty) = C_p\bar{a}_{\overline{\infty|}}$$

Hence, setting $PVC_p(\infty)$ equal to $PVB_p(\infty)$ gives

$$C_p = \frac{\left((\bar{I}\bar{a})_{\overline{20|}}\big/20 + v^{20}\bar{a}_{\overline{\infty|}}\right)\left(0.05 + 0.10v^{20} + 0.05v^{40}\right)}{\bar{a}_{\overline{\infty|}}}$$

Calculating the compound interest functions at 2% gives

$$C_p = \frac{5.8352}{50.498} = 0.1156$$

18.1.5.2 Target Fund

Having calculated the required contribution rate, we define the "target fund" at any duration as the accumulated contributions less the accumulated benefits projected up to that duration. We now use this method to calculate the target fund at $t = 60$.

In order to make use of our previous calculations, we write

$$PVB_p(60) = PVB_p(\infty) - v^{60} \times 0.20 \times \bar{a}_{\overline{\infty|}}$$

$$= 5.8352 - v^{60} \times 0.20 \times 50.498$$

$$= 2.575$$

The present value of the contribution income up to $t = 60$ is given by

$$\text{PVC}_p(60) = 0.1156\bar{a}_{\overline{60}|} = 4.058$$

And the target fund at $t = 60$ is given by

$$\begin{aligned}
\text{SF}_p(60) &= \left[\text{PVC}_p(60) - \text{PVB}_p(60)\right] \times 1.02^{60} \\
&= (4.058 - 2.757) \times 1.02^{60} \\
&= 4.269
\end{aligned}$$

As the plan is mature for $t \geq 60$, we can use the stationary fund equation to check the above result. The stationary fund equation requires that

$$\begin{aligned}
B_p(60) &= \text{SF}_p(60) \times (\delta_i - \delta_j) + C_p \\
&= 4.269 \times \ln(1.02) + 0.1156 \\
&= 0.200
\end{aligned}$$

which is equal to the projected benefit outgo for $t \geq 60$. It follows that the target fund for the mature plan could have been obtained directly from the stationary fund equation, by writing

$$\text{SF}_p(60) = \frac{B_p(60) - C_p}{\delta_i - \delta_j}$$

18.1.5.3 *Amortization of Deficits*

As the plan develops over time, its experience is likely to differ from the assumptions made by the actuary at the outset. As a result, the actual fund at any duration is likely to differ from the target fund. If a significant difference arises, the contribution rate may require modification so that the fund is returned to its target value.

In our simple model, the actual and assumed experience of the plan might diverge because the actual benefit outgo differs from the projected benefit outgo, or because actual investment returns net of salary escalation are not equal to 2% per annum. We shall focus on the second cause by examining what would happen if the net investment return were zero over a 5-year period.

We first consider the consequences of a zero net return over the 5-year period following the establishment of the plan. The target fund after 5 years is given by

$$\begin{aligned}
\text{SF}_p(5) &= 0.1156s_{\overline{5}|} - \frac{1}{20} \times 0.05(\overline{I}\overline{s})_{\overline{5}|} \qquad @2\% \\
&= 0.1156 \times 5.256 - 0.0025 \times 12.923 \\
&= 0.5753
\end{aligned}$$

To obtain the actual fund we repeat the same calculation at zero net interest:

$$F_p(5) = 0.1156 \times 5 - 0.0025 \times 12.5$$
$$= 0.5468$$

The difference between the target fund and the actual fund represents a deficit in the funding schedule that can be amortized by increasing the contribution rate over some fixed period. We shall assume that an addition, AC_p, is made to the normal contribution rate of 11.56% over a further 5-year period (between $t = 5$ and $t = 10$). To remove the deficit, this extra contribution rate must satisfy

$$AC_p \bar{a}_{\overline{5}|} = SF_p(5) - F_p(5)$$

If future net investment returns are still expected to be 2% per annum, then the annuity certain would be calculated at 2% interest, which gives

$$AC_p = \frac{(0.5753 - 0.5468)}{4.760} = 0.0060$$

Thus, the contribution rate would be increased from 11.56% to 12.16% of the payroll for a 5-year period.

It is instructive to repeat the same example for a mature plan. We assume that the actual fund at $t = 60$ is equal to the target fund, which was calculated above as 4.269 times the payroll. After 5 years of zero net interest, the fund is given by

$$F_p(65) = F_p(60) \times (1+i)^5 + (C_p - B_p) \times \bar{s}_{\overline{5}|} \qquad @0\%$$

$$= 4.269 + (0.1156 - 0.20) \times 5$$

$$= 3.847$$

The target fund for the mature plan does not change over time, so the deficit to be amortized is equal to the reduction in the value of the fund. Hence, the addition to the contribution rate is given by

$$AC_p = \frac{(4.269 - 3.847)}{4.760} = 0.0887$$

Thus, the extra contribution rate required to amortize the deficit over a 5-year period is 8.87%, making a total contribution rate of 20.43% over 5 years. This shows that it is more difficult to maintain a particular funding target for a mature plan—the large size of the fund means that any deviation from the assumed investment return requires a larger adjustment to the contribution rate to return the fund to its target value.

18.1.6 Practical Value of Cash-Flow Projections

The cash-flow approach is the most general method of formulating the funding strategy of a defined-benefit plan. It permits consideration of any funding strategy that will provide sufficient resources for the promised benefits to be paid, assuming that the plan sponsor can continue to operate the plan indefinitely. However, the cash-flow method gives no indication of the adequacy of resources if the plan were to be wound up in the near future. Nor is it helpful in assigning pension costs to individual years for the purpose of financial reporting. For these reasons, the traditional actuarial valuation is not normally based on the cash-flow approach (see Section 18.2).

The cash-flow method is, nevertheless, invaluable for understanding how the financial position of a defined-benefit plan changes as it matures. The fact that the net cash flow of a funded plan (i.e., its income less outgo) changes so significantly over its lifecycle has important implications for its investment strategy. Another feature of the cash-flow method is that future entrants to the plan can be explicitly modeled in making benefit projections. This is an important advantage over traditional "closed group" funding methods, in which new entrants can only be allowed for in an implicit manner. Lastly, a cash-flow model of a defined-benefit plan is a prerequisite for making stochastic projections to compare alternative investment and funding strategies.

18.2 The Actuarial Valuation

The funding strategy of a defined-benefit plan is determined through periodic actuarial valuations. The actuary recommends a contribution rate when the plan is established and revises the required contribution rate (if necessary) in subsequent valuations. The actuary is also required to assess the security of the *accrued benefits* of the plan, which refers to those benefits earned from service prior to the date of the valuation. The benefits payable on the wind-up of a plan are normally equivalent (or reasonably close) to the accrued benefits, so comparing the value of the fund with the present value of the accrued benefits gives an indication of the solvency of the plan if it were to discontinue.

The traditional actuarial valuation is therefore based on separate calculations for the present value of the benefits that (i) have accrued from past service and (ii) are expected to accrue from future service. The recommended contribution rate must cover the expected cost of future service benefits with a further adjustment to allow for the amortization of any surplus or deficit from past service.

18.2.1   Objectives of Actuarial Valuation

An actuarial valuation may be carried out to fulfill one (or more) of several possible objectives, which are described below.

18.2.1.1   *Funding*

The actuarial valuation used to determine the required contribution rate is sometimes called the funding valuation. The actuary calculates a target value for pension fund at the valuation date, which is referred to as the *standard fund* or *actuarial liability*. The value of this target fund depends on the choice of funding method (see Section 18.2.4). The *standard contribution rate* is the recommended contribution rate when the actual fund equals the standard fund. It also depends on the choice of funding method.

The actual fund, however, does not generally equal the standard fund. When it is less than the standard fund the plan has a *deficit*, and when it is greater the plan has a *surplus*. If there is a deficit, then a positive adjustment to the standard contribution rate may be recommended to remove the deficit over some future timeframe. If there is a surplus, it can be removed by a negative adjustment to the standard contribution rate and/or the award of benefit improvements.

18.2.1.2   *Commercial Transactions*

If the accrued liabilities of a defined-benefit plan are transferred from one party to another, e.g., as the result of a merger or acquisition, then the value of these liabilities must be determined on a basis agreed by both parties. The method and assumptions agreed upon may differ from those used in the funding valuation.

18.2.1.3   *Accounting*

Modern pension accounting standards carefully define the method and assumptions that must be used to determine pension costs charged in the financial statements of the sponsoring employer. Similar detailed requirements apply to the recognition and disclosure of surpluses and deficits. Although the approaches prescribed in these accounting standards are based on actuarial valuation methods, significant differences may exist between the pension costs charged in financial statements and the contributions paid into the fund (as determined in the funding valuation).

18.2.1.4 Regulation

Pension regulation may narrow the range of permissible funding strategies for a defined-benefit plan. The size of the fund may be subject to a minimum value (to ensure reasonable security for the accrued benefits) and a maximum value (to prevent the abuse of tax privileges). Both the upper and lower limits would be based on some definition of the accrued liabilities of the plan, using different assumptions in each case. It is likely that neither basis would coincide with that used in the funding valuation. The regulations should prescribe what action the plan sponsor must take if the fund breaches either of these limits.

18.2.2 Valuation Assumptions

The assumptions required to value the liabilities of a defined-benefit plan are either financial or demographic. The question of how actuaries select valuation assumptions is beyond the scope of this chapter. For discussion of this important topic, refer to Thornton and Wilson (1992) and Anderson (1992, chapter 6).

Most of the financial assumptions are constants for which we shall use the following symbols:

v is the discount rate used to determine present values;
j is the rate of general salary escalation;
c is the rate of increase to pensions in payment;
g is the rate of increase to pensions in deferment.

The only age-dependent financial assumption is the promotional salary scale s_x, as described in Section 18.1.2. As only relative values of promotional salary scale are important, s_x is usually set equal to one at the youngest active member age.

The demographic assumptions consist of a multiple decrement model for the active members, known as the *service table*, and a single decrement model for the deferred pensioners and pensioners. The service table decrements will be given the following symbols:

d_x is the expected number of active members aged x dying before age $x+1$;
w_x is the expected number of active members aged x leaving service before age $x+1$;
r_x is the expected number of active members aged x taking early retirement before age $x+1$;
i_x is the expected number of active members aged x taking ill-health retirement before age $x+1$.

And for the surviving lives we shall use:

l_x is the expected number of active members remaining in service at age x;
l_x^p is the expected number of deferred pensioners (or pensioners) surviving
to age x.

In order to simplify our formulae, we shall assume that all decrements occur at the start of each year of age, rather than being uniformly distributed over each year (as is normally assumed).

18.2.3   Valuation Formulae

Valuation formulae are required to determine the expected present value of benefits payable under different contingencies and arising from different periods of service. A formula for the expected present value of future pensionable salaries is also required to determine the contribution rate required to fund the plan.

The following symbols will be used for the valuation data of a final salary plan providing benefits linked to salary at the date of retirement or earlier exit (it is assumed that the data are grouped by age (nearest) and sex):

SAL_x is the pensionable salary summed for active members aged x;
$PSSAL_x$ is the pensionable salary $\times$ past service summed for active
members aged x;
PEN_x is the pensions-in-payment summed for pensioners aged x;
$DPEN_x$ is the deferred pensions (including any projected increases in
deferment) summed for deferred pensioners aged x.

For the plan under consideration, we assume that:

k is the accrual rate per year of pensionable service;
$\bar{a}_x^r$ is the expected present value of benefits on voluntary or normal
retirement per unit of pension;
$\bar{a}_x^i$ is the expected present value of benefits on ill-health retirement per unit
of pension.

The annuity factors include the value of the spouse's pension payable on death after retirement and any other post-retirement ancillary benefits (e.g., a 5-year guarantee, see Section 17.2.4).

18.2.3.1   Active Member Retirement Benefits

The benefits paid on normal retirement and voluntary early retirement can be combined in a single set of valuation formulae. These are usually the most valuable benefits provided by a defined-benefit plan. The valuation formulae given below assume that no reduction factor is applied on early retirement.

We define:

PVA^r as the present value of active member retirement benefits from accrued service;

PV1^r as the present value of retirement benefits from next year's service;

PVF^r as the present value of retirement benefits from all future service.

The formulae for these liabilities are

$$\text{PVA}^r = k \sum_{x < \text{NRA}} \text{PSSAL}_x \sum_{z=x}^{\text{NRA}} \frac{r_z}{l_x} \frac{s_z}{s_x} (1+j)^{z-x} v^{z-x} \bar{a}_z^r$$

$$\text{PV1}^r = k \sum_{x < \text{NRA}} \text{SAL}_x \sum_{z=x+1}^{\text{NRA}} \frac{r_z}{l_x} \frac{s_z}{s_x} (1+j)^{z-x} v^{z-x} \bar{a}_z^r$$

$$\text{PVF}^r = k \sum_{x < \text{NRA}} \text{SAL}_x \sum_{z=x+1}^{\text{NRA}} (z-x) \frac{r_z}{l_x} \frac{s_z}{s_x} (1+j)^{z-x} v^{z-x} \bar{a}_z^r$$

where NRA is the normal retirement age and $r_{\text{NRA}} = l_{\text{NRA}}$.

It is straightforward to incorporate early retirement reduction factors in each of the above expressions. However, when the reduction factors are such that the value of the early retirement pension is equal to the actuarial reserve (see Section 18.3.2), early retirement decrements can be ignored altogether in the service table and the valuation formulae; this is a standard result of actuarial mathematics — see Bowers et al. (1997, chapter 11.4).

18.2.3.2 Active Member Withdrawal Benefits

We assume that early leavers receive a deferred pension payable at normal retirement age, based on their salary and service at exit. The value of these benefits obviously depends on the assumed withdrawal decrements. However, the withdrawal benefit is usually worth less than the actuarial reserve, because deferred pension increases are not expected to match the salary increases for members who remain in active service. Hence, increasing the withdrawal decrements usually reduces the overall value of the active member liabilities, because the increase in the value of the withdrawal benefits is offset by a greater reduction in the value of the retirement benefits.

We define:

PVA^w as the present value of active member withdrawal benefits from accrued service;

PV1^w as the present value of withdrawal benefits from next year's service;

PVF^w as the present value of withdrawal benefits from all future service.

The formulae for these liabilities are

$$PVA^w = k\bar{a}^r_{NRA} \sum_{x<NRA} PSSAL_x v^{NRA-x} \sum_{z=x}^{NRA-1} \frac{w_z}{l_x} \frac{l^P_{NRA}}{l^P_z} \frac{s_z}{s_x} (1+j)^{z-x} (1+g)^{NRA-z}$$

$$PV1^w = k\bar{a}^r_{NRA} \sum_{x<NRA} SAL_x v^{NRA-x} \sum_{z=x+1}^{NRA-1} \frac{w_z}{l_x} \frac{l^P_{NRA}}{l^P_z} \frac{s_z}{s_x} (1+j)^{z-x} (1+g)^{NRA-z}$$

$$PVF^w = k\bar{a}^r_{NRA} \sum_{x<NRA} SAL_x v^{NRA-x} \sum_{z=x+1}^{NRA-1} (z-x) \frac{w_z}{l_x} \frac{l^P_{NRA}}{l^P_z} \frac{s_z}{s_x} (1+j)^{z-x} (1+g)^{NRA-z}$$

If the plan provides a benefit on death before retirement to early leavers, then it would be necessary to add the value of this benefit to the withdrawal liabilities. However, if this death benefit is approximately equal to the actuarial reserve, then removing the survival probability l^P_{NRA}/l^P_z from each of the above equations may be an acceptable approximation.

18.2.3.3 Active Member Ill-Health Benefits

If the ill-health benefits are based only on accrued service, then the valuation formulae take the same form as those given above for retirement benefits. However, many defined-benefit plans give a more generous benefit based on total projected service or a fixed multiple of final salary.

If the ill-health benefit does not accrue with service, then the accrued liability for ill-health retirements may be defined to be zero. Under this approach, the liability attributed to each year of future service would be the expected present value of the ill-health benefits *payable* (rather than accrued) in that year. Thus, the cost of funding ill-health benefits is calculated as a *risk premium*, i.e., the cost of meeting the benefit outgo on a pay-as-you-go basis.

We define:

PV1^i as the present value of ill-health retirement benefit payable over next year's service;

PVF^i as the present value of ill-health retirement benefit payable over all future service.

If the ill-health retirement pension is a fraction of final salary equal to f^i, then the valuation formulae are

$$PV1^i = f^i \sum_{x < NRA} SAL_x \frac{i_x}{l_x} \bar{a}^i_x$$

$$PVF^i = f^i \sum_{x < NRA} SAL_x \sum_{z=x}^{NRA-1} \frac{i_z}{l_x} \frac{s_z}{s_x} (1+j)^{z-x} v^{z-x} \bar{a}^i_z$$

An alternative approach to the valuation of ill-health retirement benefits assumes that the benefit payable at any age accrues uniformly over service up to the date of retirement. Under this approach, there is an accrued liability for ill-health benefits. The valuation formulae are similar to those used for voluntary early retirement.

18.2.3.4 Active Member Benefits on Death before Retirement

The valuation formulae for benefits payable on the death of an active member are usually similar in form to those for the ill-health benefits. If the plan provides a spouse's pension on death-in-service, the valuation formulae should allow for the probability that the active member is married at death, which is likely to vary significantly with age. The annuity factors should reflect the expected mortality of the spouse, whose average age will depend on that of the deceased member. A common approach is to assume a fixed age difference between husbands and wives (e.g., wives are 3 years younger), and calculate the spouse's annuity at the assumed age of the spouse when the member dies. If the spouse's pension is payable only until remarriage, then a further adjustment to the annuity factors would be required.

As for the ill-health retirement benefits, we obtain formulae for:

$PV1^d$, the present value of death benefits payable over next year's service;
PVF^d, the present value of death benefits payable over all future service.

If the spouse's pension is a fraction of final salary equal to f^d, then the proportion of married active members aged x is m_x and the annuity factor for the spouse of a member aged x is $\bar{a}^r_{y(x)}$, the valuation formulae for the spouse's pension are:

$$PV1^d = f^d \sum_{x < NRA} SAL_x m_x \frac{d_x}{l_x} \bar{a}^r_{y(x)}$$

$$PVF^d = f^d \sum_{x < NRA} SAL_x \sum_{z=x}^{NRA-1} m_z \frac{d_z}{l_x} \frac{s_z}{s_x} (1+j)^{z-x} v^{z-x} \bar{a}^r_{y(z)}$$

As for ill-health retirement, there is an alternative valuation method in which death-in-service benefits are assumed to accrue uniformly over service up to the date of death. This results in an accrued liability for death in service benefits.

18.2.3.5   Pensioners and Deferred Pensioners

The benefits due to pensioners and deferred pensioners are all accrued liabilities of the plan, as they have arisen entirely from service prior to the valuation date. As for the active members, the data on benefit amounts are summed for age and sex cohorts. Further subdivisions of the data may be required to separate members who are in receipt of ill-health pensions from ordinary pensioners, because of the higher mortality expected for ill-health cases.

We define:

PVA^P as the present value of benefits due to pensioners;
PVA^{dp} as the present value of benefits due to deferred pensioners.

The valuation formulae are

$$PVA^P = \sum_x PEN_x \bar{a}_x^r$$

$$PVA^{dp} = \bar{a}_{NRA}^r \sum_{x < NRA} DPEN_x \frac{l_{NRA}^p}{l_x^p} v^{NRA - x}$$

It the plan provides a benefit on death in deferment that is close to the actuarial reserve, then this can be dealt with in the manner referred to above for the active member withdrawal benefits.

18.2.3.6   Total Liabilities

To apply actuarial funding methods, we require figures for the total accrued liabilities, the total liabilities accruing from next year's service, and the total liabilities from all future service, for which we shall use the symbols PVA, PV1, and PVF respectively. These are given by summing the various liability subtotals as follows:

$$PVA = PVA^r + PVA^w + PVA^P + PVA^{dp}$$

$$PV1 = PV1^r + PV1^w + PV1^i + PV1^d$$

$$PVF = PVF^r + PVF^w + PVF^i + PVF^d$$

18.2.3.7   Active Member Salaries

As actuarial funding methods normally express the required contribution rate as a percentage of the pensionable payroll, a valuation formula is required for the expected present value of the future salaries of the current active members.

We define:

PVSAL as the present value of next year's pensionable salaries;
PVFSAL as the present value of pensionable salaries over all future service.

Making the simplifying assumptions that salaries are paid annually in advance and are increased annually in arrears gives the following valuation formulae:

$$PVSAL = \sum_{x < NRA} SAL_x$$

$$PVFSAL = \sum_{x < NRA} SAL_x \sum_{z=x}^{NRA-1} \frac{l_z \, s_z}{l_x \, s_x} (1 + j)^{z-x} v^{z-x}$$

18.2.4   Actuarial Funding Methods

Actuaries have devised a number of different funding methods for defined-benefit plans. Each one gives a general formula for calculating the required contribution rate. Most funding methods also have a formula for the target level of assets that should be held in the fund, which is called the *standard fund* or *actuarial liability*. The standard contribution rate of any funding method is the recommended contribution rate when the actual fund equals the standard fund. If they are not equal, then an adjustment to the standard contribution rate is required to amortize the surplus or deficit.

We define:

SCR as the standard contribution rate as a fraction of pensionable payroll;
SF as the standard fund.

All actuarial funding methods should result in the payment of contributions that will be sufficient to meet the cost of future benefit payments. However, the cash-flow profile of these contributions varies from one method to another. An important way of understanding different funding methods is to identify the conditions under which the contribution rate will be stable over time. A comparison of the properties of different funding methods is provided in O'Regan and Weeder (1988).

A description of several of the most commonly used funding methods is given below. For a fuller exposition, refer to Anderson (1992). It should be noted that all actuarial funding methods are consistent with the cash-flow

projection method described in Section 18.1. Each of the funding methods
described below can be derived from the general funding equation
combined with a specific population model for the plan membership (see
Khorasanee (2002)).

18.2.4.1 Projected Unit Method

The standard contribution rate and standard fund for the projected unit
method (PUM) are given by

$$SCR_{PUM} = \frac{PV1}{PVSAL}$$

$$SF_{PUM} = PVA$$

The standard contribution rate of the PUM depends on the distribution of
pensionable salary by age, as the cost of a further year's accrual of benefit
increases with the age of the active member. If the distribution of
pensionable salary by age is stable over time, and the experience of the
plan follows the actuarial assumptions, then the standard contribution rate
will also be stable.

18.2.4.2 Attained Age Method

The standard contribution rate and standard fund for the attained age
method (AAM) are given by

$$SCR_{AAM} = \frac{PVF}{PVFSAL}$$

$$SF_{AAM} = PVA$$

As the cost of benefit accrual increases with age, a higher contribution
rate is required to fund the benefits arising from total future service than
from next year's service. Hence, the standard contribution rate of the
AAM is generally higher than that of the PUM. For a plan closed to new
entrants, paying the AAM standard contribution rate until all the
active members have retired will meet the cost of benefits arising from
future service if the experience of the plan follows the actuarial
assumptions.

The AAM standard fund is the same as that for the PUM. It is equal to
the present value of the accrued benefits (including deferred pensions
and pensions in payment). Hence, if the actual fund is equal to the
standard fund, the benefits from past service will be covered by the
fund and the benefits from future service will be covered by the standard
contribution rate.

18.2.4.3 Entry Age Method

For the entry age method (EAM) we define:

$\mathrm{PVF}(x_0)$ as the present value of future benefits due to a new entrant aged x_0, based on salary projected to date of retirement (or earlier exit);

$\mathrm{PVFSAL}(x_0)$ is the present value of projected future salary of a new entrant aged x_0.

The standard contribution rate and standard fund are given by

$$\mathrm{SCR_{EAM}} = \frac{\mathrm{PVF}(x_0)}{\mathrm{PVFSAL}(x_0)}$$

$$\mathrm{SF_{EAM}} = \mathrm{PVA} + \mathrm{PVF} - \mathrm{SCR_{EAM}} \times \mathrm{PVFSAL}$$

The standard contribution rate of the EAM is generally lower than that of both the PUM and AAM, whereas the standard fund is generally higher. The EAM standard contribution rate is stable for a plan in which the average entry age is equal to the assumed entry age x_0, provided that the experience of the plan follows the actuarial assumptions. It is also stable for a plan closed to new entrants.

For a new plan with an initial population of active members, the EAM standard fund is greater than zero (even though no benefit rights have accrued). If the plan has no initial fund, extra contributions must be paid to amortize the initial deficit.

18.2.4.4 Experience Deviates from Assumptions

To determine the standard contribution rate and standard fund for any funding method requires actuarial assumptions, such as those for investment returns, salary growth, and mortality. In a deterministic actuarial valuation, these parameters are normally assumed to be constant over time, but for a real plan they will vary in an unpredictable manner. This means that the actual fund will generally differ from the standard fund, so that

$$\mathrm{F} - \mathrm{SF} \neq 0$$

The actual fund minus the standard fund is called the surplus/(deficit). Note that because the standard fund depends on the chosen funding method, so does the size of the surplus or deficit.

18.2.4.5 Amortization of Surplus/Deficit

A natural way of dealing with surpluses and deficits is to adjust the contribution rate over a temporary period so as to restore the actual fund to the standard fund. This attempt is unlikely to be entirely successful, as the experience of the plan is again likely to deviate from that assumed over the period of amortization. However, the process of driving the fund towards the standard fund ensures that the surpluses and deficits stay within reasonable limits.

The standard fund, therefore, becomes a target at which the plan is continually aiming. The fact that the plan may never actually hit the target is not important; what matters is that we have a method of deciding when the contribution rate needs to be increased or decreased relative to the standard contribution rate. It follows that the actuarial valuation should be viewed as a system for decision making and control, rather than an exercise with the sole objective of reporting on the financial position of the plan. The concept of the actuarial valuation as a control system is explored in detail in Benjamin (1999).

18.2.4.6 Methods of Amortization

Fixed Period—The simplest method of amortization is to make a fixed adjustment to the standard contribution rate that will eliminate the surplus/deficit after a period of N years. If the adjustment is ACR and the revised contribution rate is CR we obtain

$$CR = SCR + ACR$$
$$ACR = (SF - F)/(PVSAL\ddot{a}_{\overline{N|}})$$

To allow for the increasing pensionable payroll, the annuity would be calculated using a net force of interest given by $\delta_i - \delta_j$. The choice of N is somewhat arbitrary, depending on the preference of the sponsoring employer and the advice of the actuary. Too large a value of N might cause the fund to drift away from its target, resulting in excessive surpluses or deficits. Too small a value of N might cause the contribution rate to be excessively volatile. Typically, N might be anywhere between 5 and 20 years. A stochastic approach for determining optimal values of N is given in Dufresne (1988).

Remaining Service Lives—Another approach is to make an adjustment to the standard contribution rate that would eliminate the surplus or deficit if the plan were closed to new entrants. In other words, amortize over the expected remaining service lives of the current active members. The formula for ACR then becomes

$$ACR = \frac{(SF - F)}{PVFSAL}$$

If this approach is combined with the AAM we obtain

$$ACR = \frac{(PVA - F)}{PVFSAL}$$
$$CR = SCR + ACR = \frac{(PVA + PVF - F)}{PVFSAL}$$

The above formula for the contribution rate gives a special case of the AAM known as the *aggregate method*. The aggregate method gives the level contribution rate required to pay off the benefits of a closed plan, allowing for the value of the assets in the fund.

18.2.4.7 *Independent Amortization of Gains/Losses*

Surpluses and deficits continually emerge in a defined-benefit plan because the experience of the plan changes in an unpredictable way. It follows that further surpluses/deficits will arise before any previous surplus/deficit has been fully amortized.

U.S. actuaries often amortize surpluses and deficits from different periods independently of each other. The surplus/deficit that arises over any period is referred to as the actuarial gain or loss arising over that period. Each gain/loss then gives rise to its own independent adjustment to the standard contribution rate, which is maintained until the end of its own amortization period.

The alternative to treating each gain or loss independently is to re-amortize the whole surplus or deficit at periodic intervals. This approach is favored by U.K. actuaries and is much simpler to apply, as there is no need to keep a record of separate adjustments and amortization periods for each historic gain or loss. For a stochastic comparison of the properties of different amortization methods refer to Owadally and Haberman (1999).

18.2.4.8 *Discontinuance Funding*

The funding methods described above are based on the assumption that the active members will continue to accrue benefits from future service until retirement or earlier exit. These funding methods are centered on the definition of the standard contribution rate, which remains stable under certain clearly defined conditions. The definition of the standard fund can then be derived from the general funding equation given in Section 18.1.

Discontinuance funding is a different approach that aims to ensure that the plan has sufficient assets to cover the liabilities that would arise on its termination. In such an eventuality, the active members would no longer accrue benefits from future service, but would be entitled to deferred pensions based on their service at the date of termination. The benefits due to pensioners and deferred pensioners would be unchanged. The case for discontinuance funding is made by McLeish and Stewart (1987).

Discontinuance funding has some similarities to the PUM, as the required standard contribution rate is also based on the projected increase in the active member liabilities over their next year of service. The differences are as follows:

The accrued liabilities at any date are based on the benefits that would be payable if the scheme were wound up at that date.
The standard contribution rate includes the revaluation of the accrued liabilities of the active members by 1 year's projected salary growth.
Both the assets and liabilities of the plan are taken at market value. The discount rate used to value the liabilities should be based on the market yield on assets that generates a similar cash-flow profile to the accrued benefits.

If the active members receive the same benefits as early leavers on termination of the plan, then their accrued benefits would be based on current salaries rather than projected salaries, with increases in deferment up to the date of retirement as provided under the rules of the plan. In these circumstances, the *amount* of the projected accrued benefits would normally be less than assumed under the PUM. The *expected present value* of these benefits, however, depends on the market-linked discount rate, which may differ significantly from that used in a normal actuarial valuation.

18.2.4.9 Valuation Formulae for Discontinuance Funding

We now define the following variables for a final salary plan in which the active members receive the same benefits as early leavers if the plan is wound-up.

$PV1^D$ is the present value of benefits from next year's service assuming active members leave at the end of the year.
PVA^D is the present value of active members' accrued benefits assuming active members leave at the valuation date.
PVA^{D+1} is the present value of active members' accrued benefits assuming active members leave at the end of the year.

The formulae for these variables are

$$PV1^D = k\bar{a}^{\mathrm{r}}_{\mathrm{NRA}} \sum_{x<\mathrm{NRA}} \mathrm{SAL}_x \left(\frac{1+j}{1+g}\right) \frac{s_{x+1}}{s_x} \frac{l^P_{\mathrm{NRA}}}{l^P_x} (1+g)^{\mathrm{NRA}-x} v^{\mathrm{NRA}-x}$$

$$PVA^D = k\bar{a}^{\mathrm{r}}_{\mathrm{NRA}} \sum_{x<\mathrm{NRA}} \mathrm{PSSAL}_x \frac{l^P_{\mathrm{NRA}}}{l^P_x} (1+g)^{\mathrm{NRA}-x} v^{\mathrm{NRA}-x}$$

$$PVA^{D+1} = k\bar{a}^{\mathrm{r}}_{\mathrm{NRA}} \sum_{x<\mathrm{NRA}} \mathrm{PSSAL}_x \left(\frac{1+j}{1+g}\right) \frac{s_{x+1}}{s_x} \frac{l^P_{\mathrm{NRA}}}{l^P_x} (1+g)^{\mathrm{NRA}-x} v^{\mathrm{NRA}-x}$$

18.2.4.10 Current Unit Method

The current unit method (CUM) is the funding method in which the accrued liabilities of the active members are based on their salaries and service at the valuation date. This is equivalent to discontinuance funding provided that the required deferred pension increases are allowed for as in the valuation formulae given above.

The standard contribution and standard fund for the CUM are given by

$$\text{SCR}_{\text{CUM}} = \frac{\left(\text{PV1}^{\text{D}} + \text{PVA}^{\text{D}+1} - \text{PVA}^{\text{D}} + \text{PV1}^{\text{d}} + \text{PV1}^{\text{i}}\right)}{\text{PVSAL}}$$

$$\text{SF}_{\text{CUM}} = \text{PVA}^{\text{D}} + \text{PVA}^{\text{P}} + \text{PVA}^{\text{dp}}$$

Even when the experience of the plan follows the actuarial assumptions, the CUM standard contribution rate is only stable for a plan in which both the distribution of pensionable salary by age and the distribution of past service by age are stable. This could only happen for a mature plan. For a new plan, the CUM standard contribution rate tends to increase over time.

18.2.5 Example Illustrating Different Funding Methods

In the long run, all actuarial funding methods should ensure that an on-going plan has sufficient resources to pay its benefits. The general funding equation indicates that there are an infinite number of funding strategies that will meet this objective, and the actuarial funding methods described above are particular variants of the more generalized cash-flow method.

At any point in time, however, different funding methods will produce different recommended contribution rates, because the timing of the payment contributions (or the "pace of funding") is unique to each one. This is illustrated below for a hypothetical actuarial valuation in which the calculations shown in Table 18.1 have been carried out.

18.2.5.1 Projected Unit Method

The standard contribution rate of the PUM is given by

$$\text{SCR}_{\text{PUM}} = \frac{\text{PV1}}{\text{PVS}} = \frac{15}{100} = 15\%$$

The PUM standard fund is equal to the present value of the accrued benefits; hence

$$\text{SF}_{\text{PUM}} = \text{PVA} + \text{PVP} + \text{PVD} = 150 + 100 + 50 = 300 \text{ units}$$

Table 18.1 Hypothetical actuarial valuation calculations

	Symbol	Value
Present value of active member benefits[a] from next year's service	PV1	15 units
Present value of active member benefits[a] from future service	PVF	180 units
Present value of active member benefits[a] from past service	PVA	150 units
Present value of benefits due to pensioners	PVP	100 units
Present value of benefits due to deferred pensioners	PVD	50 units
Present value of pensionable salaries from next year's service	PVS	100 units
Present value of pensionable salaries from future service	PVFS	1000 units
Actuarial value of assets held in pension fund	AVF	350 units
Contribution rate to fund benefits of a new entrant aged 25	CR(25)	12%

[a]Based on salaries projected to date of retirement or earlier exit.

The actuarial value of the assets is given as 350 units. How this might have been calculated is considered in Section 18.2.6, but for the purpose of this example it will suffice to note that the plan has a surplus given by

$$\text{Surplus} = \text{AVF} - \text{SF}_{\text{PUM}} = 350 - 300 = 50 \text{ units}$$

We will assume this surplus is amortized over the expected future service of the current active members, which results in an adjustment to the standard contribution rate given by

$$\text{ACR} = \frac{-\text{Surplus}}{\text{PVFS}} = \frac{-50}{1000} = -5\%$$

Hence, the recommended contribution rate under the PUM is given by

$$\text{CR}_{\text{PUM}} = \text{SCR}_{\text{PUM}} + \text{ACR} = 15\% - 5\% = 10\%$$

18.2.5.2 Attained Age Method

The standard contribution rate of the AAM is given by

$$\text{SCR}_{\text{AAM}} = \frac{\text{PVF}}{\text{PVFS}} = \frac{180}{1000} = 18\%$$

The standard fund is the same as that for the PUM; thus

$$\text{SF}_{\text{AAM}} = \text{PVA} + \text{PVP} + \text{PVD} = 300 \text{ units}$$

And if the method of amortization of surplus is unchanged we have

$$\text{CR}_{\text{AAM}} = \text{SCR}_{\text{AAM}} + \text{ACR} = 18\% - 5\% = 13\%$$

Note that this is also the required contribution rate for the aggregate method.

18.2.5.3 Entry Age Method

With an assumed entry age of 25, the standard contribution rate of the EAM is the contribution rate required to fund the benefits of a new entrant aged 25; hence

$$SCR_{EAM} = CR(25) = 12\%$$

The standard fund of the EAM is given by

$$SF_{EAM} = PVA + PVP + PVD + PVF - SCR_{EAM} \times PVFS$$
$$= 150 + 100 + 50 + 180 - 0.12 \times 1000 = 360 \text{ units}$$

Unlike the other two funding methods, the EAM results in a deficit given by

$$\frac{\text{Surplus}}{\text{(Deficit)}} = AVF - SF_{EAM} = 350 - 360 = -10 \text{ units}$$

If this deficit is amortized, then, as before, over the expected future service of the current active members we obtain

$$ACR = \frac{10}{1000} = 1\%$$

$$CR_{EAM} = SCR_{EAM} + ACR = 12\% + 1\% = 13\%$$

This is the same answer as obtained for the AAM, which is not a coincidence. It is not difficult to show mathematically that both the AAM and EAM are equivalent to the aggregate method when the surplus/deficit is amortized over the expected future service of the current active members. For further numerical illustrations of the use of actuarial funding methods, refer to Aitken (1994).

18.2.6 Valuation of Assets

18.2.6.1 Pension Fund Assets

Pension fund assets consist mainly of quoted securities traded in the stock market or holdings in mutual funds that invest in such securities. The only significant exceptions are cash deposits and real estate, which have a smaller allocation in most pension fund portfolios.[1] It follows that the market value of most pension fund assets can be readily determined. The main exception is directly held real estate, for which an estimate from a qualified valuer would be required.

Should pension fund assets be taken at market value in an actuarial valuation? Although this is a possible approach — and one particularly

---

[1]Descriptions of all these asset classes are given in Chapter 2.

suited to certain types of valuation—actuaries have devised alternative asset valuation methods that are frequently used. Below, we outline the justifications for using "off-market" asset valuation methods and describe two such methods. We also explain how the choice of the discount rate used to value the liabilities must be consistent with the asset valuation method.

18.2.6.2 *Why Not Use Market Values?*

There are some prima facie arguments against using market values, particularly in funding valuations where the plan is assumed to be open to new entrants.

A recently established plan has no need to realize assets to pay benefits because its contribution income will exceed its benefit outgo for many years. The benefit outgo increases as the plan matures, however, and stabilizes at a level above the contribution income for a plan with a stationary population of members. The stationary fund equation (see Section 18.1.4) indicates that the gap between the contribution income and benefit outgo of a mature plan is equal to the fund multiplied by the valuation interest rate net of salary escalation. Provided that the running yield on the fund is sufficient to cover this difference (a very reasonable proposition in practice), there is still no need to realize assets in order to pay benefits. It follows that a mature plan with a stationary population of members can avoid having to sell assets to meet benefit payments. The cash flow from contributions and investment income should be sufficient to meet the benefit outgo indefinitely.

The above reasoning suggests that the market value of the pension fund is not a crucial factor in deciding upon a funding strategy for an on-going plan. What matters, instead, is the projected investment income from the assets, as it is this income stream that will be used to meet the benefit outgo. This is the idea behind the discounted cash-flow asset valuation method, as proposed by Day and McKelvey (1964).

Another problem with taking pension fund assets at market value is the potentially large short-term volatility in market values, particularly if the fund invests predominantly in equities. It is undesirable for long-term financial decisions, such as the amortization of surplus or the enhancement of benefits, to be unduly influenced by these short-term market movements. This can be avoided by applying some kind of smoothing procedure to the market value of the fund.

In the U.K., the discounted cash-flow method has worked well as a method of smoothing market values. An alternative approach, more common the U.S., is to smooth out random fluctuations directly by using a rolling average of previous market values, which is the second off-market valuation method described below.

18.2.6.3 Discounted Cash-Flow Method

The discounted cash-flow value of an asset is equal to the expected present value of the income stream from the asset. The definition of "income stream" includes equity dividends, rents, coupon payments and capital repayments from redeemable securities. The discount rate used to calculate the present value of the income stream should be the same as the discount rate used to value the liabilities, a requirement which follows from the meaning of the general funding equation presented in Section 18.1.

For a fixed-income bond, the calculation requires no valuation assumption other than the discount rate, as the interest and capital repayment are fixed monetary amounts. For equities (or real estate) the future income stream is not known in advance and is normally expected to continue over an indefinite period. Although dividends and rents can fall as well as rise, the long-term trend is generally expected to be one of positive growth correlated with the rate of price inflation. Hence, the future income stream can be estimated by assuming that the latest dividend (or rent) received will grow at some fixed compound rate. This requires an assumption for dividend (or rental) growth.

We now define the following variables for an equity stock that pays an annual dividend:

MV is the current market value of stock
DV is the discounted cash-flow value of stock
d is the latest dividend (assumed just received)
e is the projected annual rate of dividend growth
y is the current dividend yield

The expected present value of the dividend stream is given by

$$DV = d \sum_{n=1}^{\infty} (1+e)^n v^n$$

Using the formula for the sum of an infinite geometric progression gives

$$DV = \frac{d(1+e)v}{1-(1+e)v}$$

The dividend payment d is equal to the market value of the stock multiplied by its dividend yield, so substituting for d gives

$$DV = MV\left(\frac{y(1+e)v}{1-(1+e)v}\right)$$

We now define the "par dividend yield" y_0 as the dividend yield at which the discounted cash-flow and market values are equal. If the future dividend yield on the asset fluctuates around a mean value of y_0, then the market value of the asset will vary around a long-term trend equal to its discounted cash-flow value. Setting $DV = MV$ in the above equation gives

$$y_0 = \left(\frac{1+i}{1+e}\right) - 1$$

$$\Rightarrow DV = MV\left(\frac{y}{y_0}\right)$$

The above relationship gives the simplest method of calculating the discounted cash-flow value of an equity stock. The market value of the stock and its current dividend yield should be readily available for a quoted security. The par dividend yield is an assumption made by the actuary; it should equal the expected long-term average dividend yield on the stock. If we rearrange the expression given above for y_0, then we obtain

$$1 + i = (1 + e)(1 + y_0)$$

if the absolute values of e and y_0 are much less than one, this expression becomes

$$i \approx y_0 + e$$

Thus, the discount rate for the equity stock is approximately equal to the sum of the par dividend yield and the expected dividend growth. As y_0 is the expected average future dividend yield, this discount rate is the expected average return on new money invested in the stock. An important feature of the discounted cash-flow approach is the distinction made between the expected return on new money and the expected return on existing assets. The expected return on the stock at the valuation date is given by

$$i \approx y + e$$

i.e., it is based on the actual dividend yield at the valuation date rather than the par dividend yield. It follows that the market value of the stock is written up when its expected return exceeds that on new money, and written down when the reverse is true.

18.2.6.4 *Discounted Cash-Flow Method with a Notional Portfolio*

One could argue that the discounted cash-flow valuation method should be applied independently to every security in the pension fund at the valuation date. Such an approach, however, would lead to both practical and theoretical difficulties.

The practical problem is the need to estimate the future investment income from every asset in the pension fund portfolio. This would be particularly difficult for equities, where realistic dividend growth assumptions would vary greatly from one stock to another. The theoretical problem is the assumption, implicit in the discounted cash-flow method, that all of these assets will be held indefinitely by the pension fund. This is unlikely to be true, because pension fund portfolios do change over time, irrespective of whether the fund needs to realize assets to pay benefits.

A solution to these problems is to base the discounted cash-flow method on a notional portfolio, typically subdivided between market index portfolios for equities and bonds. At the valuation date, it is assumed that the actual assets of the fund are sold at market value and reinvested in the notional portfolio. The discounted cash-flow value of this notional portfolio is then taken as the actuarial value of the pension fund. As plenty of historic data are available for the market index portfolios, it should be easier for the actuary to make the assumptions required for the discounted cash-flow calculations.

The notional portfolio might be thought of as a benchmark portfolio for the future asset allocation of the pension fund. The actual portfolio will differ from the benchmark only because of active fund management in pursuit of a better return than available from the market index portfolios. The difference between the return on the actual and notional portfolios would be allowed to emerge as an item of surplus or deficit arising from the performance of the fund manager.

18.2.6.5 *Problems with the Discounted Cash-Flow Method*

The discounted cash-flow valuation method requires the actuary to assume either a par dividend yield or a rate of dividend growth. For a given discount rate, choosing the par dividend yield implies the dividend growth assumption and vice versa.

The concept of a par dividend yield implies that there is a stationary long-term mean for the dividend yield on the notional equity portfolio. Even if the historic data can be credibly interpreted in this way, estimating this mean dividend yield will require the selection of an arbitrary sampling period. As it is not clear how far back into the past one should venture when estimating the mean value of a *future* yield, the choice of par dividend yield is likely to be fairly subjective. The use of the discounted cash-flow method, therefore, is problematic for accounting and regulatory valuations, for which a standardized approach is desirable.

If the dividend yield on the equity market does not have a stationary mean, then the discounted cash-flow approach cannot be applied with any confidence. The popularity of the discounted cash-flow method in the U.K. is based on the fact that the equity market dividend yield has appeared to fluctuate around a long-term average of 4 to 5% per annum. Since the mid-1990s, however, the dividend yield on U.K. equities has remained well below its long-term average, a development partly brought about by changes in the taxation of dividends. U.K. actuaries have responded by lowering their par dividend yield assumption, so that the actuarial value of the fund does not diverge too far from its market value. This has tended to undermine confidence in the discounted cash-flow method and has led to the consideration of alternative methods, as discussed in Head et al. (2000).

18.2.6.6 Smoothed Market Value

The direct smoothing of market values in actuarial valuations is common in the U.S., where instability in the equity market dividend yield has made the discounted cash-flow method impractical. U.S. pension actuaries have devised alternative smoothing methods, which are documented in a survey of practice conducted by the Society of Actuaries (1998).

The simplest interpretation of this method is to calculate the actuarial value of the fund as a rolling average of previous market values over some fixed period. This rolling average cannot be based on historic market values of the pension fund itself, because cash flows, as well as market movements, affect the value of the fund. Instead, the rolling average must be based on notional historic market values for the assets held at the valuation date.

The smoothing method outlined below follows the approach given in Anderson (1992, chapter 5). We assume that the rolling average is calculated over N equal time intervals leading up to the valuation date.

Let:

$\bar{F}_N$ be the smoothed market value of fund at valuation date
F_n be the market value of pension fund at end of nth time period $(n \leq N)$
P_n be the price index for pension fund at end of nth time period
M_n be the new money invested at end of nth time period

The new money available for investment in each time period is equal to the investment income plus the contribution income less the benefit outgo in that period. The market value of the fund is measured immediately after the investment of new money. Values for F_n and M_n should be available from the accounting statements of the pension plan.

The smoothed market value of the fund is given by

$$\bar{F}_N = \frac{F_N}{N} \sum_{n=1}^{N} \frac{P_n}{P_N}$$

where the price index P_n can be derived from the following recurrence relationship:

$$\frac{P_n}{P_{n+1}} = \frac{F_n}{F_{n+1} - M_{n+1}}$$

A convenient scale for the price index is obtained by setting $P_N = 1$.

Unlike the discounted cash-flow method, the use of a smoothed market value involves no distinction between the expected return on fund and the expected return on new money. The valuation interest rate should be based on the expected return on the assets held over the smoothing period.

18.2.6.7 Taking the Fund at Market Value

Arguments for taking pension fund assets at market value are:

Market values are widely recognized as the true value of assets that are traded in an active market.

Market values are objective, whereas off-market values depend on the particular methodology and assumptions chosen by the actuary.

We have already seen that pension fund assets should be taken at market value when the purpose of the valuation is to assess solvency on a discontinuance basis. The case for using market values is also very strong when the purpose of the valuation is to produce pension cost and pension liability figures for the accounts of the sponsoring employer. In pension accounting standards, the need for an objective valuation method that can be understood by account users is arguably a more important consideration than stabilizing pension costs through the use of smoothing techniques.

If pension fund assets were taken at market value, then the discount rate used to value the liabilities should be linked to market investment yields at the valuation date. For a discontinuance solvency valuation, one can argue that the matching asset-class for the wind-up liabilities is long-dated bonds, so that the valuation discount rate should be equal to the market redemption yield on these bonds. Hence, the actuary should use a discount rate that is linked to the yield on long-dated bonds, irrespective of the assets actually held at the valuation date.

For pension accounting purposes, a discount rate derived from bond yields is consistent with the objective of prescribing a standardized method for the valuation of pension liabilities. On the other hand, the discontinuance valuation method has been rejected by modern pension accounting standards. If the liabilities were determined on an on-going basis, based on salaries projected to retirement, then it could be argued that equities should form a significant part of the portfolio used to cover these liabilities. If so, then it could be further argued that the discount rate should be based on the market yield on a portfolio made up of both equities and bonds.

Unfortunately, the "market yield" is less easily determined for equities than for bonds. Although the current dividend yield on an equity portfolio is known, we have shown that the expected return is approximately equal to the dividend yield *plus* expected future dividend growth. Subjective estimates for future dividend growth cannot be the basis of a standardized valuation method, so modern pension accounting standards have settled for a valuation discount rate linked entirely to bond yields. As equity returns are generally expected to exceed bond returns over long periods, it can be argued that the discount rate based on bond yields is consistent with the overriding accounting principle of prudence.

The introduction of market-based accounting standards may lead to similar approaches becoming more widespread in the funding valuation. Sponsoring employers may prefer to use the same valuation method for accounting and funding purposes, so that the contributions paid into their plan are similar to the pension costs reported in their accounts.

18.2.7   Example Illustrating Alternative Asset Valuation Methods

The calculations required to obtain both the discounted cash-flow value and smoothed market value of a pension fund will now be demonstrated for a hypothetical defined-benefit plan. We will also show how the chosen asset valuation method affects both the valuation discount rate and the ratio of the assets to the liabilities.

Quarterly financial data for a pension fund are given in Table 18.2 for a 3-year period leading up to the date of an actuarial valuation. The date of the actuarial valuation coincides with the end of the 12th quarter.

18.2.7.1   *Discounted Cash-Flow Method Using Notional Portfolio*

For the discounted cash-flow method we make the following financial assumptions:

Future rate of inflation:	3% per annum
Future dividend growth:	4% per annum
Par dividend yield:	3% per annum
Notional portfolio:	2/3 in equity index portfolio, 1/3 in 20-year bonds

To calculate the discounted cash-flow value of the fund, we assume that it is invested in the notional portfolio at the valuation date. The valuation interest rate will be taken as the expected return on investing new money in the equity index portfolio, which we have shown to be approximately equal to

$$\text{Par dividend yield} + \text{Future dividend growth}$$
$$= 3\% \text{ p.a.} + 4\% \text{ p.a.} = 7\% \text{ p.a.}$$

Table 18.2 Quarterly financial data for a pension fund

	During quarter			At end of quarter		
Quarter	Investment income	Contributions	Benefit outgo	Market value of fund	Equity index div yield (%)	20-year bond yield (%)
1	78	100	150	10 928	2.34	4.62
2	85	104	156	11 180	2.26	4.84
3	87	104	156	10 880	2.41	5.07
4	90	104	156	12 224	2.12	4.82
5	92	106	159	12 018	2.11	4.73
6	90	110	165	12 053	2.16	4.72
7	91	110	165	12 210	2.17	4.88
8	94	110	165	12 127	2.23	4.54
9	91	112	167	11 435	2.53	4.70
10	93	115	172	11 585	2.42	5.15
11	96	115	172	10 233	2.87	4.88
12	91	115	172	11 086	2.63	4.89

Although the expected return on 20-year bonds would normally be lower than the expected return on equities, the discounted cash-flow method automatically allows for this — the bonds in the notional portfolio will be given a value at which their yield will also be 7% per annum. Assuming an annual coupon rate of 5%, the discounted cash-flow value of the bond element of the notional portfolio is given by

$$\frac{\text{DCF value}}{\text{Market value}} = \frac{5a_{\overline{20}|} + 100v^{20} \, @7\%}{5a_{\overline{20}|} + 100v^{20} \, @4.89\%} = 0.777$$

For the equity part of the portfolio we have shown that

$$\frac{\text{DCF value}}{\text{Market value}} = \frac{\text{Actual dividend yield}}{\text{Par dividend yield}}$$

$$= \frac{2.63\%}{3.00\%} = 0.877$$

For the whole notional portfolio, composed of equities and bonds in the ratio $2:1$, the discounted cash-flow value is given by

$$\frac{\text{DCF value}}{\text{Market value}} = \frac{2}{3} \times 0.877 + \frac{1}{3} \times 0.777 = 0.844$$

Hence, the discounted cash-flow value of the fund at the valuation date is

$$11\,086 \times 0.844 = 9357 \text{ units}$$

Table 18.3 Smoothed market value

Quarter	New money	Market value of fund	Price index ratio	Price index
1	28	10 928	0.980	1.020
2	33	11 180	1.031	1.040
3	35	10 880	0.893	1.009
4	38	12 224	1.020	1.130
5	39	12 018	1.000	1.108
6	35	12 053	0.990	1.108
7	36	12 210	1.010	1.119
8	39	12 127	1.064	1.108
9	36	11 435	0.990	1.041
10	36	11 585	1.136	1.052
11	39	10 233	0.926	0.926
12	34	11 086	–	1.000

18.2.7.2 Smoothed Market Value

The smoothed market value will be derived from a 3-year rolling average of a historic price index of the fund. To calculate the price index for the fund over this period we note that over each quarter:

New money = Investment income + Contribution income − Benefit outgo

If we make the simplifying assumption that the new money is invested at the end of each quarter,[2] the following relationship must hold:

$$\frac{\text{Price index at start of quarter}}{\text{Price index at end of quarter}} = \frac{\text{Fund at start of quarter}}{\text{Fund at end of quarter} - \text{New money}}$$

Setting the price index at the valuation date equal to one leads to the results in Table 18.3. The mean value of the price index in the final column of Table 18.3 is 1.055. This is the adjustment factor that must be applied to the market value of the fund at the valuation date. Hence:

Smoothed market value of fund at valuation date = 1.055 × 11 086 units

= 11 696 units

This is a much higher figure than the discounted cash-flow value obtained above. However, the effect on the results of the valuation also depends on the interest rate used to value the liabilities. This interest rate should be

---

[2]It may be more accurate to assume that the new money is invested at the mid-point of each quarter and solve an equation of value to calculate the growth-rate of the fund. If the amount of new money is small compared with the size of the fund, however, the simpler method given above produces virtually the same result.

based on the expected return on the assets held over the smoothing period, which we shall assume to be close to the notional portfolio.

From Table 18.2 we can calculate the following average yields over the 3-year smoothing period:

Mean equity dividend yield $= 2.35\%$ per annum
Mean 20-year bond yield $= 4.82\%$ per annum

Our estimate for future dividend growth was 4% per annum, so

$$\text{Expected return on equities} = 2.35\% \text{ p.a.} + 4\% \text{ p.a.} = 6.35\% \text{ p.a.}$$

If we take the mean 20-year bond yield as the expected return on the bond element of the portfolio then we obtain

$$\text{Expected return on fund} = \frac{2}{3} \times 6.35\% \text{ p.a.} + \frac{1}{3} \times 4.82\% \text{ p.a.} = 5.84\% \text{ p.a.}$$

This is a much lower figure than the interest rate of 7% per annum used above for the discounted cash-flow method. It follows that the smoothed market value method would lead to a higher value for the accrued liabilities, as well as to a higher value for the fund. The possible effect on the ratio of the assets to the liabilities is explored below.

18.2.7.3 Consistent Valuation of Liabilities

We now assume that the discounted mean term of the accrued liabilities is 20 years and that the value of the accrued liabilities using a discount rate of 7% per annum is 9000 units. It follows that that ratio of the assets to the liabilities under the discounted cash-flow method is given by:

$$\text{DCF funding ratio} = \frac{9357}{9000} = 1.04$$

For the smoothed market value method, we derived a valuation interest rate of 5.84% per annum. Given that the value of the liabilities is 9000 units using a discount rate of 7% per annum, immunization theory gives us the following estimate for the value of the liabilities using a discount rate of 5.84% per annum:

$$\text{Value of liabilities @ } 5.84\% = 9000 \times \left(\frac{1.07}{1.0584}\right)^{20} = 11\,192 \text{ units}$$

Hence, the ratio of the assets to the liabilities is given by:

$$\text{Smoothed MV funding ratio} = \frac{11\,696}{11\,192} = 1.05$$

For completeness, we shall estimate the funding ratio for a valuation method based on the actual market value of the fund at the valuation date, which was 11086 units. An appropriate discount rate for the liabilities is derived as follows:

Equity dividend yield at valuation date $= 2.63\%$ per annum
20-year bond yield at valuation date $= 4.89\%$ per annum

Using an unaltered assumption of 4% per annum for future dividend growth gives

Expected return on equities $= 2.63\%$ p.a. $+ 4\%$ p.a. $= 6.63\%$ p.a.

And assuming that the actual portfolio is similar to the notional portfolio gives

$$\text{Expected return on fund} = \frac{2}{3} \times 6.63\% \text{ p.a.} + \frac{1}{3} \times 4.89\% \text{ p.a.} = 6.05\% \text{ p.a.}$$

The estimated value of the accrued liabilities using an interest rate 6.05% is given by

$$\text{Value of liabilities @ } 6.05\% = 9000 \times \left(\frac{1.07}{1.0605}\right)^{20} = 10\,757 \text{ units}$$

Hence, the ratio of the assets to the liabilities is given by

$$\text{MV funding ratio} = \frac{11\,086}{10\,757} = 1.03$$

We have shown that the consistent valuation of assets and liabilities can lead to similar funding ratios for asset valuation methods that produce very different values for the assets and liabilities separately. For this to occur, however, the assets and liabilities of the plan must have similar discounted mean terms (as in this example). If the assets and liabilities of a plan are not matched by duration, then funding ratios might vary significantly under different asset valuation methods. This is particularly likely for a recently established plan, where the duration of the liabilities is likely to exceed that of any assets available in the market.

18.2.8 Analysis of Surplus

The actuarial value of a pension fund generally differs from the target standard fund, whatever the choice of asset valuation and funding methods. This is because the experience of a pension plan inevitably differs from the assumptions used in the funding valuation. Even if the actuarial assumptions turn out to be realistic in the long term, short-term variations in the financial and demographic experience of the plan would make the emergence of surpluses or deficits unavoidable.

An analysis of surplus (or deficit) involves subdividing the surplus revealed by an actuarial valuation into various components. The reasons for doing so are

to determine the relative importance of different aspects of the plan's experience by separately evaluating their contribution to the surplus or deficit;

to determine the causes of an unexpectedly large (or small) surplus or deficit;

as a check on the results of the actuarial valuation.

Before the analysis of surplus can begin, it is necessary to adjust for those components of surplus or deficit not arising from the experience of the plan since the last actuarial valuation. These are:

Any changes to the actuarial basis since the last valuation — if these have occurred, then the previous valuation figures must be recalculated on the current basis (or vice versa).

The surplus or deficit revealed at the last valuation — this must be carried forward with interest at the valuation discount rate and deducted from the current surplus or deficit.

The amortization of surplus or deficit — the accumulated contributions in excess of the standard contribution rate must be deducted from the current surplus or deficit.

After the above adjustments have been made, we are left with the surplus or deficit that has arisen entirely from the differences between the assumed and actual experience of the plan since the last actuarial valuation.

18.2.8.1 *Main Components of Surplus or Deficit*

The experience of the plan can be subdivided into several factors, each of which can make a potentially significant contribution to the emerging surplus or deficit. The factors we shall consider are:

Investment returns — any difference between actual investment returns and the valuation discount rate is a source of surplus or deficit.

Salary growth — salary increases received since the last valuation will generally differ from the assumed general and promotional increases.

Pension increases — increases linked to an inflation index are likely to differ from the assumption made at the last valuation.

Demographic factors — these include deaths, withdrawals, and early retirements over the inter-valuation period, which will generally differ from assumed service table decrements.

We require a notation that distinguishes between the *actual* experience of each factor over the inter-valuation period and the *expected* experience of each factor at the last valuation; this is given in the following table:

	Actual experience	Expected experience
Investment return	I_A	I_E
Salary growth	S_A	S_E
Pension increases	P_A	P_E
Demographic factors	D_A	D_E

Using an extended notation, we also define:

$F(I_A, S_A, P_A, D_A) =$ fund at current valuation
$SF(S_A, P_A, D_A) =$ standard fund at current valuation

The fund at the current valuation date depends on all four factors; the investment experience obviously affects the growth in the fund, but the other factors are also relevant because they affect the benefit outgo and contribution income over the inter-valuation period.

The standard fund at the current valuation date depends on: actual salary growth, which affects the active member liabilities; actual pension increases, which affect the pensioner and deferred pensioner liabilities; the demographic experience, which affects the number members in each category. Note that the standard fund does not depend on the investment returns over the inter-valuation period. (It *does* depend on the assumed discount rate, but this is not the same thing.)

Our aim is to break down the total surplus or deficit at the valuation date into the following components:[3]

IS is the investment surplus
SS is the salary surplus
PS is the pension increase surplus
DS is the surplus arising from demographic factors

We shall use the symbol TS for the total surplus, which must satisfy

$$TS = IS + SS + PS + DS$$

And by definition:

$$TS = F(I_A, S_A, P_A, D_A) - SF(S_A, P_A, D_A)$$

---

[3]The order in which these components are calculated follows Lee (1986, chapter 25). A different order would normally have a small effect on the surplus from each component, but would not change their sum.

18.2.8.2 Investment Surplus

To remove the investment surplus from the total surplus revealed at the actuarial valuation we have to calculate the fund that would have arisen had investment returns over the inter-valuation period been equal to the valuation discount rate. It follows that:

$$TS - IS = F(I_E, S_A, P_A, D_A) - SF(S_A, P_A, D_A)$$

And the definition of TS, given above, is

$$TS = F(I_A, S_A, P_A, D_A) - SF(S_A, P_A, D_A)$$

Subtracting these two equations gives the investment surplus:

$$IS = F(I_A, S_A, P_A, D_A) - F(I_E, S_A, P_A, D_A)$$

18.2.8.3 Salary Surplus

The total surplus excluding both the investment and salary surplus is given by

$$TS - IS - SS = F(I_E, S_E, P_A, D_A) - SF(S_E, P_A, D_A)$$

And in deriving the investment surplus we used

$$TS - IS = F(I_E, S_A, P_A, D_A) - SF(S_A, P_A, D_A)$$

Subtracting these two equations gives the salary surplus:

$$SS = [F(I_E, S_A, P_A, D_A) - SF(S_A, P_A, D_A)]$$
$$- [F(I_E, S_E, P_A, D_A) - SF(S_E, P_A, D_A)]$$

18.2.8.4 Pension Increase Surplus

Removing the surplus arising from investment returns, salary growth, and pension increases gives the following expression:

$$TS - IS - SS - PS = F(I_E, S_E, P_E, D_A) - SF(S_E, P_E, D_A)$$

And in deriving the salary surplus we used

$$TS - IS - SS = F(I_E, S_E, P_A, D_A) - SF(S_E, P_A, D_A)$$

Subtracting these two equations gives the pension increase surplus:

$$PS = [F(I_E, S_E, P_A, D_A) - SF(S_E, P_A, D_A)]$$
$$- [F(I_E, S_E, P_E, D_A) - SF(S_E, P_E, D_A)]$$

18.2.8.5 Surplus from Demographic Factors

In deriving the salary surplus we used

$$TS - IS - SS - PS = F(I_E, S_E, P_E, D_A) - SF(S_E, P_E, D_A)$$

This is an expression for the surplus remaining after all the nondemographic components have been subtracted, which must be equal to the surplus arising from the demographic factors; hence

$$DS = F(I_E, S_E, P_E, D_A) - SF(S_E, P_E, D_A)$$

This equation derives the surplus arising from demographic factors as a balancing item, i.e., the surplus remaining after all the other components have been removed. It is possible, however, to calculate the demographic component directly, which is a useful check on the analysis of surplus.

18.2.9 Example of Analysis of Surplus

The procedures involved in an analysis of surplus are best illustrated by means of a simple example.

We shall assume that a final salary pension plan is established for 100 new employees. Each employee is 25 years old and has a starting salary of £10 000 per annum. The plan provides a pension of one-sixtieth final salary for each year of service at a normal retirement age of 65. On death in service, the plan provides a lump sum benefit of four times the salary at death.

It is decided that the funding method for the plan will be the EAM (using an entry age of 25). The following actuarial basis is used to determine the standard contribution rate:

Valuation discount rate:	9% per annum
General salary increases:	7% per annum
Promotional salary scale:	None
Pension increases:	5% per annum
Decrements in service:	Nil[4]
Pensioner mortality:	Standard table

---

[4]The assumption of zero decrements in service is justifiable for a plan with a relatively small number of active members; see Anderson (1996, chapter 6).

Using this basis, the actuary recommends a standard contribution rate of 16% of the pensionable payroll, payable annually in advance.

An actuarial valuation is carried out after the plan has been in operation for 3 years, on the same basis that was used when the plan was established. The changes in the pensionable payroll over the inter-valuation period were as follows:

Years since plan established	Pensionable payroll (£000)
0	1000
1	1100
2	1250
3	1400

Just before the valuation date, an active member with a salary of £20 000 per annum dies in service. All the other employees who initially joined the plan remain in service.

The results of the actuarial valuation are as follows:

		£000
(1)	Actuarial value of assets:	700
(2)	Present value of future standard contributions:	6055
(3)	Present value of total projected benefits:	6753
(4)	Standard fund $= (3) - (2)$:	698
(5)	Surplus/(deficit) $= (1) - (4)$:	2

A small surplus of £2000 is revealed by the actuarial valuation, so it appears that the experience of the plan has not deviated very far from the actuarial basis. An analysis of surplus, however, may shed further light on these figures. As the plan currently has no pensioners, the component of surplus arising from pension increases is zero and can be ignored.

18.2.9.1 Investment Surplus

The investment surplus is given by

$$IS = F(I_A, S_A, D_A) - F(I_E, S_A, D_A)$$

Working in units of £000, the valuation results indicate that $F(I_A, S_A, D_A) = 700$.

To obtain the fund that would have arisen had the investment experience been as assumed, we accumulate the standard contributions actually received at the valuation interest rate, not forgetting to deduct the lump-sum death benefit paid just before the valuation date. Thus

$$F(I_E, S_A, D_A) = 0.16 \times \left(1000 \times 1.09^3 + 1100 \times 1.09^2 + 1250 \times 1.09\right)$$
$$- 4 \times 20 = 554$$

Hence

$$IS = 700 - 554 = 146$$

18.2.9.2 Salary Surplus

Having worked out the investment surplus, we can evaluate the total surplus less the investment surplus as

$$TS - IS = 2 - 146 = -144$$

The total surplus less both the investment surplus and salary surplus is given by

$$TS - IS - SS = F(I_E, S_E, D_A) - SF(S_E, D_A)$$

Both variables on the right-hand-side must be evaluated from first principles. $F(I_E, S_E, D_A)$ is obtained by accumulating the standard contributions at the valuation interest rate and deducting the lump-sum death benefit, using the investment returns and salary escalation expected at the last valuation. Hence

$$F(I_E, S_E, D_A) = 0.16 \times 1000 \times \left(1.09^3 + 1.09^2 \times 1.07 + 1.09 \times 1.07^2\right)$$
$$- 4 \times 10 \times 1.07^3 = 561$$

$SF(S_E, D_A)$ is the standard fund that would have arisen if salary growth had been as assumed while the mortality experience remained as experienced. This can be obtained by adjusting the standard fund shown in the valuation by the ratio of the expected pensionable payroll to the actual pensionable payroll. Hence

$$SF(S_E, D_A) = \frac{698 \times \left(99 \times 10 \times 1.07^3\right)}{1400} = 605$$

It follows that

$$TS - IS - SS = 561 - 605 = -44$$

and

$$SS = (TS - IS) - (TS - IS - SS) = -144 + 44 = -100$$

18.2.9.3 Mortality Surplus

The surplus arising from demographic factors is completely explained by the death of the active member immediately before the valuation date. This is because all the other members remained in service as assumed in the valuation basis (which allows for no decrements in service). The surplus or deficit arising on the death of this member can be calculated directly as

$$\text{Expected release of reserve} - \text{expected value of death benefit}.$$

The release of reserve is equal to the standard fund held for the individual immediately before death. It can be estimated by multiplying the standard fund shown in the valuation by the ratio of the expected salary of the individual to the actual payroll at the valuation date. Hence

$$\text{Expected release of reserve} = \frac{698 \times \left(10 \times 1.07^3\right)}{1400} = 6$$

$$\text{Expected value of death benefit} = 4 \times 10 \times 1.07^3 = 49$$

$$\frac{\text{Mortality surplus}}{\text{(deficit)}} = 6 - 49 = -43$$

The mortality surplus can also be derived as the balancing item remaining after deducting the investment and salary surplus from the total surplus. In calculating the salary surplus, we showed that this was

$$\text{TS} - \text{IS} - \text{SS} = 561 - 605 = -44$$

Allowing for rounding errors, this is equivalent to the mortality surplus obtained above by the direct method.

18.2.9.4 Conclusions Drawn from Analysis of Surplus

The simple example given above shows that an analysis of surplus can be a powerful and revealing exercise. It appears from the valuation results that the experience of the plan was close to the actuarial basis, as the total surplus is relatively small. However, the analysis of surplus demonstrates that the small total surplus is made up of large components, which fortuitously tend to cancel each other out. The large investment surplus masks serious deficits arising from the salary and mortality experience of the plan.

As investment returns tend to be volatile, it might be imprudent to assume that favorable investment experience will continue to counterbalance deficits arising from the plan's salary and mortality experience. Hence, the actuary might make various recommendations based on the results of the analysis or surplus, such as (i) the use of stronger assumptions for future salary growth, including a promotional scale, and (ii) the reinsurance of death-in-service benefits by means of a group insurance policy. Further numerical illustrations of how to calculate gains or losses from different causes are given in Aitken (1994).

18.3 ' Actuarial Benefit Calculations

In respect of most contingencies, the benefits of a defined-benefit plan are calculated according to a simple formula given in the rules. The exceptions tend to occur when the plan offers optional benefits that can be accepted or rejected by the member. In these circumstances, an actuarial calculation is often required to ensure that the benefit provided is fair to the individual member without imposing a financial strain on the plan. As mentioned in Section 18.2.3, the existence of the option will have no financial consequences for the plan if the benefit provided is equal to the actuarial reserve held for the member.

Examples of benefits requiring actuarial calculations are:

Transfer values offered to early leavers in settlement of their deferred benefit rights;
Immediate pensions on voluntary early retirement;
The exchange of pension for a cash sum or an enhanced spouse's pension.

18.3.1   Transfer Values

Under U.K. pension legislation, deferred pensioners have the right to transfer a cash equivalent of their deferred benefit rights into another tax-approved pension arrangement. Even in the absence of such legislation, a pension plan may provide this option in order to reduce the administrative burden created by deferred pensioners, many of whom might have relatively small accrued benefits.

An actuarially fair transfer value would be calculated as the expected present value of the member's deferred benefits. As the member is being provided with an immediate cash sum, the discount rate used to calculate the transfer value should be based on the current market yield on assets that most closely match the term and nature of the deferred benefits. As deferred benefits are not linked to future salary growth, these assets are usually assumed to be risk-free government bonds.

If the plan increases deferred pensions at a fixed rate (or not at all), fixed income bonds would be the appropriate matching assets for the deferred benefits. Thus, the redemption yield on fixed income bonds of suitable duration would be used as the discount rate. If the increases in deferment are linked to price inflation, the appropriate matching asset would be bonds providing income and capital payments linked to an inflation index. In this case, the real redemption yield on index-linked bonds would be used as the discount rate net of deferred pension increases.

Let:

DP_x be the deferred pension at exit of a member aged x;
Y_1 be the redemption yield on fixed-income bonds;
Y_2 be the real redemption yield on index-linked bonds.

If the plan provides deferred pension increases at a fixed compound rate of g per annum, then the formula for the transfer value is

$$DP_x \left(\frac{1+g}{1+Y_1} \right)^{NRA-x} \frac{l^P_{NRA}}{l^P_x} \bar{a}^r_{NRA}$$

And when the deferred pension is increased in line with an inflation index, the formula for the transfer value is

$$DP_x \left(\frac{1}{1+Y_2} \right)^{NRA-x} \frac{l^P_{NRA}}{l^P_x} \bar{a}^r_{NRA}$$

In each case, the discount rate used to determine the value of the annuity at retirement would also be market linked. A further term would be required if a benefit were provided on death before retirement.

The method given above is often modified when the duration of the deferred benefits is greater than the duration of any bonds that are available in the market. The discount rate might then be a blend of the current market yield and an expected long-term yield estimated by the actuary. A method often used for this calculation is given below for a deferred pension subject to fixed compound increases.

Let:

N be the longest term to redemption of fixed-income bonds available in the market;
Y_1 be the current market yield on fixed-income bonds with a term to redemption of N years;
$\bar{Y}_1$ be the expected average future yield on fixed-income bonds;
$a_{\overline{N}|}$ be the annuity certain calculated at interest rate $\bar{Y}_1$;

MAF be the market adjustment factor.

The market adjustment factor is given by

$$MAF = \left[1 + (Y_1 - \bar{Y}_1) a_{\overline{N}|} \right]^{-1}$$

And the formula for the transfer value becomes

$$\text{MAF} \times \text{DP}_x \left(\frac{1+g}{1+\bar{Y}_1}\right)^{\text{NRA}-x} \frac{l^{\text{p}}_{\text{NRA}}}{l^{\text{p}}_x} \bar{a}^{\text{r}}_{\text{NRA}}$$

The discount rate used to calculate the transfer value is now the estimated future yield $\bar{Y}_1$, and the current market yield is brought into the equation by the market adjustment factor. The formula for the MAF assumes that the transfer value is invested in a long-dated bond offering a coupon rate equal to the market yield Y_1. Assuming that the coupon payments are reinvested at the future yield $\bar{Y}_1$, the present value of an investment of one unit is equal to the reciprocal of the MAF. Thus, the initial investment required to secure the deferred pension is given by the formula presented above. The same principles can be used to modify the formula involving the real redemption yield on index-linked bonds.

18.3.2 Voluntary Early Retirement

Most defined-benefit plans allow members to retire before the normal retirement age specified in the rules (see Section 17.2.3). If the early retirement pension were based on the NRA benefit formula (without reduction), then the plan would incur a financial loss. Thus, the normal benefit formula is usually subject to a reduction factor that depends on the actual age of retirement. There are two alternative principles that may be used in the calculation of actuarially fair reduction factors. These are based either on equivalence with the actuarial reserve or equivalence with the withdrawal benefit.

18.3.2.1 Actuarial Reserve

The actuarial reserve held for an active member is the standard fund for that member under the funding method used by the plan. If the expected present value of the early retirement pension is equal to the member's actuarial reserve, neither surplus nor deficit will arise from the event. The aim, therefore, is to provide cost-neutral early retirement pensions from the perspective of the funding valuation, and the assumptions used in the funding valuation should be used in the calculation of the reduction factors.

Let:

ERF_x be the reduction factor for age of retirement x;
Sal_x be the pensionable salary of active member aged x;
N_x be the past service of active member aged x.

The remainder of the notation will be as defined in Section 18.1 and Section 18.2.

Under the PUM, a simple formula for the actuarial reserve held for the active member (assuming zero withdrawal decrements) would be

$$kN_x\text{Sal}_x \frac{l_{\text{NRA}}}{l_x}(1+j)^{\text{NRA}-x}\frac{s_{\text{NRA}}}{s_x}v^{\text{NRA}-x}\bar{a}^{\text{r}}_{\text{NRA}}$$

And the present value of the early retirement pension is

$$\text{ERF}_x\, k\, N_x\, \text{Sal}_x\, \bar{a}^r_x$$

Equating these two expressions gives

$$\text{ERF}_x = \frac{l_{\text{NRA}}}{l_x}(1+j)^{\text{NRA}-x}\frac{s_{\text{NRA}}}{s_x}v^{\text{NRA}-x}\frac{\bar{a}^{\text{r}}_{\text{NRA}}}{\bar{a}^{\text{r}}_x}$$

A table of early retirement factors could be provided for the plan administrator to use in individual cases.

18.3.2.2 Withdrawal Benefit

The alternative principle is to provide members who choose to retire early with a benefit equal in value to their entitlement as early leavers. Under this approach, the funding surplus created by an early retirement would be the same as that arising from the provision of a deferred pension. The appropriate basis to use for this calculation is that used for the calculation of transfer values, so that the market value of the early retirement pension is equal to the transfer value which the member could have obtained.

Thus, for a plan providing deferred pension increases at a fixed compound rate, the equation of value for the early retirement pension is

$$\text{ERF}_x\, kN_x\, \text{Sal}_x\, \bar{a}^{\text{r}}_x = kN_x\text{Sal}_x\left(\frac{1+g}{1+Y_1}\right)^{\text{NRA}-x}\frac{l^{\text{p}}_{\text{NRA}}}{l^{\text{p}}_x}\bar{a}^{\text{r}}_{\text{NRA}}$$

$$\Rightarrow \text{ERF}_x = \left(\frac{1+g}{1+Y_1}\right)^{\text{NRA}-x}\frac{l^{\text{p}}_{\text{NRA}}}{l^{\text{p}}_x}\frac{\bar{a}^{\text{r}}_{\text{NRA}}}{\bar{a}^{\text{r}}_x}$$

The above expression would be amended accordingly for plans using alternative transfer value formulae.

18.3.3 Commutation

Commutation options give retiring members the option to exchange all or part of their pension for other benefits, such as an immediate cash sum or an enhanced spouse's pension. The commutation of pension for cash is very common in U.K. plans, where the cash sum is exempt from taxation, although only about one-quarter of the pension can be commuted under current U.K. legislation.

The actuarial calculations involved in setting the commutation terms are straightforward in principle: the expected present value of the benefit offered must be made equal to the expected present value of the pension commuted. The main issues for the actuary are:

whether to allow for adverse selection, given that lives in poor health have more to gain from exercising the commutation option;

whether the discount rate used in the present-value calculation should be taken from the funding valuation or from current market yields.

Allowing for adverse selection would require the actuary to estimate how the mortality of members taking the commutation option differs from that of the pensioner population as a whole. Whether this is necessary would depend on the proportion of members taking the option — if it is close to either one or zero, then the financial impact of adverse selection would be minimal. Another possibility would be for the plan to limit adverse selection by attaching conditions to the exercise of the option, e.g., requiring the member to provide medical evidence of reasonable health.

Using a market-linked discount rate to determine commutation terms is appealing on the grounds of fairness, as it would allow the commutation option to be exercised on approximately the same terms prevailing in the immediate annuity market. Moreover, if transfer values are calculated a using market-linked discount rate, then this would ensure that retiring members are treated consistently with those who leave the plan shortly before retirement.

It is possible, however, that the commutation option is so popular that it effectively becomes another benefit offered by the plan (e.g., commutation for cash in the U.K.). If so, then it may be beneficial for the plan to present it in this way, by offering standard commutation terms that do not vary over time. If these terms were inconsistent with the funding basis, then the actuary would treat the option as a separate benefit in the funding valuation.

18.4 Unfunded Plans

18.4.1 Pay-As-You-Go Principle

Unfunded pension plans are usually either social security plans operated by a national government or group pension plans for public employees. They operate on the principle that the contribution income over any period will equal the benefit outgo over the same period. As a result, no pension fund is accumulated and the accrued liabilities of the plan are effectively debts of the plan sponsor (i.e., the government in the case of a social security plan). Unfunded plans invariably provide defined benefits.

The concepts involved in the financial management of an unfunded plan are different from those described (above) for funded defined-benefit plans. As there are no assets to be invested, there is no need for an interest rate assumption. The present-value calculations used for funded plans have no counterpart in unfunded plans, where financial stability is assessed through long-term benefit projections using the cash-flow approach described in Section 18.1.2. The viability of an unfunded plan depends on whether the benefit outgo will increase or decrease relative to the contribution income and whether (and by how much) it will be possible to increase contribution rates.

Both employers and employees may be required to contribute to the social security plan of their country. These contributions are usually percentages of earnings that can vary over time. A strictly unfunded plan would have to change its contribution rates frequently to preserve the balance between income and expenditure. In practice, the plan is likely to have a small pool of assets so that short-term variations in benefit outgo can be funded without a change in contribution rates. The government may also subsidize a social security plan from general tax revenues if the need arises.

18.4.2 Dependency Ratio

The dependency ratio of an unfunded plan is the ratio of the number of beneficiaries to the number of contributors; this is usually approximately the same as the ratio of the number of pensioners to the number of active members.

The importance of this ratio can be illustrated by a simple example. Suppose there is an unfunded plan in which each pensioner receives a fixed annual pension of b units. If the annual contribution per active member is c units, then the pay-as-you-go principle requires that:

$$c \times \text{Number of active members} = b \times \text{Number of pensioners}$$

$$\Rightarrow c = b \times \left(\frac{\text{Number of pensioners}}{\text{Number of active members}} \right)$$

$$\Rightarrow c = b \times \text{Dependency ratio}$$

Thus, the required contribution per active member is proportional to the dependency ratio, which depends on the demographic structure of the population. In a stationary population the dependency ratio is fixed, but for a real population it is likely to vary over time. Daykin and Lewis (1999) show that the dependency ratio is projected to rise significantly in most developed countries with unfunded social security plans, which means that these plans will either have to pay lower benefits or require those in work to pay higher social security contributions.

18.4.3   Demographic Factors

The most important demographic factors that affect the future dependency ratio of an unfunded plan are:

the number of children per woman;
the mortality of the retired population.

In a country with relatively low child mortality rates, a fertility rate of approximately 2.1 children per woman is required for the population to replace itself. A lower fertility rate results in a declining population, which will have a higher dependency ratio than a stationary population, other things being equal. The opposite is true if the fertility rate is above the replacement level and the population is growing.

The mortality of the retired population will affect the number of pensioners. If pensioner mortality is falling over time, pensioners will live longer, on average, and the pensioner population will be greater as a result. Thus, the dependency ratio will be higher, other things being equal.

In most developed countries, both of these demographic factors are changing so as to cause the dependency ratio to increase. As living standards have improved, pensioner mortality has fallen, and this trend is expected to continue. Fertility rates are generally below the replacement level, and significant increases in pensioner populations are expected when the "baby boom" generation (born between 1945 and 1960) reaches retirement age. Thus, the governments of these countries will have to increase social security contributions and/or cut benefit levels to maintain their social security plans (see Daykin and Lewis (1999)).

18.4.4   Economic Factors

The main economic factors affecting the future income and outgo of an unfunded plan are:

the rate of growth in the average earnings of the working population;
the economic activity rates of the working-age population.

As social security contributions are normally linked to earnings, the growth in contribution income over time depends on the growth in average earnings, which in turn depends on the rate of economic growth in the country. If benefit rates are also linked to earnings, then the balance of income and outgo may not be affected by the rate of earnings growth. However, if benefit rates are *not* linked to earnings (e.g., a flat-rate pension linked to price inflation, as provided by the U.K. state pension plan), then a higher rate of earnings growth will result in lower social security contribution rates, and vice versa.

Economic activity rates measure the proportion of the population who are at work (or seeking work) in each age group—obviously they vary considerably by age and sex. Important factors influencing these rates are: the extent to which women participate in the labor force; the proportion of young adults in higher education; rates of long-term sickness; the number of employees who retire early or late (compared with the state pension age). Although those who are not economically active may not qualify for state pension benefits, it is possible that some of them will qualify for dependants' benefits. Furthermore, they may also receive other forms of income support from the state outside the social security plan. As nonworking individuals do not generally contribute to social security plans, lower rates of economic activity tend to increase the burden on the working population.

18.4.5 Methods of Cost Control

How can a government reduce the benefit outgo of an unfunded social security plan so as to avoid the need for unpopular increases in social security contribution rates? The main alternatives are:

increasing the pension age of future retirees;
cutting the benefits paid to future pensioners.

In both cases there are political and ethical limitations involved.

It is not likely that increasing the state retirement age of workers close to retirement would be seen as fair, as these individuals may have already made financial plans based on the existing retirement age. So any increase in the state retirement age would probably have to be phased in gradually. For example, the U.K. government has increased the female state retirement age from 60 to 65 (to make it the same as that for males), but this change will only affect women born on or after 6 April 1950. The female state retirement age will increase linearly from 60 to 65 for women born between this date and 5 April 1955.

Cutting the benefits paid to existing pensioners, or even failing to index them in line with the cost of living, is unlikely to be an acceptable alternative. It would also be difficult to reduce the accrued benefit rights of active members already earned from their past contributions. This leaves open the possibility of reducing benefits linked to future service, and the U.K. government has implemented such a change for the earnings-linked component of the state pension plan.

18.4.6 The Future of Unfunded Social Security Plans

As mentioned above, unfunded social security plans are facing financial problems because of the projected increase in dependency ratios. Many governments have taken measures to reduce future benefit payments (e.g., increasing the state pension age). This has brought home the fact that the benefits promised by such plans are not particularly secure: they depend on the willingness of the working population to pay social security contributions, which are seen as another form of taxation.

As a result, some governments have implemented reforms gradually to replace their unfunded social security plans with compulsory funded arrangements. These compulsory funded plans are typically defined contribution plans in which individual employees have some degree of control over the investment of their personal fund. It is hoped that employees will be less resistant to paying compulsory contributions if they are seen as savings rather than taxation.

However, compulsory defined contribution plans cannot replace all the features of social security plans, particularly those concerned with the redistribution of income. They also have the disadvantages of exposing workers to investment risk and relatively high administration costs. For further discussion of these issues, refer to the text of the lecture delivered to the Institute of Actuaries by the Chairman of the Pensions Commission (Turner, 2003).

References

Aitken, W. H. (1994). *A Problem Solving Approach to Pension Funding and Valuation.* ACTEX Publications.

Anderson, A. W. (1992). *Pension Mathematics for Actuaries.* ACTEX Publications.

Benjamin, S. (1989). Driving the pension fund. *Journal of the Institute of Actuaries* 117(Part III), 717–735.

Bowers, N. L., Gerber, H. U., Hickman, J. C., Jones, D. A., and Nesbitt, C. J. (1997). *Actuarial Mathematics.* Society of Actuaries.

Day, J. G. and McKelvey, K. M. (1964). The treatment of assets in the actuarial valuation of a pension fund. *Journal of the Institute of Actuaries* 90, 104–147.

Daykin, C. D. and Lewis, D. (1999). A crisis of longer life: reforming pension systems. *British Actuarial Journal* 5(Part I), 55–97.

Dufresne, D. (1988). Moments of pension contributions and fund levels when rates of return are random. *Journal of the Institute of Actuaries* 115(Part III), 535–544.

Head, S. J., Adkin, D. R., Cairns, A. J. G., Corvesor, A. J., Cule, D. O., Exley, C. J., Johnson, I. S., Spain, J. G., and Wise, A. J. (2000). Pension fund valuations and market values. *British Actuarial Journal* 6(Part I), 55–118.

Khorasanee, M. Z. (2002). A cash-flow approach to pension funding. *North American Actuarial Journal* 6(1), 137–165.

Lee, E. M. (1986). *An Introduction to Pension Schemes*. Institute of Actuaries and Faculty of Actuaries.

McLeish, D. J. D. and Stewart, C. M. (1987). Objectives and methods of funding defined benefit pension schemes. *Journal of the Institute of Actuaries* 114(Part II), 155–199.

O'Regan, W. S. and Weeder, J. (1988). A dissection of pension funding. Presented to Staple Inn Actuarial Society on 19 January 1988. www.sias.org.uk.

Owadally, M. I. and Haberman, S. (1999). Pension fund dynamics and gains/losses due to random rates of investment return. *North American Actuarial Journal* 3(3), 105–117.

Society of Actuaries (1998). Survey of asset valuation methods for defined benefit pension plans. www.soa.org/research.

Thornton, P. N. and Wilson, A. F. (1992). A realistic approach to pension funding. *Journal of the Institute of Actuaries* 119(Part II), 229–286.

Turner, A. (2003). The macro-economics of pensions. Lecture to the Institute of Actuaries, 2 September 2003. www.actuaries.org.uk.

Chapter 19

Investment Strategies for Defined-Benefit Plans

19.1 Introduction

A funded defined-benefit plan aims to meet the cost of the benefits due to each generation of members over their expected period of service. In doing so, it must accumulate contributions before their retirement age and invest those contributions in suitable assets both before and after retirement. The chosen investment strategy will affect both the security of the members' accrued benefits and the contribution rate required to fund the plan.

If investment returns are higher, on average, then this will reduce the contributions required to provide a given level of benefits. It may also be possible to use part of the surplus arising from better-than-expected investment returns to improve benefits. On the other hand, if investment returns are too volatile, then this will have a number of undesirable consequences. Excessive volatility may result in large divergences between the value of the fund and the value of the accrued benefits at different times in the lifecycle of the plan. This will increase the incidence of insolvency (on a discontinuance basis) and may cause the plan to breach regulatory funding limits (see Section 18.2.1). Furthermore, the emergence of large surpluses and deficits would make the contribution rate less stable, which generally would be an unwelcome development for the plan sponsor.

The investment strategy of a defined-benefit plan must, therefore, strike a balance between achieving high returns and limiting the size of surpluses and deficits. The appropriate balance between these objectives will depend on the maturity of the plan and the risk tolerance of the trustees and sponsoring employer. But before we can assess alternative investment strategies, we must first consider the characteristics of different pension fund assets.

19.2 Characteristics of Different Assets

A detailed description of different investments was provided in Chapter 2, with further discussion of their risk/return characteristics in Chapter 3. In this section, we summarize the relevant features of the main asset classes in which pension funds invest. Further discussion of pension fund investment is available in Lee (1986, chapter 20) and McGill et al. (1996, chapter 30). Data on the actual distribution of pension fund wealth by asset class is provided by UBS Global Asset Management (2004).

19.2.1 Equities

Equities are the securities held by the owners of a corporation. The equity stockholders have the right to attend the annual general meeting and vote on matters of company policy. Each equity stock normally carries an equal vote that can be exercised by its owner. Dividends are paid out of the corporation's profits at a discretionary rate determined by the directors, which is the same for each stock. A corporation may pay no dividends if the directors believe that reinvesting all of the profits will better serve the interests of the equity stockholders. In such cases, investors would expect the absence of dividends to be compensated by greater appreciation in the market value of the securities. Most corporations reinvest at least part of their profits; hence, capital appreciation is a significant part of the return expected from most equity stocks.

The annual return on equities is unpredictable and volatile. Over long periods, however, equities have outperformed other pension fund assets by a significant margin. Over the 50-year period ending on 31 December 2003, the real returns on representative equity market indices were 5.0% per annum for the U.S. and 5.2% per annum for the U.K.[1] Both figures are net of consumer price inflation and allow for the reinvestment of gross dividend income. The dealing costs of trading in equities include commissions paid to stockbrokers, bid–offer spreads in quoted prices, and (possibly) taxes charged by the government.

19.2.2 Government Bonds

Government bonds are securities issued by national and regional governments to finance their borrowing requirements. Most government bonds are fixed-interest securities, paying the holder a fixed percentage of the face value of the bond at regular intervals. Each stock usually has a fixed redemption date at which its face value is returned to the holder and

---

[1] Based on data provided by Barclays Capital.

the debt is repaid. If a stock has no redemption date, so that its interest payments continue indefinitely, then the holder would only be able to realize its value by selling it in the stock market. The same option is available for a redeemable bond before its redemption date.

The redemption yield on a fixed-interest bond can be calculated at issue, but the return over future periods is generally unpredictable, depending on the market price of the bond at the end of the period and the return on reinvested interest payments. Over the 50-year period ending on 31 December 2003, the real returns on representative government bond market indices were 1.3% per annum for the U.S. and 1.6% per annum for the U.K. Dealing costs are generally lower than for equities.

The real return on a fixed-interest bond can be severely eroded if inflation is higher than expected, so some national governments have issued bonds providing interest and capital payments which both rise in line with price inflation. These bonds provide a real redemption yield (net of price inflation) that can be calculated at issue. A long track record of performance is not available for index-linked bonds, but the real redemption yield has varied between 2 and 4% per annum in the U.K.

19.2.3 Corporate Bonds

Corporate bonds are fixed-interest securities issued by corporations as a form of debt financing. Loan stockholders do not have ownership rights, but their interest and capital payments are contractual liabilities of the corporation, therefore ranking above equity dividends. Debenture loan stock is secured against specific assets of the business that can be claimed if the corporation defaults on its interest or capital payments.

Unlike government bonds issued in developed countries, corporate bonds carry a significant risk of default. For "investment-grade" corporate bonds this risk is considered to be very small, but the redemption yields offered by these bonds are still somewhat higher than governments bonds of similar term to redemption. This may reflect higher dealing costs and lower marketability, as well as default risk. Sub-investment-grade corporate bonds carry a greater risk of default and, consequently, offer significantly higher redemption yields. Historic returns from investment-grade bonds have generally been close to those from governments bonds. Lower grade bonds have provided higher returns at the expense of greater volatility.

19.2.4 Real Estate and Related Assets

Direct investment in real estate normally involves the purchase of buildings that are let to tenants. Pension funds prefer to invest in commercial (as opposed to residential) buildings, which are used as offices, shops, or warehouses. These assets provide a return in the form of rents from tenants and capital appreciation. The lease agreement with the tenant specifies the amount and frequency of the rent payments and the period after which rents are reviewed. As the market for real estate is illiquid, the capital gain (or loss) on a building can only be accurately determined when it is sold.

It is difficult to construct a market index for real estate because every building is unique and market values can only be determined when buildings are traded. Historic returns on mutual funds investing in real estate have generally been lower than on equity funds, but with significantly lower volatility in returns. Real-estate dealing costs are comparatively high, and there are also significant expenses involved in managing buildings that are let to tenants.

As real estate is only available in large units, direct investment is not practicable for small pension funds. Indirect exposure can be obtained by investing in mutual funds or listed corporations specializing in real estate. As indirect investment enables a more diversified and marketable holding of assets, it may be preferred by large pension funds as well.

19.2.5   Cash

Cash refers to money held in deposits taken by banks or other financial institutions. The return on cash deposits is close to the short-term interest rate set by the central bank. Over the 50-year period ending on 31 December 2003, the real returns on cash deposits were 1.5% per annum for the U.S. and 1.8% per annum for the U.K. The volatility in annual returns is comparatively low and dealing costs are negligible. Cash is generally perceived by pension fund managers as a low-risk, low-return asset to be used as a tactical safe haven for funds that will eventually be invested elsewhere.

19.3   Objectives of Investment Strategy

A general discussion of asset allocation principles for pension funds has been presented in Section 3.5. We will now look at the objectives of the investment strategy of a defined-benefit pension fund in more detail.

19.3.1   Maximizing Expected Returns

The liabilities of most defined-benefit plans are very long term, and a plan that maintains stationary population of members can defer the realization of its assets indefinitely. For a plan open to new entrants, therefore, the average return on the fund over a long period of time may be more important than the variation in returns from year to year.

A pension fund that achieves higher long-term returns can bring about favorable consequences for both the plan sponsor and the members. If the average investment return is higher than the valuation interest rate, then the emerging surpluses could be used to reduce employer contributions or improve benefits. Another possibility would be to allow a larger fund to accumulate, so as to improve the security of the accrued benefits and provide a margin against unfavorable future experience.

It is generally accepted that the expected return on equities is higher than on other asset classes suitable for pension fund investment. A policy of maximizing expected returns, therefore, implies that the whole fund should be invested in equities. Equity returns, however, are highly volatile—the standard deviation in real returns has exceeded 20% per annum in both the U.S. and U.K. Full investment in equities is only appropriate if this volatility is genuinely unimportant to the plan sponsor and the members. This is only likely to be true for a recently established plan, where the surpluses or deficits arising from investment volatility would be small in absolute terms.

The consequences of equity market volatility cannot be dismissed as easily for a mature plan, even when the population of active members is stationary or growing. Large surpluses or deficits would result in unstable contribution rates, which the sponsor may find unacceptable. The possibility of the plan being wound up with a solvency deficit cannot be ignored, particularly when many of the beneficiaries are pensioners or active members with long past service. Mature plans, therefore, generally aim to find a compromise between maximizing returns and matching liabilities (see Section 19.3.2).

19.3.2 Matching Liabilities

A policy of investing to match liabilities can be interpreted in different ways. For a plan open to new entrants, the liabilities can be thought of as a stream of benefit payments extending indefinitely into the future. It can, therefore, be argued that a matching strategy involves investing in assets that will generate an income stream positively correlated with the projected benefit outgo. The aim of such a strategy is to make the contributions required from the plan sponsor more stable over time.

An alternative interpretation of matching is to invest in assets that will make the value of the pension fund positively correlated with the value of the accrued liabilities. The aim is to reduce the magnitude of surpluses and

Table 19.1 Correlation of U.K. real equity returns against U.K. real wage growth

Time lag (years)	0	1	2	3
Correlation	−0.18	+0.24	+0.39	+0.24

deficits emerging over time, so as to make the employer contribution rate more stable and the accrued benefits more secure.

Under either definition, a matching strategy involves selecting assets that are similar in type and duration to the liabilities of the pension plan.

19.3.2.1 Open Plans

In Section 18.1.1, it was shown that the benefit outgo of a mature plan with a stationary population of members increases in line with general salary escalation. A matching investment strategy, therefore, requires assets providing returns that are positively correlated with wage inflation.

Real wage growth in developed countries has been relatively stable over the business cycle (i.e., periods from the end of one recession to the end of the next one). It is closely linked to real growth in gross domestic product (GDP) per worker, which has averaged between 1 and 3% per annum over the business cycles of most developed countries in the second half of the 20th century. As assets providing returns explicitly linked to wage inflation are not generally available, the closest matching assets are those providing predictable returns net of price inflation, such as index-linked bonds.

Although equity returns are highly volatile, and can vary greatly from salary growth in any year, many actuaries believe that equities are a long-term match for liabilities linked to wage inflation. It is argued that equity prices should rise broadly in line with corporate profits over long periods, and that corporate profits and wages, being the two largest components of national income, should be similarly linked. Critics of this theory point to the lack of positive correlation between equity returns and wage growth measured over annual periods. Table 19.1, based on U.K. data from the 50-year period ending on 31 December 2003, may shed some light on this controversy.

Table 19.1 indicates that real wage growth appears to be positively correlated with real equity returns when time lags of 1 to 3 years are introduced. The explanation of this result follows from the behavior of profits and wages over the business cycle. When an economy emerges from a recession, the recovery in output is normally accompanied a more pronounced recovery in corporate profits.[2] The effect on the labor market,

---

[2]Following the recession of 1980–1982, for example, real corporate profits in the U.K. grew by 17% in 1983, compared with real GDP growth of 3% in the same year.

however, occurs after a time lag, as profitable businesses seek to expand by hiring more workers. As unemployment falls and the labor market tightens, the price of labor is bid up by businesses competing for workers. Thus, real growth in wages tends to lag real growth in profits. The equity market, on the other hand, tends to anticipate the recovery in corporate profits. Hence, a significant time lag would be expected between the recovery in equity prices and the time of maximum wage growth.[3]

It seems likely, therefore, that real equity returns and real wage growth are positively correlated over the business cycle. This positive correlation, however, may be of little help over a single business cycle because of the inherent volatility in equity prices. The case for equity investment as a matching asset depends on a long enough timeframe for these random price fluctuations to average out, so that the long-term connection between equity prices, profits, and wages becomes apparent. Hence, the same proviso applies to equity investment whether the intention is to maximize expected returns or to match the liabilities of an ongoing plan: it is a long-term strategy carrying significant short-term risks.

Real estate is also believed to be a long-term hedge against price and wage inflation. The capital value of a commercial building will depend on its expected future rental income, and rent increases should be positively correlated with price inflation. As with equities, however, this long-term relationship may be obscured by short-term volatility, particularly if the real-estate market is prone to slumps and booms over the business cycle.

19.3.2.2 Plan Closed to New Entrants

A pension plan closed to new entrants has a predictable lifespan. The current generation of active members will not be replaced, so the number of active members will decline as employees retire or withdraw from service. This will result in a progressive reduction in contribution income, requiring the sale of assets in order to meet benefit payments. Eventually, the plan will consist entirely of pensioners whose benefits are paid from a reducing fund.

The important differences compared with an open plan are that (i) the benefit outgo has a finite term and becomes less correlated with wage inflation as the liabilities mature and (ii) assets will have to be sold to meet benefit payments. Both these facts make equities less suitable as a matching asset. The closest matching assets for the pensioner and deferred pension liabilities, which will soon dominate the liabilities of the plan, are bonds of comparable duration. For the remaining active member liabilities, index-linked bonds would be safer matching assets than equities because of the relatively short timeframe.

---

[3]A more thorough statistical analysis of the long-term link between real equity returns and real wage growth in the U.K. is provided by Cardinale (2003), and a study of the behavior of real wage rates over the U.S. business cycle is provided by Bils (1985).

19.3.2.3   Terminated Plan

If a plan is wound-up with no further accrual of benefits, then the active members become deferred pensioners. The liabilities of the plan, there-fore, consist entirely of deferred pensions and pensions in payment. As mentioned above, the closest matching assets for pensioner and deferred pensioner liabilities would be bonds of similar duration to the liabilities.

A portfolio of fixed-income bonds can closely match pensioner liabilities, provided that the number of pensioners is sufficiently large for their mortality experience to be relatively predictable. If this is not so, then the pension plan could perfectly match its pensioner liabilities by purchasing immediate annuities from an insurance company. Long-dated bonds would also be the closest available matching assets for the deferred pensioner liabilities, although a significant mismatching risk might exist for younger deferred pensioners because of the uncertainty in future mortality trends and the unavailability of bonds of sufficiently long duration.

The option of buying out the liabilities of a terminated plan through the purchase of immediate and deferred annuity contracts should be given serious consideration. The main reason for not doing so would be a belief that the members' benefits could be improved by retaining control over the investment policy. When the number pensioners has fallen to a level at which the mortality risk can no longer be self-insured, however, the purchase of immediate annuity contracts is the only prudent option.

19.3.3   Portfolio Diversification

Portfolio diversification involves investing the pension fund in a wide spread of assets, so that the specific risk associated with any one asset does not have an excessive influence on the performance of the fund. The principles of portfolio diversification can be applied both within a given asset class, such as equities, and across different asset classes, such as equities and bonds. An introduction to modern portfolio theory was given in Section 5.3, and its application to an institutional investor with liabilities was covered in Section 5.4. The main points of relevance to a defined-benefit pension fund are summarized below.

The theory of portfolio diversification decomposes the risk of investing in any asset into "specific risk" and "systematic risk." Specific risk is unique to that particular asset and results in price movements that are uncorrelated with any other asset. Systematic risk arises from price movements caused by general economic factors that influence a large number of assets in a similar way. Portfolio diversification reduces the consequences of specific risk but does not have any effect on systematic risk. Therefore, it is most important for assets that have high specific risks, such as equities and real estate.

The principles of portfolio diversification are expressed mathematically in mean – variance portfolio theory. The systematic risk on any asset can be measured by the covariance between its return and the return on a portfolio representing the whole market. The lower the systematic risk of a particular asset or asset class, the greater its diversifying effect on the portfolio. If means, variances, and covariances can be estimated for the returns on all the assets in the market, then it is possible to derive an efficient set of portfolios. These portfolios have the lowest possible standard deviation of return for any given mean return.

Mean – variance portfolio theory needs to be modified for a pension fund because of its liabilities. For a pension plan, risk is represented by the standard deviation of its surplus or deficit, not the standard deviation of the return on the fund. This means that assets which are positively correlated with the liabilities are most effective in reducing risk, which is another way of saying that matching liabilities is an important investment objective. Liabilities can be included in mean–variance portfolio theory by treating them as a "negative asset." The mathematical precision of the theory masks problems with its practical application, however, because of the uncertainty involved in parameter estimation. For more on the application of portfolio theory to pension funds, refer to Sherris (1992) and Panjer et al. (1998, chapter 8).

A particularly important aspect of risk diversification for a pension plan arises in connection with investment in securities issued by the plan sponsor, or "self-investment." This is because the bankruptcy of the plan sponsor would lead to these securities becoming devalued at a time when the pension plan was being wound up. As one of the main reasons for funding a defined-benefit plan is to provide security for the accrued benefits on termination of the plan, self-investment is generally seen as bad practice and may be prohibited or restricted by legislation.

19.3.4 Liquidity

Liquidity refers to ease with which assets can be converted into cash. Although quoted securities can normally be bought and sold whenever the stock exchange is open for business, variations in liquidity arise because dealing costs are not uniform and because the size of a stock issue affects its liquidity. In general, government bonds are the most liquid quoted securities, with low dealing costs and large volumes of stock in issue. The most liquid corporate securities are those issued by "blue-chip" companies, for which large volumes of stock are traded daily. The securities of smaller companies are less liquid because of larger bid–offer spreads and lower amounts of stock in issue. If an issue is too small, then a pension fund may not be able to buy or sell the required amount of stock without affecting its price. Real estate is the least liquid pension fund asset, as it often takes several months to find a buyer for a building and transaction costs are significantly higher than for quoted securities.

The importance of liquidity for a pension fund depends on the maturity of the plan. It could be argued that liquidity is relatively unimportant for a newly established plan, because its contribution income will provide sufficient cash to meet the benefit outgo for decades. For a closed plan, on the other hand, assets must be sold off progressively to meet benefit payments, making liquidity very important. Given that the possibility of termination or closure is a reality for all pension plans, the issue of liquidity can never be entirely disregarded. If the sponsor of a recently established plan went bankrupt, for example, then it would probably be necessary to liquidate the pension fund in order to buy out the accrued liabilities through the purchase of deferred annuity contracts.

In practice, most pension fund assets are quoted securities and, therefore, sufficiently marketable to deal with the circumstances of a closure or termination. It might be reasonable for a large, mature pension fund to invest a significant proportion of its assets directly in real estate, on the assumption that it could be run as a closed fund after its closure or termination. This would allow sufficient time for the property assets to be sold.

19.4   Stochastic Asset/Liability Modeling

Stochastic asset/liability modeling is a quantitative approach to optimizing the investment portfolio of a pension fund. It involves projecting the assets and liabilities of the pension plan to some future point in time, which may be the date of the next actuarial valuation or later. In making these projections, one or more of the elements of the future experience of the plan are assumed to be random variables. The projected surplus or deficit, therefore, is also a random variable. A general discussion of asset/liability modeling for institutional investors was provided in Section 5.6.

The initial objective of a stochastic asset/liability modeling exercise is to obtain a probability distribution for the projected surplus or deficit. Only in very simple models is it possible to obtain an explicit distribution function. More usually, the distribution is obtained by randomly simulating a large number of outcomes with the aid of a computer program. This procedure is carried out for a range of different investment strategies, each one producing its own distribution for the projected surplus or deficit.

The second stage of the exercise is to calculate carefully chosen performance measures for each distribution and use them to compare the alternative investment strategies. Some performance measures are described below.

19.4.1 Performance Measures

A general discussion of investment risk measures was provided in Section 4.3. In a defined-benefit pension plan, all aspects of the future experience of the plan contribute to the uncertainty in its projected surplus/deficit, including investment returns, salary growth, pension increases, service decrements, pensioner mortality, and new entrants. Investment experience, however, is normally the most significant risk factor, because market volatility tends to dominate other elements of the plan's experience, at least in the short term.

The performance measures outlined below are similar to those used for pure investment problems, although one should bear in mind that they apply to the emerging surplus rather than the return on the fund and, therefore, are influenced by noninvestment factors.

19.4.1.1 *Mean and Variance*

The mean and variance of the projected surplus can be calculated for each investment strategy and used to derive "efficient" investment strategies, as in standard mean–variance portfolio theory.

The first step is to plot the position all possible asset portfolios in a graph of mean against standard deviation. If there are more than two assets in the portfolio, then these points form a region referred to as the "opportunity set." The principle of mean–variance optimization states that a portfolio is efficient if no other portfolio offers a higher mean for the same standard deviation or a lower standard deviation for the same mean. Applying this principle leads to an "efficient frontier" of portfolios along the upper edge of the opportunity set.

Although the efficient frontier eliminates many portfolios from consideration, it still contains a wide range of portfolios from which the plan must choose. A further constraint is required if this choice is to be made using the results of the asset/liability projections. One possible type of constraint is to specify an upper limit for the probability of a deficit and maximize the mean surplus subject to this limit. The probability of a deficit can be estimated directly from the simulated output for any particular investment strategy. Applying this type of constraint would lead to a unique investment portfolio.

19.4.1.2 *Mean and Semi-Variance*

Using the variance (or standard deviation) as a measure of risk has been criticized because both positive and negative outcomes are treated as being equally undesirable. For a pension fund, a large deficit is normally a more serious problem than a large surplus. A possible "one-sided" risk measure is the semi-variance, which differs from the variance in that positive deviations from the mean are ignored. The portfolio selection method is otherwise the same as in the mean–variance approach described above and would produce similar results if the distribution for the projected surplus/deficit were approximately symmetric.

19.4.1.3 *Conditional Tail Expectation*

The conditional tail expectation, like the semi-variance, is a one-sided risk measure. It is defined as

$$E[-S|S<0] \times \text{Probability}\{S<0\}$$

where S is the projected surplus.

Thus, the conditional tail expectation is the expected value of the projected deficit, treating surpluses as deficits of zero. Therefore, it can be estimated directly from the simulated output for any particular investment strategy. There is no need to consider the mean value of the projected surplus/deficit when using the conditional tail expectation, because this risk measure is already sensitive to the mean — clearly, if the mean surplus is higher while the shape of the distribution remains the same, then the conditional tail expectation will be lower. The optimal investment portfolio can be taken simply as the one producing the lowest conditional tail expectation.

19.4.2 Modeling Tools

Projecting the assets and liabilities of a pension plan requires a number of modeling tools. As investment returns are generally the largest source of uncertainty, a stochastic investment model enabling the simulation of returns from different asset classes is often the dominant feature of an asset/liability study. However, an investment model that generates only returns may be inadequate because of the need to simulate yields as well. As explained in Section 18.2.6, dividend yields and bond yields are required to calculate the discounted cash-flow value of a pension fund. In valuations where the assets are taken at market value, these yields would be required to determine the discount rate used to value the liabilities. A cash-flow projection model of the kind described in Section 18.1 would be required to project the value of the fund.

Projecting the liabilities of the pension plan may require a population model that allows for both service table decrements and new entrants. A sophisticated model would allow for stochastic variability in the demographic experience, so that the future number of members in each age/sex cohort was a random variable. For a large plan, however, it may be sufficient to assume that the demographic experience is deterministic on the grounds that it is relatively predictable compared with the financial experience. It would be more important to have a stochastic model for future salary escalation, which is normally the most important variable driving both the liabilities of the plan and the active member payroll.

19.4.2.1 Investment Returns and Salary Escalation

A description of stochastic investment models was presented in Section 5.5, and further discussion of alternative models would be beyond the scope of this chapter. All such models, however, must enable the return on each asset class to be projected as a random variable over an arbitrary timeframe. The simplest type of model would assume that the real returns on any asset are independent and identically distributed over time, and the same assumption could be made for the rates of salary escalation. Over any single period, however, the model should allow for correlations between returns on different asset classes (and between returns and salary escalation). The construction of such a model is outlined below for a pension fund that can invest either in equities or bonds.

Let:

$r_e(t)$ be the real annual return on equities between time t and $t+1$
$r_b(t)$ be the real annual return on bonds between time t and $t+1$
$j(t)$ be the real annual salary escalation between time t and $t+1$

The lognormal distribution is often used to model compound growth rates because it generates returns in the correct range. A multivariate lognormal model for the variables defined above would have the form

$$\ln(1 + r_e(t)) = \mu_e + \sigma_e Z_1(t)$$
$$\ln(1 + r_b(t)) = \mu_b + b_1 Z_1(t) + b_2 Z_2(t)$$
$$\ln(1 + j(t)) = \mu_s + c_1 Z_1(t) + c_2 Z_2(t) + c_3 Z_3(t)$$

where $Z_1(t)$, $Z_2(t)$, and $Z_3(t)$ are independent unit normal random variables. The parameters μ_e, μ_b, and μ_s are the expected real forces of growth for equities, bonds, and salaries respectively. They would be estimated from past financial data, possibly modified by judgements about the future. These data would also be used to estimate:

σ_e, the standard deviation of real force of return on equities
σ_b, the standard deviation of real force of return on bonds
σ_s, the standard deviation of real force of salary growth

σ_{eb}, the covariance between real returns on equities and bonds
σ_{es}, the covariance between real equity returns and real salary growth
σ_{bs}, the covariance between real bond returns and real salary growth

The constants b_1, b_2, c_1, c_2, and c_3 are then derived from the following equations:

$$\sigma_{eb} = b_1 \sigma_e$$
$$\sigma_{es} = c_1 \sigma_e$$
$$\sigma_b^2 = b_1^2 + b_2^2$$
$$\sigma_{bs} = b_1 c_1 + b_2 c_2$$
$$\sigma_s^2 = c_1^2 + c_2^2 + c_3^2$$

The main drawbacks of the simple model outlined above are its failure to allow for correlations between returns over different time periods and the difficulty in deriving running yields on equities and bonds from simulated returns. More sophisticated stochastic investment models have attempted to address these problems (see Section 5.5). However, the multivariate lognormal model has the advantages of being easy to understand and simple to use. It should also give similar results to other models over a reasonably short timeframe. For further information about this model, including reasonable parameter values for U.K. asset classes, refer to Kemp (1996).

19.4.2.2 Demographic Experience

Projecting the liabilities of a pension plan requires a model for simulating the number of active members, their pensionable salaries, and their past service. Such a model requires an assumption to be made for the number and distribution of future entrants to the plan. Applying the appropriate service table decrements to the active member population can then generate the pensioner and deferred pensioner populations. The output from this model would be used to project the benefit outgo, contribution income, and accrued liabilities of the plan.

We now define the following discrete-time variables:

$N(x, t)$ is the number of active members aged x at time t
$g(x, t)$ is the number of new entrants[4] aged x at time t
$SAL(x, t)$ is the pensionable salaries summed for active members aged x at time t
$PS(x, t)$ is the average past service of active members aged x at time t

---

[4]Defined as active members who had not joined the plan at time $t - 1$.

PSSAL(x, t) is the past service times pensionable salary summed for active
   members aged x at time t
s_x is the promotional salary scale
q_x is the probability of an active member aged x leaving service (for any
   cause) before attaining age $x + 1$
$j(t)$ is the general salary escalation between time t and $t + 1$

The projection formula for the number of active members is

$$N(x + 1, t + 1) = N(x, t) \times (1 - q_x) + g(x + 1, t + 1)$$

If we make the simplifying assumption that new entrants have the same
average salary as active members of the same age, then the pensionable
salaries of the active members can be projected as follows:

$$\text{SAL}(x + 1, t + 1) = \text{SAL}(x, t)\frac{s_{x+1}}{s_x}(1 + j(t))\frac{N(x + 1, t + 1)}{N(x, t)}$$

If we assume that the members who leave the plan (for any cause)
have the same average past service of those who remain in service,
then the average past service of the active members can be projected as
follows:

$$PS(x + 1, t + 1) = PS(x, t)(1 - q_x)\frac{N(x, t)}{N(x + 1, t + 1)}$$

Hence, the product of the past service and pensionable salary is given by

$$\text{PSSAL}(x, t) = PS(x, t)\text{SAL}(x, t)$$

In principle, all of the age-dependent parameters in the above equations
could be modeled stochastically, allowing both the number of active
members and their average past service to be projected as random
variables. For a large plan, however, it may be acceptable (as noted earlier)
to assume that general salary escalation is the only stochastic variable, in
which case the projected future pensionable salaries would be the only
random output.

The new entrant function $g(x, t)$ is not available from the service table
and may be difficult to estimate. A possible simplification would be
to assume that $g(x, t)$ is such that the population of active members
will remain stationary, which may again be a reasonable assumption for a
large plan. If so, then the active member function $N(x, t)$ would be
proportional to the survival function of a model where mortality is the only
decrement.

19.4.3 A Simple Example Based on Mean–Variance Analysis

19.4.3.1 Model Plan

We shall now derive formulae for the mean and variance of the projected surplus of a simple model plan. It is assumed that the plan is mature with a stationary population of members and that its benefit outgo and contribution income are each a constant fraction of the pensionable payroll. The only stochastic elements of the experience are the return on the fund and the rate of salary escalation. These can be reduced to a single random variable by using currency units that increase in line with salary escalation. Hence, we define:

$F(t)$ is the value of fund at time t
SF is the standard fund
$r(t)$ is the return on fund net of salary escalation between times t and $t+1$
M is the new money, assumed to be invested at times $t=1$, $t=2$, $t=3$, etc.

The standard fund SF and the new money invested M are constant multiples of the pensionable payroll, which rises in line with general salary escalation. It follows that both SF and M are constants when measured in salary-linked currency units. The return on the fund net of salary escalation is assumed to be an independent and identically distributed random variable such that

$$E[1 + r(t)] = U$$
$$E[(1 + r(t))^2] = V$$

where U and V are constants.

The projected surplus at time t is given by

$$\text{Surplus}(t) = F(t) - \text{SF}$$

As the standard fund is a constant in our salary-linked currency units, the mean and variance of the projected fund can be used as a proxy for the mean and variance of the projected surplus.

19.4.3.2 Mean and Variance of Projected Fund

We assume that the timeframe for the projection is from $t=0$ to $t=T$. The recurrence formula for the fund is

$$F(t + 1) = F(t)(1 + r(t)) + M$$

Under our chosen stochastic investment model, $r(t)$ is independent of $F(t)$; hence

$$E[F(t+1)] = E[F(t)]U + M$$

which can be rewritten as

$$E[F(t+1)]U^{-(t+1)} - E[F(t)]U^{-t} = MU^{-(t+1)}$$

We now sum both sides of the above expression from $t=0$ to $t=T-1$, noting that the summation on the left-hand side collapses into just two terms. This gives

$$E[F(T)]U^{-T} - F(0) = Ma\frac{U}{T|}$$

$$\Rightarrow E[F(T)] = F(0)U^T + Ms\frac{U}{T|} \tag{19.1}$$

To obtain the variance of the projected fund, we square both sides of the recurrence formula for $F(t)$ to obtain

$$F(t+1)^2 = F(t)^2(1+r(t))^2 + 2MF(t)(1+r(t)) + M^2$$

Taking the expected value of both sides gives

$$E[F(t+1)^2] = E[F(t)^2]V + 2ME[F(t)]U + M^2$$

A formula for the expected value of the fund at time t can be inferred from Equation (19.1), which allows the above expression to be rewritten as

$$E[F(t+1)^2]V^{-(t+1)} - E[F(t)^2]V^{-t} = 2MUV^{-(t+1)}\left(F(0)U^t + Ms\frac{U}{t|}\right) + M^2V^{-(t+1)}$$

Substituting the standard actuarial formula for $s\frac{}{t|}$ gives

$$E[F(t+1)^2]V^{-(t+1)} - E[F(t)^2]V^{-t} = 2MUV^{-(t+1)}\left[F(0)U^t + M\left(\frac{U^t - 1}{U - 1}\right)\right]$$
$$+ M^2V^{-(t+1)}$$

$$\Rightarrow E[F(t+1)^2]V^{-(t+1)} - E[F(t)^2]V^{-t} = \left(2MF(0) + \frac{2M^2}{U - 1}\right)\left(\frac{U}{V}\right)^{(t+1)}$$
$$- \left(\frac{2M^2U}{U - 1}\right)V^{-(t+1)} + M^2V^{-(t+1)}$$

If we sum both sides of the above expression from $t = 0$ to $t = T - 1$, then the left-hand side again collapses into two terms to give

$$
\begin{aligned}
E[F(T)^2]V^{-T} - F(0)^2 &= \left(2MF(0) + \frac{2M^2}{U-1}\right)\left(\frac{U}{V}\right)\left[\frac{(U/V)^T - 1}{(U/V) - 1}\right] \\
&\quad - \left(\frac{2M^2U}{U-1} - M^2\right)a_{\overline{T}|}^{\frac{V}{}} \\
\Rightarrow E[F(T)^2] &= F(0)^2 V^T + 2MU\left(F(0) + \frac{M}{U-1}\right)\left(\frac{U^T - V^T}{U - V}\right) \\
&\quad - M^2\left(\frac{U+1}{U-1}\right)s_{\overline{T}|}^{\frac{V}{}}
\end{aligned}
\tag{19.2}
$$

And the variance of $F(T)$, which is the same as the variance of the projected surplus, is then given by

$$
\mathrm{Var}[F(T)] = E[F(T)^2] - (E[F(T)])^2
$$

19.4.3.3 Deriving the Moments of the Return on a Portfolio of Assets

We now show how the parameters U and V can be derived for an arbitrary portfolio. Let $r_n(t)$ be the return net of salary escalation on the nth asset class between times t and $t + 1$. The first and second moments of the return on the asset classes are estimated from historic data and given the following symbols:

$$
\begin{aligned}
E[1 + r_n(t)] &= U_n \\
E[(1 + r_n(t))^2] &= V_n
\end{aligned}
$$

And for each pair of asset classes, we estimate from the same historic data:

$$
E[(1 + r_n(t))(1 + r_m(t))] = U_{nm}
$$

It is assumed that the pension fund is realigned at the start of each period in a portfolio given by

$$
\sum_{n=1}^{N} X_n = 1
$$

where X_n is the proportion invested in nth asset class and N is the total number of asset classes in which the pension fund can invest.

It follows that the return on the fund $r(t)$ is given by

$$1 + r(t) = \sum_{n=1}^{N} X_n(1 + r_n(t))$$

And the moments of the return on the fund are given by:

$$E[1 + r(t)] = U = \sum_{n=1}^{N} X_n U_n$$

$$E\left[(1 + r(t))^2\right] = V = \sum_{n=1}^{N} \sum_{m=1}^{N} X_n X_m U_{nm}$$

where $U_{nn} = V_n$. Hence, the parameters U and V can be determined for any portfolio.

19.4.3.4 Numerical Example for a Two-Asset Portfolio

We shall consider a simple numerical example for a pension fund restricted to two asset classes (e.g., equities and bonds). The pensionable payroll is defined as one currency unit and the unit of time is 1 month. The projection timeframe of 36 months corresponds to an inter-valuation period for the plan. The parameters are as follows:

$F(0) = 4$
$\quad M = -0.005$
$\quad T = 36$
$\quad U_1 = 1.003$
$\quad U_2 = 1.001$
$\quad V_1 = 1.010$
$\quad V_2 = 1.003$
$\quad U_{12} = 1.005$

The data above indicate that the expected monthly return net of salary escalation is 0.3% for asset 1 and 0.1% for asset 2. The higher expected return on asset 1 comes at the price of greater volatility, as seen in the fact that V_1 exceeds V_2. We can also see that the returns on these two assets are positively correlated, because U_{12} is greater than the product of U_1 and U_2. As this is a mature plan, in which the benefit outgo exceeds the contribution income, the new money available for investment at the end of each month M is negative.

Modern Actuarial Theory and Practice

Table 19.2 Statistical properties of projected fund after 36 months

X_1	X_2	$E[F(36)]$	$E[F(36)^2]$	$\sqrt{\mathrm{Var}[F(36)]}$
0	1.0	3.963	16.310	0.776
0.1	0.9	3.993	16.570	0.793
0.2	0.8	4.022	16.872	0.833
0.3	0.7	4.052	17.218	0.895
0.4	0.6	4.082	17.610	0.974
0.5	0.5	4.112	18.051	1.069
0.6	0.4	4.142	18.544	1.177
0.7	0.3	4.173	19.093	1.296
0.8	0.2	4.204	19.701	1.425
0.9	0.1	4.234	20.373	1.563
1.0	0	4.266	21.114	1.708

Based on these parameters, the moments of the return on the fund are given by

$$U = U_1 X_1 + U_2 X_2 = 1.003 X_1 + 1.001 X_2$$

$$V = V_1 X_1^2 + V_2 X_2^2 + 2 U_{12} X_1 X_2 = 1.01 X_1^2 + 1.003 X_2^2 + 2.01 X_1 X_2$$

where $X_1 + X_2 = 1$.

The statistical properties of the projected fund after 36 months, as given by Equation (19.1) and Equation (19.2), are given in Table 19.2.

Table 19.2 shows the trade-off between the expected value of the projected fund and its standard deviation. When there are only two assets in the portfolio, the opportunity set is a curve rather than a region. Hence, all of the portfolios tabulated above are efficient in the mean–variance sense.

To enable further comparison of these portfolios, we shall calculate for each one the ratio of the mean projected surplus to its standard deviation,[5] which is given by

$$\frac{E[\mathrm{Surplus}(36)]}{\sqrt{\mathrm{Var}[\mathrm{Surplus}(36)]}} = \frac{E[F(36)] - \mathrm{SF}}{\sqrt{\mathrm{Var}[F(36)]}}$$

[5]This is analogous to the ratio used by Sharpe (1966) to compare the performance of investment portfolios.

Table 19.3 Ratio of mean to standard deviation of projected surplus

X_1	X_2	Surplus(0) = 0	Surplus(0) = 0.2	Surplus(0) = 0.4
0	1.0	−0.047	0.211	0.469
0.1	0.9	−0.009	0.243	0.495
0.2	0.8	0.027	0.267	**0.507**
0.3	0.7	0.058	0.282	0.505
0.4	0.6	0.084	**0.289**	0.495
0.5	0.5	0.105	0.292	0.479
0.6	0.4	0.121	0.291	0.461
0.7	0.3	0.133	0.288	0.442
0.8	0.2	0.143	0.283	0.423
0.9	0.1	0.150	0.278	0.406
1.0	0	**0.156**	0.273	0.390

As the value of SF does not change over time in our constant salary units, we can write

$$SF = F(0) - \text{Surplus}(0)$$

Hence

$$\frac{E[\text{Surplus}(36)]}{\sqrt{\text{Var}[\text{Surplus}(36)]}} = \frac{E[F(36)] - F(0) + \text{Surplus}(0)}{\sqrt{\text{Var}[F(36)]}}$$

If we make the assumption that the projected surplus is approximately normally distributed, then the ratio of the mean to the standard deviation is inversely related to the probability of a deficit occurring. Hence by maximizing this ratio, we minimize the probability of a deficit. These ratios are presented in Table 19.3 for values of the initial surplus equal to 0, 0.2 and 0.4. The maximum value of the ratio for each value of the initial surplus is highlighted in bold, corresponding in each case to the portfolio that minimizes the probability of a deficit in 36 months' time.

Table 19.3 indicates that the portfolio that minimizes the probability of a deficit depends greatly on the initial surplus in the fund. This portfolio consists entirely of the more risky asset if the initial surplus is zero (or negative). However, the optimal allocation to the risky asset falls quickly as the initial surplus increases.

19.4.4 Combining Funding and Investment Decisions

The illustrative example in the previous section assumed that the contributions paid to the plan were a stable fraction of the payroll over the timeframe for the projection. This is only a reasonable assumption if the timeframe does not extend past the date of the next actuarial valuation (or whenever the contribution rate is next reviewed). For these relatively short-term projections, changing the asset allocation of the fund is the only method by which the plan can influence the distribution of the projected surplus/deficit.

It is often desirable, however, to use longer projection timeframes (e.g., 10 or 20 years). In doing so, an assumption must be made about how the contribution rate would be revised at future actuarial valuations. A simple approach would be to assume that the standard contribution rate does not change, but that the projected surplus or deficit at each future valuation is amortized over some fixed period. As the projected surplus or deficit is random, a different series of contribution rates would arise for each simulation.

For a long-term projection, therefore, the asset allocation of the fund is no longer the only control variable. The results now also depend on the period over which future surpluses and deficits are amortized, and simulations can be carried out for different amortization periods in order to compare their effect on the distribution of the projected surplus/deficit. In principle, therefore, the asset/liability study can be used to optimize both the investment portfolio of the fund and the method of amortizing surpluses and deficits.

If we carry these ideas to their logical conclusion, then there is no reason why the asset/liability study should not examine the effect of different standard contribution rates as well as amortization periods. In doing so, it could effectively replace the deterministic funding valuation as the method of evaluating the required contribution rate for the plan. The main advantages of using an asset/liability study in this way are:

The required margins for prudence can be determined by evaluating quantifiable risk measures rather than by making arbitrary judgments about what constitutes a prudent actuarial basis in a deterministic valuation.

The funding and investment decisions are made together on the basis of their combined financial impact on the plan.

In order to include the funding strategy in the optimization process, performance measures are required for the contribution rate paid over the projection timeframe. As for the projected surplus, possible measures could include the mean, variance, or semi-variance of the contribution rate. Alternatively, the following definition might be used for the conditional tail expectation of the contribution rate:

$$\text{Conditional tail expectation} = \sum_{t=1}^{T} v^t E_0[\max(\text{CR}_t - \text{SCR}, 0)]$$

where T is the number of periods in projection timeframe, v is the discount rate for interest net of salary escalation, E_0 is the expected value of simulated output at start of projection timeframe, CR_t is the total contribution rate paid at end of period t, and SCR is the standard contribution rate.

The conditional tail expectation, as defined above, is a measure of how frequently and by how much the future contribution rate will exceed the standard contribution rate. This is an important risk measure for the plan sponsor, which quantifies the risk of having to pay extra contributions in order to amortize a funding deficit. One of the reasons for the plan sponsor agreeing to pay a higher standard contribution rate might be to reduce this risk. However, other factors, such as the mean contribution rate and the risk of insolvency, would also have to be considered. The question of how an optimal trade-off between different performance measures can be achieved is beyond the scope of this chapter. A stochastic approach to decision making that allows for multiple objectives is presented in Haberman et al. (2003).

19.5 Practical Issues

The investment policy of a defined-benefit pension plan is influenced by common practice in the country where it operates and how the local legal environment affects this practice. Pension legislation may have a direct effect on investment policy by placing explicit restrictions on the asset allocation of a pension fund. Even when such legal restrictions are absent, regulations dealing with matters such as taxation and the use of surplus assets are likely to have a material influence on investment strategies.

19.5.1 Who Controls the Investment Policy?

The control of investment policy is an issue of fundamental importance. It could be argued that if the plan sponsor has a legal obligation to pay defined benefits, then it should be free to finance these benefits in any way it chooses. The opposing argument is that the main function of the pension fund is to provide security for the members' accrued benefits, so the plan sponsor should leave its investment policy to the members or their representatives.

In practice, the responsibility for investment policy may be shared in some way between the plan sponsor and the members. The trust law status of plans in the U.K. and U.S. requires that control of the pension fund is ceded to trustees who are legally required to act in the best interests of the members. But many of the trustees might also be senior employees, appointed by the plan sponsor, who would not ignore the interests of their employer.

Whether there is any real conflict of interest between the plan sponsor and the members as regards investment policy is a debatable point. It has been argued that if the plan sponsor is concerned primarily with protecting the interests of the equity stockholders, then this should lead to an investment strategy based either on matching the liabilities or maximizing tax efficiency, which may not be in the members' interests (see Section 19.5.2 and Section 19.5.3).

19.5.2   Who Owns the Surplus?

In defined-benefit plans where the plan sponsor meets the balance of the cost (after fixed member contributions), it can be argued that any surplus assets are the result of earlier overfunding by the sponsor. By this reasoning, the plan sponsor would appear to have a strong claim for the entire surplus, which could be reclaimed by a reduction in future employer contributions.

In practice, the issue may not be this clear cut because (i) legislation may give the members entitlement to part of the surplus in certain situations and (ii) the plan itself may have provisions for the use of surplus assets to improve benefits. If surplus assets *can* be used to improve benefits, then it can be argued that the members and the equity stockholders would have conflicting interests in relation to investment policy. If the members can take a share of surpluses while avoiding any liability for deficits, then they effectively have a call option (in addition to their defined benefits). In order to maximize the value of their option, they would prefer the fund to be invested in volatile assets. The equity stockholders, on the other hand, would prefer the assets to be matched (as far as possible) to the accrued liabilities.

The validity of these theoretical arguments depends on the extent to which the employees perceive benefit improvements as an increase in their own remuneration, thereby offsetting other pay awards or improving their motivation. As there is no simple answer to the general question of whether increasing employees' remuneration is in the interests of the equity stockholders, the implications of benefit improvements would have to be judged on a case-by-case basis.

A further important point is that the members' accrued benefits are not absolutely secure because of the possibility of default, i.e., the sponsoring employer's business may fail when the plan has a solvency deficit. In effect, the limited liability of the equity stockholders allows them to exercise a put option against the plan members in the event of bankruptcy. If this default risk is judged to be significant, then a matching investment strategy might then be in the members' best interests.

19.5.3   Tax Considerations

Pension funds usually enjoy a more favorable tax regime than other forms of saving. These tax concessions are granted in order to encourage pension provision, and might be necessary to compensate pension plans for the additional regulations and restrictions that apply to them. The tax concessions that are relevant to the investment policy of a pension fund relate to the taxation of investment income and capital gains. A simple and common form of tax relief is to exempt all income and gains within a tax-approved pension fund from taxation.

The relevant question for the investment policy of the fund is how these tax concessions affect comparative returns from different assets relative to tax-paying investors. If investment income is generally taxed more heavily than capital gains, then a pension fund that is wholly tax exempt would obtain a relative tax advantage by investing in high-income assets, such as fixed-interest bonds.

It might appear that these tax issues are quite minor because of the very different gross returns expected from different asset classes. In a plan where both surpluses and deficits belong to the sponsor, however, it can be argued that the plan sponsor effectively owns the pension fund assets. Ignoring the effects of taxation, the investment strategy of the pension fund should not matter to the equity stockholders, who could nullify any change in the pension fund portfolio by adjusting their own private portfolios. This argument is a version of the "irrelevancy proposition" of corporate finance put forward by Modigliani and Miller (1958).

If certain assets have a relative tax advantage within the pension fund, however, then the equity stockholders would clearly prefer to hold them via the pension fund rather than in their own private portfolios. Thus, if investment income has a relative tax advantage, then it can be argued that shareholder value is maximized by investing the pension fund entirely in fixed-interest bonds. It follows that in a somewhat simplified model in which pension fund assets are treated as belonging to the plan sponsor, it can be argued that taxation is the deciding factor in determining which investment strategy is in the interests of the equity stockholders.

An introduction to the treatment of defined-benefit plans in corporate finance models based on the Modigliani and Miller proposition was presented in Section 3.5.2. The case for this approach is made by Exley et al. (1997), and a critique can be found in the written comments submitted by Haberman et al. (2003) to the discussion following their paper. For a discussion of the relevance of corporate finance concepts to the investment strategy of a defined-benefit plan, refer to Blake (2001).

19.5.4 Legal Constraints on Investment Policy

The investment policy of a tax-approved pension fund is normally subject to legal constraints, although the scope of this legislation varies greatly from country to country. In lightly regulated countries, such as the U.S. and U.K., the legal constraints do not go much beyond requiring the fund to be invested as a prudent person would do so, although certain inappropriate assets have also been explicitly prohibited. In more heavily regulated countries, such as Japan and France, there may be limits on the proportion of the fund that may be invested in particular asset classes. A brief survey of pension fund investment regulations in developed countries is available in Turner and Watanabe (1995, chapter 6). For a discussion of the effect of investment restrictions on pension fund performance, refer to Srinivas and Yermo (1999). The main types of restriction that exist are outlined below.

19.5.4.1 Restrictions on Self-Investment

The meaning of "self-investment" was explained in Section 19.3.3. It may be prohibited or restricted to ensure that the value of the pension fund is largely independent of the plan sponsor's business. As the pension plan would be terminated if the plan sponsor were to go out of business, it is not in the members' interests for a significant proportion of the fund to be invested in securities that may become worthless in such an eventuality.

In the U.K., a special relaxation to the self-investment rules applies to pension plans with fewer than 12 members who include directors of the sponsoring employer. These plans may use a proportion of the pension fund to makes loans to the plan sponsor (at commercial rates of interest) or to purchase an office building that is leased to the plan sponsor. These special rules recognize the particular difficulties that a small business may have in raising capital.

19.5.4.2 Restrictions on Foreign Assets

A pension fund may wish to invest in foreign assets in order to secure the advantages of international portfolio diversification. It may also benefit from access to more liquid and diverse capital markets than exist in its home country.

Many countries, however, restrict or prohibit investment in foreign assets in order to prevent the depletion of foreign exchange reserves or protect the exchange rate. Another possible motive might be to limit the currency risk in pension fund portfolios, although the benefits of international diversification may more than compensate for such risks.

19.5.4.3 Regulation of Investment Portfolios

The most intrusive type of regulation may specify limits for various asset classes in the pension fund, such as a maximum proportion of the fund that may be invested in equities. The aim of such regulation would be to limit portfolio risk and ensure adequate diversification across asset classes. Such restrictions would be most justified in countries with relatively immature capital markets, where corporate securities might be subject to unusually high volatility or low liquidity.

Another type of regulation may require that a minimum proportion of the fund is invested in government bonds, so that pension plans are obliged to use part of their assets to finance the national debt. This type of regulation would be most justified in countries moving the burden of their pension provision from unfunded social security plans to funded private sector plans, when heavy government borrowing may be required to finance transitions costs.

References

Bils, M. (1985). Real wages over the business cycle: estimates from panel data. *Journal of Political Economy* 98, 666–689.

Blake, D. (2001). UK pension fund management: how is asset allocation influenced by the valuation of liabilities? The Pensions Institute. www.pensions-institute.org/wp.

Cardinale, M. (2003). Cointegration and the relationship between pension liabilities and asset prices. Watson Wyatt LLP Technical Paper 2003–RU01.

Exley, C. J., Mehta, S. J. B., and Smith, A. D. (1997). The financial theory of defined benefit schemes. *British Actuarial Journal* 3(Part IV), 835–996.

Haberman, S., Day, C., Fogarty, D., Khorasanee, M. Z., McWhirter, M., Nash, N., Ngwira, B., Wright, I. D., and Yakoubov, Y. (2003). A stochastic approach to risk management and decision making in defined benefit schemes. *British Actuarial Journal* 9(Part III), 493–618.

Kemp, M. (1996). Asset/liability modeling for pension funds. Presented to Staple Inn Actuarial Society, 15 October 1996. www.sias.org.uk.

Lee, E. M. (1986). *An Introduction to Pension Schemes*. Institute and Faculty of Actuaries.

McGill, D. M., Brown, K. N., Haley, J. J., and Schieber. S. J, (1996). *Fundamentals of Private Pensions*, 7th edition. University of Pennsylvania Press.

Modigliani, F. and Miller, M. H. (1958). The cost of capital, corporation finance and the theory of investment. *American Economic Review* 48, 261–297.

Panjer, H. H. (Ed.), Boyle, P. P., Cox, S. H., Dufresne, D., Gerber, H. U., Mueller, H. H., Pedersen, H. W., Pliska, S. R., Sherris, M., Shiu, E. S., and Tan, K. S. (1998). *Financial Economics with Applications to Investments, Insurance and Pensions*. The Actuarial Foundation.

Sharpe, W. F. (1966). Mutual fund performance. *Journal of Business* 39(1, Part 2), 119–138.

Sherris, M. (1992). Portfolio selection and matching: a synthesis. *Journal of the Institute of Actuaries* 119(Part I), 87–105.

Srinivas, P. S. and Yermo, J. (1999). Do investment regulations compromise pension fund performance? — Evidence from Latin America. World Bank Discussion Paper No. 19562. www-wds.worldbank.org.

Turner, J. and Watanabe, N. (1995). *Private Pension Policies in Industrialized Countries*. W.E. Upjohn Institute for Employment Research.

UBS Global Asset Management (2004). Pension fund indicators. www.ubs.com.

Chapter 20

Individual Pension Choices

20.1 Introduction

Pension actuaries have traditionally specialized in advising the sponsors and trustees of defined-benefit pension plans. In many countries, this role has been reinforced by legislation requiring defined-benefit plans to commission actuarial valuations and to obtain actuarial certification of their compliance with solvency and funding regulations.

In much of the industrialized world, however, there is a trend towards greater pension provision through individual accumulation policies or defined-contribution group plans. In both types of arrangement, the individual member is usually responsible for important funding and investment decisions affecting his or her benefits, and bears the risk associated with these decisions.

Pension actuaries may, therefore, be called upon to provide advice on individual pension choices, which might be delivered efficiently through computer software that members could use without the need for face-to-face consultations. Some of the decisions that individuals might require help on are:

the contribution rate required to achieve a target pension at a particular retirement age;
the investment strategy to follow at different durations from retirement;
the type of annuity product to buy at retirement.

For a fuller discussion of the role of the pension actuary in the design and management of defined-contribution plans, refer to Stocker et al. (1999).

20.2   Projection of Money-Purchase Benefits

Money-purchase benefit projections can give members an idea of what benefits they can expect at retirement for a given future rate of contribution. These benefit projections are normally an essential part of the process of selling individual pension policies and may also be a legal requirement in defined-contribution group plans.

An important problem associated with such projections is the misleading impression created by the use of nominal currency units. The actuarial assumptions used to project benefits contain implicit allowances for future inflation rates, which, when compounded over long periods, have a significant effect on the purchasing power of money. Thus, the "money illusion" created by projecting benefits in nominal currency units can give members an exaggerated sense of the value of their future benefits.

If members are to receive useful information, then benefit amounts should be projected in units that will give them a realistic idea of their standard of living at retirement. This can be done in two ways:

using currency units which increase in line with price inflation, so that the purchasing power of each unit does not change over time;
projecting future benefits as a fraction of the member's final salary.

The second type of projection might also be useful when an employer is seeking to replace a final salary plan with a defined-contribution plan offering benefits of comparable value.

The formulae used to project money-purchase benefits are very similar to the cash-flow equations for defined-benefit plans given in Section 18.1. The main differences are that the periodic cash flows are contributions alone (rather than contributions and benefit payments) and that the projected fund at retirement must be divided by an annuity factor to obtain the projected pension.

The variables that the member can potentially control are the retirement age, the future contribution rate, the investment strategy, and the type of annuity purchased at retirement. Therefore, it would be useful to design a software package in which all these variables could be altered at the discretion of the member.

The uncertainty in future benefits is an important issue. Some indication of risk can be given by calculating the projected benefits on both optimistic and pessimistic assumptions, so that the member has an idea of the potential range of outcomes. A more rigorous approach would involve making stochastic projections, so that a probability distribution can be obtained for the projected benefits. This could be communicated to the member by tabulating the projected benefits at different percentiles.

20.2.1 Deterministic Projections

20.2.1.1 *Projected Pension in Real Currency Units*

To make projections in real currency units (adjusted for price inflation) we define the following variables in discrete time:

$f(t)$ is the projected fund after t periods in real units
$c(t)$ is the real contribution paid at end of period t
i is the projected real investment return per period
a_R is the projected value of annuity at retirement age
T is the number of periods up to retirement age
e_1 is the expense fraction for contributions
e_2 is the expense loading for annuity purchase

The projected fund at retirement in real currency units is given by

$$f(T) = f(0)(1 + i)^T + (1 - e_1) \sum_{t=1}^{T} c(t)(1 + i)^{T-t}$$

For some individuals, it may be reasonable to assume that the contribution will increase broadly in line with price inflation. If so, then $c(t)$ will be a constant in real currency units; hence, we can write

$$c(t) = c$$

The formula for the projected fund then simplifies to

$$f(T) = f(0)(1 + i)^T + (1 - e_1) c s_{\overline{T}|}^i$$

And the projected pension at retirement is given by

$$\text{Real projected pension} = \frac{f(0)(1 + i)^T + (1 - e_1) c s_{\overline{T}|}^i}{a_R(1 + e_2)} \tag{20.1}$$

The value of the annuity at retirement may include the cost of any spouse's pension required by the member. As the fund is being projected in real currency units, it would be logical to assume that the pension is also measured in these units, in which case the value of the annuity should be determined using a real rate of interest. This real interest rate may differ from the assumed real return on the fund prior to retirement, because the investment strategy adopted during the accumulation phase would probably be more aggressive than that of a typical annuity fund.

20.2.1.2 *Projected Pension as a Fraction of Final Salary*

To project the member's benefits as a fraction of final salary we require the following additional variables:

$s(t)$, the member's real salary after t periods
j, the real rate of salary escalation (including promotional increases)
$cr(t)$, the contribution as a fraction of salary paid at end of period t

The projected fund as a fraction of final salary is given by

$$\frac{f(T)}{s(T)} = \frac{f(0)(1+i)^T}{s(T)} + \sum_{t=1}^{T} cr(t)\frac{s(t)}{s(T)}(1+i)^{T-t}$$

We assume that the member's salary grows at the fixed real compound rate j, so that

$$s(t) = s(0)(1+j)^t$$

Using the above formula to substitute for $s(t)$ and $s(T)$ in the expression for the projected fund gives

$$\frac{f(T)}{s(T)} = \frac{f(0)}{s(0)}\left(\frac{1+i}{1+j}\right)^T + (1-e_1)\sum_{t=1}^{T} cr(t)\left(\frac{1+i}{1+j}\right)^{T-t}$$

In group defined-contribution plans, the contribution rate is often a constant fraction of salary, in which case we can write

$$cr(t) = cr$$

The formula for the projected fund as a fraction of final salary then simplifies to

$$\frac{f(T)}{s(T)} = \frac{f(0)}{s(0)}(1+i^*)^T + (1-e_1)\,crs_{\overline{T}|}^{i^*}$$

where i^* is the expected investment return net of salary escalation given by

$$1+i^* = (1+i)/(1+j)$$

The projected pension as a fraction of final salary is obtained, as before, by dividing the projected fund by the value of the annuity loaded for expenses; thus

$$\frac{\text{Projected pension}}{\text{Final salary}} = \frac{(f(0)/s(0))(1+i^*)^T + (1-e_1)\,crs_{\overline{T}|}^{i^*}}{(1+e_2)a_R} \tag{20.2}$$

20.2.1.3 *Projections for Comparing Investment Strategies*

Individuals with money-purchase funds are normally advised to vary the asset allocation of their fund with their duration from retirement.

Individuals who are far from retirement have a long investment time-scale and a relatively small fund. They have less to fear from volatile asset prices because: (i) it will be many years before they have to realize their assets; (ii) the financial impact of market volatility on a small fund is relatively minor. Such individuals are usually advised to maximize expected returns by investing in asset classes such as equities.

On nearing retirement, however, the position changes. As the member's fund grows, the consequences of a fall in asset prices become more serious. This is because there is less time left for prices to recover and less time to make up for the reduced value of the fund by paying extra contributions. As a result, individuals are usually advised to switch their fund into less risky asset classes, such as bonds and cash, as they get nearer to retirement. A reason for preferring bonds to cash is that long-dated bonds can be used to hedge the uncertainty in the cost of an annuity at retirement (because annuity funds invest in these bonds).

Although the most rigorous methods of comparing different investment strategies involve stochastic projections, some simple insights can be obtained from a deterministic approach. Assume that a money-purchase fund is invested wholly in an equity portfolio for which the following parameters are estimated:

y_0, the current dividend yield
g, the projected real dividend growth
i_0, the projected real return on fund

Based on the arguments presented in Section 18.2.6, the projected real return on the fund is given by

$$i_0 \approx y_0 + g$$

Hence, modifying our earlier notation, the projected real fund at retirement is given by

$$f_0(T) = f(0)(1 + i_0)^T + (1 - e_1)cs\frac{i_0}{T|}$$

Now, suppose that a re-rating of equity prices occurs over the period up to retirement, causing the dividend yield over this period to be higher than the current dividend yield. For this hypothetical scenario we define

y_1, the dividend yield over the remaining period up to retirement
i_1, the projected real return on future contributions over the remaining period up to retirement

If we assume that the re-rating of equity prices has no effect on dividend growth,[1] then we can write

$$i_1 \approx y_1 + g$$

The re-rating of equity prices will cause the market value of the fund to fall in the same proportion to the increase in the dividend yield; thus, the projected retirement fund under this scenario is given by

$$f_1(T) = f(0)(y_0/y_1)(1 + i_1)^T + (1 - e_1)cs_{\overline{T}|}^{i_1}$$

The relative magnitudes of $f_0(T)$ and $f_1(T)$ will depend on the parameters in both formulae. However, the term to retirement T has a particularly important effect on how they compare.

In the trivial case when $T = 0$ we have

$$f_1(T) = f_0(T)(y_0/y_1)$$

At $T = 0$, the fall in equity prices would inevitably result in a lower retirement fund. This will also be true for small values of T, although the gap between $f_0(T)$ and $f_1(T)$ will narrow as T increases. Eventually, we come to a critical value T_c at which

$$f_1(T_c) = f_0(T_c)$$

Hence, under this simple deterministic model, it would be reasonable to invest wholly in equities when $T \geq T_c$. As T falls below this critical value, a more diversified portfolio might be advisable.

A further useful projection can be made on the basis that the fund and future contributions are switched into assets offering a guaranteed real return r. The projected fund would then be given by

$$f_2(T) = f(0)(1 + r)^T + (1 - e_1)cs_{\overline{T}|}^{r}$$

We would ordinarily expect $f_2(T)$ to be less than $f_0(T)$, because the equity risk premium means that $i_0 > r$. However, $f_2(T)$ might well be greater than $f_1(T)$, as investing in risk-free assets would avoid the capital loss caused by the fall in equity prices. Thus, for certain values of T we would have

$$f_0(T) > f_2(T) > f_1(T)$$

For these values of T, a reasonable investment strategy would involve splitting the fund between equities and the risk-free asset.

---

[1] Although this assumption is inconsistent with simple interpretations of market efficiency, Campbell and Shiller (2001) show that it is supported by historic data from the U.S. equity market.

20.2.2 Examples of Deterministic Benefit Projections

20.2.2.1 *Example 1*

An individual 30 years from retirement is about to start contributing to a single-member pension plan. It is assumed that future contributions will increase broadly in line with price inflation. The member requires projections for the pension at retirement based on the following assumptions:

$T = 30$ years
$c = 1000$ units per annum
$e_1 = 0.05$, $e_2 = 0.02$
Optimistic scenario: $i = 0.06$, $a_R = 16$
Pessimistic scenario: $i = 0.02$, $a_R = 18$

Although we have chosen the unit of time to be 1 year, it is more likely that contributions would be paid monthly. To simplify matters we shall assume a continuous rate of payment, so that Equation (20.1) becomes

$$\text{Real pension at retirement} = \frac{c(1 - e_1)\bar{s}_{\overline{30}|}}{a_R(1 + e_2)}$$

The optimistic and pessimistic bases give

$$\text{Real pension } \{i = 0.06,\ a_R = 16\} = 1000 \times 0.95 \times 81.407/(16 \times 1.02)$$
$$= 4739 \text{ units per annum}$$

$$\text{Real pension } \{i = 0.02,\ a_R = 18\} = 1000 \times 0.95 \times 40.972/(18 \times 1.02)$$
$$= 2120 \text{ units per annum}$$

The wide gap between the pensions projected for the two scenarios is typical for projections made at long durations from retirement. It reflects the high degree of uncertainty in the pension from a money-purchase plan.

20.2.2.2 *Example 2*

An individual 10 years from retirement has an accumulated fund of 40 000 units. What real contribution rate is required to obtain a pension of 5000 units per annum? The parameters for the projection (apart from T) are the same as for Example 1.

Rearranging Equation (20.1) to solve for the contribution rate gives

$$c = \frac{a_R(1 + e_2) \times \text{Pension} - f(0)(1 + i)^T}{\bar{s}_{\overline{T}|}(1 - e_1)}$$

The optimistic and pessimistic bases give

$$\text{Lower contribution rate} = \frac{16 \times 1.02 \times 5000 - 40\,000 \times 1.06^{10}}{13.572 \times 0.95}$$

$$= 773 \text{ units per annum}$$

$$\text{Higher contribution rate} = \frac{18 \times 1.02 \times 5000 - 40\,000 \times 1.02^{10}}{11.059 \times 0.95}$$

$$= 4097 \text{ units per annum}$$

The higher contribution rate is more than five times the lower one, which presents a disturbingly unclear picture for an individual attempting to target a particular pension. This uncertainty arises because changes in the future contribution rate are less important than changes in the projected value of the fund for an individual who is close to retirement. The only way to reduce the uncertainty is to invest in assets that offer relatively predictable real returns.

20.2.2.3   Example 3

The individual described is Example 2 is considering a change in the asset allocation of the fund, which is currently invested wholly in equities. It would be possible to invest all or part of the fund in a risk-free asset providing a guaranteed real return over the 10-year period up to retirement. Investigate whether there is a strong case for diversification based on the following assumptions:

Current annual divided yield on equities is 3%
Projected real dividend growth is 1% per annum
Possible fall in equity prices is 25%
Real return on risk-free asset is 2% per annum

If the dividend yield remains at 3%, the projected real return on equities is given by

Dividend yield + Real dividend growth = 3% + 1% = 4% per annum

In this case, the projected real retirement fund is given by

$$f_0(10) = 40\,000 \times 1.04^{10} + 0.95 \times 4000 \times s_{\overline{10}|}^{4\%} = 105\,739 \text{ units}$$

If equity prices fell by 25% then the dividend yield would increase by the ratio 100 : 75; thus

$$\text{New annual dividend yield} = 3\% \times (100/75) = 4\%$$

New projected real return on equities $= 4\% + 1\% = 5\%$ per annum
The projected real retirement fund after the fall in equity prices is given by

$$f_1(10) = 40\,000 \times 1.05^{10} \times \frac{3}{4} + 0.95 \times 4000 \times \bar{s}_{\overline{10|}}^{5\%} = 97\,848 \text{ units}$$

As the projected fund after the hypothetical fall in equity prices is lower, there might be a case for a more diversified investment strategy. To investigate further, we calculate the projected real retirement fund that would arise from investing the fund in the risk-free asset, which is given by

$$f_2(10) = 40\,000 \times 1.02^{10} + 0.95 \times 4000 \times \bar{s}_{\overline{10|}}^{2\%} = 90\,783 \text{ units}$$

As this projected fund is lower than the fund projected from investing in equities even after the fall in equity prices, the case for switching part of the fund into the risk-free asset appears to be weak at this duration from retirement. However, the relative magnitudes of $f_0(T)$, $f_1(T)$, and $f_2(T)$ will change as the individual gets closer to retirement; at some stage it will certainly be advisable to switch part of the fund into the risk-free asset.

The results from this simple type of analysis are dependent on the parameters assumed. We would, for example, obtain significantly different answers if the scenario for the equity market re-rating involved a 50% reduction in market prices. However, the approach presented above is very simple to use and would be comparatively easy to explain to an individual member. It is also consistent with basic actuarial principles (the idea that a fall in asset prices can be redressed by an increase in future yields is a version of Redington's immunization concept; see Section 5.2).

20.2.3 Stochastic Projections

In a stochastic projection, both the rate of return on the fund and the interest rate used to price the annuity would be modeled as random variables.
We define:

$i(t)$ to be the random return on the fund in period t
$y(t)$ to be the random interest rate used to price annuities after t periods

The random interest rate $y(t)$ would normally be the simulated yield on government bonds of suitable duration for matching whole-life annuities. The cash-flow projection formula used to simulate the retirement fund has the form

$$f(t+1) = f(t)(1 + i(t)) + (1 - e_1)c(t)$$

And the projected pension at retirement is given by

$$\text{Pension } (T) = \frac{f(T)}{(1 + e_2)a_R^{y(T)}}$$

A stochastic investment model that could simulate both $i(t)$ and $y(t)$ would be required. Although the return on the fund and the annuity yield are believed to be the largest sources of uncertainty in projecting a money-purchase pension, other elements of the projection formula could also be modeled stochastically. In particular, if the contribution $c(t)$ is expected to be a constant fraction of salary, then there might be a case for projecting the member's future salary as a stochastic variable. Another refinement worth considering, which might be more difficult to apply in practice, would be to allow for the uncertainty in future mortality rates.

The required output from the projection model would be probability distributions for the projected pension, either in real currency units or as a fraction of final salary. These distributions would depend on various factors under the control of the individual, such as the investment strategy, contribution rate, and retirement age. The most useful application is likely to be guiding the choice of investment strategy at different durations from retirement.

20.2.4 Example Illustrating Stochastic Benefit Projections

The projected pension of an individual 10 years from retirement is simulated stochastically for different investment strategies.[2] Each strategy involves splitting the fund between equities and bonds in fixed proportions, which are realigned annually. The simulations are used to estimate the following points of the distribution function for the pension at retirement.

Table 20.1 suggests that the preferred investment strategy will largely depend on the individual's own personal evaluation of the potential risks and rewards involved. If, for example, the individual's main concern is the

---

[2]For details of the stochastic model used to produce these simulations, refer to Khorasanee (1996).

Table 20.1 Simulated distributions for pension at retirement

Equity proportion (%)	Bond proportion (%)	Probability pension at retirement $<P$ units per annum			
		$P = 4000$	$P = 5000$	$P = 6000$	$P = 7000$
0	100	0.085	0.256	0.472	0.664
10	90	0.070	0.224	0.431	0.626
20	80	0.062	0.201	0.395	0.587
30	70	0.059	0.187	0.367	0.550
40	60	0.061	0.181	0.347	0.518
50	50	0.066	0.180	0.333	0.492
60	40	0.073	0.183	0.325	0.471
70	30	0.082	0.190	0.322	0.456
80	20	0.093	0.198	0.323	0.446
90	10	0.105	0.208	0.324	0.438
100	0	0.118	0.218	0.328	0.434

risk of ending up with a pension below 4000 units per annum, then this probability is minimized by a portfolio split between equities and bonds in the ratio 30:70. If, on the other hand, the individual wants to maximize the chances of getting a pension above 6000 units per annum, then the optimal asset allocation is reversed to 70:30. It follows that simulations of this kind should be used to help individuals understand the risks and rewards of different investment strategies and make decisions consistent with their own subjective preferences.

An unrealistic feature in the distributions presented above is the assumption that asset allocations remain fixed until retirement. Unless the individual is very close to retirement, this is unlikely to be so. As mentioned above, there are sound practical reasons for switching the fund into less risky assets as the duration from retirement reduces. Although it would be relatively straightforward to obtain distributions for investment strategies where the asset allocations change over time in a predetermined manner, this does not quite reflect the way in which decisions can and will be made. The individual can change future asset allocations at regular intervals up to retirement in a manner (yet to be decided) that will depend on the accumulated fund (and possibly other factors) at the time that the decision is made.

The consequences of unknown future investment decisions cannot easily be incorporated into the simulation method described above. It is clear, however, that being able to change the future asset allocation has a radical effect on how the fund should be invested now. A risky investment strategy is more appealing if it can be modified at future durations, according to circumstances, than if it cannot. Allowing for the periodic optimization of asset allocations up to retirement requires an algorithm for deciding what is "optimal" and a mathematical method for iterating the effect of multiple decision points. This type of problem is well known in financial economics, where a technique known as dynamic programming has been used to optimize decision making based on the principle of maximizing expected utility. An introduction to this work and its possible practical applications will be presented in Section 20.4. For further information on how dynamic programming can be applied to decision making in defined contribution plans, refer to Haberman and Vigna (2002).

20.3   Annuity Choices for Pensioners

The purchase of a whole-life annuity can be viewed as one possible investment option for individuals retiring with a money-purchase fund. Hence, there is a case for allowing members the freedom to decide whether or not to buy an annuity and, if they choose to do so, to select the type of annuity and its provider.

In the U.K., however, an individual retiring from a tax-approved plan is legally obliged to use at least 75% of the fund to buy a whole-life annuity. The reason for this requirement is to ensure that tax-subsidized pension plans are used primarily to provide pensions, i.e., incomes guaranteed until death. Otherwise, it is argued, some pensioners would exhaust their assets before death and require income support from the state. As a result, the U.K. has a mature and competitive market for whole-life annuity products. Subsequent legislation has allowed retiring members to delay the purchase of an annuity until age 75 while drawing income from the fund.

For the purpose of this section, we shall assume that pensioners are free to decide whether and when to buy an annuity. We shall also assume that they are free to choose between an annuity offering a level monetary income and an annuity offering an income linked to an index of price inflation. Our purpose will be to outline various actuarial techniques that could help pensioners with these difficult decisions. For more detailed information on these modeling techniques, refer to Khorasanee (1996).

20.3.1   Whether and When to Buy an Annuity

A life office annuity fund receives single premiums from its policyholders and provides each annuitant with an income until death. Unless the annuitant pays an extra premium, there is no return of capital on death. For conventional types of annuity contract, the annuity fund is invested in "safe" assets matching the duration of its liabilities, such as government bonds. However, the income received by each annuitant significantly exceeds the interest that could have been obtained by investing the single premium directly in the bond market. This is because each annuity payment includes a repayment of capital as well as interest. In effect, the reserves released in respect of annuitants who die are used to subsidize the income of the survivors.

The alternative to buying an annuity is to continue to invest the fund, while drawing a regular income from it. If the income drawn is to match that from a whole-life annuity, however, then the investment return on this personal fund must exceed that achieved on the assets of a life office annuity fund. This is because the individual member receives no income subsidy from the reserves released for annuitants who die. Another way of thinking about this is to note that the capital held in the personal fund is not lost on the death of the pensioner — it will become part of the pensioner's estate. Compared with a whole-life annuity, a pensioner drawing income from a personal fund has a life assurance benefit that must be paid for in some way.

It is useful to estimate the return required on the fund that will allow the individual, over some specified period, to:

draw the same income that could have been obtained from a whole-life annuity;

buy a whole-life annuity at the end of the period that will provide the same income.

Assume that the individual retires at age x and intends to draw a continuous level income for a period of T years. Retaining our earlier notation, we define:

$y(t)$ is the yield used to price annuities after t years
i is the constant return on fund
$f(t)$ is the fund remaining after t years

The income that could have initially been obtained from an annuity is given by

$$\text{Income at start of period} = \frac{f(0)}{\bar{a}_x^{y(0)}}$$

If the same income is drawn from the fund, then the assets remaining after T years are given by

$$f(T) = f(0)(1+i)^T - \left(\frac{f(0)}{\bar{a}_x^{y(0)}}\right)\bar{s}_{\overline{T}|}^i$$

If the remaining fund is used to buy an annuity, then the income then obtained is given by

$$\text{Income at end of period} = \frac{f(T)}{\bar{a}^{y(T)}_{x+T}} = \frac{f(0)(1+i)^T}{\bar{a}^{y(T)}_{x+T}} - \frac{f(0)\bar{s}^i_{\overline{T}|}}{\bar{a}^{y(0)}_x \bar{a}^{y(T)}_{x+T}}$$

Equating the expressions for the income at the start and the end of the period gives

$$\bar{a}^{y(0)}_x - \bar{a}^i_{\overline{T}|} = v^T \bar{a}^{y(T)}_{x+T} \qquad (20.3)$$

where $v = 1/(1+i)$.

Solving for i in the above equation gives the "break-even" return, i.e., the return required on the fund over T years to provide the same income that could have initially been obtained from a whole-life annuity. The instantaneous break-even return can be derived from the limiting form of Equation (20.3) as $T \to 0$. In this limit we can write

$$y(T) \to y(0)$$

$$\bar{a}^i_{\overline{T}|} \to T$$

$$\bar{a}^{y(0)}_x \to T + \exp(-\delta_x T)\big(1 + y(0)\big)^{-T} \bar{a}^{y(0)}_{x+T}$$

where δ_x is the force of mortality at age x.

Inserting these limiting values into Equation (20.3) gives

$$1 + i = \big(1 + y(0)\big) \exp(\delta_x)$$

$$\Rightarrow \delta_i = \delta_{y(0)} + \delta_x$$

where $\delta_i = \ln(1+i)$ and $\delta_{y(0)} = \ln(1 + y(0))$.

Thus, the instantaneous break-even return is approximately equal to the sum of the force of interest and the force of mortality used to price the annuity. It follows that the break-even return will increase with the age of the pensioner.

20.3.2 Example Illustrating Calculation of Break-Even Return

A 60-year old man is free to invest his retirement fund in any way he chooses. The most competitive immediate annuity rate available in the market would provide a continuous annual income equal to 7.2% of his fund. This income would rise to 10.7% of the fund for a 75-year old man.

Estimate the break-even return required if the man decides:

never to buy an annuity;
to delay the purchase of an annuity until age 75.

If the man intends never to buy an annuity, then we require the limiting form of Equation (20.3) as $T \to \infty$, which is

$$\bar{a}_x^{y(0)} - \bar{a}_{\overline{\infty}|}^i = 0$$

For a 60-year old man this gives

$$\bar{a}_{60}^{y(0)} - 1/\delta_i = 0$$
$$\Rightarrow \delta_i = 0.072$$

Hence, the force of return required on the fund is the same as the income yield from the whole-life annuity. To match this income over an indefinite period, the capital value of the fund must be preserved while the required income is provided entirely by the investment return.

If the man intends to defer the purchase of an annuity until age 75, then we must solve

$$\bar{a}_{60}^{y(0)} - \bar{a}_{\overline{15}|}^i = v^{15}\bar{a}_{75}^{y(15)}$$

The value of the annuity at age 75 is unknown because it will be calculated using mortality and investment yields applicable 15 years into the future. However, if we take the current annuity price for a 75-year old man as our best estimate of its price in 15 years, then the expression becomes

$$(0.072)^{-1} - \bar{a}_{\overline{15}|}^i = v^{15}(0.107)^{-1}$$

By trial and error, the solution for the required return is 6.0% per annum. This is significantly lower than the return required if no annuity is ever bought, because the purchase of an annuity at age 75 allows the capital value of the remaining fund to be converted into income. It follows that delaying the purchase of an annuity, rather than avoiding it altogether, might be a more realistic strategy for improving retirement income.

20.3.3 Level or Increasing Annuity

It might appear that a whole-life annuity providing an income increasing in line with an inflation index would be the natural option for a pensioner with a fund to invest. The effect of inflation on the purchasing power of a fixed monetary income can be devastating, so a pensioner who buys a level annuity is apparently taking a serious risk.

The protection provided by an index-linked annuity, however, is reflected in its price; the initial income provided by such an annuity will be significantly lower than that from a level annuity. Given the price of each type of annuity, it is a simple matter to estimate the duration at which the index-linked income will "overtake" the level income for any future inflation rate. To gain advantage from purchasing an index-linked annuity, the pensioner must survive well beyond this duration, so that the aggregate value of the income from the index-linked annuity exceeds that which a level annuity could have provided.

It is useful, therefore, to estimate the break-even lifespan for an individual considering whether to buy an index-linked or level annuity. This is the survival duration at which the income stream from each type of annuity will have the same present value at retirement. If it is reckoned that the probability of surviving to this end of the break-even lifespan is small, then the individual may be willing to accept the risk of a level annuity.[3]

We define the following variables for an individual retiring at age x:

N is the break-even lifespan
y is the nominal yield used to price level annuities
r is the real yield used to price index-linked annuities
g is the expected future inflation rate
i is the interest on savings held in retirement

For every unit invested at retirement:

$1/\bar{a}_x^y$ is the income from a level annuity
$1/\bar{a}_x^r$ is the initial income from an index-linked annuity

At the break-even lifespan, the two income streams have the same present value; hence

$$\frac{\bar{a}^i_{\overline{N}|}}{\bar{a}^y_x} = \frac{\bar{a}^{i-g}_{\overline{N}|}}{\bar{a}^r_x}$$

Solving for N in the above expression gives the break-even lifespan.

20.3.4 Example Illustrating Calculation of Break-Even Lifespan

An individual retiring at age 65 can choose between a level annuity providing a continuous income equal to 8.4% of the retirement fund and an

---

[3]It would be prudent, of course, to save a proportion of earlier payments from a level annuity in order to build up a fund that can later be used to offset the effects of inflation.

index-linked annuity providing a continuous income initially equal to 6.0% of the retirement fund. Estimate the break-even lifespan using the assumptions

$$\delta_i = \ln(1 + i) = 0.04$$

$$\delta_g = \ln(1 + g) = 0.03$$

The present value of the income stream from the level annuity is given by

$$0.084\,\bar{a}^i_{\overline{N}} = \frac{0.084(1 - e^{0.04N})}{0.04}$$

The present value of the income stream from the index-linked annuity is given by

$$0.060\bar{a}^{i-g}_{\overline{N}} = \frac{0.060(1 - e^{0.01N})}{0.01}$$

These two functions are tabulated below for selected values of N.

N	Level annuity	Index-linked annuity
10	0.69	0.57
15	0.95	0.84
20	1.16	1.09
25	1.33	1.33
30	1.47	1.56

This indicates that the break-even duration is 25 years, which implies that a pensioner would have to survive beyond age 90 to obtain a more valuable income-stream from the index-linked annuity. The individual may decide that the probability of surviving to this age is too small to warrant the purchase of an index-linked annuity.

20.3.5 Stochastic Approaches

Stochastic projections can be used to quantify the risks of different investment choices at retirement. If maintaining a guaranteed standard of living until death is the pensioner's prime objective, then the purchase of an index-linked annuity can be taken as the risk-free option.[4] The alternatives of buying a level annuity or drawing income from the fund both involve

---

[4]Although this is the nearest thing to a risk-free option, there is still some uncertainty because the pensioner's own expenses may escalate at a different rate from the inflation index.

risks that can be measured relative to this risk-free option. In either case, a stochastic approach allows us to estimate the probability that the pensioner would fail to reproduce the real income stream that an index-linked annuity would have provided.

For an individual retiring at age x, we define the following random variables in discrete time:

$f(t)$ is the pensioner's fund t years after retirement
$i(t)$ is the return on fund between t and $t+1$
$G(t)$ is the inflation index at time t
$\ddot{a}_{x+t}^{y(t)}$ is the value of a level annuity at time t using nominal yield $y(t)$
$\ddot{a}_{x+t}^{r(t)}$ is the value of an index-linked annuity at time t using real yield $r(t)$

The random variables $i(t)$, $G(t)$, $y(t)$, and $r(t)$ would be simulated directly from a suitable stochastic model. To enable cash flows to be projected in discrete time, it is assumed that the income from the annuities is paid annually in advance.

20.3.5.1   Income Drawdown

Suppose, first, that the pensioner wishes to draw income from the fund for a period of T years. For the purpose of comparison with the risk-free option, we shall assume that the pensioner draws the same income that an index-linked annuity would have provided and uses the fund remaining at time T to buy an index-linked annuity. It follows that the fund must obey the following recurrence formula:

$$f(t+1) = \left[f(t) - \left(\frac{f(0)}{\ddot{a}_x^{r(0)}}\right)\left(\frac{G(t)}{G(0)}\right)\right](1+i(t))$$

At time T, we define S_T as the assets remaining after the purchase of the index-linked annuity. The formula for S_T is

$$S_T = f(T) - \left(\frac{f(0)}{\ddot{a}_x^{r(0)}}\right)\left(\frac{G(T)}{G(0)}\right)\ddot{a}_{x+T}^{r(T)}$$

In reality, it is not possible for S_T to be less than zero. Thus, when $S_T < 0$ in a stochastic simulation, we can infer that the pensioner would not have been able to maintain the real income stream. The probability of not being able to match the income from an index-linked annuity can therefore be estimated as

$$Pr\{S_T < 0\} = \text{(Frequency of simulations for which } S_T < 0)/$$
$$\text{(Total number of simulations)}$$

20.3.5.2 Level Annuity

Turning to the comparison with a level annuity, we again assume that the pensioner only spends the income that could have been obtained from an index-linked annuity. As the income from a level annuity would always initially be higher, the balance of the income can be invested in a sinking fund.

We now define sf(t) to be the value of sinking fund t years after retirement, where $sf(0) = 0$. The recurrence formula for this sinking fund is

$$sf(t+1) = \left[sf(t) + f(0)/\ddot{a}_x^{y(0)} - \left(\frac{f(0)}{\ddot{a}_x^{r(0)}} \right) \left(\frac{G(t)}{G(0)} \right) \right] (1 + i(t))$$

If the pensioner survives to durations at which the real income stream exceeds the income from a level annuity, then money is drawn from the sinking fund to bridge this gap. This strategy will only fail to reproduce the real income stream if the sinking fund is exhausted before pensioner's death. Hence, for all values of t, we use stochastic simulation to estimate:

$$Pr\{sf(t) < 0\} = \text{(Frequency of simulations for which } sf(t) < 0)/$$

$$\text{(Total number of simulations)}$$

The probability of not matching the income from an index-linked annuity is the same as the probability that the pensioner will die after the sinking fund is exhausted. Assuming that the mortality of the pensioner is independent of all the financial variables, this probability is given by

$$\sum_{t=1}^{\infty} Pr\{sf(t) < 0\} \frac{d_{x+t}}{l_x}$$

where d_x and l_x are taken from a suitable life table.

20.3.6 Example Illustrating Stochastic Approach

A man retires at age 60 with a money-purchase fund of 100 000 units. The most competitive index-linked annuity available in the market offers an initial income of 5500 units per annum in exchange for his fund. He decides, instead, to invest the fund in equities while drawing an income from it that will rise in line with inflation. He asks for advice on how much income he should draw and for how long he should defer the purchase of an annuity.

Table 20.2 Probabilities of failing to maintain real income stream

Initial income per annum	Duration until purchase of annuity		
	$T = 5$	$T = 10$	$T = 15$
4500	0.12	0.13	0.18
5000	0.28	0.27	0.34
5500	0.44	0.43	0.50

A stochastic investment model is used to estimate the probability of failing to maintain various real income streams for a range of durations until the purchase of an index-linked annuity. These probabilities are tabulated in Table 20.2, which indicates that if the income drawn from the fund matches that from an index-linked annuity, then the probability of failing to maintain the real income stream is close to 0.5 over a range of durations. Drawing a somewhat lower income significantly reduces this probability.

20.4   Utility-Maximization Models

20.4.1   Expected Utility Theory

An introduction to utility functions and their application to portfolio theory was presented in Section 4.2. In this section, we assume that the fundamental axioms underpinning the use of utility functions and expected utility theory are valid. For further information on utility theory, refer to Eichberger and Harper (1997, chapter 1).

Expected utility theory is used to explain how individuals make financial decisions when future outcomes are uncertain. It is assumed that every individual has a utility function, i.e., a mathematical function that measures the satisfaction provided by any level of wealth, income, or consumption.

Although utility functions are subjective, they are believed to have some common properties for all individuals. If the utility function for the wealth W of an individual is given by $u(W)$, then it would normally be assumed that

$u'(W) > 0$, because individuals prefer more wealth to less
$u''(W) < 0$, because individuals exhibit diminishing marginal utility

Expected utility theory postulates that individuals seek to maximize the expected value of their utility when their future wealth is random. To calculate this expected value, the probability distribution of the individual's future wealth must be known. In order to optimize decision making, the way in which this distribution depends on choices made by the individual must also be known.

For example, suppose that a person's future wealth is a random variable that depends on the proportion of current wealth invested in equities, so that

X is the proportion of wealth invested in equities
$f(W, X)$ is the probability density function for future wealth

The optimal value of X is given by finding

$$\max_{X} \left(\int_0^{\infty} u(W)f(W, X)\, dW \right)$$

The assumption that utility functions are increasing and concave implies that individuals are "risk averse," i.e., they will prefer a certain amount of wealth to a random amount of wealth with the same expected value. This is why a risky asset must offer a "risk premium," which is the difference between its expected return and the return on a risk-free asset. The market prices risky assets so that their risk premium is always positive; otherwise no one would buy them.

The degree of risk aversion, however, can vary significantly from one individual to another. It depends on the exact shape of the utility function. This means that the optimal amount of wealth to invest in risky assets, such as equities, depends on the utility function (and risk aversion) of the particular individual.

20.4.2 Utility of Lifetime Consumption

One kind of utility-maximization model is based on the utility of lifetime consumption. The amount consumed by an individual is defined as the money spent on goods and services, which can vary over the lifecycle. It is assumed that individuals borrow or save over different periods in their lives so as to maximize the expected utility of their lifetime consumption. This is known as the *lifecycle permanent income hypothesis* (refer to Bateman et al. (2001) for further discussion of this concept). A refined version of this model also allows for the bequest motive, i.e., the desire of individuals to pass on wealth to their next of kin.

We now define the following variables:

S_t is the salary payment received at time t
B_t is the amount consumed at time t
W_t is the wealth accumulated to time t

R_t is the random return on savings between t and $t+1$
k is the subjective rate of time preference
I_x is the utility of future consumption of an individual aged x

Assuming that mortality is independent of financial variables, Kapur and Orszag (1999) postulate that the utility of lifetime consumption can be formulated as

$$I_x = \sum_{t=0}^{\infty} {}_tp_x u(B_t)\, e^{-kt}$$

where $u(B_t)$ is the utility function for consumption at time t.

The relationship between earnings, consumption, investment returns, and wealth is given by

$$W_{t+1} = (W_t + S_t - B_t)(1 + R_t)$$

At time $t=0$, the individual is free to choose both the initial amount of consumption B_0 and the manner in which the accumulated wealth is invested. The investment decision, which we assume can be defined by a single parameter X_0, will determine the probability distribution for the investment return R_0. The objective is to optimize these decisions by maximizing the expected utility of lifetime consumption. Hence, we must determine

$$\max_{B_0, X_0}\{E_0[I_x]\}$$

where E_0 denotes the expected value at $t=0$.

In maximizing I_x, however, we must allow for the fact that the individual will be able to change the consumption and investment decisions at the start of each future period. The problem is thus to optimize the decision at time zero given that the individual will be able to make further optimizing decisions at the start of all future time periods. The required solutions are of the form

$$B_t^* = f_1(W_t, t)$$
$$X_t^* = f_2(W_t, t)$$

where B_t^* and X_t^* are the optimal consumption and investment decisions at time t.

Closed-form solutions for B_t^* and X_t^* only exist for certain types of utility function and for relatively simple stochastic processes for R_t. In the general case, a continuous-time formulation of the model leads to a nonlinear partial differential equation for the maximum expected value of I_x, which might be solved by numerical methods. This leads directly to solutions for B_t^* and X_t^*. For further details, refer to Merton (1990).

The strength of the utility of consumption model is that investment and consumption decisions are optimized together. In theory, it could be used to advise an individual both on how much to contribute and how to invest the fund at different durations from retirement. However, the practical problems involved in the use of such a model are considerable. For example, it would be necessary to determine a utility function for consumption, valid at all future ages, and the subjective parameter k, which measures the time preferences of the individual. For further discussion of these problems, refer to Khorasanee (2002).

A major limitation of the model presented above is its failure to allow for the utility of leisure (or the disutility of work). Although some theoretical work has been done with functions allowing for the utility of both consumption and leisure, the difficulties of eliciting such a function for a particular individual may be insurmountable. For an appraisal of portfolio choice models based on the maximizing utility of lifetime consumption, refer to Campbell and Viceira (2002).

20.4.3   Utility of Retirement Wealth

A simpler type of utility-maximization model is based on utility of wealth at retirement. In this type of model it is assumed that future savings are predetermined, so that only investment decisions can affect the distribution of retirement wealth. The retirement age is also taken as predetermined, which sidesteps the problem of allowing for the utility of leisure. This might be a reasonable representation of the position of some individuals in defined-contribution pension plans. For details of how to apply this model to investment choices in a defined-contribution plan using simulation, refer to Thomson (1998). In this section, we adopt a simple stochastic investment model for which analytical solutions are possible.

We now define the following variables:

T is the duration until retirement
$u(W_T)$ is the utility of retirement wealth
C_t is the contribution to fund at time t
X_t is the parameter defining investment strategy between t and $t+1$
    (as before)
R_t is the random investment return between t and $t+1$ (as before)

In this model, we must determine

$$\max_{X_0}\{E_0[u(W_T)]\}$$

given that $W_{t+1} = (W_t + C_t)(1 + R_t)$.

The aim is to obtain a solution for the optimal investment strategy of the form

$$X_t^* = f_t(W_t)$$

In maximizing $u(W_T)$, we must again allow for the fact that the individual will be able to modify the investment strategy at all durations up to retirement. This problem can be solved by starting the optimization procedure one period before retirement, at time $T-1$. The utility-maximizing investment strategy at $T-1$ can then be determined to give

$$X_{T-1}^* = f_{T-1}(W_{T-1})$$

This expression for X_{T-1}^* enables us to determine the maximum expected utility of retirement wealth at $T-1$ as function of W_{T-1}. We write this function as

$$u_{T-1}(W_{T-1}) = E_{T-1}^*[u(W_T)|W_{T-1}]$$

The function $u_{T-1}(W_{T-1})$ is the implied utility function for the wealth at $T-1$. It allows us to apply the same algorithm at $T-2$ to determined X_{T-2}^* and $u_{T-2}(W_{T-2})$. This procedure is repeated, working backwards in time, until the optimal investment strategy at the required time t is obtained. As with the utility of lifetime consumption model, closed-form solutions for X_t^* only exist for certain types of utility function and stochastic process for R_t.

20.4.4   Example Based on Utility of Retirement Wealth

A woman can split her retirement fund between a diversified equity portfolio and a risk-free asset. The annual real return on the equity portfolio R_t is an independent and identically distributed random variable. The risk-free asset provides a fixed annual real return r at all durations from retirement. The following parameters are provided for these investment returns:

$$r = 0.02$$

$$E_t[R_t] = \mu = 0.06$$

$$E_t[(R_t - r)^2] = \sigma^2 = 0.20^2$$

The woman's utility function for real retirement wealth is given by

$$u_T(W_T) = \ln(W_T - 40) \quad \text{for} \quad W_T > 40$$

Derive a formula for the optimal proportion of her fund that should be invested in the equity portfolio, if her contribution to the fund is one unit per annum.

20.4.4.1 Nature of the Utility Function

If we take first and second derivatives of the utility function, then we find that

$$u_T'(W_T) = (W_T - 40)^{-1}$$

$$u_T''(W_T) = -(W_T - 40)^{-2}$$

Hence, the utility function is increasing and concave for $W_T > 40$. As W_T tends to 40 units the woman's utility tends to minus infinity. The shape of the utility curve is such that her aversion to risk becomes infinite at this level of wealth. We can infer from this that 40 units is the minimum amount of wealth that she would be willing to accept at retirement; anything less would leave her below her subsistence level.

20.4.4.2 Solving the Single-Period Problem

Let x_t be the proportion of wealth invested in equities at time t, where $0 \le t \le T$. Assuming that the contributions of one unit per annum are paid annually in advance, the recurrence formula for the woman's wealth is

$$W_{t+1} = (1 - x_t)(W_t + 1)(1 + r) + x_t(W_t + 1)(1 + R_t)$$

$$\Rightarrow W_{t+1} = (W_t + 1)(1 + r) + x_t(W_t + 1)(R_t - r)$$

The implied utility function for the wealth at time t must satisfy

$$u_t(W_t) = \max\{E_t[u_{t+1}(W_{t+1}|W_t)]\} = E_t^*[u_{t+1}(W_{t+1}|W_t)]$$

A time t, the expected value of the implied utility function at time $t+1$ is given by

$$E_t[u_{t+1}(W_{t+1}|W_t)] = E_t[u_{t+1}((W_t + 1)(1 + r) + x_t(W_t + 1)(R_t - r))]$$

We now define W_t^c as the wealth that can be guaranteed in 1 year's time by investing entirely in the risk-free asset; hence

$$W_t^c = (W_t + 1)(1 + r)$$

Expressing $u_{t+1}(W_{t+1}|W_t)$ as a Taylor expansion about W_t^c gives

$$u_{t+1}[W_t^c + x_t(W_t + 1)(R_t - r)] = u_{t+1}(W_t^c) + x_t(W_t + 1)(R_t - r)u'_{t+1}(W_t^c)$$
$$+ \frac{1}{2}x_t^2(W_t + 1)^2(R_t - r)^2 u''_{t+1}(W_t^c) + \ldots$$

Ignoring terms of higher order than second, the expected value of the above expression is given by

$$E_t[u_{t+1}(W_{t+1}|W_t)] = u_{t+1}(W_t^c) + x_t(W_t + 1)(\mu - r)u'_{t+1}(W_t^c)$$
$$+ \frac{1}{2}x_t^2(W_t + 1)^2\sigma^2 u''_{t+1}(W_t^c)$$

The optimal investment strategy at time t corresponds to the value of x_t that maximizes the above expression, which we give the symbol x_t^*. Setting the derivative with respect to x_t equal to zero gives

$$x_t^* = -\left(\frac{\mu - r}{\sigma^2}\right)\left[\frac{u'_{t+1}(W_t^c)}{u''_{t+1}(W_t^c)}\right]\left(\frac{1}{W_t + 1}\right) \qquad (20.4)$$

Inserting this formula back into the expression for the expected utility gives

$$u_t(W_t) = E_t^*[u_{t+1}(W_{t+1}|W_t)] = u_{t+1}(W_t^c) - \frac{1}{2}\left(\frac{\mu - r}{\sigma}\right)^2 \frac{[u'_{t+1}(W_t^c)]^2}{u''_{t+1}(W_t^c)} \qquad (20.5)$$

This concludes the solution for the single-period problem, which is extended below to the multi-period case.

20.4.4.3 *Solving the Multi-Period Problem*
Setting $t = T - 1$ in Equation (20.5) gives

$$u_{T-1}(W_{T-1}) = E_{T-1}^*[u_T(W_T|W_{T-1})] = u_T(W_{T-1}^c) - \frac{1}{2}\left(\frac{\mu - r}{\sigma}\right)^2 \frac{[u'_T(W_{T-1}^c)]^2}{u''_T(W_{T-1}^c)}$$

The function $u_T(.)$ is the utility function for retirement wealth specified above; thus

$$u_T(W_{T-1}^c) = \ln(W_{T-1}^c - 40)$$

$$u'_T(W_{T-1}^c) = (W_{T-1}^c - 40)^{-1}$$

$$u''_T(W_{T-1}^c) = -(W_{T-1}^c - 40)^{-2}$$

where $W^c_{T-1} = (W_{T-1} + 1)(1 + r)$.

Substituting these expressions into the formula for $u_{T-1}(W_{T-1})$ gives

$$u_{T-1}(W_{T-1}) = \ln[(W_{T-1} + 1)(1 + r) - 40] + \alpha \qquad (20.6)$$

where

$$\alpha = \frac{1}{2}\left(\frac{\mu - r}{\sigma}\right)^2$$

To derive the implied utility function at time $T-2$ is now straightforward, because the implied utility function at time $T-1$ is similar in form to the utility function at retirement.

Setting $t = T - 2$ in Equation (20.5) gives

$$u_{T-2}(W_{T-2}) = E^*_{T-2}[u_{T-1}(W_{T-1}|W_{T-2})] = u_{T-1}(W^c_{T-2})$$
$$- \frac{1}{2}\left(\frac{\mu - r}{\sigma}\right)^2 \frac{[u'_{T-1}(W^c_{T-2})]^2}{u''_{T-1}(W^c_{T-2})}$$

Substituting the function for $u_{T-1}(.)$ obtained in Equation (20.6) gives

$$u_{T-2}(W_{T-2}) = \ln[(W^c_{T-2} + 1)(1 + r) - 40] + 2\alpha$$
$$= \ln[W_{T-2}(1 + r)^2 + \ddot{s}^r_{2|} - 40] + 2\alpha$$

Going back an arbitrary number of years to time t gives

$$u_t(W_t) = \ln[W_t(1 + r)^{T-t} + \ddot{s}^r_{\overline{T-t|}} - 40] + \alpha(T - t)$$

from which we can derive

$$u'_t(W_t) = \frac{(1 + r)^{T-t}}{W_t(1 + r)^{T-t} + \ddot{s}^r_{\overline{T-t|}} - 40} = \frac{1}{W_t + \ddot{a}^r_{\overline{T-t|}} - 40(1 + r)^{-(T-t)}}$$

$$u''_t(W_t) = \frac{-1}{[W_t + \ddot{a}^r_{\overline{T-t|}} - 40(1 + r)^{-(T-t)}]^2}$$

The formula for the optimal equity proportion, given in Equation (20.4), was

$$x^*_t = -\left(\frac{\mu - r}{\sigma^2}\right)\left[\frac{u'_{t+1}(W^c_t)}{u''_{t+1}(W^c_t)}\right]\left(\frac{1}{W_t + 1}\right)$$

Substituting the formulae derived above for the first and second derivatives of the implied utility function gives

$$\Rightarrow x_t^* = \left(\frac{\mu - r}{\sigma^2}\right)\left(\frac{W_t + \ddot{a}^r_{\overline{T-t}|} - 40(1+r)^{-(T-t)}}{(W_t + 1)(1+r)}\right) \qquad (20.7)$$

Equation (20.7) consists of two quotients multiplied together. The first quotient is a function of the investment parameters alone—it equals the equity risk premium divided by a parameter approximately equal to the variance of the equity returns. Thus, the proportion of the fund that should be invested in equities increases with the assumed equity risk premium and falls with assumed volatility in equity returns. These are intuitively sensible results.

The second quotient is a function of variables that depend on the age and accumulated wealth of the woman. The numerator is the present value of her projected wealth at retirement minus the present value of her subsistence wealth, each evaluated at the risk-free interest rate. If this surplus wealth is zero, then the woman should take no risks and invest everything in the risk-free asset. The denominator is the projected risk-free wealth at the end of the year. If the woman is far from retirement, then the denominator is likely to be much smaller than the numerator, leading to a high optimal equity proportion.

20.4.4.4 Portfolio Isoquants

The optimal equity proportion can be viewed as a function of accumulated wealth and duration from retirement. Its relationship to these variables can be illustrated by plotting *portfolio isoquants*, i.e., curves mapping out points at which the optimal equity proportion is a constant. These curves are plotted on a graph in which accumulated wealth and duration from retirement are shown on the perpendicular axes.

Figure 20.1 shows portfolio isoquants for the woman in this example. The isoquants meet at a specific point on the horizontal axis, 30 years from retirement. At this duration, the woman should always invest 100% of her wealth in equities, whatever her accumulated wealth. The optimal equity proportion is always less than 100% at smaller durations from retirement and increases with her accumulated wealth at these durations. Note that her wealth can never fall below the $x_t = 0$ isoquant in this model, because she would always switch her entire fund into the risk-free asset on reaching this isoquant.

When the duration from retirement exceeds 30 years, the optimal equity proportion can be greater than 100%, which implies that the woman should

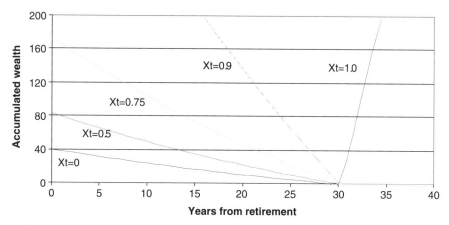

FIGURE 20.1 Portfolio isoquants

borrow at the risk-free rate to gear up in equities. Although such a strategy would not normally be possible in a pension fund, it may be feasible if the model is applied to the entire wealth of an individual who can borrow against other assets (e.g., the mortgage on a private residence). Note that at durations greater than 30 years the optimal equity proportion reduces with accumulated wealth.

References

Bateman, H., Kingston, G., and Piggott, J. (2001). *Forced Saving: Mandating Private Retirement Incomes*. Cambridge University Press.

Campbell, J. Y. and Shiller, R. J. (2001). Valuation ratios and the long run stock market outlook. Cowles Foundation Discussion Paper 1295, Yale University. http://cowles.econ.yale.edu.

Campbell, J. Y. and Viceira, L. M. (2002). *Strategic Asset Allocation: Portfolio Choice for Long-Term Investors*. Oxford University Press.

Eichberger, J. and Harper, I. R. (1997). *Financial Economics*. Oxford University Press.

Faculty and Institute of Actuaries (2002). TM1: statutory illustrations of money purchase benefits. www.actuaries.org.uk.

Haberman, S. and Vigna, E. (2002). Optimal investment strategies and risk measures in defined contribution pension schemes. *Insurance: Mathematics and Economics* 31, 35–69.

Kapur, S. and Orszag, M. (1999). A portfolio approach to investment and annuitisation during retirement. Birkbeck College Working Paper, Birkbeck College, London.

Khorasanee, M. Z. (1996). Annuity choices for pensioners. *Journal of Actuarial Practice* 4(Part 2), 229–255.

Khorasanee, M. Z. (2002). Can utility-maximization models assist with retirement planning? *Journal of Actuarial Practice* 10, 97–129.

Merton, R. C. (1990). *Continuous-Time Finance*. Blackwell.

Stocker, M. A., Dudley, S. D., Finlay, G. E., Fisher, H. J., Harvey Wood, O. C., Kemp, M. H. D., Lumb, W., Miles, M. W., and Wasserman, S. L. (1999). The role and responsibility of actuaries in the defined contribution environment in the United Kingdom. *British Actuarial Journal* 5(Part IV), 763–800.

Thomson, R. J. (1998). Investment channel choices in defined contribution retirement fund: the use of utility functions. In: *Proceedings of the 8th International AFIR Colloquium*.

Part V
Health Insurance

Chapter 21

An Introduction to Health Insurance

21.1 Background

Health insurance products provide useful cover for individuals and employers who wish to insure themselves against financial losses that might be incurred as a result of sickness or disability. In Part V we discuss the four main products on offer within the health insurance market. These are:

income protection (IP)
critical illness (CI)
long-term care (LTC)
private medical insurance (PMI)

In Chapter 22 to Chapter 25 we describe the main features of each of the above health products in turn. We also discuss the underwriting, pricing, reserving, and claims management procedures associated with running each line of business. The reader will see that these processes vary with each product; they also vary within the same product line across different markets. As Laüter (1999) comments, the products can vary widely, from country to country, as a result of differences in state health provision and attitudes to health care.

21.2 Actuarial Control Cycle

The actuarial control cycle was mentioned in Section 11.2.1, in the context of the management of a life insurance company. The cycle has been customized for health insurance by Elliott (2000). This is depicted in Figure 21.1.

The control cycle is a series of management procedures carried out in sequence. It applies to all forms of health insurance in a similar manner to other insurances. For the actuary, the profitability of the health portfolio is

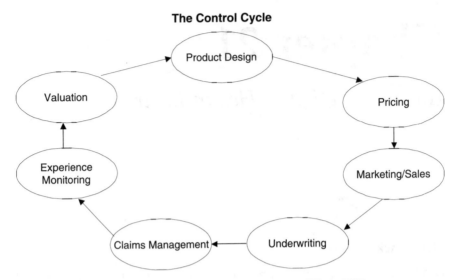

FIGURE 21.1 The actuarial control cycle for health insurance (Elliott, 2000)

only achieved and maintained by the systematic application of the control cycle. This does not mean to say that the actuary performs all of the operations, rather that the actuary oversees some of the operations and has management access and influence on the others.

Actuarial involvement in health insurance goes beyond the mathematically based role that traditionally typifies the work of an actuary. There is a multi-disciplinary function to health business (as with general insurance) within which the actuary will need to participate. In order to carry out their role, actuaries will need to liaise with, inter alia, medical professionals, underwriters, claims assessors, sales and marketing teams, and reinsurers. For this reason, Chapter 22 to Chapter 25 cover aspects of the whole cycle, from product design through to the valuation of the product.

21.3 Basic Contract Structures of Health Insurance Products

The four main types of health insurance contract covered in Part V are briefly described below.

21.3.1 Income Protection Insurance

IP insurance (formerly known as permanent health insurance in the U.K. and called disability insurance or group long-term disability insurance in the U.S.) provides benefits on sickness or disability. The form of benefit is usually an income, the amount of which is usually related to the loss of earnings suffered by policyholders due to their being unable to continue with their normal occupation. Benefit entitlement is strictly linked to the continuation of "sickness" as defined by the policy. Policies are usually issued over a long term, e.g., up to the expected retirement age of the policyholder.

21.3.2 Critical Illness Insurance

CI insurance (also known as dread disease) pays a lump sum on the diagnosis of certain terminal or highly debilitating illnesses, such as cancer, stroke, and heart disease.

21.3.3 Long-Term Care Insurance

Benefits under this policy are designed to meet the costs of caring for the insured life when he or she becomes unable to perform some or all of the normal activities of daily living. This would include the costs of community care in the home, or of staying in a residential or nursing home.

21.3.4 Private Medical Insurance

PMI reimburses the insured in whole, or in part, for the expense of any private medical treatment that satisfies the terms and conditions of the policy.

21.4 Comparison between Life and Health Insurance

Health insurance products are often offered by life offices, but it should be noted that there is a significant difference between life and health insurance policies.

Life insurance benefits are paid on the basis of a defined event, i.e., the death of the insured person. The benefit is normally fixed in advance as a cash sum. The death triggers the payment of the benefit to the policyholder's beneficiaries. By contrast, health insurance is the generic term used to describe a variety of insurance products where the benefits are dependent upon the "health status" of the insured person.

The health insurance event may be defined as a single event or it may be the result of a change in health status over a period of time. The event may activate a stream of benefit events, which vary in severity and cost. Cost may be expressed as the delivery of a service or as a monetary amount. The benefit is often subject to a degree of limitation on the scope and/or amount of cash or services rendered.

This difference between life and health insurance is fundamental. When a life insured dies, the death is confirmed and the benefit is paid; whereas the position for health insurance is much more complicated. A health insurance policy normally requires that an event takes place, either an accident or the onset of an illness. This event must lead to the insured person qualifying for benefit within the terms and conditions of the health insurance policy. The individual's state of health may lead to their seeking medical care or being unable to work for medical reasons. The terms of the policy will include a number of conditions that help the insurer to control the size of the claim. For example, policyholders may be required to contribute towards the cost of claims through co-insurance or through the payment of an excess.

21.5   External Factors Affecting Health Insurance

In order to understand the nature of the health products available within the insurance market, it is necessary to look at the context within which each product is set. This is because health products, more so than most other insurance products, are sensitive to a number of external factors. These factors are discussed more fully in the following chapters. However, for illustrative purposes, three factors are outlined below:

demography
economic climate
state provision of health and care services

21.5.1   Demography

The likelihood of suffering from ill health or disability increases with age. Most developed countries are experiencing a demographic shift such that their populations are getting older. As this trend continues, there is an increasing burden on the working population to support the elderly. This support may come in the form of direct care or through taxation to fund the costs of state health provision.

21.5.2   Economic Climate

In times of economic hardship, those in work are likely to feel the stress caused by financial insecurity. For some people, this stress can lead to ill health, resulting in, for example, IP claims. There is also a moral hazard risk that employees are more likely to make a fraudulent claim when they feel that their job is in jeopardy. It is to be expected, therefore, that the claims experience for products linked to employment (i.e., IP business) will deteriorate during a recession. This is what occurred in the

1980s and 1990s in various countries, including South Africa, the U.S. and the U.K.

Where products are perceived as a luxury form of cover, rather than a basic necessity, premiums may be too expensive for policyholders to maintain in a financially insecure climate (e.g., PMI in the UK). Take-up rates for nonessential insurance policies may, therefore, fall during a recession.

21.5.3 State Provision of Health Care and Services

Health insurance business, by its very nature, is concerned with the design of products to aid sick or disabled people. In any country, the insurance market will offer policies that complement the health and care services provided by the state. In countries where the level of state provision is very low, the insurance market tends to be well established (e.g., U.S.). By contrast, the health insurance market tends to be less developed in countries where state intervention is greater (e.g., Germany).

There has been a noticeable shift in many countries towards the privatization of health care since the 1980s. A description of how this has taken place in Mexico, Chile, Columbia, and Brazil can be found in Hierro and Wrede (1999).

21.6   Features of Health Insurance

Health products vary from each other as they are designed to fulfill different needs. These will be discussed more fully in Chapter 22 to Chapter 25. However, there are some key features that are common to all of the health products. These are described below.

21.6.1   Deferred Period

This refers to the period for which sickness needs to last before a claim commences. Usually, the period is between 1 week and 1 year, depending on the type of health insurance product. Deferred periods are most commonly associated with IP and LTC insurance.

Often, policyholders are allowed to choose their deferred period from a range of options. In making the choice, they will wish to take into account any period for which they will receive state benefits and/or employer disability benefits. Also, they need to bear in mind that the shorter the deferred period, the higher the premium that they will be required to pay for the policy.

From the insurer's perspective, the deferred period reduces the claims cost and the administration costs associated with paying small claims.

Both of these savings could be passed on to policyholders in the form of lower premiums.

21.6.2   Guaranteed and Reviewable Premiums

It is possible that an insurer will issue a health insurance contract with guaranteed premium rates throughout the term. However, in view of the unpredictability of the claims experience for health insurance, it is unlikely that that the insurer will offer such guarantees. Guarantees would only be given if insurers have built in a substantial contingency margin to the premium. This, in turn, could make the premiums unattractive to potential policyholders compared with competitors' rates that are not guaranteed. Therefore, most health insurance products are issued with premium rates that are reviewable or renewable.

If the rates are reviewable then they will be guaranteed for, say, the first 5 years of the term and altered (upwards or downwards) if justified by the claims experience during that period. If the rates are renewable, then they are, effectively, guaranteed for only 1 year. PMI premiums are almost always renewable due to the risk that medical expense inflation will be much higher than expected over a period of years.

Whenever insurers increase their premium rates, they need to be aware of the risk of "selective lapses." This occurs when lives who are less likely to claim (i.e., healthy lives) allow their policies to lapse, and the other (less healthy) lives remain with the insurer. The lives remaining in the insurer's portfolio would then represent poorer risks, on average, than the lives covered by the insurer before the premiums were increased.

21.6.3   Claims Management

When a claim is made on a life insurance contract, the decision over whether or not to pay is usually a straightforward one for the insurer to make; in most cases, the death of a policyholder is easy to prove. However, within health insurance, the decision over the validity of a claim is potentially much more subjective. This is because the claim has to satisfy more complicated criteria. For example, in IP business, a claims assessor might have to decide whether or not an individual is too disabled to carry out, not just his or her own, but any occupation. Clearly, this cannot always be determined objectively, and the insurer will rely on the experience and technical knowledge of the assessor in making the decision.

Claims assessors must be mindful of the fact that, since the claims are health related, they must be particularly careful to treat all claimants sensitively. In addition, they need to be aware that any claims that they refuse may lead to negative publicity for the insurer.

21.7 Risks to the Insurer

There are certain risks to the insurer that are common to the range of health products offered by companies. The main ones are described below.

21.7.1 Adverse Selection

With health insurance, the insurer needs to be aware that there is a significant risk of adverse selection. In particular, individuals may decide to purchase a health insurance product when they have reason to believe that they have a higher probability of making a claim than that assumed in the premium basis. For example, an individual who begins to suffer severe stomach cramps may take out PMI insurance in case an operation is thought to be necessary.

The insurer will use underwriting to try to minimize the risk of adverse selection. For example, in the above case, any pre-existing medical conditions are likely to be excluded from the cover, at least for a specified period of time (see "moratorium" in Section 25.4).

As mentioned in Section 10.4.1 in the context of life assurance, an issue that has recently become a concern to the insurance industry is that of genetic testing. In countries where applicants are not required to disclose the results of any genetic tests that they have undergone, it is possible that adverse selection will become a bigger problem. For example, individuals who know, from taking a genetic test, that they are predisposed towards developing Huntington's chorea are likely to purchase CI insurance. For a discussion on social policy issues related to genetic testing, see Daykin et al. (2003). For a full discussion of the impact that genetic testing could have on health insurance in the U.K., see Brett and Hopkins (2001).

21.7.2 Moral Hazard

There is always the risk that an individual's behavior might be influenced by the fact that they have health insurance cover. For example, an individual receiving IP benefits might be reluctant to return to work if these benefits, together with any state and employer disability income, compare favorably with their pre-disability earnings. This is known as "overinsurance". It is an example of "ex post" moral hazard, since it is only after the claim has been made that the policyholder's attitude alters.

This risk would be reduced by the use of financial underwriting. This means that, at the application stage, the insurer obtains details of all income (net of tax) that the applicant would receive if they were to claim in the future. The insurer could then ensure that the IP benefit is set at a low enough level for the applicant to have a sufficient incentive to return to work upon recovery.

21.7.3   Mismatch of Claim Definitions

It is important that the definitions of medical conditions used by the claims assessors should be the same as those used in the premium basis and those understood by the underwriters. Any mismatch could result in unprofitable business being written. For example, in CI business, if the definition of cancer used by the claims assessors is significantly wider than that assumed in setting the premiums, then the experience for cancer claims could be worse than expected, resulting in losses being made.

The risk is reduced if the insurer ensures that all members of staff involved in the claims management, underwriting, and pricing processes have received sufficient training or appropriate advice from medical experts.

21.7.4   Medical Advances

Insurers need to be aware that improvements in medical science and technology will affect the risks associated with health insurance business. Examples of the way this can occur are given below.

Example 1: the early diagnosis of a type of cancer could bring forward the claim from a CI policy.
Example 2: the use of new and expensive medical procedures could increase the cost of PMI claims.
Example 3: advances in medical treatment could enable a policyholder to live longer than they might have previously done, but in a severe state of disability. This could increase the cost of both LTC and IP claims.

21.8   Reinsurance

The reinsurance requirements for IP, CI, and LTC business tend to be similar to those described in Section 10.4.3 for life insurance business. Most portfolios will be reinsured, often a significant amount on a quota share basis, although some insurers may use a combination of surplus and quota share.

The reinsurance of a PMI portfolio is normally confined to excess of loss insurance on an aggregate and/or a specific loss basis. Catastrophic portfolio loss reinsurance is also required. Smaller portfolios will reinsure on a quota share basis or co-insure with another direct insurer. Further details on these types of reinsurance can be found in Chapter 15.

As well as providing reinsurance, reinsurers provide important support services to direct insurers. The fact that reinsurers are able to collect a substantial amount of data from various insurers writing health insurance business means that they can act as consultants to the direct insurers.

In particular, they can offer advice on underwriting, marketing, pricing, and reserving.

References

Brett, P. and Hopkins, J. (2001). Genetics and underwriting — health care module. Faculty and Institute of Actuaries.

Daykin, C. D., Akers, D. A., Macdonald, A. S., McGleenan, T., Paul, D., and Turvey, P. J. (2003) Genetics and insurance — some social policy issues. *British Actuarial Journal* 9(IV), 787–874.

Elliott, S. (2000). Income protection — the role of the actuary. In: *UK Actuarial Healthcare Conference*, University of Warwick, England.

Hierro, H. and Wrede, P. (1999). Privatization of health care systems in Latin America. *GeneralCologne Re Risk Insights* 3(5).

Laüter, G. (1999). Different markets — different health insurance products. *GeneralCologne Re Risk Insights* 3(5).

Chapter 22

Income Protection Insurance

22.1 Introduction

The inability to work has financial consequences for both the individual and for his or her employer. From the individual's perspective, loss of income will mean that their financial commitments may not be met. For the employer, the incapacity of employees, particularly if they are key workers, could affect the profitability of the business. Taylor (2001) indicates how costly it is to lose staff through disability. He cites the direct and indirect costs to the U.S. economy as being over $300 billion dollars (equating to approximately 9% of total payroll).

The concept of making up for the loss of income is not a new one. Indeed, as early as the 14th century there is evidence that German miners were given money from a specially designated fund if they were unable to work as a result of injury (Munich Re, 1998). The modern forerunners to current products were available from the 1880s onwards in the U.K. (through the Friendly Societies) and in the U.S. (through similar organizations catering for the professional classes).

The term income protection (IP) insurance applies to a family of contracts under which income is paid to the insured whilst they are unable to work, by reason of illness or injury. IP is known as disability insurance or long-term disability insurance in the U.S. and was formerly known as permanent health insurance in many countries, including the U.K. It has gained popularity, particularly in group business form, across the developed world. However, uptake amongst employers can be mixed. For example, in the U.S., 57% of employers with more than 100 employees do not offer IP benefits (Taylor, 2001).

Many countries provide state disability benefits to members of the population who are unable to work through illness or injury. Therefore, IP insurance can be thought of as bridging the gap between the benefits payable from the state (if any) and the income required by the recipient.

The biggest danger to insurers of IP business is the risk of moral hazard. That is, that if the disability income (from all sources) payable to an individual is relatively high, they might be inclined to make a claim; furthermore, they might have little incentive to return to work. Since the payment of an IP claim is linked to an individual's inability to work, the IP market is extremely sensitive to a country's economic circumstances. This has been shown to be the case in a number of countries, and is explored in more detail in Section 22.2. IP business is also heavily influenced by other external factors. Factors that Taylor (2001) has identified as having become important in recent years include the following.

22.1.1   Changing Demography

In many industrialized countries, the workforce is aging and, therefore, more prone to suffer from disability. A return to the workforce following recovery can be harder to achieve for an older person, particularly if the individual has to retrain or find another job. The length of time a benefit is paid to an older claimant may be greater, as their recovery time may be more prolonged than that of a younger person.

22.1.2   Changing Attitude to Disability

There is now less of a stigma attached to being unable to work through disability. Employers used to be less keen to employ disabled people, due to a mix of prejudice and reticence to make any necessary changes to the work environment. In some countries, these issues have been tackled, to a degree, through legislation. For example, the Disability Discrimination Act (1995) in the U.K. prohibits companies from discriminating against disabled job applicants and existing disabled staff.

22.1.3   The Nature of Disability

In recent years, insurers have accepted a range of occupational-related disabilities that were not previously acknowledged as being grounds for making a claim. These disabilities include chronic fatigue syndrome, repetitive strain injury, and work-related stress.

22.2   The Evolution of the Product

IP business is particularly sensitive to external economic and cultural influences. A number of countries, including the U.S., Canada, the U.K., and South Africa, have suffered economic downturns during the last two decades. These unfavorable conditions have resulted in significant increases in IP claims.

As discussed later, the poor claims experience has led to heavy losses in the insurance industry. This has been partly as a result of weaknesses in policy design inherent in these policies. For example, Ainslie (2001) describes how the U.K. insurance market concentrated on developing product lines which have a large savings element incorporated within them (e.g., pensions products); IP insurance, not having a savings element within it, was only considered as a secondary line. As such, it lacked a coherent policy development strategy until the industry took stock, having incurred significant losses on the business.

Globally, many developed countries underwent a shift from a manufacturing-based economy to a service-providing economy. The recessions encountered during this shift were characterized by a loss of jobs amongst manual workers that were not replaced by similar roles in the emerging economy. A significant increase in claims incidence resulted. In particular, group business claims increased as employers shifted the cost of redundancy to the insurance sector. Insurers subsequently reacted by increasing their premium rates for group business (Coetzee et al., 2001).

As well as the unfavorable economic conditions, the profitability of IP business suffered as a result of increased competition between insurers, causing them to offer overgenerous terms. For example, in pricing, insurers would sometimes treat occupations as being less risky than they were actually believed to be in order to make the premiums attractively low. An additional problem was that once a claim commenced it tended not to be reviewed as frequently as it should have been. Indeed, the initial validation of the claim was often undertaken by poorly trained staff, with no expertise in the areas of occupational disability, treatment, and rehabilitation. Consequently, insufficient incentives to return to work were written into the policies and few, if any, independent medical views were sought at any time during the claims management process (Ainslie, 2001).

As a result of this financially painful experience, insurers have made significant changes to the way in which they write and manage IP business. Not least, the value of well-conceived product design has been an important step forward in turning IP business into a profit-making venture once more (Taylor, 2001). Additionally, the claims management procedures that were deficient in the past have been tightened considerably. For example, insurance companies have invested heavily in the recruitment of appropriately qualified health professionals to their claims departments. Claims are now regularly assessed whilst in payment and there is a much greater emphasis on encouraging the claimants to seek employment as they begin to recover (Coetzee et al., 2001). The claims process is discussed in more detail in Section 22.6.

Section 22.3 considers the current designs of policies that reflect the lessons that have been learnt from the experience of the last few decades.

22.3 Product Design

IP product design is discussed as it relates to individual business in Section 22.3.1 and Section 22.3.2. Additional design features that relate specifically to group business are described in Section 22.3.3.

22.3.1 Benefit Types

IP insurance provides a benefit to an individual during a period of disability, with the aim of replacing some of the income lost whilst they are unable to work. The benefit is normally paid as an annuity in monthly installments until recovery or expiry of the policy term, whichever is the sooner. The policy term could be a fixed number of years, the period until the individual reaches a certain age (e.g., normal retirement age), or it could be for life. The benefit only commences after a disability has lasted for longer than the deferred period (see Section 21.6.1). Deferred periods for IP business are typically 3 to 12 months.

Whilst it is possible for the benefit to be paid in the form of a lump sum rather than an annuity, the former suffers from the following drawbacks:

The fact that an individual would receive a lump sum may increase the risk of moral hazard. In other words, it may be more tempting for the individual to make a fraudulent claim if they know that they will receive a large up-front cash benefit if such a claim were successful. This risk increases as the expiry of the policy term is approached. Moral hazard is discussed in more detail by Haberman (1987).

It does not appear to be in keeping with the main objective of IP insurance, which is to replace a proportion of the income lost due to the onset of a disability.

It would be virtually impossible for the insurer to reclaim any of the lump sum benefit if the individual subsequently were to recover.

The amount of IP benefit payable should be high enough to provide a decent standard of living but low enough to provide an incentive for the insured to return to work. Table 22.1 shows, for the U.S., the probability of eventual return to work by gross replacement ratio, where the latter is defined as the ratio of gross post-disability income to gross pre-disability income. It demonstrates that, as expected, the higher the gross replacement ratio, the lower the probability of eventual return to work. For example,

Table 22.1 Probability of eventual return to work by gross replacement ratio

Gross income replaced (%)	Proportion disabled for ≥ 5 months who returned to work (%)
>26	84
26–50	70
51–75	52
76–100	38
101–150	26
>150	6

Source: Meilander and Simbro (1993)

if the gross replacement ratio is over 150% (i.e., the insured lives are receiving substantially more in IP income than they were as salary when working) then very few will return to work.

Table 22.1 highlights the importance of ensuring that IP benefits are constrained so that the income received from all sources is not over-generous. In the U.K., for example, the amount of benefit payable from an IP contract is often limited to a maximum of 50% of gross pre-disability earnings. The significance of such a benefit limitation clause being included within product design is demonstrated by the fact that the ceiling applies in approximately 40% of claims made in the U.K. (Elliott, 2001).

Sometimes a claimant returns to work but suffers a relapse and once more is absent from work. In order not to discourage a claimant from returning to work, there will be a policy provision linking the spells of absence and treating the two claim periods as a single period, with no second deferred period being imposed. The maximum period given for a linked claim to occur will normally be either 3 or 6 months.

IP insurance could be offered as a rider to a life assurance policy, or it could be issued as a stand-alone contract. Practice varies from country to country. For example, IP benefits in France tend to be written as rider benefits, whereas IP insurance in the U.K., North America, and the Netherlands tends to be stand alone (Munich Re, 1998).

It is common for certain causes of disability to be excluded from IP policies. These might include HIV/AIDS, war, self-inflicted injury, drugs, hazardous pursuits, failure to follow medical advice, and complications arising from pregnancy.

22.3.2 Definitions of Disability

Since the objective of IP insurance is to replace lost income, a natural way to define disability is in terms of the ability to carry out an occupation. There are three common types of definition. These are discussed below.

22.3.2.1 Own Occupation

The person is defined as disabled if they can satisfy the insurer that they are unable to carry out their job as a result of injury or illness.

22.3.2.2 Own or Similar Occupation

The person is defined as disabled if they can satisfy the insurer that, as a result of injury or illness, they are unable to carry out their job or any similar job for which they would be qualified, by reason of their training, education, and experience.

22.3.2.3 Any Occupation

The person is defined as disabled if they can satisfy the insurer that, as a result of injury or illness, they are unable to carry out any gainful employment.

To satisfy the "any occupation" definition of disability, the insured must be severely disabled. Therefore, adopting that definition will lead to the insurer accepting fewer claims than under either of the other two definitions. As a result, this definition would result in the lowest premiums being charged.

Some insurers use an "own occupation" definition for a certain period of disability (e.g., 2 years) after which the definition is switched to "any occupation". This enables the insurer to reassess the claim after the initial period and to terminate payment if the claimant is found to be capable of carrying out another occupation.

Clearly, a policyholder's particular occupation is a significant factor in determining the likelihood that the individual will ultimately make an IP claim. In practice, insurers tend to use a few broad occupation categories to classify the policyholder's occupational risk. Typically, four to seven bands of occupation are used. An example, from the U.K., is given in Table 22.2.

The purpose of these bands is to group occupations according to common risk profiles, thereby enabling insurers to calculate premiums consistently

Table 22.2 Example of occupation classes used in the U.K.

Class	Occupation class
One	Professional and managerial
Two	Clerical
Three	Manual
Four	Heavy manual

Source: Elliott (2001)

across similar occupations. However, the use of broad occupation classes has not always correctly assessed the risk inherent within a particular occupation. For example, teaching would be placed in class one, since it is carried out by professionals. This would suggest that a teacher's risk of claiming IP benefits should be low; however, in recent years, this has not been the case. In the U.K., for example, a number of teachers have taken sick leave due to work-related stress (Campbell, 2001). Mackay (1993) suggests that teachers, doctors, and nurses would all be more accurately placed in class three rather than class one.

22.3.2.4 *Alternative Measures of Disability*

In view of the heterogeneity within the occupation classes, some insurers have moved away from occupation-based definitions of disability. Instead, disability is measured by the difficulty that the insured has in carrying out the typical functions or tasks within their job. These tasks are commonly referred to as activities of daily work (ADWs) and functional assessment tests (FATs). The principles involved are similar to those used in long-term care insurance, where, for many years, activities of daily living (ADLs) have commonly been used to measure disability (see Chapter 25).

An example of this task-based approach would be to pay the disability benefits if the insured fails two or more out of six occupational tasks. Such tasks might include the following:

Walking 200 meters on a flat surface
Lifting up an object weighing 1 kilogram and carrying it 5 meters
Using a pen or keyboard

For more details on the use of work task definitions, the reader is referred to Campbell (2001).

Although occupational ability and functional ability are both common ways to determine whether a claim is payable, not all countries use these approaches. In France, for example, the claimant's level of disability is determined from a combination of two factors: the level of economic disability (which is based on the resulting loss of income) and the degree of physical disability (as advised in the medical report). Often, the full annual benefit is paid if the level of disability exceeds 67%; a proportion of the benefit is paid if the level is 33% or more, and nothing is paid for a lower level of disability (Munich Re, 1998).

22.3.3 **Group Business**

As mentioned in Section 22.1, many employers set up group IP insurance schemes for their employees.

As with individual business, a ceiling is imposed on the benefits that a person can receive. For example, in the U.K., an employee's benefit entitlement is often limited to a maximum of approximately 70% of gross salary. In addition to covering lost salary, the IP insurance may also cover any pension contributions due in respect of the employee from both the employer and the individual.

As with individual policies, claims criteria will be based on the employee's inability to work. However, particularly in the case of large schemes, the insurer is unlikely to have much detail with regard to a specific employee's duties. Therefore, the definition of disability tends to be based on the member's "normal occupation" (Elliott, 2001).

An important feature of group business is the "free cover limit." This is the maximum amount of benefit for which any employee in the group can be covered, without the need for underwriting to take place. Provided that a substantial proportion of the employees decide to participate in the scheme, the insurer can assume that the risk of adverse selection is minimal. The insurer then saves the costs associated with underwriting whilst the employer avoids having to collect detailed personal information from the employees.

22.4   Underwriting

Underwriting is required to ensure that the risks associated with the policyholder have been suitably taken into account when pricing the product. There are two main components: medical underwriting and financial underwriting.

With medical underwriting, the insurer needs to consider the information provided by the applicant (and their doctor, if required) that relates to the individual's current state of health and previous medical history. The insurer will also need to take into account the individual's family history, together with other risk factors such as hypertension, body build, cholesterol level, and smoking status.

Under financial underwriting, the insurer will consider the ratio of the level of benefit for which the individual has applied to the individual's current level of income. The higher this replacement ratio, the lower the incentive the policyholder has to return to work following a claim. This moral hazard risk was discussed in Section 22.3.1.

22.5   Pricing

In this section we discuss the pricing process for IP insurance under seven headings: risk factors, data sources, trends, guarantees, variations, and pricing methodology.

22.5.1 Risk Factors

The major risk factors are

Age
Sex
Smoking status
Occupation
Medical factors

22.5.1.1 Age

As people get older, they are more likely to suffer a disability or injury. Hence, the disability incidence rate would be expected to increase with age. Many studies have shown this to be the case, including CMIR 12 (1991) and Renshaw and Haberman (2000).

22.5.1.2 Sex

There are marked differences in both the incidence rates of disability and the duration of claims between the sexes. Females tend to have a far higher claims experience than males (CMIR 20, 2001). This means that different morbidity tables need to be used for males and females. In practice, in the U.K., often the female rates are calculated on the basis of male rates increased by 50%.

22.5.1.3 Smoking Status

Smokers have a much higher propensity to make claims than nonsmokers do. This needs to be reflected in the pricing basis.

22.5.1.4 Occupation

Occupation is a key determinant in risk assessment. As mentioned in Section 22.3.2, occupations are usually classified by the insurer, into four or more broad classes. Typical premium loadings for the U.K. that are associated with the occupation classes shown in Table 22.2 are given in Table 22.3.

Table 22.3 Typical premium loadings for occupation classes in the U.K.

Class	Occupation class	Premium level (%)
One	Professional and managerial	100
Two	Clerical	175
Three	Manual	300
Four	Heavy manual	400

Source: Elliott (2001)

In effect, the extent to which claims experience varies by socio-economic class and geographical area of residence is implicitly incorporated within the occupation loadings.

22.5.1.5 Medical Factors

Medical factors are discussed in connection with underwriting in Section 22.4.

22.5.2 Data Sources

In order to price the product, the insurer will need to collect relevant morbidity and mortality data. The starting point would be for the insurer to consider its own office data. However, as this data set is likely to be too small to be credible, the insurer will also wish to make use of any inter-office data available (i.e., data from other insurers).

In the U.K. and the Republic of Ireland, very useful inter-office claims experience data have been regularly collected and analyzed by the Continuous Mortality Investigation (CMI) Bureau for many years. Such data have enabled insurers to identify trends in IP business in the U.K., and to monitor the extent to which their own experience is out of line with that of U.K. IP business as a whole. Even if the insurer's own experience is very small, it should still be compared with the inter-office results as there can be a wide variation in claims experience between offices.

Other useful data can come from population statistics and data collected in other countries. For example, the Society of Actuaries in the U.S. produces inter-office data that is helpful to, amongst others, U.K. insurers. It must be borne in mind that data from one country will need to be adjusted before it could be applied in another, since the types of IP insurance product sold in any two countries are likely to be different. Indeed, even countries that have a similar profile with regard to demography, social structure, and economy may exhibit quite different claims experience. This is demonstrated for Australia and New Zealand by Fabrizio and Mak (2001).

There are other external sources of data, namely

Competitors' prices
Regulatory returns
Reinsurers

All of these sources may give relevant information, but it will be of varying quality and should be used with circumspection. Margins should, therefore, be included in the pricing assumptions to allow for difficulties with the data.

22.5.3 Trends

It is important that the insurer takes the effects of relevant trends into account when setting the pricing basis. An example of an important trend affecting IP business is that of improved screening and diagnosis from new medical techniques. As a result, some illnesses are now detected at an earlier and, therefore, less critical stage. For example, some cancers (in particular, prostate cancer) and diseases such as multiple sclerosis have far better detection rates now than in the past. This trend could affect IP business through, for example, claims tending to commence earlier over time. The associated increase in claims cost would then need to be allowed for in the assumptions used to calculate the premiums.

Medical advances may also affect IP business through a reduction in the mortality rate for policyholders with illnesses that were previously likely to be terminal. An individual who dies from a disease does not qualify for an IP benefit; however, a person who survives a disease but is left incapacitated may well be able to make an IP claim. Therefore, a decrease in mortality rates can be correlated with an increase in morbidity rates; the latter may well influence future IP business.

22.5.4 Guarantees

Some IP insurance business is written with the premium rate guaranteed throughout the term of the policy. This has proved to be a dangerous strategy to adopt, since insurers have no scope to increase the premiums if the terms prove to be too generous. Where guarantees were offered in the past, actuaries often allowed for the cost in pricing the product by assuming a deterioration of a set amount per annum in future claims experience. However, this deterministic approach is now regarded as being deficient for assessing the impact of guarantees, and a stochastic approach would be used instead.

In recent years, it has become accepted practice for premiums to be reviewable. This means that they are reassessed periodically (e.g., every 5 years) and can be increased (or decreased) if the claims experience, for the portfolio as a whole, warrants such action.

22.5.5 Variations

There can be significant variations in claims experience between insurance companies writing IP business. The experience level will be determined by a variety of factors, including underwriting levels, product design, sales methods, claims control, and the target market for sales. An office needs to build the actual level of experience into its pricing and product design. However, as mentioned in Section 22.5.2, it is possible that the data available may not be enough to be credible.

22.5.6 Pricing Methodology

As noted by Haberman and Pitacco (1999, chapter 3), there are a number of calculation procedures that can be used for calculating premiums for IP contracts and their analogues. Using a multiple state approach as the basis, they demonstrate that procedures common in a number of countries (in particular the U.S., the U.K., Denmark, Norway, Sweden, Finland, Netherlands, Germany, Austria, and Switzerland) are essentially derivatives of this fundamental approach. Often these approaches have arisen from attempts to make best use of the data being routinely collected by the insurance provider.

In this section, we will consider three of the most important approaches used for pricing:

Multiple state modeling approach
Inception rate and disability annuity method
The Manchester Unity method of costing (the "Friendly Society approach")

In many countries, IP contracts have deferred periods of 3 months or more. Since the Manchester Unity method is unable to deal satisfactorily with contracts that have such long deferred periods, the first two methods listed above are the approaches most often used.

Whichever method is used, the premium rates should be investigated further by constructing cash flow models and analyzing the profit emerging (see Chapter 7 for a full discussion of profit testing).

22.5.6.1 Multiple State Modeling Approach

In the multiple state model, we consider three states:

Healthy
Sick
Dead

and the four possible transitions presented in Figure 22.1: death of a healthy life, death of a sick or disabled life, becoming sick or disabled, recovering from being sick or disabled.

The probabilistic structure is defined by the underlying transition intensities. If we assume that all of the transition intensities depend only on attained age, then the underlying model is Markov in nature. An implication is that conditional probabilities of transition, which are needed for

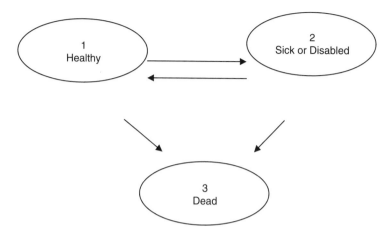

FIGURE 22.1 IP insurance: multiple state model.

determining expected present values (and hence premiums and reserves), are the solutions to a series of differential equations. For example, the expected present value of a continuous n-year temporary annuity of 1 p.a. payable to a life aged x at the start of a policy (who is then in state 1) while the life is in state 2 could be written:

$$\bar{a}_{x:\overline{n}|}^{12} = \int_{0}^{n} v^{t}{}_{t}p_{x}^{12}\, dt$$

where ${}_{t}p_{x}^{12} = \text{Prob}\{(x) \text{ is in state 2 at age } x + t|(x) \text{ is in state 1 at age } x\}$.

Values of ${}_{t}p_{x}^{12}$ would be obtained from the solution of second-order ordinary differential equations, where the coefficients depend on the transition intensities.

It is often assumed that the transition intensities from state 2 depend both on the attained age and the duration since entry to state 2 (i.e., duration since the onset of sickness or disability). This complication means that the underlying model is semi-Markov in nature and the required transition probabilities are now the solution of a more complex series of equations — which are a mixture of differential and integro-differential equations.

In both cases we require relevant observational data that would allow us to estimate the transition intensities; this is best done by maximum likelihood estimation. These estimates can then be used to feed into the equations in both the Markov and semi-Markov cases. These equations would then need to be solved numerically in order to provide us with solutions in terms of transition probabilities that would be used for pricing. Haberman and Pitacco (1999) provide full details of each step of these procedures.

This approach is very demanding of data, and it was first implemented practically by the U.K.'s CMI Bureau in the early 1990s: see CMIR 12 (1991). The next two methods are approximate versions of this approach, and depend on a range of approximating assumptions.

22.5.6.2 *Inception Rate and Disability Annuity Method*

This method originated in North America and is now the most common approach adopted in the U.K., U.S., Austria, Germany, and Switzerland for pricing IP insurance.

As noted in Appendix A, which illustrates how the method may be derived from the multiple state modeling approach, there are three important components involved:

claim inception rate w_{x+h}

expected present value of an annuity payable to a disabled life which ceases on death or recovery $a^*_{x+h+\frac{1}{2}}$

probability of being in the healthy state just before claim commences $_hp_x^{11}$.

In a contract with a deferred period, the definitions of these components will need to be modified. Other features, e.g., rates of escalation to claim payments, could also be incorporated.

22.5.6.3 *The Manchester Unity Method*

A major sickness experience investigation was undertaken for the years 1893–97 for the Manchester Unity of Oddfellows Friendly Society. One of the major occupational classes, Occupation Group AHJ, formed the standard table for this business for many years.

The methodology used to present the results of this investigation (and others) is based on the proportion of the insured population that are sick, rather than multiple state probabilities. The results are tabulated for $z_x^{a/b}$, which is the average number of weeks of sickness at age x for the period between a and $a+b$ weeks of disability. Risk premiums are then calculated by applying the appropriate $z_x^{a/b}$ to the exposed to risk to obtain the expected claims.

One weakness of this approach is that the $z_x^{a/b}$ factors are only tabulated for subintervals of sickness up to 104 weeks, with periods of sickness lasting more than 104 weeks being aggregated together. However, since many IP contracts written by Friendly Societies had short deferred periods, (e.g., 1 week), the method was reasonably suitable for pricing and reserving, provided that the underlying probabilities remained stable.

A second weakness is that the approach is not presented in probabilistic terms, and so it is not possible to allow for information about which state a policyholder currently occupies. These (and additional) weaknesses mean that other methods are much more suitable for modern IP contracts.

Appendix B demonstrates how the Manchester Unity method can be derived from the more general multiple state model described in Section 22.5.6.1.

For more detail on the Manchester Unity method, the reader is referred to Neil (1977).

22.6 Claims

Claims management is a particularly important aspect of IP business since, with this type of product, there is always a substantial risk that once a claim commences it will continue until the expiry of the policy term. The issues involved are discussed in more detail in Section 22.6.1. Section 22.6.2 describes the data that will be required once a claim is made.

22.6.1 Claims Management

As discussed in Section 22.2, the claims experience of insurers across the world deteriorated significantly during the 1980s and 1990s. This was, in part, due to a number of factors that highlighted deficiencies in the claims management process and fueled the change to a more active management of IP business.

22.6.1.1 Causes of Claims Deterioration: The U.K. Example

Pitcher (2001) gives a full account of the causes of claims deterioration in the U.K. market. These causes include the following.

The Role and Attitude of Claimants' Own Doctors (General Practitioners) — In the 1990s, GPs were actively opposed to the notion of providing their patients' disability information to external bodies such as government agencies, insurance companies, and employers. This was because they felt that their overriding duty to their patients was being compromised. The GPs believed that, in effect, they were being asked to "whistleblow" on their patients' state of health to organizations that would not necessarily use the information in the patients' best interests.

The Role of the National Health Service (NHS), as the Main Provider of Healthcare to the U.K. Population — The NHS primarily focuses on the most appropriate treatment for the patient. Any requirements for rehabilitation are a secondary concern. Furthermore, the waiting time to receive help with rehabilitation can be long.

Fraudulent Claims Following the 1980s Recession — The 1980s recession led to increased unemployment and a general fear amongst workers that their jobs were not secure. Inevitably, this increased the temptation amongst policyholders to make fraudulent IP claims.

Changing Attitudes to Disability — The stigma attached to taking time off work through ill health has decreased substantially (see Section 22.1.2). Consequently, employees are more likely to claim for a period of sickness or take early retirement on the grounds of ill health.

22.6.1.2 *Improvements to the Claims Process*

Claims management was traditionally seen as, primarily, an administrative process. Now it is seen as being integral to the risk assessment process; as such, it is given much more credence by insurers (Coetzee et al., 2001). Indeed, the claims management process is now accepted as being a key determinant of how successful an insurer's IP business becomes.

It has also been recognized that claims assessors of IP business require different skills than those who manage lump-sum benefit claims (e.g., critical illness insurance; see Chapter 23). As such, there is now a greater emphasis on the need for technical and medical knowledge relating to IP claims, as well as strong interpersonal and communication skills (Dorter, 2001).

The claims assessor acts as the liaison between the claimant, medical professionals, the claimant's employer, and any other interested parties. They are now well trained in the areas of disability and rehabilitation, with some recruited from the relevant health professions (Coetzee et al., 2001).

It has become common for the insurance companies to have specialist claims assessment teams to deal with just one type of claim, e.g., mental illness claims (Pitcher, 2001). Insurance companies also now make a much greater use of their own in-house medical staff to aid claims assessment. This would be through, for example, confirming the diagnosis of the illness or injury and interpreting the results of tests carried out by independent medical professionals. There is much less emphasis placed on information obtained from the claimants' own doctors for reasons discussed in Section 22.6.1.1.

Early intervention is crucial in managing a successful claim. This helps the insurer to make initial claims assessments, e.g., through home visits soon after the claim is made. Also, regular follow-up visits help the assessor to manage a program of treatment and rehabilitation for the claimant that should remain under constant review.

There is now a strong emphasis placed on rehabilitating claimants as much as possible, so that they might re-enter the workforce. Consequently, insurance companies will go to greater lengths in order to provide financial assistance for rehabilitation services if, for example, there is a long waiting time to receive them through state health provision.

22.6.2 Claims Data

In order for a claim to be processed efficiently, certain data need to be collated from a variety of sources. Data will be required not just in respect of the claimant's medical condition, but also in respect of his or her financial position. The latter is required in order to ascertain that any benefit limit stated in the policy will not be breached. It also helps to determine the level of motivation that the policyholder is likely to have in returning to work.

The claimant's medical condition will be described on the claims form by the insured, and confirmed in a separate report by his or her doctor. In addition, other medical information might be required. This could include the following:

A specialist report from a hospital consultant
Information collected from a visit to the claimant by the insurer's health
 claims assessor

Regarding the collection of financial information, this will include the following:

Full details of the claimant's taxable earnings before the disability occurred
Details of any state disability benefits being received
Details of income being received from other IP policies that the insured
 might have taken out
Details of any other income (e.g., pension in respect of early retirement due
 to ill health).

In order to investigate further the likelihood that the claim will eventually cease, the insurer will also wish to collect information directly related to the claimant's occupation. This might include details of the individual's current duties, their level of job satisfaction, the employer's attitude to the claimant, and the claimant's previous employment history. The latter would be useful in ascertaining which skills might enable the claimant to retrain (Elliott, 2001).

22.7 Reserving

Reserves at a valuation date can be considered under three headings

Reserves in respect of active lives
Reserves in respect of claimants
Contingency reserves

22.7.1 Reserves in Respect of Active Lives

Reserves need to be set up at a valuation date in respect of policy-holders who are not currently making a claim. An unearned premium reserve will be required to cover the liability in respect of the period from the valuation date to the date when the next premium is due (see Chapter 16).

An additional reserve may be required if level premiums are payable throughout the term of the contract. This is usually known as the active life reserve, and this will be the present value of the benefits due under the contract less the present value of the future level premiums. This is similar in concept to the reserve set up in respect of a life insurance contract (see Chapter 7).

22.7.2 Reserves in Respect of Claimants

Reserves need to be set up in respect of policyholders who satisfy the claims criteria as at the valuation date. There are three categories of claim for which reserves are required:

Claims currently in payment
Claims incurred but not reported
Claims due and unpaid.

22.7.2.1 Claims Currently in Payment

If the number of claims in payment at the valuation date is very small it may be appropriate to set up the reserve from case estimates. In this situation, the claims department estimates the expected future duration of each claim. The reserve would then be the present value of all the claim payments assuming that these duration estimates are correct.

If the number of claims in payment at the valuation date is too large for case estimates to be used, then a statistical approach will be required. In effect the current claims will be valued as impaired life annuities, taking into account the expected impaired life mortality and termination rates.

22.7.2.2 Claims Incurred but Not Reported

It is common for policyholders only to notify the insurer of a claim just before the end of the deferred period. Consequently, at any valuation date, the insurer will have a liability in respect of claims of which it is unaware. It is important, therefore, that insurers set up appropriate reserves in respect of such incurred but not reported (IBNR) claims. See Chapter 16 for more details on IBNR reserves.

22.7.2.3 Claims Due and Unpaid

A reserve may be required in respect of claims that have been accepted by the insurer as at the valuation date but for which no payments have yet been made. In this situation, a case estimate could be made following the approach described in Section 22.7.2.1.

22.7.3 Contingency Reserves

An additional reserve may be required to cover the shortfalls in the reserves described above if experience proves to be far worse than expected. Elliott (2001) suggests that, in the U.K., it is common for this reserve to amount to 50% of the total annual premium.

22.8 Conclusion

The success of IP business has been shown, historically, to follow the general economic cycle. It was only after the global recessions of the 1980s and 1990s that insurers realized just how unprofitable this product line had become. Previously, the losses incurred were hidden by the profits made in more successful areas of business. With the poor claims experience and associated losses exposed, insurers were involved in a greater effort to redesign IP contracts and manage the claims process more actively.

One of the most important developments in recent years has been the move away from occupation-based claims criteria towards measures based on the functions and tasks that people carry out in their work. This has the advantage to both insurer and policyholder of ensuring that the claims management process is more transparent, objective, and consistent.

Appendix A: Inception Rate and Disability Annuity Method

We consider

$$\bar{a}^{12}_{x:\overline{n}|} = \int_0^n v^t \, {}_t p^{12}_x \, dt$$

Using integration by parts and considering the range of integral as comprising a partition of adjacent 1-year subintervals, Haberman and Pitacco (1999) show that we can rewrite this expression as

$$\bar{a}^{12}_{x:\overline{n}|} = \sum_{h=0}^{h-1} \int_0^1 {}_{h+u} p^{11}_x \, \mu^{12}_{x+h+u} \, v^{h+u} a^*_{x+h+u} \, du$$

where μ^{12} is the transition intensity from state 1 to state 2 and a^*_{x+h+u} is the expected present value of a continuous annuity payable to a disabled life which ceases on death or recovery.

We then use the approximation, based on the mean-value theorem for integrals:

$$\bar{a}^{12}_{x:\overline{n}|} \cong \sum_{h=0}^{n-1} v^{h+\frac{1}{2}} a^*_{x+h+\frac{1}{2}} \int_0^1 {}_{h+u}p^{11}_x \mu^{12}_{x+h+u} \, du$$

We now assume the further approximation

$$ {}_{h+u}p^{11}_x \cong {}_{h}p^{11}_x \, {}_{u}p^{11}_{x+h}$$

and consider

$$w_{x+h} = \int_0^1 {}_{u}p^{11}_{x+h} \mu^{12}_{x+h+u} \, du$$

We note that this integral represents a good approximation to the probability of entering state 2 between ages $x+h$ and $x+h+1$, on the basis that the probability of two or more transitions during the year is very low.

Thus, we obtain the final approximation

$$\bar{a}^{12}_{x:\overline{n}|} \cong \sum_{h=0}^{n-1} {}_{h}p^{11}_x w_{x+h} v^{h+\frac{1}{2}} a^*_{x+h+\frac{1}{2}}$$

Appendix B: Manchester Unity Method

We consider how the Manchester Unity method may be regarded as a simplification of the multiple state modeling approach.

As with Appendix A, we start with

$$\bar{a}^{12}_{x:\overline{n}|} = \int_0^n v^t {}_{t}p^{12}_x \, dt$$

Let $j_{[x]+t}$ denote $\text{Prob}\{(x)$ is in state 2 at age $x+t \mid (x)$ is in state 1 at age x and (x) is alive at age $x+t\}$.

If we let $S(y)$ denote the state of the underlying process at any age y, then

$$j_{[x]+t} = \Pr\left[S(x+t) = 2 \mid S(x) = 1 \text{ and } \{S(x+t) = 1 \text{ or } S(x+t) = 2\}\right]$$

To simplify matters we use A, B, C to denote the respective events

$$S(x+t) = 2 \qquad S(x) = 1 \qquad S(x+t) = 1 \text{ or } S(x+t) = 2$$

Then

$$j_{[x]+t} = \Pr[A|B \text{ and } C] = \frac{\Pr(A \text{ and } B \text{ and } C)}{\Pr(B \text{ and } C)}$$

$$= \frac{\Pr(A \text{ and } B)}{\Pr(B \text{ and } C)} = \frac{\Pr(A|B)}{\Pr(C|B)} \quad \text{using the laws of probability}$$

Hence

$$j_{[x]+t} = \frac{{}_t p_x^{12}}{{}_t p_x^{11} + {}_t p_x^{12}}$$

Thus, we can rewrite the original single premium as

$$\bar{a}_{x:\overline{n}|}^{12} = \int_0^n v^t ({}_t p_x^{11} + {}_t p_x^{12}) j_{[x]+t} \, dt$$

$$\cong \int_0^n v^t {}_t p_x j_{[x]+t} \, dt$$

where we approximate the multiple state model transition probability $({}_t p_x^{11} + {}_t p_x^{12})$ by a simple survival probability ${}_t p_x$ which is independent of the initial health status at age x.

This approximation has been widely used in Norway; see Haberman and Pitacco (1999).

We now introduce $\bar{z}_{x+t}$, the probability that an individual is in state 2 at age $x+t$ given that he is alive at age $x+t$. Then, ignoring the state at age x at the issue of the policy:

$$\bar{z}_{x+t} = \Pr\left[S(x+t) = 2 \middle| S(x+t) = 1 \text{ or } S(x+t) = 2\right]$$

so that we remove the conditioning on $S(x) = 1$.

We replace $j_{[x]+t}$ by $\bar{z}_{x+t}$ and obtain the following approximation, which forms the basis of the Manchester Unity method:

$$\bar{a}_{x:\overline{n}|}^{12} \cong \int_0^n v^t {}_t p_x \bar{z}_{x+t} \, dt$$

References

Ainslie, R. (2001). Income protection experience in the UK market—recovery at last? *GeneralCologne Re Risk Insights* 5(1).

Campbell, R. (2001). Using work task definitions of disability for income protection policies. *GeneralCologne Re Risk Insights* 5(1).

Coetzee, S., Temple, P., and Crause, S.-M. (2001). Financial performance of disability products is significantly improved by new product design and enhanced claims management. *GeneralCologne Re Risk Insights* 5(1).

CMIR 12 (1991). The analysis of permanent health insurance data. Continuous Mortality Investigation Bureau. Faculty and Institute of Actuaries.

CMIR 20 (2001). Sickness experience 1995–98 for individual and group income protection policies. Continuous Mortality Investigation Bureau. Faculty and Institute of Actuaries.

Dorter, J. (2001). Case management: transferring the skills of claims management. *GeneralCologne Re Risk Insights* 5(1).

Elliott, S. (2001). Income protection insurance — healthcare module. Faculty and Institute of Actuaries.

Fabrizio, E. and Mak, A. C. T. (2001). Dangers of using international experience for pricing — a comparison of Australian and New Zealand individual disability experience. *GeneralCologne Re Risk Insights* 5(2).

Haberman, S. (1987). Long term sickness schemes and disability insurance — forecasting and other actuarial problems. *Journal of the Institute of Actuaries* 114, 467–533 (with discussion).

Haberman, S. and Pitacco, E. (1999). *Actuarial Models for Disability Insurance.* Chapman & Hall/CRC Press, Boca Raton.

Mackay, G. (1993). Permanent health insurance — overviews and market conditions in the UK. *Insurance: Mathematics & Economics* 13, 123–130.

Meilander, R. and Simbro, D. (1993). The impact of replacement ratios. *Disability Newsletter* (June).

Munich Re. (1998). *The International Situation of Disability Insurance.* Munich Reinsurance Company.

Neill, A. (1977) Life Contingencies. Heinemann, UK.

Pitcher, D. (2001). Managing disability claims in the United Kingdom. *GeneralCologne Re Risk Insights* 5(1).

Renshaw, A. and Haberman, S. (2000). Modeling the recent time trends in the UK PHI recovery, mortality and claim inception transition intensities. *Insurance: Mathematics & Economics* 27, 365–396.

Taylor, R. (2001). Group long-term disability insurance in the United States: at a crossroads. *GeneralCologne Re Risk Insights* 5(1).

Chapter 23

Critical Illness

23.1 Introduction

Critical illness (CI) insurance provides a benefit that is payable upon the first occurrence or diagnosis of one of a number of medical conditions. These conditions will be specified in the policy document and can vary from insurer to insurer.

The first CI product was issued in South Africa in 1983, where it was known as "dread disease". Although the product was popular in South Africa, success was not universal. The unpleasant nature of illnesses covered such as cancer deterred many from purchasing the insurance; they did not wish to dwell on the consequences of suffering from such serious conditions (Watthey, 2001). The product has since been renamed as "Critical illness" and remarketed as a modern lifestyle product. Its aim, therefore, is to give people the peace of mind that, should they suffer a CI, they will be able to cope with the financial consequences (Munich Re, 2001). It has proved to be a very popular form of health insurance in many countries, including the U.K., Australia, Canada, East Asia, and Israel. Interestingly, however, it has had limited success in the U.S. (Watthey, 2001). CI products are also sold under a variety of different names, including "crisis cover," "trauma cover" and "living insurance".

The policy typically pays a lump sum to the insured. The benefits are not designed to reimburse the insured for the costs of treating the disease. Rather, the benefits are intended to help the insured and their family to cope with the financial consequences of the disease in the way they see fit. This flexibility makes the product attractive since, for example, they may need the lump sum to pay medical bills, repay loans, or provide financial security for their family. The product can be considered as providing cover for elements of the total range of health conditions, but it is not intended to provide cover for all medical conditions or procedures. The common medical conditions that are covered are described in Section 23.3.

For a condition to be considered for inclusion in a policy, the insurer would wish the following criteria to be met:

The condition or illness is considered by the public to be life (or lifestyle) threatening.

The condition can be clearly defined so that it is clearly distinguishable from any other condition.

A sufficient amount of data is available to be able to price the product.

It is usually only the first of these criteria that is fully met in practice. The second and third criteria are discussed in more detail in Section 23.2 and Section 23.5.2 respectively.

23.2   Product Structure

There are two main types of CI product on offer, the acceleration benefit and the stand-alone benefit; however, other variations have also been designed to service a range of more targeted markets. These variations are discussed in Section 23.2.3.

23.2.1   Acceleration Benefit

The acceleration (or "prepayment") product provides a combination of a death benefit and CI cover. A payment is made when either the policy-holder dies, or he or she is diagnosed as having one of the conditions specified in the policy. Typically, a proportion of the sum assured is paid when a CI is diagnosed and the balance is paid on death. This type of CI product is very popular in the U.K. and Irish markets (Reynolds, 1999). The design can be beneficial to insurers, since it reduces some of the uncertainty surrounding the pricing of the product. For example, when a CI benefit is attached to whole-life assurance or endowment assurance, a CI claim would simply be bringing forward the payment that would ultimately have been made on death.

23.2.2   Stand-Alone Benefit

The stand-alone product provides a sum assured if the policyholder is diagnosed with one of the conditions specified in the policy, provided he or she survives for a period (often 28 days) after diagnosis. This period is known as the "survival period." It is important for policyholders to understand that if they die during the survival period, there is no payment made to their beneficiaries under this type of cover.

23.2.3   Other Types of Critical Illness Product

As mentioned above, there are a wide variety of CI products that are aimed at a number of different markets. Some of these are described below.

23.2.3.1   *Buy-Back Benefits*

One of the drawbacks to the standard acceleration product is that the CI claim reduces the life cover element of the policy. The insured could find it very difficult to purchase further life cover following a CI claim, since their life expectancy will have been substantially reduced. For those policyholders wishing to maintain their level of life cover, a buy-back policy could be attractive. This product combines an acceleration benefit with an option to buy back some, or all, of the death benefit after a diagnosis of CI. Following such a diagnosis, the insured can reinstate the original death benefit provided that he or she has survived for a specified period (usually at least a year). The policyholder is not required to provide any further medical information to reinstate the life cover.

23.2.3.2   *Scaled Benefits*

Some insurance companies link the amount of benefit they offer to the level of severity of the policyholder's illness. For example, 20% of the sum insured may be paid on a heart attack and 30% in respect of heart by-pass surgery. This enables the cover to target need more effectively. It also reduces the opportunity for a policyholder to receive a substantial lump sum in respect of a relatively minor ailment. This is sometimes known as a "windfall payment" and is discussed in more detail in Section 23.3.2.

23.2.3.3   *Annuity Benefits*

It is possible for the CI benefit to be paid in the form of an annuity as opposed to a lump sum. This would ensure that the policyholder is guaranteed an income for a fixed period of time, or until death. However, policyholders generally prefer the flexibility that a lump sum can offer. In the U.K., for example, it is the attraction of receiving a lump sum that has made CI insurance more popular than income protection insurance (Reynolds, 1999). See Chapter 22 for income protection insurance.

23.2.3.4 Terminal Illness Benefits

Most acceleration products pay a proportion of the death benefit on the diagnosis of a terminal illness. However, providing such cover in respect of a stand-alone policy can lead to difficulties. Stand-alone products are not designed to provide death benefits; however, if death occurs soon after the onset of a terminal illness, then effectively a death benefit is being provided. Even requiring the insured to survive a certain period after onset will not necessarily solve this problem, as it can be difficult to ascertain precisely when the terminal illness began.

23.2.3.5 Gender-Specific Benefits

There are clearly some conditions that either only affect one sex (e.g., testicular cancer) or are far more common in one sex (e.g., breast cancer). CI products can, therefore, be designed to suit gender-specific markets. For example, in Asia, where CI is popular, female-specific CI products have been successfully launched. These typically cover three main areas:

Female diseases
Pregnancy complications
Congenital abnormalities of the insured's children

More details on female-specific products can be found in Liu (2002). Products aimed at males have been less successful (Kroll, 2002).

23.2.3.6 Children's Benefits

Some insurers offer a modest benefit on the lives of the children of the insured as part of the adult's policy. In this situation, a child's claim would not terminate the policy. The policy would only cease if the adult made a claim during the term.

There are, however, also products designed for children as the main policyholders. These usually cover conditions that are also suffered by adults, together with childhood diseases such as poliomyelitis and autism. In addition, there are products that specifically cover infants from birth to 3 years old. For these policies, most conditions covered relate to congenital abnormalities, such as Down's syndrome and spina bifida (Kroll, 2002).

23.3 Policy Content

23.3.1 Conditions Covered

Insurers vary considerably with regard to the number of diseases covered in a CI policy. Some will include an extensive list, whilst others will offer a basic cover, which includes only cancer, heart attack, and stroke. Policy-holders who wish to be covered for more medical conditions are usually then given the opportunity to take out an augmented version of the policy.

In order to obviate the need to include a more extensive list of medical conditions than necessary, many insurers will include cover for total and permanent disability (TPD) and/or loss of independent existence (LIE). TPD could be defined as the inability to carry out any occupation in the future. LIE could be defined as the inability to carry out at least three out of six activities of daily living (ADLs) on a permanent basis. ADLs are more usually associated with long-term care insurance (see Chapter 24). It should be noted that these kinds of definition are often poorly understood by the policyholders. For example, in the U.K., around half of the TPD claims made during 2002 were declined (GE Frankona, 2003).

23.3.2 Defining the Conditions

The insurance company will need to consult its senior medical advisers in order to ensure that the definitions used for the medical conditions are clear and precise. Claimants need to be dealt with consistently and need to understand exactly what cover they have. From the insurer's perspective it is vital that a definition does not encompass more conditions than originally envisaged in designing and pricing the policy. For example, Fabrizio and Mak (1999) describe how the definition of a stroke that was commonly used in the past has been found to be so wide that it inadvertently included other conditions affecting the brain, e.g., those resulting from head injuries.

In addition to ensuring that the policy wording is sufficiently tight, the provider needs to be aware that advances in medical science may mean that conditions which are considered serious at one point in time may be viewed as being relatively minor at a later date. Historically, for example, pneumonia was routinely considered a terminal illness, whereas today it is not necessarily so, since effective treatments are in use.

In order to mitigate the risk that the insurer may be asked to pay substantial lump sums in respect of what are now considered to be minor ailments, the insurer may state in the policy document that they reserve the right to review the definitions. An example of such wording is given by Molesworth (1999):

"Due to continuing advances in medical treatment and diagnostic techniques, we may need to review the definitions used in your policy document to ensure that in the future they:

remain appropriate with regard to medical terminology and classi-fication,

take into account effective cures, vaccines and modern diagnostic
    procedures,
include some diseases considered appropriate in the future,
exclude diseases which are found to have become minor in the future.

We reserve the right, therefore, to adjust your critical illness definitions
and/or premium rates, but only if the changes apply to all policies of this
class and are approved by the actuary having access to information and
data which are reasonable and reliable in the circumstances."

23.3.3 Standardization of Policy Wording: The U.K. Example

As mentioned in Section 23.3.2, it is imperative that the policy document
contains precisely worded definitions. During the 1990s, insurers in the
U.K. collectively decided to standardize the definitions of the conditions
covered in their policies. This led to a Statement of Best Practice being
issued by the Association of British Insurers (ABI) in 1999, which was
updated in 2003 (ABI, 2003). It has proved to be very successful, as
policyholders no longer need to examine the illness definitions in detail
when comparing different insurers' products. In addition, it means that
any inter-office claims experience that is collected is made more valuable,
since the definitions used by each insurer are the same.

The Statement of Best Practice defines model wordings for seven
"core" medical conditions, thirteen "additional" medical conditions and
nine "common exclusions." The conditions are listed below. The reader
can obtain the suggested policy wordings from the ABI (2004).

23.3.3.1 *Core Medical Conditions*

CI contracts offered by ABI members in the U.K. will usually include
heart attack, cancer, stroke, coronary artery bypass surgery, kidney failure,
major organ transplant, and multiple sclerosis.

23.3.3.2 *Additional Medical Conditions*

Additional conditions defined by the ABI are as follows:

Aorta graft surgery
Benign brain tumor
Blindness
Coma
Deafness
Heart valve replacement or repair
Loss of limbs
Loss of speech

Motor neurone disease
Paralysis/paraplegia
Parkinson's disease
Terminal illness
Third-degree burns

23.3.3.3 Common Policy Exclusions

Model policy exclusion wordings have been provided for:

Aviation (excluding nonroutine aviation), e.g., spraying of agricultural crops
Criminal acts
Drug or alcohol abuse
Failure to follow medical advice
Hazardous sports and pastimes
HIV/AIDS
Living outside U.K.
Self-inflicted injury
War and civil commotion

23.4 Underwriting

CI insurance is highly susceptible to adverse selection. For example, in South Africa, in the first few years after CI insurance was introduced in 1983, there were a substantial number of claims made for cancer and multiple sclerosis soon after policy commencement. As a result, a waiting period (or moratorium) clause was introduced into the policy, which stipulated that claims could not be made in respect of conditions that began during this period. Waiting periods tend to be either 3 or 6 months (Munich Re, 2001).

Whilst waiting periods help to reduce the risk of adverse selection, they also, however, reduce the attraction of the product to potential policy-holders. For example, individuals may wish to take out CI insurance to cover a specific loan and would effectively be exposed to the risk of being unable to repay the loan if they were to suffer a CI during the waiting period. Furthermore, such a design gives policyholders a disincentive to visit their doctor during the period for fear of being diagnosed as having one of the excluded CIs.

The risk of adverse selection is higher for stand-alone products than for acceleration products (Grossman, 1999). This is because an individual who knows that he or she is predisposed towards a CI is more likely to purchase a stand-alone product than an acceleration product, since the premiums are lower for stand-alone policies but give the same level of CI benefit.

As Grossman (1999) suggests, an important way to try to reduce the risk of adverse selection is to establish precisely why the policyholder has chosen to take out CI insurance, especially since such a product tends to require relatively large premiums. Probing in this way can be very helpful in identifying applicants who might intend to make fraudulent claims in the future.

The underwriting process is focused on the applicant's prior history of illness, family history, and the presence of other risk factors (including smoking status, hypertension, body build, cholesterol, diabetes, existing and proposed insurance benefit levels). In order to obtain such detailed information, the insurer will issue an application form that is more extensive than that required for standard life insurance (see Chapter 10). In addition, in order to substantiate the data provided, the insurer will require a report from the prospective policyholder's doctor.

As an alternative to carrying out detailed underwriting, some insurers accept all applicants but exclude from cover all pre-existing conditions. This negates the need for underwriting at the application stage; however, following a claim, underwriting then becomes part of the claims assessment procedure (Watthey, 2001).

23.5 Pricing the Risk

In this section, pricing is considered under the following headings:

Risk factors
Data sources and difficulties
Trends
Guarantees
Variations in claims experience
Pricing methodology

23.5.1 Risk Factors

The major risk factors are:

Age
Sex
Smoking status
Socio-economic class
Occupation
Geographical area of residence and birth
Family history
Medically related factors

A CI contract covers a number of different medical conditions. The total risk premium for a policy will be the sum of the individual risk premiums in respect of each of the conditions. The influence that each risk factor has on the likelihood of a particular condition occurring will vary by condition. Therefore, assessing the overall impact that a risk factor has on the whole policy is complicated. This is different to a life insurance contract, where there is only one decrement, i.e., death.

We comment on each risk factor in turn. For more details, the reader is referred to Watthey (2001).

23.5.1.1 Age

For most of the CIs under consideration, the incidence rate increases with age. However, it should be borne in mind that some of the illnesses covered are usually only diagnosed amongst people in certain age groups. For example, the onset of multiple sclerosis tends to occur mostly between the ages of 20 and 45.

23.5.1.2 Sex

Separate tables of morbidity rates are required for males and females. There are marked differences between the sexes in the incidence rates of the various conditions covered. For example, men are more likely to suffer a heart attack than women are.

23.5.1.3 Smoking Status

There is a significantly higher claims incidence rate amongst smokers than nonsmokers, particularly with regard to certain diseases such as cancer. It is important, therefore, to rate the two groups separately. For this purpose, a nonsmoker would typically be defined as someone who has not smoked any tobacco for the previous 12 months.

23.5.1.4 Socio-Economic Class

Socio-economic differences most commonly manifest themselves in the education and diet of the population. As a result, people in the higher socio-economic classes tend to be less susceptible than others are to most of the diseases covered. The difference in the incidence rates between the classes decreases with age.

23.5.1.5 Occupation

A policyholder's occupation is likely to have a bearing on their propensity to suffer a CI. For example, people in stressful jobs are more likely to suffer a heart attack.

When pricing policies that include a total and permanent disablement benefit, consideration must be given to the significant variations in incidence of TPD by occupation. For example, builders will tend to be at risk of suffering a serious injury at work.

23.5.1.6   Geographical Area of Residence and Birth
As with socio-economic factors, people living in a deprived geographical area may be more likely to suffer from certain medical conditions. For example, they are more at risk if they live in an area that is highly polluted with toxic waste.

23.5.1.7   Family History
This is an important factor, as it relates to the genetic makeup of the individual and, therefore, the likelihood that they will contract certain conditions in the future. For example, there may be a history of breast cancer in the family.

23.5.1.8   Medically Related Factors
Medically related factors include for example, hypertension and cholesterol levels. To illustrate, an individual who is hypertensive is more at risk of having a stroke.

23.5.2   Data Sources and Difficulties

23.5.2.1   Data Sources
As CI contracts cover a variety of medical conditions, a number of data sources need to be found in order to cost each of these components. However, as CI insurance is a relatively new product, for most countries very little insurance claims experience is available.

A useful source of insured life data can be found in Dinani et al. (2000). In this work, claims experience that relates to approximately 60% of the U.K. CI insurance market is described for the years 1991–1997. In addition, a base table for CI business in the U.K. is constructed from population data, for comparison purposes. The report also includes the results of a survey of reserving methods being used by U.K. insurers during 1997. The study was the first undertaken for the U.K. CI market. Other countries for which earlier studies have been made include Australia and New Zealand (Collins et al., 1997), and Ireland (Jeffery, 1995).

Since most insurers are unlikely to have sufficient insured life data to be credible, they tend to rely heavily on population-based data when pricing the product. Such data need to be adjusted to reflect assumed differences in incidence rates between the general population and insured lives. This was done in constructing the base table in Dinani et al. (2000).

A useful list of general population data sources is provided in Fabrizio and Mak (1999). These sources are summarized below.

Cancer Societies—Most countries carry out a considerable amount of research into cancer. Therefore, statistics related to incidence rates and types of cancer are usually readily available. These data are usually broken down by age, sex, location of the cancer in the body, and sometimes by size of tumor.

In England and Wales, for example, the Office of National Statistics (ONS) Cancer Registrations provides data in respect of individuals who were diagnosed with cancer during the year being investigated (Watthey, 2001).

Hospital Episode Statistics—Most hospitals in developed countries record patient information, including duration of stay, conditions suffered, and treatment provided. Such data will not give a complete picture; for example, when patients die soon after admission to hospital their full medical condition may not be noted.

Health Interest Groups—Many countries have health interest groups that are committed to collecting data related to specific illnesses (e.g., multiple sclerosis, stroke, and circulatory disorders). They often compile statistics that are useful for pricing different components of the CI insurance benefits.

Specific Population Studies—Some countries undertake specific studies usually based on population registers. However, care is needed in the interpretation of the results, since the studies are usually only based on a small population sample. The conclusions will be unreliable if the sample is unrepresentative of the general population.

Other Sources—Countries may have motor accident statistics, morbidity statistics collected from doctors' surgeries, and general population studies. In addition, academic papers published in medical journals can be a very useful source of data related to specific diseases.

23.5.2.2   Difficulties with the Data

There are difficulties with the data sources. Although the published data sources are reasonable, considerable work is needed to derive meaningful statistics for use in pricing. In particular, it should be borne in mind that if the published data are population-based, then they will need to be adjusted. This is to reflect the fact that the lives under consideration are insured lives and may exhibit a different claims experience to that of the general population, because of temporary initial selection (see Benjamin and Pollard (1993)).

Although a few inter-company surveys exist (see Section 23.5.2.1), the exposure tends to be of short duration and there is, as yet, no hard evidence on the shape of the total experience over the first generation of business. Consequently, margins should be included in the pricing to allow for these data difficulties, and the trends discussed in the next section.

23.5.3 Trends

There is considerable uncertainty about the likely future level of CI claims. The major change in the past few years has been the improvement in screening and diagnosis through the introduction of new tests. This has led to changes being made in the wording of definitions for some medical conditions. This trend is likely to continue.

The improvement in screening has led to earlier detection of some cancers (in particular prostate cancer) and multiple sclerosis. This may not have changed the overall incidence of the particular disease, but earlier diagnosis will result in claim payments being brought forward.

23.5.4 Guarantees

It is possible for CI insurance to be written with the premium rate guaranteed throughout the term of the policy. With all of the uncertainty surrounding future claims experience, it is important that such guarantees are allowed for in the pricing basis. This can be done, for example, by using a stochastic model for changes in future claims experience.

23.5.5 Variations in Claims Experience

There are significant variations in claims experience between insurers in this class of business. The range of experience is wider than for life products. The experience level will be determined by a variety of factors including underwriting levels, product design, sales methods, claim control, and the target market for sales. An insurer always needs to build its own level of experience into its pricing and product design.

23.5.6 Pricing Methodology

There are two main approaches to calculating premiums for CI products: the traditional approach and the more modern multiple state model approach. Both approaches are described below.

23.5.6.1 Traditional Approach

As mentioned in Section 23.5.2.1, it is possible that the insurer will have insufficient of its own insured life data to be able to produce credible CI incidence rates. In this situation, the insurer might choose to derive such rates approximately from population-based incidence rates. The premiums would then be based on these estimated incidence rates.

To derive the insured life incidence rates, it is assumed that the relationship between insured life and population-based incidence rates is approximately the same as that between insured life and population-based mortality rates. Fabrizio and Mak (1999) give the formulae for the premium rates for the acceleration and stand-alone products under this assumption.

The formulae are set out below, together with a general reasoning explanation for each result. In each case, the formula for the risk premium is given for a term of 1 year, unit sum assured, and policyholder age x. For completeness, the common approach used to derive the result for the acceleration benefit as described by Dash and Grimshaw (1990) is shown in Appendix A.

(1) Pricing formula for acceleration product:

$$\text{Risk premium} = i_x + q_x(1 - k_x)$$

where i_x is the incidence rate of CI, k_x is the proportion of deaths caused by CI, and q_x is the mortality rate. The i_x term covers the cost of the CI element of the benefit. The $q_x(1 - k_x)$ term covers the cost of the mortality element of the benefit, but only in respect of deaths due to a cause other than CI.

(2) Pricing formula for stand-alone product:

$$\text{Risk premium} = i_x(1 - q_x^i)$$

where i_x is the incidence rate of CI and q_x^i is the proportion of lives who die during the survival period following a CI (see Section 23.2.2).

The deduction of $i_x q_x^i$ from the incidence rate i_x is required since no payment is made on this policy if the policyholder dies during the survival period.

23.5.6.2 Multiple State Model Approach

CI benefits can be priced using a multiple state model. Full details of this approach are described in Haberman and Pitacco (1999, chapter 5).

Figure 23.1 shows the basic model for CI insurance, which may be used to build up the necessary actuarial models for determining premiums and reserves. Here, the states are

active a
suffering from a CI, i
death due to a CI, $d(D)$
death due to other causes, $d(o)$

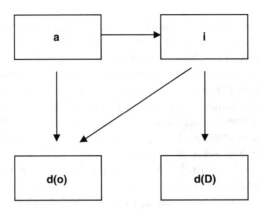

FIGURE 23.1 A multiple state model for CI insurance

In this structure, sudden deaths due to a CI are represented by a pair of transitions (from state a to state i and then from state i to state $d(D)$) so that all such deaths are represented by transitions that end up in the $d(D)$ state. We note that there are no permitted transitions from state i to state a — this arises from the policy design feature that insurance cover ends after diagnosis and payment of the lump sum.

The four transition intensities define the structure of the continuous time model:

$$\mu_x^{ai}, \quad \mu_x^{ad(o)}, \quad \mu_x^{id(o)}, \quad \mu_x^{id(D)}$$

where x is the attained age. It would be more theoretically correct to assume that the latter pair of intensities, for transitions out of state i, depend not only on x but also on r, the time elapsed since the CI was diagnosed. This feature has been suppressed here because, in practice, data restrictions mean that such bivariate transition intensities are difficult to estimate.

Given the absence of any reversible transitions in Figure 23.1 (unlike the version of the multiple state model used for income protection — Section 22.5.6.1), the transition probabilities needed for the calculation of premiums and reserves are relatively simple functions of the transition intensities. This is due to the simple form of the underlying differential equations.

The first step is to define the occupancy probabilities:

$$_t p_x^{\overline{aa}} = \exp\left[-\int_0^t \left(\mu_{x+s}^{ai} + \mu_{x+s}^{ad(o)}\right) ds\right]$$

and

$$_t p_x^{\overline{ii}} = \exp\left[-\int_0^t \left(\mu_{x+s}^{id(o)} + \mu_{x+s}^{id(D)}\right) ds\right]$$

We will use these to illustrate the form of single premiums for two simple benefit structures derived from the equivalence principle.

Consider a stand-alone policy providing 1 unit at diagnosis of a CI within n years, thus payable if the $a \to i$ transition occurs. The single premium is equal to the expected present value of the benefits, i.e.

$$\bar{A}_{x:n}^{\,\text{CI}} = \int_0^n {_u p_x^{\overline{aa}}} \, \mu_{x+u}^{ai} v^u \, du$$

This would also relate to an additional benefit provided, e.g., as a rider to a temporary insurance.

Consider an acceleration benefit whereby, for a given amount of insurance, the proportion λ is payable on diagnosis of a CI while the remaining proportion, $1 - \lambda$, is payable on death, if this occurs within the policy term n. The single premium can be shown to be given by

$$\bar{A}_{x:n}^{\text{CI}:\lambda} = \int_0^n {_u p_x^{\overline{aa}}} \left\{ \mu_{x+u}^{ad(o)} v^u + \mu_{x+u}^{ai}[\lambda v^u + (1-\lambda)B_{x+u}] \right\} du$$

where

$$B_{x+u} = \int_0^{n-u} {_s p_{x+u}^{\overline{ii}}} \left(\mu_{x+u+s}^{id(o)} + \mu_{x+u+s}^{id(D)} \right) v^{u+s} \, ds$$

Haberman and Pitacco (1999) consider more complex structures and also discuss formulae for the reserves.

It is clear that estimates for the four sets of transition intensities are needed so that the integrals in the above expressions can be evaluated numerically. In practice, a number of simplifying assumptions are made in order to reflect the nature and quality of the data available. These assumptions include the following:

Introduce a factor $1 + m_x$ to represent the ratio of the force of mortality from CI to the force of mortality from causes other than the CI. Hence, $\mu_x^{id(0)} = \mu_x^{ad(o)}(1 + m_x)$.

Use population-based disease incidence data for μ_x^{ai}.

Use the approximation $\mu_x^{ad(o)} + \mu_x^{id(o)} = \mu_x(1 - k_x)$, where μ_x is the population force of mortality from all causes and k_x is the population-based proportion of deaths at age x that are attributable to the CI.

This approach underpins the widely used approximation based on Dash and Grimshaw (1990) that is discussed in Appendix A and which is presented via a discrete-time version of the model.

23.6 Claims

23.6.1 Claims Management

When a claim is made, the policyholder needs to complete a detailed claims form, and his or her doctor will be required to confirm that the information provided is correct. The claims assessor must then ascertain that the condition is covered under the policy and that it did not commence during the waiting period, if any (see Section 23.4).

The claims assessment process should be straightforward and objective, given that the policy covers particular illnesses that have been tightly defined in the policy documents.

In practice, a CI provider needs to recognize that the definitions of the conditions covered will be written using very technical medical terminology. Consequently, it is unlikely that a policyholder will understand precisely what the product does and, importantly, does not cover. For example, individuals may believe that they are covered for all types of cancer when, in fact, mild skin cancer might not be included in the policy. It is quite possible, therefore, for claims decisions to be disputed and for claims assessment staff to spend a significant amount of time defending their decisions. Generally, in dealing with the claimants, staff need to provide a fast and sympathetic service; by the very nature of the type of claim, the claimant will have undergone significant traumas. Poorly handled claims could result in adverse publicity and a significant loss of business for the insurer.

23.6.2 Causes of Claim

Table 23.1 illustrates the typical distribution of causes of claim by sex and territory during the late 1990s. It highlights the marked differences in the prevalence of claims due to cancer and heart attack between the different territories. It also illustrates the differences in causes of claim between the sexes, particularly in the U.K. and Australia.

23.7 Reserving

The reserving methods used will be similar to those used for pricing. Prudence dictates that the assumptions used to calculate the reserves

Table 23.1 Distribution of claims

		Cancer	Heart attack	Stroke	Other
U.K.	Male	42	30	8	20
	Female	75	2	6	17
Australia	Male	46	17	7	30
	Female	84	1	4	11
South East Asia	Male	77	7	4	12
	Female	89	1	2	8

Source: Munich Re (2000)

should be more cautious than those used to calculate the premiums for a product. This is particularly true of CI insurance, given the paucity of data as regards claims experience. In effect, the valuation should be undertaken on a conservative pricing basis. Due allowance needs to be made for guarantees on products by using an appropriate stochastic model for changes in future claims experience.

It is important that likely future trends in morbidity rates are allowed for in the valuation basis. For example, in the U.K. it is common for the morbidity assumptions used in the pricing basis to be loaded by around 25% for use in the reserving basis (Dinani et al., 2000).

Mortality is another important component of the valuation basis. For stand-alone CI insurance business, however, it should be remembered that the conservative assumption is to assume lighter mortality, since no benefit is paid on death with this product.

23.8   Conclusion

CI insurance is a relatively new product in the health insurance market. The first policies preyed on people's fears of being struck by certain terminal diseases, such as cancer. The early take up rates of CI policies were mixed internationally. Since the early days of the product's existence, the insurance industry has redesigned and remarketed CI policies to much greater effect. CI is now a significant product in most well-established health insurance markets.

The ongoing issues that need to be considered are those of policy design. Medical advances in the diagnosis and treatment of conditions result in the need to review the policy documents on a regular basis. In particular, the definitions of the conditions covered are crucial. Failure to keep pace with medical science can have a significantly detrimental effect on the profitability on this line of business. Other countries could benefit from following the approach to definitions that has been adopted in the U.K., i.e., to use industry-wide, standardized definitions.

Appendix A

A.1 Derivation of Approximate Risk Premium Formula for an Acceleration Product

We follow the approach described in Dash and Grimshaw (1990) to derive the approximate risk premium rate for an acceleration product. The approach is approximate, in that it uses a discrete-time representation of the continuous-time model discussed in Section 23.5. The cover for this product is for 1 year and the benefit involves a unit sum assured.

The discrete time model is shown in Figure 23.2, where I_x is the number of incidences of CI between age x and $x+1$, $(dh)_x$ is the number of deaths amongst healthy lives from a cause other than CI between age x and $x+1$, $(dc)_x$ is the number of deaths due to CI amongst lives suffering from CI between age x and $x+1$, and $(do)_x$ is the number of deaths due to a cause other than CI amongst lives suffering from CI between age x and $x+1$.

It can be seen that the three states are healthy, CI sufferer, and dead. The number of CI incidents, I_x, is the aggregate of the number of incidents for each CI covered. Also, the number of deaths amongst CI sufferers needs to be subdivided between those where death was caused by CI and those where death was attributable to another cause.

The total number of claims between age x and $x+1$ on an acceleration CI policy is given by

$$I_x + (dh)_x \tag{23.1}$$

The difficulty is then to find suitable data to calculate these quantities and, in particular, $(dh)_x$. We can find a good approximation to Equation (23.1) by making suitable assumptions as follows:

d_x is the number of deaths in the population between age x and $x+1$
k_x is the proportion of deaths in the population between age x and $x+1$
 which are due to CI

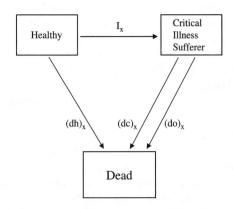

FIGURE 23.2 Multiple decrement model used to price CI benefits

k_x should exclude deaths that occur immediately following the onset of a CI, since such lives will not enter the population of CI sufferers. Then

$$(\text{dc})_x = k_x d_x \tag{23.2}$$

and since $d_x = (\text{dh})_x + (\text{dc})_x + (\text{do})_x$, it follows that the number of deaths not due to CI is

$$(\text{dh})_x + (\text{do})_x = (1 - k_x)d_x \tag{23.3}$$

If we now assume that the mortality of CI sufferers from causes other than CI is the same as the mortality of healthy lives, then

$$\frac{(\text{do})_x}{(\text{lc})_x} = \frac{(\text{dh})_x}{l_x - (\text{lc})_x} \tag{23.4}$$

where l_x is the number of lives aged x exact and $(\text{lc})_x$ is the number of lives aged x exact suffering from a CI.

From Equation (23.3) and Equation (23.4), we can eliminate $(\text{do})_x$ to obtain

$$\frac{(\text{dh})_x l_x}{l_x - (\text{lc})_x} = (1 - k_x)d_x \tag{23.5}$$

Now Equation (23.1) is an expression for the number of claims between age x and $x+1$ on an acceleration CI policy. To convert this into an incidence rate, we must divide by the healthy population at age x (i.e., divide by $l_x - (\text{lc})_x$).

Hence, the incidence rate for the acceleration product is

$$\frac{l_x}{l_x - (\text{lc})_x} + \frac{(\text{dh})_x}{l_x - (\text{lc})_x} \tag{23.6}$$

From Equation (23.5), this can be rewritten as

$$i_x + (1 - k_x)q_x \tag{23.7}$$

where the incidence rate $i_x = I_x/(l_x - (\text{lc})_x)$ and the mortality rate $q_x = d_x/l_x$.

Hence, Equation (23.7) represents a simple expression for the risk premium rate for an acceleration CI product.

References

ABI (2003). *Critical Illness Insurance: ABI Code of Practice (revised 2003)*. Association of British Insurers, London.

ABI (2004). www.abi.org.uk.

Benjamin, B. and Pollard, J. H. (1993). The analysis of mortality and other actuarial statistics. Faculty and Institute of Actuaries.

Collins, E., Howes, S., and Tsui, L. (1997). Crisis and TPD experience in Australia and New Zealand in 1994 and 1995. Centenary Convention, Institute of Actuaries of Australia, vol. 5.

Dash, A. and Grimshaw, D. (1990). Dread disease — an actuarial perspective. Presented to the Staple Inn Actuarial Society, London.

Dinani, A., Grimshaw, D., Robjohns, N., Somerville, S., Spry, A., and Staffurth, J. (2000). A critical review: report of the critical illness healthcare study group. Presented to the Staple Inn Actuarial Society, London.

Fabrizio, E. and Mak, A. (1999). Actuarial aspects of dread disease products. *GeneralCologne Re Risk Insights* 3(1).

GE Frankona (2003). Critical illness claims survey. In-house publication.

Grossman, E. (1999). Dread disease — underwriting and claim considerations. *GeneralCologne Re Risk Insights* 3(1).

Haberman, S. and Pitacco, E. (1999). *Actuarial Models for Disability Insurance*. Chapman & Hall/CRC Press, Boca Raton.

Jeffery, A. J. (1995). Reserving for critical illness guarantees. 25th International Congress of Actuaries, 3, 383–411.

Kroll, K. (2002). Product variety in dread disease insurance. *GeneralCologne Re Risk Insights* 6(1).

Liu, S. Y. (2002). Ode to the fairer sex. *GeneralCologne Re Risk Insights* 6(1).

Molesworth, M. (1999). The future of critical illness definitions. *GeneralCologne Re Risk Insights* 3(1).

Munich Re (2000). The marketing of critical illness insurance. In-house publication.

Munich Re (2001). Critical illness insurance. In-house publication.

Reynolds, N. (1999). Current issues in critical illness product design. *GeneralCologne Re Risk Insights 3(1)*.

Watthey, D. K. (2001). Critical illness cover — healthcare module. Faculty and Institute of Actuaries.

Chapter 24

Long-Term Care

24.1 Introduction

Many countries in both the developed and the developing world have been experiencing a marked demographic shift towards an older population. This has been due to two main factors: increasing life expectancy and decreasing fertility rates. The social structure, particularly in developed countries, has also changed. Decreasing marriage rates and increasing divorce rates have shattered the traditional concept of a nuclear family. More people are living on their own and in isolation from other family members. As they get older and more incapacitated, it is becoming less likely that the elderly will be looked after in their own home by other family members. This has been further exacerbated by the fact that women, who have generally been the main carers, are increasingly in paid employment and have less time to act as a carer. Therefore, the provision of long-term care (LTC) for the elderly has become a very topical issue.

In this chapter, we define LTC to be care provided to people who are unable to look after themselves without some sort of support. We are focusing on the elderly population who are the significant cohort requiring LTC. The care could be provided in the person's own home or within an institution offering a care facility. LTC is either provided informally by family and friends or, formally, on a paid basis.

Care requirements and services vary greatly from country to country, since they are a function of a country's demographic structure and its social attitude to care. It is important to consider what care services are expected by the elderly population and their families. For example, Rappaport (1998) describes how, in the U.S (as in many other countries), it is women who greatly outnumber men in old age, and who are therefore more likely to require formal care. It is also necessary to consider the extent to which state services are provided and, by implication, the extent to which private care provision is required to top up those state services to the level desired by each individual. It is in connection with private care provision that the actuarial profession is most commonly involved, and this will be the main emphasis of this chapter.

In the following sections, although other countries are mentioned, the focus will be on LTC provision in the U.K. and the U.S. The reason for this is that the latter has a well developed LTC insurance market; the U.K., by contrast, is a good example of a country whose LTC insurance market is still in the developmental stage.

24.2 Funding Long-Term Care

24.2.1 Background

The means by which LTC is funded varies from country to country. National resources through taxation, the use of private assets, and the insurance market are three examples of the mechanisms that might be used in the financing of care requirements. The emphasis on each of these mechanisms will change from country to country, leading to a complex picture of care provision.

LTC insurance is fairly new to the global insurance industry. It originated in the U.S. during the 1970s; it was first issued in Germany and France in 1985 and in the U.K. in 1991. Insurance products are also being developed in other European countries, Israel, South Africa, and Asia (Stracke, 1998).

The insurance market tends to provide LTC solutions through the following vehicles:

Purchasing a pre-funded LTC insurance product (see Section 24.3).
Purchasing an impaired life annuity at the time of entry into institutional care (see Section 24.4.2).
Realizing the value of the owner-occupied property by an equity release mechanism (see Section 24.4.2).

24.2.2 Brief International Overview of Long-Term Care Funding

We have already suggested that the approach to care provision in any country is, in part, a function of the interplay between the state and the private funding of services. In order to present a flavor of the complexities involved, the main features of LTC provision are given for the U.S., the U.K., Sweden, Germany, and Japan. For a fuller description of care provision in these countries, the reader is referred to Karlsson et al. (2004).

24.2.2.1   The U.S.

The American system is characterized by personal responsibility and private provision mainly obtained through the purchase of LTC insurance. In addition to this national Medicare and Medicaid programs were introduced in 1965. These were designed to increase the general population's access to medical services; they are briefly described below.

Medicaid is available to those on low income. Although initially designed to cover the cost of institutional care, it now also covers the cost of care received in the individual's home. The scope of the services offered and the conditions of eligibility vary from state to state.

Medicare is a national insurance program designed to cover the disabled and the elderly population for relatively low levels of benefits. For example, Medicare only covers the medical costs incurred during a stay in hospital; it does not cover the accommodation costs. Furthermore, hospital stays are only covered for the first 60 days, and nursing home stays are only fully covered for the first 20 days. Contributions are paid in the form of a special tax by employees and their employers, and special premiums by pensioners.

24.2.2.2   The U.K.

The U.K. has a rather complex system for LTC, as the funding of medical care (through the National Health Service) is different to that of the social aspects of care (through local authority Social Services departments). The NHS is responsible for covering the cost of nursing and medical care wherever the individual resides (i.e., in their own home, a residential home, or a hospital). Any services that are paid for by the local authority are means tested. This means that anyone with assets over a certain threshold becomes ineligible for free care, and is obliged to liquidate the assets in order to pay for the care. Assets, unlike in the U.S., include an individual's home. Consequently, many people have had to sell their homes in order to pay for care. This is covered more fully in Section 24.4.2.

24.2.2.3   Sweden

Sweden provides LTC mainly through the state. Local authorities provide care, primarily financed through local income tax, but some central funding is also available through the government. Additionally, people may be asked to contribute to the services they receive, but the amount that they are required to pay varies across local authorities. There is no means testing in Sweden; anyone in need of care is entitled, by law, to receive it. However, there is no standardized method of assessing entitlement across the local authorities, and so there is some regional variation.

24.2.2.4   Germany

A compulsory LTC social insurance scheme was set up in Germany in 1995. The scheme provides three types of benefits that depend on need: a cash benefit (which can be used at the discretion of the recipient), home care, and institutional care. Approximately 10% of the population have been allowed to opt out of the social insurance scheme provided that they purchase LTC insurance instead. They have this option when their earnings exceed a certain, specified threshold.

24.2.2.5   Japan

The provision of LTC has become a key issue in Japan in recent years, as it has the fastest aging population in the world. In 2000, Japan introduced a compulsory social insurance LTC scheme that covers the whole population. It is funded through a combination of taxation and earnings-related private insurance contributions. People aged over 65 are entitled to obtain care for physical or mental disabilities; people aged between 40 and 65 are entitled to care for mental conditions only. Benefits are not means tested and the availability of informal carers does not affect the entitlement to formal care.

24.3   Pre-Funded Long-Term Care Insurance Products

Various types of pre-funded LTC insurance products are available in the insurance market. They are aimed at individuals who are currently in good health but who are concerned that, at some stage in the future, they will require formal care. Such arrangements are either stand alone or written as riders to other health or life insurance products. These are discussed separately in Section 24.3.2 and Section 24.3.3.

24.3.1   Claim Triggers

In many countries, an insurance company will agree to pay a claim if the insured life is either sufficiently cognitively impaired or unable to carry out a certain number of tasks, known as activities of daily living (ADLs). ADLs are used as policy triggers, since failing some or all of them indicates the loss of independent living through chronic illness. The most commonly used ADLs are as follows:

Washing — the ability to wash in the bath or shower (including getting into and out of the bath or shower) or wash satisfactorily by other means.

Dressing — the ability to put on and take off all garments and/or braces, artificial limbs, or other surgical appliances, and to secure and unfasten the garments or devices.

Feeding — the ability to feed oneself once food has been prepared and made available.

Toileting — the ability to use the lavatory or otherwise manage bowel and bladder functions, so as to maintain a satisfactory level of personal hygiene.

Mobility — the ability to move indoors from room to room on level surfaces.

Transferring — the ability to move from a lying position in a bed to a sitting position in a chair or wheelchair (and vice versa).

It is important that the claims criteria used by an insurance company should be objective, measurable, and consistent if the insurer is to price and control its LTC business successfully. ADLs have, on the whole, managed to fulfill these requirements.

In addition to ADLs, insurance companies will also often use instrumental ADLs (IADLs). Examples include using the telephone and shopping. IADLs are not used as policy triggers as ADLs are. Rather, they help to predict an individual's decline in mobility and independence, both of which are signs of the need for increased care (Campbell, 1998).

It is, of course, possible for an insurance company to have a different arrangement in place in order to determine when a claim should be paid. For example, in the early days of writing LTC business, insurance companies in the U.S. only paid claims when a policyholder entered a nursing home, following a period of hospitalization.

24.3.2 Stand-Alone Insurance Products

The benefit payable in respect of a stand-alone product could be an annuity that ceases after a fixed period (e.g., 3 years). More usually, however, it is an annuity that is payable until death or recovery. A deferred period normally applies. This could be only a few months, or it might be several years. In the U.K. it is usually 3 months.

Regarding the level of benefit payable, it is common in the U.K. for 50% of the benefit to be paid upon failing two of the six ADLs. This would increase to the full amount upon failing the third ADL. The full amount would always be payable if the claim is in respect of cognitive impairment, rather than ADL failure. The benefit level is the same regardless of whether the policyholder resides in their own home or in a residential/nursing home.

In the U.S., on the other hand, the benefit might be paid at a lower rate if policyholders receive the care in their own home. The rationale for the latter is that, in general, policyholders will not choose to reside in an institution unless their health was so poor that they had to.

24.3.3 Long-Term Care Insurance as a Rider Benefit

LTC insurance could be written as a rider to a whole-of-life policy. With such an arrangement, if policyholders satisfy the claims criteria then they are paid an accelerated death benefit in installments. Usually, 2% of the sum assured would be paid each month, for a maximum of 50 months. This rider has been popular in the U.S., but less so in the U.K.

LTC cover could also be provided as a rider to critical illness insurance, where "LTC" would effectively be treated as one of the specified illnesses. As with the rider to whole-of-life assurance, the benefit would normally be paid in installments.

It is also natural to consider offering LTC cover as a rider to income protection insurance. One way of doing this is to offer income protection insurance to policyholders while they are under, say, normal retirement age and LTC cover after that age. As a result, the claim criteria would switch from being occupation related (as for income protection) to being based on ADL/cognitive ability (as for LTC) at normal retirement age.

24.3.4 Underwriting

The underwriting of an LTC contract requires specialist knowledge of the conditions which are commonly faced by the aging population. In order to make an accurate prediction of life expectancy and independent living, both the medical and social status of the individual need to be taken into account (Campbell, 1998).

There are two approaches to underwriting new applicants:

Minimal underwriting so as to decline as few applicants as possible.
An active approach to underwriting so that only those applicants of a
    reasonable standard of health are accepted.

It should be recognized that the underlying morbidity experience arising from the first approach is likely to be higher than that from the second. In the U.S., approximately 15% of applicants are rejected due to poor health. The corresponding percentage in the U.K. is around 10% (Werth, 2001).

Whichever approach is adopted, the underwriting assessment needs to concentrate on three key factors:

Risk of death before claim
Likelihood of claim
Likely length of claim

24.3.5 Pricing the Risk

In this section, the pricing of pre-funded LTC insurance products will be considered under the following headings:

Risk factors
Data sources
Guarantees and experience monitoring
Pricing methodology

24.3.5.1 Risk Factors

The major risk factors that need to be taken into account in calculating LTC insurance premiums are:

Age
Sex
Occupation
Geographical area of residence and birth
Family history
Medically related factors

24.3.5.2 Data Sources

One of the biggest problems with writing LTC insurance is the difficulty in finding satisfactory data, both in terms of quantity and quality, in order to underpin the premium calculations. The ideal data source would be the insurer's own internal price and claims experience. Obtaining accurate data in this way is a significant task, but it is vital if the sales of profitable contracts are to be achieved.

In the early years of an LTC insurance portfolio, the internal database will not be sufficient in itself to price the product. Therefore, it is important to use any external, relevant data sources that may be available. It should be borne in mind, however, that claims experience could vary considerably from one insurance company to another, as there is often an element of subjectivity about whether or not a claim is accepted. Consequently, considerable margins should be included in the pricing assumptions to allow for the data difficulties and possible adverse trends.

24.3.5.3 Guarantees and Experience Monitoring

LTC products often include a guarantee that the premiums will not increase significantly over the term of the contract. The product, therefore, needs to be costed with significant rate guarantees. In reality, it would be difficult to increase premiums substantially even on reviewable contracts. This is because companies would risk damaging their reputations if they were to ask aging policyholders to increase significantly the premiums that they pay during the contract.

It is necessary to monitor the claims experience on a continuing basis so that trends and changes can be factored into the repricing of the product.

24.3.5.4   Pricing Methodology

There are two main approaches used for pricing the current generation of LTC contracts:

Multiple state model
Inception rate and disability annuity value

For a full mathematical analysis of the formulae that should be used to price LTC contracts, the reader is referred to Haberman and Pitacco (1999, Chapter 6).

24.3.5.4.1   Multiple State Model

Figure 24.1 presents a typical multiple state model for LTC benefits, with three levels of care: homecare, residential home, and nursing home. In this

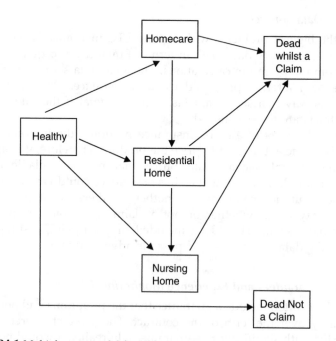

FIGURE 24.1 Multiple state model for LTC insurance

representation, it is assumed that there are no reversible transitions. Therefore, it is not possible for the individual to recover such that they can re-enter the healthy state from one of the care states; nor is it possible to re-enter the homecare or residential home states.

The absence of reversible transitions leads to considerable modeling simplifications. As in Chapter 23 for critical illness insurance, this means that the derivation of the transition probabilities from the transition intensities is relatively straightforward.

The key step is to be able to estimate these underlying transition intensities from data, and in Figure 24.1 we note that there are nine sets of intensities. Relevant LTC data are scanty, and this is the main obstacle to the implementation of this approach.

However, if the transition intensities could be estimated and graduated so that the transition probabilities could be obtained, then we could use the equivalence principle to calculate the single premium or annual premium (payable while the policyholder is in the healthy state) for a contract, represented by Figure 24.1, which pays annuity-type benefits of:

b_H per annum while the policyholder is in the homecare state
b_R per annum while the policyholder is in the residential home state
b_N per annum while the policyholder is in the nursing home state

24.3.5.4.2 *Inception Rate and Disability Annuity Value*

General principle — An alternative method to the multiple state modeling approach is to consider the risk premium in a similar manner to the inception rate/disability annuity approach used in income protection business (see Chapter 22). The risk premium will be constructed from two components:

The incidence rate
The disability annuity value at the time of claim

The incidence rate is i_x at age x and will be tabulated by sex and age. The disability annuity value will be a_x using an appropriate rate of interest and an appropriate mortality table for claimants. It is unlikely that the insured going into care will recover and cease claiming. The use of the disability annuity approach enables a recovery factor to be incorporated if the liability of the insurer fundamentally changes. For example, this could be in the event of the insured's health changing to a higher or lower level of disability.

Haberman and Pitacco (1999, Chapter 3) demonstrate that results obtained from using the inception/annuity approach can, in fact, be obtained from using a multiple state approach with certain assumptions being adopted. The two approaches are therefore consistent.

Obtaining incidence rates — In some countries, for example the U.K., disability prevalence rates are publicly available and are the prime source of data. These would need to be converted into disability incidence rates for use in the pricing of a product. For a description of approximate ways in which this can be done, the reader is referred to Gatenby (1991), Dullaway and Elliott (1998), and Werth (2001).

24.3.6 Sales and Marketing

The immature nature of the LTC market in the U.K. was referred to in Section 24.1. At the end of 2002 there were only about 44 000 LTC policies in force in the U.K. (FSA, 2003). Reasons for this poor uptake may include:

Belief that the state will take care of all citizens from birth through to death
Reluctance to face the issues surrounding aging and consequent frailty
Perceived high cost of the insurance premiums
Concern that premiums will be wasted if care is not required
Lack of understanding of the products on offer by potential buyers
Lack of understanding of the products by those selling them

The U.K. market has therefore remained very small. By contrast, in the U.S., the market has grown significantly in recent years, increasing by an average of 18% each year between 1987 and 2001. In total, 1.4 million policies were sold in both 2000 and 2001 (HIAA, 2003).

When marketing and selling these products, insurance companies need to be aware of the sensitive nature of the issues involved. The products are often being sold to elderly people who are more vulnerable, with little opportunity to rebuild finances if the product and/or advice fails. They are less likely to be financially aware and, therefore, might need to rely heavily on the advice of the salesperson.

24.3.7 Claims

24.3.7.1 Claims Management

The claims management process is a key determinant in the profitability of LTC insurance business. Effective policy design, policy wording, and underwriting controls should ensure that the claims procedure runs efficiently.

The claims assessment process consists of collecting the relevant data to ascertain that the policy conditions are met and that it is appropriate for the claim to be accepted. As suggested by Dullaway and Elliott (1998), the information might include some of the following:

A report on the insured's medical history and current state of health by his or her doctor.

Physical and cognitive assessment tests performed by occupational therapists in the insured's normal residence (e.g., own home or nursing home).

Evidence from a geriatrician.

Invoices for the care received if the benefit entitlement is the reimbursement of the policyholder's care costs.

An important aspect of the claims assessment procedure will be to determine whether any assistive devices could be provided to the insured which would help them to regain some independence (e.g., installing stair lifts, grab handles, and emergency alarms).

It needs to be recognized that, at the ages at which most claims will be made, nearly all policyholders will have suffered some loss of functionality and some degree of mental impairment. These disabilities can be intermittent. It can be difficult, therefore, to distinguish between the cases where a policyholder has genuinely suffered a permanent loss of function (and, therefore, is entitled to claim) and cases where the policyholder has simply become more frail, but not to the point of triggering a claim. Furthermore, given that the claimants are likely to be old and frail, it is particularly important that claims are handled sensitively.

24.3.7.2 Causes of Claims

The breakdown of causes of claim in respect of pre-funded LTC policies for a major U.K. insurer as at December 2001 is given in Table 24.1. It shows that the majority of claims have arisen due to dementia (including Alzheimer's disease). A similar picture has emerged in the U.S., where the

Table 24.1 Distribution of causes of LTC claims for a major U.K. insurer

Condition	%
Dementia	39
Stroke and neurological	21
Orthopedic and arthritis	16
Cardiac and circulatory disorder	11
Cancer	6
Other (frailty)	7

Source: PPP Lifetime Care (2001)

Society of Actuaries' Long Term Care Intercompany Study for 1984–1999 shows that Alzheimer's disease has become the biggest cause of claim in recent years (Corliss et al., 2002).

24.3.8 Reserving

Since LTC insurance is a relatively new product, with little claims experience available, it is important that the valuation basis is cautious. This is particularly true when the product includes substantial guarantees. Even when guarantees are not included, the valuation assumptions should be conservative. This is because, as mentioned in Section 24.3.5.3, marketing pressures will make it difficult to increase significantly the premiums if the claims experience is worse than that assumed in the pricing basis.

Separate reserves will need to be held in respect of in-force policyholders and current claimants. The reserves in respect of in-force lives need to be prudent. However, since surrender values are not usually payable on early termination of the policy, it is important to ascertain what a prudent assumption would be with regard to lapses. In the early years, lapses are likely to lead to a loss, due to the impact of initial expenses and commission. By contrast, lapses occurring in the later years are likely to lead to a profit. The reserves in respect of existing claimants will depend on the cause of claim. For example, claims resulting from dementia (including Alzheimer's disease) tend to last longer than most other causes of claim.

In view of the paucity of data, a significant margin should be included in the reserve. Dullaway and Elliott (1998) suggest that in the U.K., a margin of approximately 30% should be added to the reserve calculated on the basis of best estimate assumptions.

24.4 Alternatives to Pre-Funded Insurance Products

Section 24.3 described pre-funded LTC insurance products. In this section, alternative types of funding and care provision are briefly outlined.

24.4.1 Noninstitutionalized Settings for Long-Term Care

Although care institutions may give perfectly acceptable standards of medical and social care, residing in these places will inevitably lead to at least some loss of independence. It is generally agreed that elderly people would wish to live in their own homes for as long as possible (Dullaway and Elliott, 1998).

Where possible, the integration of services and an individual's accommodation (be it rented or owned) has been the goal of a number of schemes in Europe. These have been particularly successful in the Netherlands and Scandinavia. The Netherlands has actively sought to deinstitutionalize the elderly since the 1970s and has increased the availability of sheltered housing and other home and community-based services to that end (Gibson et al., 2003).

Continuing Care Retirement Communities (CCRCs) are another illustration of the trend towards the deinstitutionalization of elderly people. They are specialized residential complexes where varying degrees of care are available on site. They are well established in the U.S., with an estimated 220 000 residents as at 1995 (Humble and Ryan, 1998). Interest in CCRCs has been growing in other countries such as the U.K.

The CCRC is, in effect, a specially designed village for the elderly, with facilities incorporated into the site. For example, there will be communal dining facilities, forms of entertainment, social activities, and shops. The care for residents can range from minimal (where people are living independently) to full nursing home care for severely disabled residents. The main reason for their success is their flexibility. For example, couples where one member is healthy and the other is disabled can continue to live in the same complex, even as they both become more frail with age.

CCRCs are usually funded by members paying a lump sum entry fee on joining, together with monthly charges. The entry fee tends to be large and is usually paid out of the proceeds of the sale of the member's home on joining the CCRC. In keeping with the community ethos, CCRCs work on the principle of "risk pooling." This means that those members who do not require expensive nursing care subsidize those who do. However, costs of any acute medical care which an individual receives normally have to be met by the recipient, in addition to their normal CCRC fee (Gatenby, 1991).

An interesting actuarial analysis of both the financial structure and the protocols surrounding the development of a CCRC can be found in Humble and Ryan (1998). Jones (1997) provides a detailed mathematical analysis of how CCRCs operate. For a discussion of the risks faced by CCRCs, see Moorhead and Fischer (1998).

24.4.2 Use of Equity in Property to Fund Long Term Care: The U.K. Example

As with all developed countries, the proportion of elderly people in the U.K. is increasing. There is also an increasing tendency for the elderly to own their own homes. Recent surveys in England have shown that over 70% of the elderly have little or no savings (Home Improvement Trust, 2003). Their major asset is, therefore, the equity in their property and this may need to be accessed in order to fund LTC. The position in the

U.K. is unusual in that the value of property needs to be included in an individual's assets when ascertaining whether he or she qualifies for state funding of LTC (i.e. means testing). It has therefore become common in recent years for a person living alone, with no pre-funded insurance, to have to sell their home in order to meet their full care costs. This has been a very controversial issue and it was discussed in detail in the Royal Commission report into LTC provision in 1999 (The Royal Commission on Long Term Care, 1999).

Since the amount of equity in the home is often a key component of a homeowner's capital, the U.K. insurance market has devised some innovative products to exploit this equity. The main products are described briefly below. All except the immediate-needs annuity would be described as equity-release products. A detailed analysis of the issues involved in using the value of the home to fund LTC can be found in Benjamin (1992). For a full discussion of issues surrounding equity release mechanisms, the reader is referred to Le Grys (2001) More information regarding equity release products available in the U.K. can be obtained from Council of Mortgage Lenders (2004).

24.4.2.1 Immediate-Needs Annuity

The property could be sold and the proceeds could be used to purchase an impaired life annuity at the time that care is required. The annuity amount is related to the degree of disability. The reduced life expectancy of the policyholder means that he or she will receive an enhanced annuity compared with standard terms. This feature makes such an annuity attractive, especially since, by definition, it becomes available just at the time that a substantial amount of formal care is required.

The product is individually underwritten and is priced on an individual basis, allowing for age, sex, and level of disability.

24.4.2.2 Home Income Plan

The most common form of home income plan involves using a fixed interest-only loan, which has been secured against a proportion of the value of the house, to purchase a lifetime annuity. Part of the income from the annuity is used to meet the interest repayments of the loan and the balance is available to the individual. The plans tend to be limited to those over age 70, as it is only at the older ages that the income from the annuity exceeds loan interest charges by a margin sufficient to be attractive to a potential borrower. On the death of the borrower, the house is sold and the proceeds are used to pay off the loan. Any remaining money can be passed on to the beneficiaries.

24.4.2.3 Sale of Property on a Reversionary Basis

Under a home reversion scheme, an elderly homeowner sells all (or part) of his or her property to an investor at a discount to the full market value of the property. In return, the individual is allowed to continue to live in the property rent free. He or she can then use the lump sum raised to purchase the LTC requirements. The drawback to the individual of this arrangement is that the difference between the full market value of the property and the lump sum offered tends to be substantial.

24.4.2.4 Reverse Mortgage (or "Interest Roll-Up") Arrangement

Under a reverse mortgage arrangement, the homeowner obtains a loan against a percentage of the value of the home in a similar fashion to the home income plan. The interest on the loan is not paid to the lender immediately, but allowed to accumulate. The interest rate may be fixed or variable. The loan plus the accumulated interest is then repaid from the sale of the property on the death of the borrower. Most lenders offer a guarantee that the total amount repaid cannot be more than the value of the home.

The disadvantage of this arrangement is that the amount owed can mount up quickly. The scheme is, therefore, only really appropriate for people aged over 75.

A variation on the reverse mortgage product is the protected appreciation mortgage. Under this arrangement, interest is not paid on the loan. Instead, the total amount repaid on death is the original loan, together with an agreed percentage of the amount by which the value of the home has increased since the loan was taken out.

24.4.2.5 Reversion Scheme

Under a reversion scheme, the homeowner sells a proportion of the property to a reversion company. When the homeowner dies, that proportion of the proceeds belongs to the company. The drawback to this scheme as far as the individuals are concerned is that the terms are very unattractive. For example, if the homeowner decides to sell all of the property, he or she may only receive between 20 and 50% of the property's market value.

24.5 Conclusion

The prospect of becoming old and infirm can be unpleasant. However, given that state provision may only partially cover care requirements, it is important that this issue is not ignored by the population.

Many countries are experiencing a demographic shift towards an aging population. There is also a greater propensity for elderly people to live by themselves and without access to informal care. The need to provide health and care services to greater numbers of people is becoming a significant financial issue. The level of state-provided care varies from one country to another. For example, Japan is now providing substantial state-funded care, whilst the U.S. provides very little.

In this chapter, we have considered various solutions to the problems surrounding the funding and receiving of care. Whilst the pre-funded LTC policy is an obvious area for development within the insurance industry, insurers could become involved in more imaginative solutions. These could include greater involvement with CCRCs, immediate-need annuities, and equity release products. These alternatives could be attractive to individuals, as they do not require the payment of premiums until the individuals' care needs are imminent.

References

Benjamin, S. (1992). Using the value of the home to fund long term care for the elderly. In: *24th International Congress of Actuaries*, vol. 4, pp. 19–34.

Campbell, R. (1998). New tools for underwriting long term care. *GeneralCologne Re Risk Insights* 2(5).

Corliss, G., Gagne, R., Murphy, M., Newton, M., and Tillmann, K. (2002). Society of Actuaries Long Term Care Experience Committee Intercompany Study 1984–1999.

Council of Mortgage Lenders (2004). Unlocking the value of your home — a guide to equity release. In-house publication.

Dullaway, D. and Elliott, S. (1998). Long term care insurance. Presented to the Staple Inn Actuarial Society, London.

FSA (2003). Regulation of long-term care insurance. Financial Services Authority. http://www.fsa.gov.uk/pubs/cp/cp200.pdf

Gatenby, P. L. (1991). Long term care. Presented to the Staple Inn Actuarial Society, London.

Gibson, M. J., Gregory, S. R., and Pandya, S. M. (2003). Long-term care in developed nations: a brief overview. AARP Public Policy Institute.

Haberman, S. and Pitacco, E. (1999). *Actuarial Models for Disability Insurance.* Chapman & Hall/CRC Press, Boca Raton.

HIAA (2003). Long-term care insurance in 2000–2001, executive summary research findings. Health Insurance Association of America, Washington, DC.

Home Improvement Trust (2003). http://www.improvementtrust.fsbusiness.co.uk/.

Humble, R. A. and Ryan, D. G. (1998). Continuing Care Retirement Communities. In: *26th International Congress of Actuaries. Transactions* vol. 6, pp. 227–256.

Jones, B. L. (1997). Stochastic models for continuing care retirement communities. *North American Actuarial Journal* 1(1), 50–73.

Karlsson, M., Mayhew, L., Plumb, R., and Rickayzen, B. (2004). An international comparison of long-term care arrangements. Actuarial Research Paper 156, Faculty of Actuarial Science and Statistics, Cass Business School, London.

Le Grys, D. J., (2001). Report of equity release mechanisms. Presented at the Institute of Actuaries, London.

Moorhead, E. J. and Fischer, N. H. (1998) Analyzing the dynamic financial condition of a Continuing Care Retirement Community. In: *26th International Congress of Actuaries. Transactions* vol. 6, pp. 257–270.

PPP Lifetime Care (2001). Long term care: the experience that counts. An in-house publication

Rappaport, A. M. (1998). Challenges of an aging society. In: *26th International Congress of Actuaries. Transactions* 6, 37–54.

The Royal Commission on Long Term Care (1999). With respect to old age: long term care — rights and responsibilities. HMSO.

Stracke, A. (1998). The challenge of designing and pricing long term care products. *GeneralCologne Re Risk Insights* 2(5).

Werth, M. (2001). Long term care — healthcare module. Faculty and Institute of Actuaries.

Chapter 25

Private Medical Insurance

25.1 Introduction

The amount of state healthcare provision offered to a population will vary considerably from country to country. The use of drugs and operative procedures can put a strain on a country's healthcare budget. This strain is made worse by the aging nature of many populations. As a result, in many countries, the resources required to provide state healthcare are being stretched and the insurance market is growing in order to accommodate any gaps that occur. Where public provision is lacking, private funds may be used to pay for healthcare. The insurance market, through its offering of private medical insurance (PMI) policies, is an important route by which healthcare is financed when it is not publicly funded. PMI products reimburse the insured, in whole or in part, for the cost of receiving medical treatment.

How the product is taken up amongst the population is also affected by state provision. For example, where the state plays a minor role in providing healthcare, only those on the lowest incomes will be given access to publicly funded treatment. The majority of the population will rely on PMI to cover most, if not all, of their health costs. This is the case in the U.S.

In countries where state provision is offered more widely to the population, people have a choice over whether or not to buy private insurance. The design of policies will, therefore, be focused on giving the policyholder a faster or superior level of care to that which they would expect to receive if they relied solely on state provision. In such countries, PMI is likely to be purchased only by those in the higher socio-economic classes. For example, this occurs in the U.K.

Figure 25.1 shows how the use of publicly and privately funded healthcare varies across a range of countries. For example, Japan's healthcare provision is dominated by government intervention, whereas in the U.S. the onus is placed fully on the individual to ensure that their healthcare

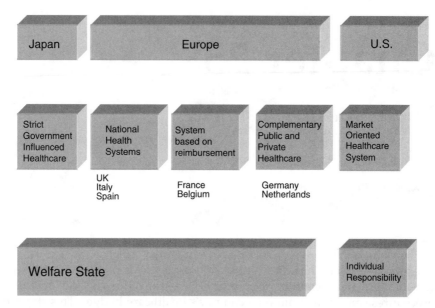

FIGURE 25.1 Overview of healthcare provision in various countries
Source: Munich Re (2003)

requirements are met. Only the elderly, the disabled, and those who are unable to meet the financial costs of insurance use public healthcare in the U.S. (Laüter, 1999).

Many people obtain PMI cover through employer-sponsored group schemes. Although the benefits are often similar to those available under individual policies, the pricing and underwriting considerations are quite different. This is discussed more fully in Section 25.3.

25.2 Product Design

25.2.1 General Design Principles

The policy design of PMI products is heavily influenced by the level of public healthcare provision and by the attitudes of the target market. Where state provision is very scarce, the design must give prospective policy-holders the confidence that such a product will enable them to undergo virtually any medical treatment, knowing that the costs will be met. In markets where state provision plays a more prominent role, the design must give confidence that access to the highest level of healthcare (often perceived as private provision) will be attainable.

Specific policy design features are described in Section 25.2.2, using the U.K. market as an example. However, the main features of any standard

PMI policy can be summarized as follows:

It is a form of general insurance product (see Chapter 12).
The benefit is simple and specific in its nature, i.e., to cover the costs of medical treatment. However, complications arise from the use of exclusions and policy limitations.
It can be offered as either an indemnity product (where benefits of a fixed monetary amount are paid), or it can be a reimbursement product (where the full cost of treatment is covered).
It is renewable (often annually) and is a short-tailed product, since claims are usually settled quickly.
It does not generally use guarantees, as medical costs are unpredictable.
An insurer's claims experience typically comprises a high frequency of small claims, and a small number of large claims.
The policy can be written in individual or group form.

25.2.2 Private Medical Insurance Products: The U.K. Example

The U.K. provides its state healthcare through the National Health Service (NHS); treatment is free to all at the point of use. However, owing to increased demands placed upon the NHS and a lack of government resources, the public may see public healthcare as falling below their expectations. PMI, which is often viewed as a "luxury" insurance product, can be used to upgrade the medical services on offer to the policyholder. The upgrade relates to all parts of the medical event, from diagnosis through to treatment. Policyholders would therefore expect to receive prompt attention from senior professionals (maybe even staff of their choice) as soon as a medical problem has been identified.

As well as the medical aspects of care, the policyholder would expect to be given hospital accommodation that is superior to that offered on a typical NHS ward. Expectations would include a private room with better facilities (for example, en suite bathroom) and a wide choice of menu for meals.

We describe below the main types of PMI product available in the U.K. Many of these products are also sold in other countries.

25.2.2.1 Comprehensive Cover

Table 25.1 lists the typical benefits provided in a comprehensive PMI plan. It can be seen that most costs would be covered in full, with a few fees covered up to a monetary limit. The cover includes consultations with specialists, out-patient treatments, in-patient treatments (i.e., where an overnight stay in hospital is required) and accommodation.

Table 25.1 Example of benefits provided in a comprehensive PMI plan

A. Hospital Costs	
1. Accommodation, nursing care, operating theater, diagnostic procedures, surgical dressings, and drugs for in-patient treatment	Full cover
2. Specialist consultations and physiotherapy received as an in-patient	Full cover
3. Accommodation, nursing care, operating theater, diagnostic procedures, surgical dressings, and drugs for day-care operation not involving an overnight stay	Full cover
4. Accommodation for one parent to stay in hospital with an insured dependent under 12 years old	Full cover
B. Specialist Fees	
1. Surgeons' and anesthetists' fees for in-patient and day-care operations	Full cover
2. Physicians' fees for in-patient treatment	Full cover
3. Out-patient treatment:	
(i) Specialist consultations	Full cover
(ii) Diagnostic procedures, such as radiology and pathology	Full cover
(iii) Physiotherapy	Up to £500 per policy year
4. Radiotherapy, chemotherapy, and scanning	Full cover
C. Other Features	
1. Private ambulance	£150 per policy year
2. Recuperative care (including nursing and domestic services) immediately following in-patient surgical treatment where the stay in hospital was for a minimum period of seven nights	Up to 3 hours care per day for a maximum of 7 days
3. Maximum overall limit	None

Source: Alexander et al. (2001)

The cover does not usually include pre-existing medical conditions, chronic conditions, or accident and emergency admission. In addition, the following general policy exclusions would normally apply (ABI, 2003):

Drug abuse
Self-inflicted injuries
Outpatient drugs and dressings
HIV/AIDS
Infertility
Normal pregnancy
Cosmetic surgery
Organ transplants and kidney dialysis
Hazardous pastimes

In view of the fact that the fully comprehensive cover described above may only be affordable to the wealthy, insurers also offer more limited (and, therefore, cheaper) versions of the cover. Examples of variations of the fully comprehensive cover are as follows.

Preferred provider plan — Fully comprehensive cover is provided, but the hospital must be on an approved list drawn up by the insurer. This is known as a "preferred provider" arrangement.

Waiting-list plan — Fully comprehensive cover is provided but only if policyholders are unable to receive the treatment from the NHS within a certain time period (usually 6 weeks) and within a reasonable distance of where they live. This is known as a "waiting list" plan.

Excess plan — Fully comprehensive cover is provided, but policyholders must pay the first part of every claim. This amount, known as the "excess" (or "deductible" in the U.S.), will be a fixed sum of money. Insurers often allow policyholders to choose the level of excess, within specified limits. Policyholders who wish to keep their premiums as low as possible will opt for the highest level of excess possible.

The main advantage to policyholders of having an excess, therefore, is that it enables them to have some control over the level of premiums that they pay. The main advantage to the insurer is that the excess acts as a disincentive for policyholders to make a PMI claim, if the NHS could be used instead.

25.2.2.2 Major Medical Expenses

In the U.K., "major medical expenses" is a product that provides a lump sum to defray the costs of undergoing surgery. The intention is to provide a lump sum that not only covers the in-patient expenses of the procedure but also makes a contribution towards the recuperation costs as well. In order to keep premiums low, it does not cover out-patient treatment; however, this point is not always fully appreciated by policyholders at the time when the product is purchased, and it can lead to aggrieved claimants (Faculty and Institute of Actuaries, 2004).

The term "major medical expenses" in the U.K. must not be confused with the same term in the U.S., where it means that cover is provided for both acute and chronic medical conditions.

25.2.2.3 Cash Plans

With a cash plan, policyholders pay level premiums, usually weekly. In return, they are entitled to fixed cash payments if any of a number of specified medical events occur during the period of cover. These events might include, inter alia, hospitalization, recuperation, hearing aids, visual aids, and physiotherapy.

Policyholders purchase the benefits in units. This enables them to choose the level of cash sum that would be paid to them in the event of a claim. To reduce the risk of adverse selection, policyholders do not usually become eligible to make a claim until the policy has been in operation for a certain period. This period is known as the "waiting period," and is often 6 months. In order to control further the claim costs, the cash sum is usually limited to a proportion of the medical bill (often 50%).

25.2.2.4 Dental Plans

Under this type of cover, the costs of dental treatment are partly or wholly indemnified. Pre-existing conditions will be excluded, as will the cost of any work identified as being required imminently after policy commencement.

In view of the aversion that many people have to visiting the dentist, there is a significant moral hazard risk associated with dental insurance. Policyholders might be tempted to ignore any dental problems until they require emergency treatment. In order to encourage policyholders to visit the dentist for regular checkups (which might avoid such expensive treatment) the insurer may impose an excess on any claim (see Section 25.2.2.1). They may also require the individual to pay a proportion of any claim; this is known as "co-insurance."

25.2.2.5 Optical Plans

An optical plan covers the cost of eye tests, together with the cost of visual aids, such as spectacles and contact lenses. The cover is usually incorporated within cash plans or other PMI products. However, it can also be purchased on a stand-alone basis.

As with other health products, the insurer needs to be aware of the risks of moral hazard and adverse selection. These risks can be significantly reduced by the imposition of waiting periods, co-insurance, and the exclusion of pre-existing conditions.

25.3 Group Business

Group business is an important part of PMI provision. Employers recognize that PMI cover is seen as a valuable benefit to employees. The provision of PMI will also benefit the employer with regard to reducing employee absenteeism through illness. Private cover will ensure that employees receive treatment quickly and are, therefore, likely to make a faster recovery. Approximately 60% of individuals with PMI cover in the U.K. are in employer-sponsored group arrangements (Alexander et al., 2001). Approximately two-thirds of Americans under pension age have access to health insurance cover through their employers (Dauser, 1999).

The risk of adverse selection in writing PMI business is reduced considerably if the take up rate of the group PMI cover amongst the employees is high. Indeed, provided nearly all of the employees opt to participate in the scheme, and that there are more than 100 employees involved, it is likely that the insurer will not underwrite the policy. As with group IP business, a "free cover" limit is likely to apply (see Section 23.3.3). The insurer will usually provide automatic cover only to those employees who are at work on the day that cover begins. This is known as an "actively at work" condition. Employees who are not at work on that day and wish to be covered may be required to complete a certain period of work, uninterrupted by illness, when they return to the workplace. Alternatively, they may need to be underwritten on an individual basis.

The benefits and exclusions in group PMI schemes tend to be similar to those in individual business. However, in recognition of the fact that the risk of adverse selection is lower, cover is much more likely to be offered for pre-existing conditions.

With regard to pricing, experience rating is likely to be used, at least to a certain extent. This is where an insurer takes into account the past claims experience of the particular group. The greater the exposure, the greater the weight that the insurer will give to this experience compared to the claims experience of the insurer's portfolio as a whole. Many insurers will base the premium entirely on the group's own experience if the exposure is at least 100 man-years (Faculty and Institute of Actuaries, 2004). For an overview of the technical and practical aspects of using an experience-rating model in pricing group PMI cover, the reader is referred to Price (2003).

Group schemes that contain at least 50 employees are often flat rated. This means that the premium is a fixed amount per person as opposed to being related to the age profile of the membership.

For a more detailed account of the differences between group and individual PMI business, see Crause (2003).

25.4 Underwriting

As with other types of insurance contract, underwriting is an important component of the insurer's risk management system and is used specifically to manage the risk of adverse selection. Thus, the purpose of the underwriting function is to assess the policyholders' risk profiles and place them, accordingly, into homogeneous groups. The premiums charged should then reflect the risk characteristics of each group. In the context of PMI business, underwriting is used to estimate the risk of pre-existing medical conditions returning and the risk of related medical conditions occurring.

There are various ways in which underwriting can be approached. The following methods are often used:

Declared medical history
Moratorium
Premium loading

25.4.1 Declared Medical History

This is the most common method of underwriting used in the U.K. The applicants provide details of their full medical history and current state of health on an application form. Their doctor may be asked to confirm the information provided.

The insurer then uses this information, together with other risk factors including age, sex, and occupation, to assess the risk posed by any pre-existing medical conditions that the applicant may have. Even if the underwriter decides to exclude the conditions (and any linked ailments), it is possible that they will review the decision after a certain period of time. If, during this period, an applicant has not had cause to visit the doctor with any symptoms related to the condition, then it is possible that the exclusion clause would be lifted.

25.4.2 Moratorium

The concept of the moratorium clause was mentioned in connection with critical illness insurance (see Section 23.4). Under this approach, the applicant's PMI cover commences immediately, without a detailed medical history being required. It is only at the point that a claim is made that the insurer collects the medical information necessary to determine whether the claim stems from a condition which the policyholder previously had (i.e., prior to policy commencement).

The insurer will usually not pay claims in respect of any conditions that the applicant had during a certain number of years before the policy started (often 5 years). However, if the applicant has no symptoms during a fixed period after the policy commences (often 2 years) then these conditions would be covered in the future. This latter period is known as the "moratorium period."

With this method, the underwriter saves the time and expense of collecting detailed medical information at the application stage. However, it is this apparent simplicity which can cause problems. The policyholders may not appreciate that, even though their application for PMI has been accepted, not all conditions will be covered for at least the moratorium period.

25.4.3 Premium Loadings

Under this approach, the insurer agrees to pay any claims that arise from an applicant's pre-existing conditions, but charges the applicant higher premiums to reflect the extra risk.

Policyholders may prefer this approach, as they will be confident that they are fully covered. Indeed, this approach is commonly used in many territories, including the U.S. However, the drawback is that insurers may have difficulty in collecting sufficient data for them to determine the appropriate premium loadings. For this reason, this approach is rarely used in the U.K., for example.

25.5 Pricing

In this section, consideration will be given to the way in which premiums are determined for individual PMI business. The pricing of group business was mentioned in Section 25.3.

25.5.1 Data Sources

In order to calculate the premium, the actuary needs to be able to estimate the future cost of claims for each policyholder over the term of the contract.

Useful data sources for this information include the following:

The insurer's own data
Inter-company data
Reinsurers
Hospitals
Returns to the insurance supervisor
International sources

25.5.1.1 The Insurer's Own Data

The most suitable source of information will be the insurance company's own claims experience. However, for the insurer to be able to use its own data it must be confident that the claims experience is large enough to be credible. For small insurers and insurers new to the PMI market, this point is particularly pertinent.

25.5.1.2 Inter-Company Data

Insurers will wish to make use of any inter-company data available. However, they need to be aware that any industry-wide data will be based on policies that will not have exactly the same characteristics as their contracts. For example, other insurers' policies may differ with regard to underwriting, claims definitions, and claims management. Therefore, some adjustment will be required to premium rates derived from such external sources of data.

25.5.1.3   Reinsurers

Since reinsurers advise many direct insurers that operate in the PMI market, they collect a significant amount of industry data. Reinsurers are, therefore, ideally placed to assist insurance companies in setting their premium rates.

25.5.1.4   Hospitals

Research staff attached to teaching hospitals often produce statistical reports that might be helpful to insurance companies writing PMI business. For example, research could be published on the average time spent in hospital by patients admitted with conditions covered by PMI policies.

25.5.1.5   Returns to the Supervisor

In most countries, insurers are required to submit data to the regulator in order to demonstrate that they are solvent. Notwithstanding the fact that the information is being provided solely for regulatory purposes, insurers may find the data provided by their competitors (which is usually publicly available) very informative.

25.5.1.6   International Sources

When PMI products are offered in a country with no prior history of PMI business, there may not be any insurance data on which to base premium levels. Insurers, therefore, will need to look to international insurance markets in order to obtain appropriate information. However, it has to be remembered that data from an insurance market in one country will not necessarily transfer easily to another. As already discussed in Section 25.1, the products designed for PMI business will depend greatly on a number of factors, e.g., a country's public health provision, cultural attitude to healthcare, and legislative regime. Thus, the premiums derived from international data sources will need suitable adjustment before they can be used.

In addition to overseas insurance markets, international organizations such as the World Health Organization and the United Nations also publish useful data.

25.5.2   Pricing Methodology

The premium for a PMI product will comprise the risk premium, together with a suitable loading for expenses, contingency margins, and profit.

Table 25.2 Risk cost model for typical U.K. policy

Risk premium component	Risk premium (%)
Hospital accommodation	35.40
Theater fees	5.90
Other hospital charges	3.20
Surgeon's fees	16.00
Anesthetist's fees	6.90
Drugs and dressings	3.60
Total in-patient charges	**71.00**
In-patient physician's services	4.50
Out-patient consultations	17.00
NHS cash benefits	3.00
Claims outside the U.K.	1.00
Psychiatric illness	3.50
Total risk premium	**100.00**

Source: Orros and Webber (1988)

As noted in Faculty and Institute of Actuaries (2004), the formula for the risk premium will be

$$\text{Risk premium (age, sex)} = \sum_k i_k \times AC_k$$

where i_k (the incidence rate for claims cost component k) and AC_k (the average claims cost for claims cost component k) depend on the age and sex of the policyholder, and where claims costs are subdivided into separate components (e.g., hospital treatment, outpatient treatment, dental care, etc.).

Table 25.2 shows a specimen breakdown of the risk premium into its different cost elements for a typical U.K. PMI policy.

25.5.3 Medical Expense Inflation

In many territories, over the last few years, medical expenses have been increasing at a much faster rate than price inflation. Therefore, careful consideration must be given to the assumptions made in estimating future levels of medical expenses. This is a crucial aspect to pricing policies appropriately.

Carroll (2004) has identified various components to medical expense inflation. These are set out below, together with an illustrative example in each case. Not all of these factors will apply in any particular market.

25.5.3.1   Quality

This term refers to the improvement of the treatment for a given complaint. This is often due to the introduction of new technological techniques. For example, in a hip replacement operation, a titanium-based hip socket is now used instead of a stainless-steel socket. Titanium is generally accepted as being a better material to use, but it is more expensive than stainless steel.

25.5.3.2   Utilization

This refers to the increase in the amount of treatment used. For example, increasing the number of pills taken daily in order to control post-operative pain.

25.5.3.3   Addition

An additional medical procedure might be added to the existing regime for a patient. For example, giving the patient physiotherapy to aid recovery following an operation.

25.5.3.4   Substitution

This is where new treatment replaces the usual treatment. For example, whereas previously medication may have included two separate drugs with different functions, a newer, more expensive drug might be administered that combines the two functions in a single pill.

25.5.3.5   Medical Negligence

This refers to the capital cost of settling medical negligence claims. The cost may be in the form of direct compensation to the patient or as a premium paid to a reinsurer to cover the risk of such compensation payments being required. An example is the rising cost of medical negligence insurance premiums, particularly in the U.S., in the area of gynecology.

25.5.3.6   Inflation

This refers to the rise in costs of existing services due to general inflation. For example, where the cost per night of private hospital beds increases in order to cover increases in staff salaries.

25.5.3.7   Cost Shifting

This is the process of shifting the burden of cost from one payer to another. In the U.S., for example, Medicare pays the doctors and dentists. However, Medicare stipulates a maximum amount that can be paid on any medical procedure; therefore, doctors and dentists make up their income by charging their private payers a larger fee than would otherwise have been the case.

25.5.4 Risk Factors

The medical history of a policyholder will clearly be crucial to the insurer when assessing the risk that the individual poses. As well as taking this into account, as suggested by Alexander et al. (2001), an insurer typically considers the following risk factors when pricing a PMI product:

Age
Sex
Number of people covered
Geographical area
Occupation
Hospital band
Smoker status

25.5.4.1 Age

As individuals get older, their probability of making a PMI claim in a particular year increases. The premiums need to reflect this, together with the fact that certain conditions are only likely to occur at particular ages. For example, it is unusual to require heart surgery before the age of 40.

25.5.4.2 Sex

An analysis of claims experience demonstrates a clear difference in PMI claim rates between the two sexes. The difference depends on the benefits included in the PMI contract. For example, the high cost of gynecology treatment, and other related female conditions, tends to cause claim rates to be higher amongst females than males at younger ages. The converse tends to be true at older ages.

25.5.4.3 Number of People Covered

Some PMI contracts automatically cover policyholders' children regardless of how many children there are and how old they are. Insurers should assume in their premium rates that the families which they cover are larger than average if there are other insurers offering PMI cover to families for which the premiums take into account the full details of the children. In other words, they should assume that a certain amount of adverse selection will take place (Alexander et al., 2001).

25.5.4.4   Geographical Area

The costs associated with hospitals in larger cities tend to be higher than those associated with hospitals in more rural areas. The PMI premium, therefore, will be affected by where the policyholder lives.

25.5.4.5   Occupation

The type of work an individual does may make them more susceptible to suffering an illness or injury for which a PMI claim could be made. For example, people in stressful jobs are more likely to suffer from duodenal ulcers.

25.5.4.6   Hospital Band

The premium needs to reflect the quality of hospital care and accommodation that is being provided under the cover. For example, private rooms will be more expensive than a bed on an open ward.

   In addition, some PMI contracts offer policyholders the right to choose where they wish to receive their treatment, whereas others offer a much more restricted choice. In general, the wider the choice, the higher the premium.

25.5.4.7   Smoker Status

Smokers are generally charged higher premiums than nonsmokers due to the increased risks of claims related to, for example, heart and lung conditions.

25.5.5   Guarantees

As mentioned in Section 25.1, PMI products are usually created to offer those aspects of healthcare which are unavailable from the state system, or which are available at an inferior level to that expected by the policyholder. Since state healthcare is easily affected by political change, PMI claims experience can be unpredictable. As a result, few PMI insurers offer substantial guarantees in their terms. Consequently, PMI tends to be written as one year renewable business.

   In cases where an insurer does choose to guarantee benefits for a period longer than one year, suitable allowance must be made in the premium charged. In this situation, guarantees are usually priced using a stochastic model.

25.5.6   No-Claims Discount

Some insurers offer a no-claims discount system on individual PMI business (see Chapter 14). Effectively, policyholders who make no claims in a year are rewarded by receiving a discount on their renewal premium. This encourages loyalty amongst policyholders who have made no claims in the previous year. It also discourages policyholders from making small claims.

25.6   Claims

25.6.1   The Nature of the Claim

As mentioned previously, the size and nature of a PMI claim will depend on the relationship between what the state healthcare system will provide and what the private insurance industry will cover. Where the main infrastructure for healthcare provision falls within the state system, the insurance industry will generally offer cover for routine, acute treatments. More complex or chronic procedures would be the responsibility of the state sector. In the U.K., for example, the NHS is better placed to carry out organ transplant operations than the private sector. As a consequence, the average claim size for a PMI policy in the U.K. is generally much lower than in other countries in Europe. Indeed, very few PMI claims in the U.K. exceed £5000 (Alexander et al., 2001).

25.6.2   Claims Control

It is important that the claims management process for PMI business is as cost efficient as possible, since the insurer will expect to pay a significant proportion of incoming premiums as outgoing claims. Nevertheless, efficiency must be balanced with public faith. If the management process is too harsh, then the insurer will suffer from negative publicity. Fairness and objectivity are, therefore, as crucial as cost efficiency.

In order to assess a claim properly, the policy itself must have been written clearly and concisely. Both the claimant and the claims manager should be able to tell what the policy covers and, importantly, what the policy excludes. Any benefit limitations and exclusions should be transparent in the policy document.

As mentioned in Section 25.2 and Section 25.5.6, there are various features of product design that help the insurer to contain claim costs:

The general exclusions referred to in Section 25.2, together with specific exclusions related to pre-existing conditions, will restrict the cover to a level that the insurer is happy to offer.

Excesses ensure that policyholders pay the first part of any claim. They also
     reduce the number of small claims made.
Co-insurance will mean that policyholders must share the cost of each
     claim. This will act as a deterrent for the policyholder to choose to
     be treated privately if free treatment is available from the state as an
     alternative.
The use of a no-claims discount system tends to discourage policyholders
     from making small claims.

Alexander et al. (2001) describe the following two additional elements
to claims management:

Prospective claims management
Retrospective claims management

25.6.2.1   Prospective Claims Management

It is desirable for an insurer to have in place agreed fee schedules with
core providers. This involves negotiating, in advance, prices for hospital
accommodation, consultants' fees, and treatment costs. Particularly for
larger insurers, the costs can be reduced by block-booking business with
certain care providers. The savings may then be passed on to policyholders
in the form of competitively priced premiums.

25.6.2.2   Retrospective Claims Management

When a claim is settled, the insurer will wish to review the fees charged by
the healthcare provider in order to ensure that they accord with those
charges agreed in any prospective arrangement. The insurer will also wish
to verify that the treatment given was efficient and effective (e.g., that the
time the claimant stayed in hospital compared favorably with published
statistics for the condition). Large claims will be of particular interest to the
insurer in the review process.
   Insurers might insist on claimants getting their treatment and place
of treatment approved by the insurer before treatment is carried out.
However, this may not seem to be in keeping with the flexible character
of PMI policies often presented in the marketing literature.

25.6.3   Claims Procedures

When a claim is made, the following information needs to be supplied
to the insurer:

Details of the claimant's medical condition;

A description of the medical diagnosis and recommended treatment;
Confirmation from the consultant that the treatment is appropriate.

The claims department will wish to ascertain that the condition for which the claim is being made is covered. It must not, for example, be a pre-existing condition occurring within the moratorium period (see Section 25.4.2). The claims assessors will then obtain a detailed medical history from the policyholder's doctor. This is to ensure that the policyholder did not omit details of a relevant condition at the time the policy commenced. The risk of such nondisclosure is particularly high for claims made during the first 6 months of the policy (Alexander et al., 2001).

The insurer needs to establish that the condition is acute rather than chronic, since claims are only allowed in respect of an illness or injury that is acute. However, it should be noted that sometimes an acute condition becomes chronic, and it can be difficult to ascertain precisely when this change occurred.

25.7 Reserving

PMI is an example of short-term general insurance business. Consequently, the methods used to determine appropriate levels of reserves are those described in Chapter 16.

The main reserves will be in respect of:

Claims reported but not settled
Claims incurred but not reported (IBNR)
Unearned premium reserve (UPR)
Unexpired risk reserve (URR)
Policy specific reserve

25.7.1 Claims Reported but Not Settled

Usually, insurers set up these reserves only in respect of claim payments outstanding at the valuation date for which treatment has already been received. This means that the reserve would not include payments due in respect of treatment to be received after the valuation date (e.g., the part of a hospital stay that takes place after the valuation date).

It is important to note that treatment costs will only be reimbursed if the policyholder is covered whilst the treatment is given. In other words, it is not sufficient for the individual to have PMI cover for the period during which the diagnosis leading to treatment took place; the individual must also have PMI cover for the period when the treatment is actually received. This feature is unusual when compared with other general insurance products.

25.7.2 Claims Incurred but Not Reported

The standard methods described in Chapter 16 will be used to estimate the IBNR reserve.

25.7.3 Unearned Premium Reserve

It is common for the UPR to be estimated on a 24ths basis (see Chapter 16).

25.7.4 Unexpired Risk Reserve

An insurer may decide that it is necessary to hold a URR in addition to the UPR. This would be, for example, if recent claims experience has been significantly worse than expected.

25.7.5 Policy Specific Reserves

It is possible that a reserve will be required to deal with particular features of the PMI policies being offered. For example, reserves will be required in respect of any guarantees contained within the products.

25.8 Conclusion

In this chapter we have described how PMI contracts will vary in design and uptake, depending on a country's state healthcare provision. Where the public perceive the latter to be incomplete, insurers can capitalize on the deficiencies by devising suitable PMI policies in order to plug the gaps.

Of the four health products described in Part V, PMI is the only one that follows general insurance principles, as opposed to those of life insurance. The product tends to lead to many small claims, some very large claims, and settlement is generally completed quickly.

The claims experience for PMI is significantly affected by new medical treatments, as expensive procedures and drugs will increase claims costs. As a result of this, and general medical expense inflation, contracts are almost always renewed annually. Insurers are constantly aiming to control claims costs by using prospective claims management procedures.

This is where the insurers negotiate the fee schedules for treatment and hospital accommodation in advance with healthcare providers.

References

Alexander, D., Hilary, N., and Shah, S. (2001). Private medical insurance — healthcare module. Faculty and Institute of Actuaries.

ABI (2003). A guide — are you buying private medical insurance? Association of British Insurers, London.

Carroll, H. D. (2004). Information provided by VectorRisk LLC, Minneapolis, USA.

Crause, S.-M. (2003). Principles of group insurance, part 1. *GeneralCologne Re Risk Insights* 7(4).

Dauser, A. M. (1999). Cost trends for large health claims in the United States. *GeneralCologne Re Risk Insights* 3(5).

Faculty and Institute of Actuaries (2004). Health insurance core reading. Faculty and Institute of Actuaries.

Laüter, G. (1999). Different markets — different health insurance products, *GeneralCologne Re Risk Insights* 3(5).

Munich Re. (2003). International healthcare trends. http://www.munichre.com/pdf/praes_migdal_health_insurance_trends_e.pdf.

Orros, G. C. and Webber, J. M. (1988). Medical expenses — an actuarial review. *Journal of the Institute of Actuaries* 115, 169–253.

Price, R. (2003). Experience rating in group life insurance. *GeneralCologne Re Risk Insights* 7(4).

Index